TECHNOLOGY
TODAY AND TOMORROW

BRITANNICA
SCIENCE
AND THE
FUTURE
LIBRARY

VOLUME
3

Encyclopædia Britannica, Inc.

CHICAGO • GENEVA • LONDON • MANILA • PARIS • ROME • SEOUL • SYDNEY • TOKYO • TORONTO

Library of Congress Catalog Card No.: 81-71792
International Standard Book Number: O-85229-402-6
No part of this work may be reproduced or utilized in any form or by any means,
electronic or mechanical, including photocopying, recording, or by any information storage and retrieval system,
without permission in writing from the publisher.

Printed in U.S.A.

BRITANNICA
SCIENCE
AND THE
FUTURE
LIBRARY

MILITARY TECHNOLOGY
PART FIVE

ELECTRONICS IN ACTION
PART SIX

MEDICAL SCIENCE
PART SEVEN

CONTENTS

VOLUME THREE

MILITARY TECHNOLOGY
PART FIVE

INTRODUCTION TO PART FIVE

From the very beginning science and technology, knowing very well on which side their bread is buttered, have placed themselves at the disposal of the powerful, the ambitious, and the cruel. Such men are almost irresistibly tempted to blow each other up, and if they don't get the tools for the job from one scientist they will get them from another. It is a somber prospect, and yet there is a bright side to military technology, too. Not only do we need it to defend ourselves from the Other Fellow, but the discoveries we stumble upon while trying to find out how to kill our enemies more efficiently often make our own lives better.

This part of *Britannica Science and the Future Library* is a veritable cornucopia of death-dealing machines and devices—but of course all of them are also life-giving in certain circumstances. There are rifles for infantrymen, tanks and antitank weapons, mines and minefields, killer copters, fighter planes of the future, and the A-10 killer plane that is already on the horizon. Then there are missiles of many kinds, SAMs and MIRVs and MX systems. Finally, of course, there are nuclear weapons.

Most frightening of all, perhaps, are the various kinds of chemical warfare: "death drop by drop," as the article says. Would you rather die instantly from poison in the water supply, or be vaporized in a nuclear explosion? Perhaps it doesn't matter much.

There are things one can do. Watching for war with every electronic eye open is one thing. Another is camouflage, an art that all should learn. Weapons are not always turned only on enemies, or, rather, the Enemy is not always without— sometimes he is one of us. Then we need riot equipment to control him, and interrogation techniques to find his secrets. Or a deadly fence to keep him from "voting with his feet" for freedom, as in Berlin. Not a pleasant creation, that. It reminds us of how well we live, after all.

The last article in this part is perhaps the most interesting of all. Despite the technological advances, the exquisite skill with which one can now kill, in the end it is courage that counts most. "Who dares, wins." So it always was, and so it always will be.

Guns for today's infantryman

The infantry are the most flexible and adaptable of land forces. Forests, jungles, swamps and mountains may all thwart armoured forces and even make life difficult for artillery; but a foot soldier can engage an enemy wherever he chooses to maintain himself.

The infantryman is the keystone of military power for, as all commanders know, you only hold the ground upon which your infantry are positioned. Other areas may be swept by fire, crossed by armour and dominated by patrols, but it is still possible for an enemy to establish himself within them. But he cannot take infantry positions until he has driven your men out of them.

Although they fulfill such a basic and all-purpose role, infantry are specialist soldiers who use a variety of technical equipment to tackle their tasks. Some of them are trained to reach their objectives in particular ways: the paratroops from the air, the marines and commandos from the sea. But any unit whose soldiers fight on foot with personal weapons is an infantry unit, and that includes the misleadingly named US Air Cavalry and elite squads such as the British Special Air Services.

There are considerable variations in the composition of the basic infantry unit, but the standard formation is the battalion. This is organized not only into companies of infantry, but also contains a support company which can deploy comparatively heavy weapons. The smallest unit in such a battalion is the infantryman himself and he is, most typically, armed with a rifle capable of automatic fire.

The outstandingly successful rifle since World War 2 has been the Soviet AK-47 which was designed by Mikhail Kalashnikov and fires 7.62 mm ammunition with acceptable accuracy. It is an extremely rugged and easily manufactured piece of equipment, and as many as 20 million may be in use worldwide with any number of variants and copies. Its mechanical loading and firing is operated as is usual with automatic weapons,

THE ACTION OF THE ARMALITE RIFLE

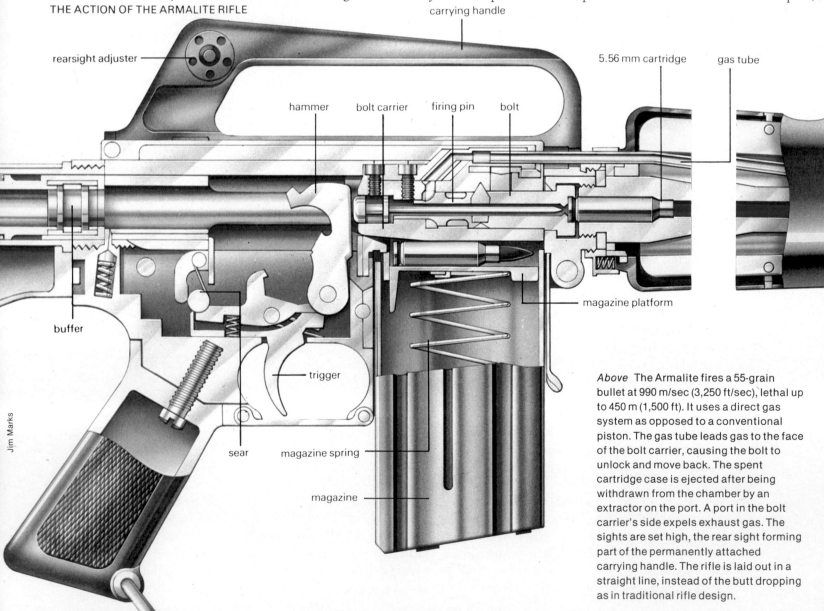

Above The Armalite fires a 55-grain bullet at 990 m/sec (3,250 ft/sec), lethal up to 450 m (1,500 ft). It uses a direct gas system as opposed to a conventional piston. The gas tube leads gas to the face of the bolt carrier, causing the bolt to unlock and move back. The spent cartridge case is ejected after being withdrawn from the chamber by an extractor on the port. A port in the bolt carrier's side expels exhaust gas. The sights are set high, the rear sight forming part of the permanently attached carrying handle. The rifle is laid out in a straight line, instead of the butt dropping as in traditional rifle design.

by using the explosive force which propels the bullet forward to force the bolt to recoil.

When a bullet is fired the pressure of released gases which drive it down the barrel is as great backward as forward. In the AK-47 and most other types of automatic rifles, this backward pressure is used to push the bolt back, eject the empty bullet case and allow a new round into position. This can then be pushed into the barrel by the returning spring-loaded bolt and fired. The AK-47 is fed by a 30-round, detachable magazine and can be used to fire single shots as well as bursts of fire.

The AK-47 has not been without rivals. The armies of the Western industrialized nations have used a bewildering diversity of automatic rifles, some of which, like the G3 produced by the German firm of Heckler and Koch, were hardly inferior to their Soviet competitor. However, the most significant development occurred some 20 years ago with the design and introduction of Armalite rifles by the American Eugene Stoner.

One of the chief difficulties in equipping an infantryman is to keep down the weight he has to carry into action. The Armalite AR-15 solved some of the problems by being designed to fire a lightweight bullet at such high velocity that it was as lethal and accurate as heavier rounds. Lightweight materials such as nylon, plastic and metal alloys were used in manufacturing the AR-15 and it overcame initial teething problems to prove itself reliable.

By 1980 the Armalite revolution was just beginning to reach the world's best equipped armies and the change to small-calibre rifles was under way. The Soviet Union had begun to issue a 5.45 mm Kalashnikov called the AKS 74 and Israel had the 5.56 mm Galil. The NATO nations were conducting various trials of a wide range of rifle types and, although it seems nearly certain that they would adopt 5.56 mm calibre ammunition, nothing else was decided—except by the Americans, who pre-empted the result of trials by adopting the Armalite M16.

Sub-machine guns

The rifle is the most typical and useful personal weapon of an infantryman, but for special situations he may use a pistol or sub-machine gun. These are both comparatively inaccurate weapons, firing low velocity rounds, so they are used for work at close range. The sub-machine gun in particular gives impressive fire power and may be useful in house clearing—for example when resolving incidents of hostage taking. Both pistols and sub-machine guns are cheap and easy to produce, so they are manufactured all over the world and there are no absolutely outstanding makes.

Because the infantryman is called upon more and more to fight in guerrilla wars and policing actions he has also been given some protection in the form of body armour. Layers of heavy-weave nylon cloth have been shown to have a resistance to low-velocity projectiles: the Hardcorps I Armour made by

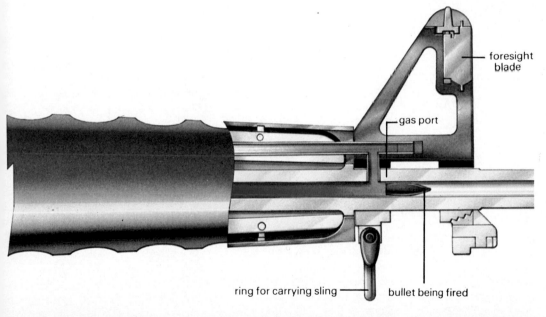

foresight blade

gas port

ring for carrying sling

bullet being fired

Below left Afghan guerrillas display captured Russian weapons. The anti-tank launcher (foreground) is the RPG-7V. For greater accuracy the projectile ignites an internal rocket as it leaves the muzzle.
Below The Israeli Uzi 9 mm SMG is only 44.45 cm (17.5 in) without the butt, making it ideal for troops in armoured vehicles.

Gunshots

necessity when the gun is used in its heavy, sustained fire role but, in that case, its barrel overheats and needs changing as it spits out between 750 and 1,000 rounds per minute.

In spite of all this it is difficult to see what sort of weapon can provide a replacement for the GPMG. It is obvious that a section machine gun must use the same ammunition as the section rifle, and the move to lighter 5.56 mm bullets by NATO has rather limited the range of experimental machine guns which have been evolved for it. The Israelis have simply fitted the ARM version of their Galil rifle with a bipod and a choice of magazine sizes up to 50 rounds. Whether this provides their infantry sections with the exceptional range and accuracy of a genuine light machine gun is a matter for debate.

The single man anti-tank weapon which is an infantry section's other support weapon is usually a shoulder-held rocket launcher. Unfortunately for NATO soldiers their US-designed M72, which launches a 66 mm rocket, is not good enough to harm modern tanks unless it is used against the thin armour at the back of the vehicle.

The best weapon of the type in service is the Soviet RPG-7. This 40 mm launcher fires an 85 mm rocket-propelled grenade to a maximum target range of 500 m (1,600 ft). The shaped explosive charge in the missile causes it to be focused into a high-temperature, high-velocity gas jet on contact which will penetrate 30 cm (12 in) of armour and squirt this hot gas with the molten armour into the tank's interior.

LAW

This very serious deficiency in the infantryman's anti-tank capability may soon be ended on the NATO side by a new British light anti-armour weapon (LAW). As far as is known LAW is a throwaway launcher which fires a single highly lethal round. It can destroy any armoured vehicle (including the formidable Soviet T-72 MBT) at ranges up to 300 m (1,000 ft). No doubt the Warsaw Pact armies will soon be issued with a replacement for the obsolescent RPG-7.

Infantry faced by an armoured threat do not rely completely on hand-held weapons to combat tanks. Heavier weapons are generally handled by a support company in each battalion. A fairly typical organization would have sections grouped in threes with a headquarters section to make up a platoon, and for three platoons with a company headquarters to form each company. The main structure of the full infantry battalion would be for three companies of this sort (usually

the Second Chance Corporation of America gives remarkable protection against high-velocity pistols, sub-machine guns and shell, mortar and grenade fragments. Protection against high-velocity rifle fire is claimed for Hardcorps I when it is reinforced by plate inserts which have the disadvantage of making it more cumbersome.

Because most infantrymen are sceptical about the ability of any body armour to protect them against the high-velocity rifle and machine-gun fire which they fear most, it is hard to convince them of its value. Statistically, however, the greatest cause of battlefield casualties is shrapnel and, as that can be resisted, it seems that body armour may soon find its way from special operations to full-scale battle.

Heavier weapons

Behind their personal weapons and protection is a long line of increasingly heavier weaponry to support infantry in battle. The smallest unit of riflemen is normally a section, which contains about ten men and would usually be equipped with a machine gun and a one-man anti-tank weapon. As most modern armies are equipped with automatic rifles, it has been argued that machine guns are unnecessary at section level; but most soldiers who have been in battle have no such doubts. Because of its heavier barrel and bipod or tripod rest, the machine gun fires the same ammunition as an automatic rifle considerably further and more accurately.

Above The British Army's 4.85 mm LMG (light machine gun). It has a 30-round magazine and an optical × 4 Trilux sight. It is recoilless and weighs only 4.08 kg (9 lb). A rifle version is expected to be in service by 1984.

Even those who accept the value of section machine guns are undecided as to which is the best type. During World War 2 it was common for sections to be equipped with light machine guns with box magazines and for belt-fed heavier guns to be used when sustained firing was required. This meant that each used a variety of weapons, with resulting strains in manufacture and maintenance, plus the additional disadvantage that soldiers were rarely familiar with more than one type. The answer seemed to be a general purpose machine gun (GPMG), which could perform all the necessary tasks. Most armies now have something similar in service.

Fairly typical of a GPMG is the British Army's L7A2, which is based on the Belgian MAG. While it performs all its tasks adequately it is not entirely satisfactory in any of them. As a light machine gun, which is what an infantry section requires, it suffers from the handicap of weighing more than 12 kg (28 lb) when loaded with a belt of 30 rounds. That is a lot for an infantry man to run across country with.

An additional disadvantage is the belt feed, because the trailing edges of the belt can become caught in undergrowth and hedgerow as the gunner doubles to a new firing position. The belt feed becomes a

referred to as rifle companies) to be backed up by a support company commanded from battalion headquarters.

The support company's weapons are normally of two sorts—heavier anti-tank systems and mortars. Until a few years ago, recoilless rifles and guns of various types requiring a crew of two or three men were the backbone of infantry anti-tank capability, but these are largely being replaced by guided missile systems. However, the recoilless gun has not been completely abandoned and the Soviet 73 mm SPG-9 fires a rocket-assisted round which does a very efficient job at well over 1,000 metres.

The missile systems use a number of methods of guidance, from radio control to wire fed out from the missile to the operator. For the operator himself there are basically two ways of issuing commands to correct the missile's flight. The first is by manual control, using levers or a joystick; and the second is by line-of-sight, in which the

Right The British General Purpose MG L7A1 in action. This gas-generated tipping bolt gun is a slightly altered version of the FN Mitrailleur à Gaz (MAG) weighing 10.89 kg (24 lb) and firing 7.62 mm NATO ammunition.

MOD/MARS

Royal Ordnance Factories

Above The General Purpose MG L7A2 is a slightly heavier model of the L7A1, weighing just over 12 kg (28 lb). The belt feed can make the gun unwieldy, but in its sustained fire role it is capable of firing up to 1,000 rounds of 7.62 mm ammunition per minute.

operator simply keeps the target in his optical sights for automatic commands of correction to be sent to the missile.

Because the line-of-sight method is more modern soldiers tend to consider it superior, but this view is difficult to justify. Such excellent weapons as the British 'Swingfire' and the Soviet 'Swatter' (the name is a NATO designation) are as accurate and effective as many line-of-sight controlled rivals despite their reliance on manual controls.

For practical purposes, a more significant division of anti-tank guided missiles might be between those which are portable by two or three men and those which really need to be mounted on vehicles. Of the more easily portable type, the French-designed MILAN (Missile d'Infanterie Leger Anti-char) is a line-of-sight guided weapon which has become popular with a number of NATO armies. Its makers claim that it gives a 98 per cent chance of striking targets 250–2,000 m (800–6,500 ft) away—and it hits hard. Of the heavy, vehicle-mounted weapons, the

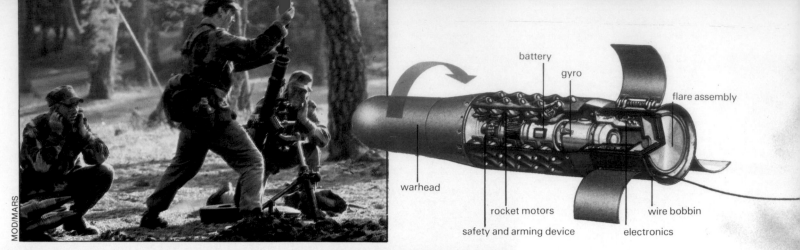

battery

gyro

flare assembly

warhead

rocket motors

safety and arming device

electronics

wire bobbin

MOD/MARS

Photri

US-made TOW (tube-launched, optically-tracked, wire-commanded) is easy to operate and battle tested.

The mortar, which is often found in the support company alongside these battalion anti-tank weapons, is an exceptionally useful weapon which, in effect, gives the infantryman his own light artillery section. It is basically a steel tube into which a bomb is dropped, to be shot out over a high trajectory arc by an explosive charge. Because of their simplicity, very light mortars can be carried by soldiers at platoon level. These are placed on the ground and hand held, usually for the purpose of providing concealing smoke or illuminating flares in an emergency —but their high-explosive bomb is not to be despised.

The medium mortars of the support company are more sophisticated and accurate, with an exceptional rate of fire and a very lethal high-explosive bomb. The new British–Canadian ML 81 mm L16 is a highly successful example of the type, which can

throw out 15 rounds a minute to a maximum range of 3,200 metres (2 miles). Just a few of these three-man weapons can provide a fearsome barrage when needed.

The weapons of the support company complete the inventory of a standard infantry battalion as it might be recognized world wide, but not absolutely everywhere—in Germany for instance such a battalion would be considered to be hardly equipped at all. There, with NATO and the Warsaw Pact facing each other on the Central Front, and to a lesser extent in the Middle East, where Israel borders the Arab states, the threat of tank warfare in the fast-moving *blitzkrieg* style have forced infantry to adopt expensive and advanced transport.

The basic doctrine of this sort of warfare is that a fast-moving armoured column should punch through the enemy's defences and quickly overwhelm the bases and depots beyond. With the punctured front-line now rendered powerless, the victorious armour can flood forward over the enemy's

heartland. The established counter to this is for the defenders to have equally mobile armoured forces, which will slow up the advancing column and harry its flanks.

For both attack and defence, infantry are as necessary as tanks. Defenders know that infantry can dig in where tanks are at a disadvantage—behind tree lines or in built-up areas—and attackers know that the only way to clear such places is by infantry assault. Consequently, infantry battalions are broken up and dispersed among the tank formations in combat teams and battle groups. To give these units the mobility and speed of the tank formations they accompany, each section is normally conveyed in a tracked armoured vehicle known as an armoured personnel carrier (APC).

By the early 1970s it had occurred to many soldiers that these sturdy APCs were capable of carrying heavy weapons to support the infantry in assault and defence. At first they simply mounted machine guns by the commander's hatch, but now some extremely powerful weapons are coming into service—particularly on Warsaw Pact vehicles. While NATO seem to favour a 20 mm cannon, the Soviets prefer a single-shot gun which fires a larger round. Such armed APCs are known as infantry fighting vehicles (IFVs) and the most lavishly equipped with offensive weaponry is the Soviet BMP-1, which has a 73 mm gun with a coaxially mounted 7.62 mm machine gun. A launcher for the 'Sagger' anti-tank guided missile is positioned over the gun and an anti-aircraft SA-7 is carried inside it.

As infantry weapons have enabled foot soldiers to accompany or oppose the most

error sensor   trigger

launch tube   tracker battery

initiator squib   propellant sticks

Sarson/Bryan

telescopic sight

aft end cap

propulsion canister

DRAGON XM47 SURFACE ATTACK GUIDED MISSILE

support stand

Below The 'Blowpipe' ground-to-air missile. A complete one-man anti-aircraft system, it is transported and launched from the same container, and weighs in all 18 kg (40 lb).

Short Brothers

sophisticated tanks, they have also begun to give them some defence against the potent threat of hostile airforces. Suitably mounted machine guns have always given infantry at least a slim chance of rebuffing low level attack from aeroplanes or helicopter gunships, but the age of the guided missile may strengthen their hand. The best known of these is undoubtedly the Soviet SA-7 which is a shoulder-fired, heat-seeking missile launcher which has seen wide combat service—particularly with guerrilla armies.

There are not many shoulder-held, anti-aircraft missile systems available, but the Swedish firm of Bofors produce the RBS 70 which is laser-beam guided. This may well provide a deterrent to any aircraft up to an altitude of 3,000 m (10,000 ft). Certainly the Swedish Army, which demands a very high equipment standard, has put it into service.

Even in some of the most modern armies, anti-aircraft missile systems are not normally used at the level of the infantry battalion. With recent advances in electronics, however, the 1980s should see light missiles with exceptionally accurate guidance systems that may well give dug-in infantry a crucial advantage over armour and aircraft.

But, just as infantry weapons improve, so do armour and aircraft defence. As the technological advantage sways from one military arm to the other, the only reasonable certainty is that the basic job of a foot soldier will be unchanging. In the final analysis all his sophisticated support weapons are designed to take him and his personal weapon—his rifle—to victory over similarly equipped enemies: to take or hold ground that is important to his commanders.

Ammunition, the sharp end of warfare

Effective small arms ammunition is crucial to any army. Today's soldier may carry into battle cartridges which fire not bullets, but several thin metal darts in a single shot. The cartridge case itself may be made of propellant rather than metal and so be consumed when the shot is fired. These and other developments augment continuing research into more conventional ammunition.

In the hundred years or so since the metallic cartridge appeared, its basic form has changed remarkably little. Basically, the metallic cartridge has four components; the cartridge case, primer, propellant charge and bullet. In *rim fire* cartridges, the primer is within the rim of the case whereas in *centrefire* cartridges, it is in a centrally located cap in the base of the case.

Since their introduction in the middle of the last century, rim fire and centrefire cartridges have undergone two major modifications, both dating from about 1886. The first was the substitution of smokeless powder for the original black powder and the second, largely dependant upon the first, was the progressive reduction in service rifle calibres from the usual 10–11 mm to 6.5–8 mm. These lower calibres remained unchanged until World War 2.

In Germany, a new development occurred that was to change post-war design and development considerably. Germany alone recognised that infantry fighting ranges were no longer about 900 m (3,000 ft) or more, for which the first metallic cartridges were designed, but were, in most instances, about 350 to 550 m (1,000 to 1,800 ft) only. For such reduced range, and for use in a new type of automatic rifle, known in Germany as the 'Sturmegewehr' (assault rifle), Germany produced a new 7.92 mm cartridge with a shortened case.

This new short-cased cartridge was ideal for the new class of rifle; it was lighter and produced less recoil. After World War 2 several countries, including the USSR, Britain and the USA, experimented along similar lines and, as a result, the old breed of rifle cartridges first introduced in various countries in the 1890s was largely phased out, although a few have remained in service for use with medium machine guns. The most important consequence of this post-war

development based upon the German short cartridge was the introduction of two 'treaty cartridges'—firstly the Soviet 7.62 mm short cartridge, with a 390 mm case length and secondly the NATO 7.62 mm cartridge with a 51 mm case length. The former cartridge was adopted as standard by the whole of the Warsaw Pact armies and the latter by NATO and similar Western alliances.

Several development trends in rifle cartridges are now apparent. Research conducted largely in the USA focused on the problems of 'hit probability' with the rifle

fired by an ordinary soldier. This work included an analysis of battle casualties, and the result confirmed that hit probability fell off sharply once the engagement range increased beyond about 100 m. It was also ascertained that hits on human targets were random and often the result of unaimed fire. An infantryman armed with an automatic or self-loading rifle had a better chance of securing hits than one armed with an ordinary magazine rifle. This investigation in the early 1950s resulted in the first of a series of experimental small-calibre, high-velocity car-

Daily Telegraph Colour Library

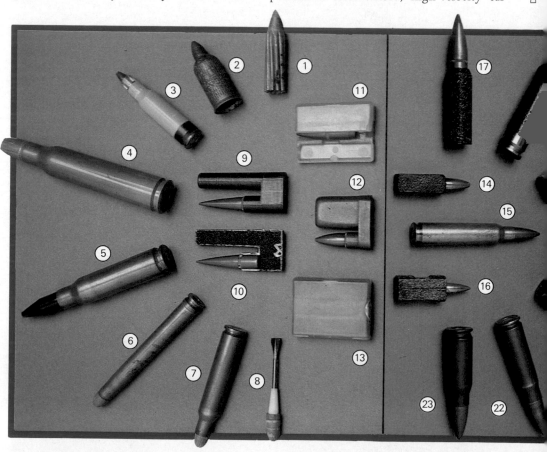

For more than a century, basic design of small arms ammunition changed little. But even in the age of missiles, the soldier's equipment is crucial. Continual research into conventional ammunition has produced some novel developments, for example, the consumable case cartridge. (1) and (2) caseless projectile after and before firing; (3) 4 mm projectile in sabot, in plastic case with steel head; (4) 8.35 mm multi-flechette; (5) 7.62 mm with depleted uranium slug in plastic sabot (NATO); (6) 5.6 mm flechette; (7) and (8) 5.6 mm flechette, whole and without case; (9) and (10) folded case, whole and sectioned; (11) and (13) Hughes, fully telescoped, sectioned and whole; (12) folded case; (14) 5.56 mm consumable case (USA); (15) 5.56 mm Armalite; (16) 4.7 mm consumable case (W Germany); (17) 7.62 mm consumable case (USA); (18) 7.62 mm consumable case with steel head (USA); (19) 7.62 mm (NATO); (20) 7.62 mm (USA); (21) 7.92 mm Mauser (Germany, 1898); (22) 7.62 mm Kurz (German).

tridges that, 25 years later, are having a direct influence on the re-arming of NATO and the Warsaw Pact armies with new weapons and ammunition.

In the US experiments, small calibre —5.56 mm (0.22 in)—was combined with a relatively long case—about 45 mm (1.7 in)—to produce an extremely lethal, high-velocity cartridge. The bullet for this class of cartridge was reduced to nearly a third of the original weight of the 7.62 mm NATO cartridge, so that it weighed only 55 grains, compared with the 144 grains of the NATO ball bullet.

This reduction in weight helped to increase velocity, but at the expense of range, and the bullet became far less lethal beyond a range of about 450 m (1,500 ft). This was not a disadvantage, however, because the current tactical doctrine in the USA was veering towards short-range engagements for infantry. The muzzle velocity of the new bullet was about 970 m/sec (3,100 ft/sec), a considerable advance on the 820 m/sec (2,700 ft/sec) of the 7.62 mm NATO bullet. But most importantly, the new cartridge could be fired from a new, light-weight automatic rifle. Such a rifle was then developed in the USA, originally called the Armalite, and eventually known by its US Army designation of M16.

The American experiments with 5.56 mm cartridges led to widespread imitation elsewhere, with various calibres and case sizes, but all the experimental cartridges that stemmed from the American research were of 5.56 mm calibre or smaller, and all behaved in the same ballistic fashion. In the next ten years, many countries adopted the American cartridge and the Armalite rifle. The USSR, Communist China and their various satellites continued to favour the original Soviet 7.6 mm cartridge. But this scene has recently changed dramatically.

For some time NATO have adopted the American 5.56 mm calibre lightweight cartridge for both rifle and light machine gun. Non-aligned countries, such as Austria, Switzerland and Sweden, have already decided individually to adopt a similar cartridge, so that by the mid-1980s, most of western Europe will have switched to the new lightweight cartridge.

In the late 1970s it became clear that the Soviet Union was also contemplating a

Soon, law-enforcement officers may carry weapons that fire not bullets but several thin, metal darts in a single shot—so lethal are the weapons of modern technology.

757

switch to a new calibre, but details were obscure. It is now apparent that the USSR has adopted a 5.6 mm cartridge, with a 39 mm case length, to be fired from a modified Kalashnikov-type rifle. The new rifle was first used by front-line troops of Afghanistan.

It is estimated, therefore, that between 1980 and 1985 the major armies of the world will be adopting rifles that fire one of two main types of 5.56 or 5.6 mm cartridges, and it is likely that these will remain in service, once adopted, for a considerable period. It should be stressed that, although representing a considerable break with the past, in terms of calibre and weight, both the Soviet and the American–NATO cartridges are conventional in design: they comprise case, primer, propellant and bullet.

Since about 1950, attempts have been made to improve the hit ratio (the number of hits divided by the number of shots fired) by using more than one projectile in each cartridge case. The largest practical number of bullets in a normal case is three mounted in tandem—known as a *triplex cartridge. Duplex loads*—two bullets mounted in tandem —have also been tried.

Salvo squeezebore (SSB) is another attempt to improve hit ratio. It is a variation on the duplex or triplex theme and consists of mounting several projectiles in a plastic sheath projecting from the mouth of the cartridge case. As many as eight projectiles have been fired by this method, but the usual maximum has been five. These rounds can be fired from rifles, machine guns or pistol calibre weapons, the barrels of which have either specially reduced bores at the muzzle, or muzzle attachments with specially reduced diameters. When fired, the projectiles are swaged into a reduced diameter form, and are propelled at a high velocity with fair accuracy and satisfactory dispersion. Although used in combat in Vietnam, SSB

cartridges have not yet been fully developed.

As a variation on the multi-projectile cartridge, *multi-flechette* cartridges consist of thin steel arrows, initially mounted in a discarding carrier or *sabot,* which on account of their extremely light weight have an extremely high velocity—about 1,400 m/sec (4,600 ft/sec). Although lethal at short range, flechettes rapidly lose accuracy as range increases.

In the USA and in Europe, extensive work is being done to reduce the number of cartridge components to reduce weight and cost, and conserve strategical raw materials by the exclusion of copper-containing brass cases. The commonest and best favoured solution to this problem is the *consumable-cased cartridge,* in which the case is made of solid propellant.

The bullet is secured in a solid cylinder of propellant, fashioned in the form of a normal cartridge case. In the base of the cylinder is secured a pellet of priming composition. When the primer is struck by the firing pin, the solid propellant case ignites and burns, being totally consumed in the process and produces a hot, expanding gas which expels the bullet through the rifle barrel.

Major problems encountered with consumable-cased cartridges relate to breech blocking, overheating, and safety. With a consumable-cased cartridge, the heat can be transferred only to the weapon itself, causing serious problems of weapon heating.

Another solution to the problem of the cartridge case makes use of a bell-shaped or elongated projectile with a hollow base. Within this hollow base is secured propellant and a primer pellet. When the pellet is ignited, and in turn ignites the propellant, the whole moves forwards. The hollow rear end is dragged through the barrel, and is swaged into shape in the process. The entire mass then forms the bullet in subsequent flight. Problems with this type include the breech-

ROF Radway Green

Bullet design is continually being reviewed in many countries to provide better bullets and to meet the changing demand, not only of armed forces but also of civil defence forces throughout the world. Distinct advantages have been found in the use of depleted uranium as a core material but its deployment has been discouraged on moral grounds. Other materials, such as wood, plastic and aluminium, have been used, generally in attempts to shift the bullet's centre of gravity to suit various designs.

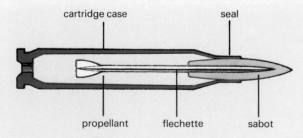

FLECHETTE CARTRIDGE

cartridge case   seal

propellant   flechette   sabot

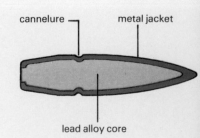

LEAD-CORED RIFLE BULLET

cannelure   metal jacket

lead alloy core

blocking problem found with consumable-cased types, and limited effective range.

Weapon designers have had to cope with breech actions capable of handling the long, thin, form of the conventional cartridge. The ability to feed such a round through the mechanism imposes considerable design constraints and results in a heavy weapon. In recent trials, it has been claimed that these problems are solved by a cartridge that retains the conventional number of basic components, but has these arranged in a different layout. Various designs exist of *folded* or *encapsulated* cartridges, some of which have cases of plastic rather than metal. These cartridges have not only been made in rifle calibres, but also have been tried in cannon calibres up to 30 mm (1.2 in).

Most cartridge cases are made of brass, but in some countries, notably Germany, the USSR, China and several of the eastern European communist states, steel cases are standard. In the past, most countries have attempted to use aluminium instead of brass, mainly to save weight, but aluminium has proved to be prone to rupture with consequent gas escape. Nevertheless, efforts are still being made to use aluminium and, as a compromise, cases with steel heads and aluminium bodies have also been tried. Similar savings in weight and in strategic

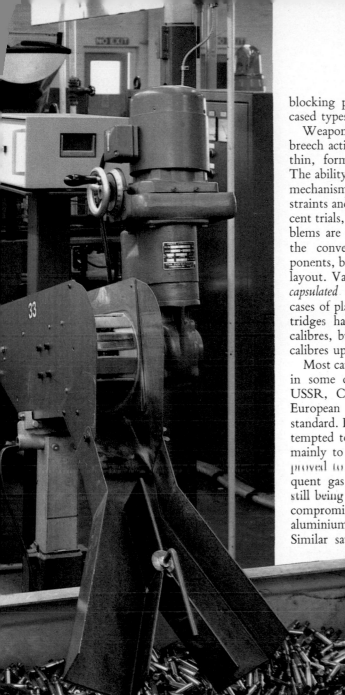

Left In a modern factory for making cartridge cases, automation is an efficient means of increasing output and reducing costs. The conventional cartridge case is made of brass but experiments with caseless cartridges are under way.

raw materials of cartridge cases (mainly copper) were expected from the use of plastic instead of brass or steel but, as yet, this expectation has not been realized for ball ammunition, although plastic-cased training ammunition (blanks) and grenade propelling ammunition are used in several countries.

The original solid lead projectiles, common when metallic cartridges were first introduced, gave way in the 1880s to composite bullets having metal envelopes with, usually, lead alloy cores. Round nosed bullets have given way to pointed or *spitzer* bullets at about the turn of this century, and until recently little basic change occurred in ball bullet design.

An efficient method of causing a bullet to transfer its energy quickly to the target has come from Germany. Experimental bullets with *Löffelspitz* have recently been produced. Such bullets have the area near the tip scooped out, as if by a spoon—hence Löffel, the German word for spoon.

The asymmetrical bullet is designed to tumble more readily than a normally shaped bullet when the target is struck, thus increasing wound effect. Another, similar, more recent development is the double Löffelspitz, in which there are two dissimilar spoon-shaped depressions on opposite sides of the tip.

The velocity of a bullet is an important factor that determines its accuracy, resistance to gravity, and wounding effect. In general, the recent trend has been towards lighter bullets moving extremely fast. The use of a sabot around the bullet so that a smaller diameter bullet may be fired from a larger capacity case has been one way of achieving high velocities. This system has already been described with flechettes but with conventional amunition, bullets of about 3 mm diameter have been fired at enhanced velocities from 5.56 mm cases using sabots.

In the past, projectile cores have usually

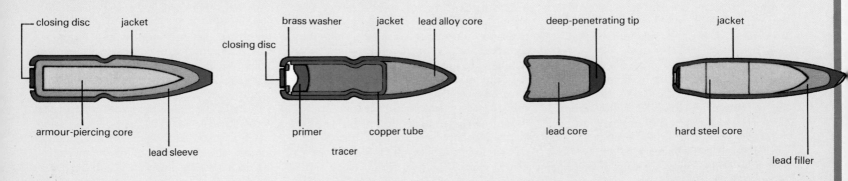

ARMOUR-PIERCING BULLET — closing disc, jacket, armour-piercing core, lead sleeve

TRACER BULLET — brass washer, jacket, lead alloy core, closing disc, primer, copper tube, tracer

PISTOL BULLET — deep-penetrating tip, lead core

LÖFFELSPITZ — jacket, hard steel core, lead filler

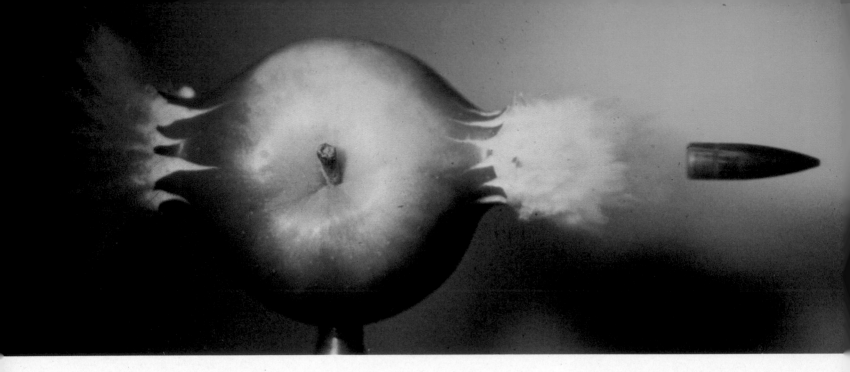

been made of lead alloy, but cones of hardened steel or of tungsten carbide have often been used to achieve armour piercing capability.

In addition to the work being done on rifle-calibre weapons, there are other developments that apply more to larger calibre weapons. It is likely that liquid propellants will replace conventional nitrocellulose powder propellants, and electrical ignition will supplant the percussion primer system.

Of the various developments, some are far more likely to represent the future real trend than others. The NATO small arms calibre trials ending early in 1980 included, as a West German entry, a consumable-cased 4.7 mm cartridge. This was finally withdrawn, but was entered with the main objective of gaining further experience under tough trial conditions of this type of cartridge. It seems likely that further sustained efforts will be made to perfect this type so that, for the next generation of small arms, conventional ammunition can be partly replaced. It is also likely that development of flechette will continue in the USA, together with the special weapons required for this ammunition, for some years ahead, and this class of ammunition could, therefore, become adopted, at least in a limited role.

The ammunition for revolvers, self-loading pistols and sub-machine guns has changed less (and will probably change even less in the future) than rifle calibre ammunition. Pistols and even submachine guns have a lower value and lower priority to the military than rifles or machine guns. Pistols and submachine guns are usually of larger calibre, usually between about 7.62 mm and

A bullet passes through an apple virtually unchecked. A high-velocity bullet is particularly lethal because its shock waves damage tissues beyond the immediate wound.

11.25 mm than rifles. Most rounds in police or military use are based upon cartridge cases that date back to before World War 1, but a few revolver cartridges in police use are of more modern vintage. The main development in pistol calibre ammunition is now in the bullet design, and here there are two main requirements.

A normal army requirement for pistol ammunition would be met with conventional lead-cored, jacketed bullets, but, increasingly, the military is becoming involved with anti-terrorist activity in urban areas, where the requirements are different. Here the army requirement overlaps with police requirements, to the extent that the police also have a similar duty against terrorism. For this kind of deployment, special bullets are being developed and used.

For anti-terrorist use in urban, built-up areas, the target may be fleeting and there will not be opportunities for a second shot. This is particularly important if hostages are being held. Accuracy and killing power are of prime importance. A further important consideration is that, especially when hostages are involved, or with bystanders at risk, the bullet, although being lethal to the terrorist target, should ideally not be able to continue and injure or kill the innocent.

Some designs—many originating in West Germany—have been produced to cope with this requirement. They are known under the generic term of *effect geschoss*. These appear normally in 9 mm Parabellum calibre, which

is probably the most widely used calibre in the world for self-loading pistol or sub-machine guns, but could easily be made in any other large pistol calibre. The basic effect geschoss design has a nose cavity in the bullet. This cavity is either covered with a thin metal shroud so that the bullet seems to be full jacketed, or it is filled with a plastic plug. The intention is that the cavity causes the bullet to expand upon impact. Effect geschoss bullets are lighter than normal ball bullets and have higher velocity.

Pistol bullets that are highly penetrating have been produced in the USA and in various parts of Europe and Scandinavia. Normal ball bullets are ineffective in stopping cars, and several designs exist to cope with such targets. In one form, the bullet has a hard steel core inserted as a separate component. In a more recent design, the bullet has a clad steel envelope with a normal lead core, but the envelope at the nose is specially thickened to form an armour-piercing cap.

In the USA, it was discovered that all-steel bullets coated with Teflon had special piercing qualities. The Teflon provided lubrication and, therefore, improved penetration significantly.

Further development work on pistol calibre ammunition is likely to include special bullets suitable for use by security guards in hijacked aircraft. Such ammunition has already been produced but has not yet been perfected. The design of a small-calibre bullet with sufficient accuracy and power to injure or kill a person, but which will not seriously damage the pressurized hull of an aircraft should it miss the hijacker, is a difficult problem to solve.

Tanks: speed versus weight

Since their first spectacular appearance in the battlefields of World War 1, tanks have played a decisive role in numerous hard-fought actions. During 60 years of development, particular effort has gone into increasing their mobility.

The first tanks were designed to overcome the stalemate of trench warfare. The basic concept was of a mobile armoured gun platform which could overcome the barbed wire and machine guns proving so deadly to infantry. It would open a gap in the enemy trench system for a cavalry breakthrough. When used properly in 1916 and 1917, the tank performed its task adequately—but the unmotorized cavalry and infantry of the day proved unable to make the desired breakthrough.

Perceptive soldiers in Britain and later in Germany realised that its mobility made the tank itself the ideal instrument for exploiting any gaps it made in the line. By the outbreak of World War 2 the Germans had created an armoured force to try out this theory. But while its cutting edge was the tank, the force was not composed solely of tanks. The Germans considered that the tank needed support from the other military arms to achieve maximum results, and that these would need to be mechanized so that they could keep up

with the tanks and not deprive them of the priceless asset of mobility. So motorized infantry, artillery workshops, and engineers rolled forwards with the tanks in highly manoeuvrable armoured columns to achieve notable successes.

This idea of warfare has changed little since World War 2, but there have been great technical advances in the construction of the whole range of vehicles that make up the armoured mass. Indeed the reconnaissance vehicles, infantry carriers and self-propelled guns in service today are so much like tanks in appearance and some specifications that they are easily confused with the real thing. In this rather obscure situation the narrowest and best definition of the tank is of an armoured vehicle that is designed to be used only in the role of a main battle tank (MBT) and not as a battlefield taxi for infantry or as a reconnaissance aid.

If the first characteristic of a tank is mobility, it comes as some surprise to find that this is achieved today by substantially the same methods as 65 years ago. The first tanks to lumber across the shell craters of

Top A tank in its element—a German Leopard 1 churning up the mud on manoeuvres. Leopard is a classic post-War battle tank designed for high mobility to break through enemy lines.

northern France did so on caterpillar tracks, and their most modern descendants have yet to improve upon that system of carrying heavy vehicles across difficult ground.

Today, however, the strain placed upon these tracks is vastly greater. The earliest tanks moved at scarcely above walking pace, but Germany's latest MBT—the Leopard 2—has a road speed in forward gear of 68 km/h (42 mph). Of course, the Leopard's cross-country speed is considerably less, for its crew could hardly exceed 35 km/h (20 mph) without serious risk of injury from being jolted about in the hard armoured turret and hull. Furthermore, a tank manoeuvring at 35 km/h across country will soon shed a track if it is not correctly adjusted.

A caterpillar track is made up of 100 or so pin-jointed links and is prone to stretching with wear. It is kept at the correct tension by adjustment to the idling roller or, occasionally, the removal of a link. This process is not endless; tracks have a fairly short life and need regular replacement. The length and breadth of the tank critically affects the length and width of the tracks that it can use, but, in general, the best way to give a tank agility is to put as much track under it as possible all the time.

The ratio of the tank's weight to the area covered by its tracks—its ground pressure

761

—is one of the two important equations that will decide how manoeuvrable it will be. The other is the ratio of the power developed by its engine to its weight. If we look at two MBTs currently in service we can see from the figures that there will be a vast difference in performance. The British Chieftain has a massive combat weight of 54 tonnes, a power-to-weight ratio of 13.49 hp/tonne and a ground pressure of 0.84 kg/sq cm (12 psi). In contrast, the French AMX-30 has a combat weight of 36 tonnes, a power-to-weight ratio of 20 hp/tonne and a ground pressure of 0.77 kg/sq cm (11 psi). A quick glance at these figures makes it a mechanical certainty that the AMX-30 will be by far the more agile vehicle—but this is not to be attributed to British incompetence in tank specification and construction.

The value of armour

The reason why the Chieftain gives these rather unimpressive figures for manoeuvrability is that the British believe that a tank's capacity to survive depends upon its having armour and striking power superior to its enemies. Armour is heavier if it is thicker, and that is the cost which the British are prepared to pay; they regard the very much thinner protection of the AMX-30 as controversial to say the least. In fact both the British and the French are due to replace their MBT fleets during the 1980s and it seems likely that the AMX-32 will have increased armour protection. The British MBT-80 is certain to have a more powerful engine.

Engine power has long been recognized as the most serious, yet most easily remedied, defect of the Chieftain. In the days of its development it was considered desirable for NATO armoured fighting vehicles (AFVs) to have engines that could use any fuel—petrol, diesel or jet fuel —that might be available in wartime. The British attempted to develop such an engine, but the result was so weighty and complicated that they eventually settled for the Leyland L-60 in the Mark 1 Chieftain. This was the usual diesel engine; diesel is a favoured fuel for AFV's because it carries a lower combustion risk than petrol.

The first L-60s developed a puny 585 hp, which was clearly inadequate, and even an

improved version which managed 650 hp was not good enough. Development has continued; indeed it was hoped that an engine providing at least 750 hp would be in service from 1979. However, Britain had sold about 750 Chieftains to Iran before the revolution and when the Shah ordered new tanks based on the Chieftain he showed a preference for Rolls-Royce engines, culminating in the CV1 2CA. This power pack is a 60° Vee-form 12 cylinder, direct injection diesel engine which can develop 1,200 hp. Some development of it may well power the next British tank—the MBT-80—if it is not superseded by a new American design.

In spite of the technological sophistication

that the Americans normally bring to weapon development, they have recently been one step behind so far as tanks are concerned. All this was changed with the new XMI MBT introduced in 1980. Among the many advances incorporated in this remarkable tank is an Avco-Lycoming AFT-1500C gas turbine engine which can use diesel, petrol or jet fuel and develops 1,000 hp at 3,000 rpm. The obvious advantage of this powerpack should make it the first choice for most of the new MBTs being considered by NATO countries, but it has had a number of teething troubles during trials. Although the engine is now said to be reliable, it may have to be in use some years before the last doubt is dispelled.

120 mm smoothbore gun

gunsight with integrated thermal image unit

bore evacuator

steering column

driver's seat

The German Leopard 2 typifies the modern approach to tank design with its immensely powerful engine, high-energy shock absorbers, and heavy armour. It is also highly mobile and has heavy fire power.

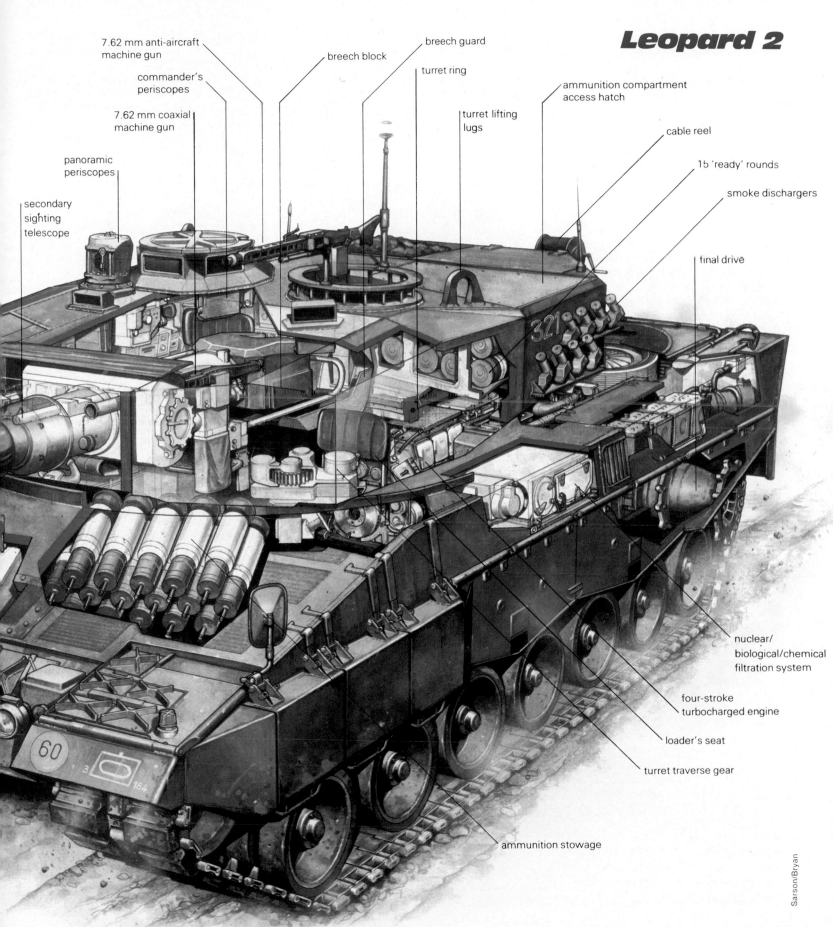

Leopard 2

7.62 mm anti-aircraft machine gun

commander's periscopes

breech block

breech guard

turret ring

breech block

7.62 mm coaxial machine gun

panoramic periscopes

turret lifting lugs

ammunition compartment access hatch

cable reel

15 'ready' rounds

smoke dischargers

secondary sighting telescope

final drive

nuclear/ biological/chemical filtration system

four-stroke turbocharged engine

loader's seat

turret traverse gear

ammunition stowage

60

321

3

184

Sarson/Bryan

763

Even with its gas turbine engine the XMI will have a slower road speed than the most modern Soviet tanks. With armoured forces numerically superior to any deployed by a potential adversary, the Soviets are believers in speed and mobility, which should give them the initiative in war. All their recent MBTs have shown an impressive road speed, with the T-72, first seen with their forces in October 1977, capable of 80 km/h (50 mph). The T-80 which is following it is very unlikely to be slower.

However, these road sprinting capabilities may not give Soviet tanks any advantage through superior manoeuvreability in battle. The fact is that the crew of an MBT must have a reasonably stable platform from which to do their jobs. While tank suspension is designed to give as smooth a ride as possible, it also has to be rugged. Modern suspension systems generally consist of between four and seven rubber-tyred wheels over which the track runs; on a number of MBTs, shock

Inset A display of Russian armoured might in Moscow. The elegant lines are indicative of the traditional Soviet emphasis on speed and mobility in their tanks.

Below A tank advance can be brought to a halt by a river or deep ditch, so bridge-laying tanks like this Chieftain have been developed.

absorbers have also been fitted to some of the road wheels. But it still seems even Soviet tanks are limited to a practical maximum of 35 km/h (20 mph) across country and considerably less in rough conditions. Any greater speed would injure the crew.

Manoeuvreability, however, is not simply a question of speed: a tank must be handy at turning. Early tanks were guided simply by applying a brake to one track, which slowed it down in relation to the other and caused the machine to turn—the harder the brake was used the sharper the turn would be. With this rather primitive method the braking caused an unacceptable wastage of power, so the modern technique, known as *regenerative steering*, was devised. In this system the power is subtracted from one track and transferred to the other through a differential gear. It has been so highly developed that the driver can now slew the tank around on a point.

However high-powered, well suspended and manoeuvrable a tank may be, it is generally necessary to make some modification for special obstacles. In some cases the whole concept of an army's MBT will be influenced by the sort of country it is expected to encounter. One example is provided by the TAM (Tanque Argentino Mediano), built for Argentina by the West German

Powered by a 1200 hp Rolls-Royce engine, the British P4030/3 has a power-to-weight ratio of 19.5 hp/tonne enabling it to travel at speeds of up to 60 km/h (37 mph).

firm of Thyssen Henschel. The TAM is a medium tank in the 30 tonne class, instead of the more usual 40 to 50 tonnes, because the Argentine Army has taken the realistic view that many roads and bridges in South America are unable to take the heavyweights. In fact the TAM also has an effective Argentine-designed 105 mm gun, which gives it the fire power of many heavier tanks.

Crossing the river

Even where bold, radical decisions in tank design are necessary, it is often found that straightforward off-road agility is not enough. On the Central Front in Europe, where NATO and Warsaw Pact MBTs are deployed, it has long been axiomatic that armoured forces should have some amphibious capability. It is obvious at first sight that heavy metal objects such as tanks do not make natural amphibians. The established ways of making a tank take to water are either to erect screens around the tank hull until it displaces so much water that it floats, or simply to make the vehicle watertight, erect a snorkel and drive it across underwater.

The Soviet Union, with its habitual emphasis on mobility, makes quite a point of having an amphibious armoured force and favours the snorkel method—as does France. Although it must be said that the Soviet MBTs are as good amphibians as any other, the idea of fixing a flotation collar around a vehicle, as opposed to using a snorkel, has many advantages—not the least being that

the crew are more likely to escape in the event of an accident. No tanks are really secure crossing deep, fast moving rivers and all have to cross in carefully selected places. River banks can often become effective anti-tank ditches—a tank trying to get out of a river with its nose up can lay very little track on the ground and so loses traction.

Because tanks are not at their best when wading or swimming, considerable thought has been given to the swift bridging of obstacles. Military science relies heavily upon statistics and one statistical survey revealed that 90% of all obstacles that need to be bridged in 'normal' terrain are less than 9 m (30 ft) across so the provision of a standard, swiftly installed 12 m (40 ft) bridge will mean an important improvement in mobility of an armoured force. All modern armies employ bridgelaying tanks to overcome these lesser obstacles.

Basically, a bridgelaying tank consists of the armoured hull of a tank topped by a bridge and hydraulic lifting gear. The armour means that the bridging can take place under fire, when the hydraulic apparatus will lift the bridge and push it forward into position. Then the bridgelaying tank disengages and backs away, allowing fighting vehicles to cross. Often, as in the case of the German Leopard bridgelayer, the bridge itself is in two interlocking sections that give it added length—in the Leopard's case 22 metres (72 ft)—which will cross most streams.

The bridgelayer is only one of a family of adapted tanks which perform special roles to keep MBTs mobile. In order to simplify production most of these specialist vehicles are constructed on the chassis of a nation's standard MBT. The highly successful German

Leopard 1 which has been adopted by so many armies, provides illustration of this. Besides the bridgelayer there is the Leopard MBT armoured recovery vehicle, which is equipped with a crane, bulldozer blade, electric wrench and welding systems and a spare Leopard MBT engine. The MBT itself is so designed that the engine, transmission and cooling system have couplings that can be rapidly disconnected, allowing the entire engine to be lifted out and replaced in the field within 20 minutes. The vehicle's gun barrel can be changed in the same time.

Obviously this sort of ready mechanical help in the field can restore many crippled tanks to the battle with little delay.

Supporting roles

As a companion to the armoured recovery vehicle, there is an armoured engineer vehicle which carries explosives for demolition work, has an auger instead of a spare power-pack and can have its bulldozer blade fitted with scarifiers to rip up road surfaces. This family of support vehicles will obviously develop with Germany's MBT fleet and their successors will, in time, be mounted on the chassis or hulls on the Leopard 2.

The existence of support vehicles has proved significant in battle and it points to the last and most delicate aspect of tank mobility. This is known as RAM-D (reliability, availability, maintainability and durability). The effort to find the correct balance between protection, mobility and firepower has called on the most advanced technical accomplishment. Yet it is essential that designers do not step over the frontier between improvement and complication. Ideally, all equipment must be both utterly reliable and easily reached for replacement when failure or battle damage make that necessary. There is no point in having a sophisticated MBT if it spends more time in the workshop than on the field.

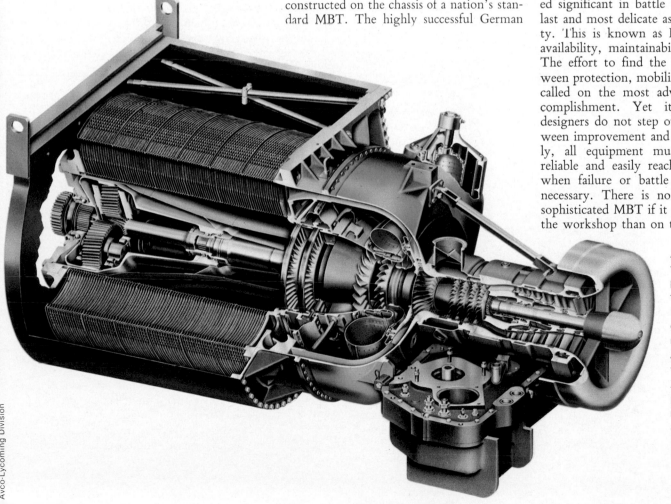

A powerful engine that will run on diesel, petrol or even jet fuel—the Avco-Lycoming AFT-1500C gas turbine from the remarkable new American main battle tank, the Chrysler XM1. Although not perfected yet, its massive 1,000 hp output should make it popular with NATO.

Tanks: hunter and hunted

No one who has seen a tank at close range can fail to be impressed by the fearsome spectacle. Although the true power of the modern battle tank is seldom witnessed by civilians, examination of its design does give some idea of its strengths and weaknesses.

A tank is essentially a mobile, armoured gun platform, its first characteristic—mobility —being greatly affected by its second—armour. Strong armour generally means heavy armour and that in turn means a lower top speed, so armies have had to decide whether they need a fast, manoeuvrable tank or a well armoured one. The British 'Chieftain' is a tank which sacrifices mobility for protection, as is the Swedish 'Stridsvagn 103'—an armoured vehicle designed to dig itself in when it encounters resistance. The development of more powerful engines in recent years had made the most modern tanks both highly mobile and heavily armoured, but these 'supertanks'—such as the German Leopard 2 and the US XM1—are not yet ready for full scale conflict.

'Hull down'

Armies that expect to fight on the defensive, as the British and Swedish do, seem correct to value protection above mobility. This has been borne out by the wealth of recently acquired experience of the Israeli Armoured Corps. They found that highly manoeuvrable attacking main battle tanks (MBTs) have been outmatched by more heavily armoured vehicles in the 'hull down' position—that is, to make as small a target as possible by placing the tank on a reverse slope so that the hull is under cover but the turret and gun are still clear and usable.

The value of the hull down position is so well known that there is a long history of effort to give MBTs a lower profile. When the Chieftain first appeared in the late 1950s, it introduced the idea of the driver—who usually sits in the hull in front of the turret—carrying out his duties in a semi-reclining position, because a few vital inches had been lopped off the height of the hull. In the Stridsvagn 103, this principle was taken even further, by dispensing with a turret

altogether and fixing a retractable bulldozer blade on the front of the vehicle so that it could dig itself in.

The Stridsvagn's gun is worked from inside the hull by a crew of three, using periscopic sights. The gun is moved horizontally by traversing the whole vehicle, and it is elevated by altering the pitch of the entire hull, using its remarkable suspension. Although the Stridsvagn has several disadvantages, compared with more modern MBTs—for example, it has to stop to change gear—the concept of a turretless tank is still alive. The German firms of Thyssen Henschel, MaK and GST have suggested a design for the development of Kampfpanzer 3 (Germany's future MBT), which has only the gun and mount above the hull, with the crew and ammunition below.

One of the best modern tanks in service is the Israeli Merkava, which has been designed especially to fight from the hull down position. Design work on the Merkava project was begun in 1969, and the first production tanks were delivered to the Israeli Army in 1978. Of the three tank characteristics, armour was made top priority with firepower second and mobility third.

The Merkava is an unusual design which resulted from lessons learnt in combat. In the usual tank layout, the driver's seat is in the front of the hull, the turret and fighting compartment in the middle and the engine in the rear. In the Merkava, the engine is in the front on the right, with the driver on its left, and the fighting compartment is in the rear. The Israelis have also built a turret with a distinctive shape that, evidently, presents an even smaller target when the rest of the tank is in the hull down position. From its considerable weight—estimated at between 58 and 62 tonnes—the Merkava also seems to carry an unusual thickness of armour.

It is not just its thickness, however, that gives armour its high resistance. For a long time anti-tank weapons used high velocity shots to pierce armour. To defeat this kind of attack, armour has usually been constructed from nickelchrome steel, which does not

crack easily. And the armour is well sloped to deflect high-velocity rounds, which always arrive on a flat trajectory. Experience in battle has shown that the most hits are received on the front of the hull and turret. Consequently, the heaviest armour and the most steeply sloping glacis are at the front.

But increased protection is limited not only to retain mobility but also by space. The thickness and slope of armour both contribute to reducing the inner dimensions of hull and turret, and although a tank crew is always expected to work in confined quarters, there is a point beyond which constriction seriously impairs efficiency.

Chemical-energy warheads

Sloping steel armour may be the most effective protection against high velocity shot but it is vulnerable to chemical-energy warheads. These are of two main types: high explosive anti-tank (HEAT)—a shaped charge explodes close to the armour in such a way that the hot gases from the explosion are focused into a jet which lances the armour, squirting molten metal out of the other side; and high explosive squash heads (HESH) which are exploded against the armour by a slightly delayed action fuse causing shock waves—these fragment the armour and send scabs of metal flying off the inside at great velocity. Because chemical-energy rounds attain their best effect only when fired at low velocity, they arrive on

Bottom right Ideal for keeping a low profile—with no turret and equipped with a bulldozer blade for digging in, the Swedish Stridsvagn 103 presents a small target when 'hull down'.

Right Lurking behind this heavy camouflage is a German Leopard 2 MBT. With its special Chobham-type armour, however, it has very little need to hide—this armour with its granular filling sandwiched between two layers of heavy plate is resistant even to chemical-energy warheads. Tanks using Chobham armour can be recognized by their vertical sides.

Zefa

Soldrs AE

target from a curved trajectory, which nullifies the deflection from a sloping glacis.

Although chemical-energy warheads are deadly against conventional armour of nickelchrome steel plates, some armour does give protection against them. A double thickness of armour, separated by an air gap—like the cavity wall of a building—has been proved to be fairly effective. In this arrangement, the energy of the chemical warhead can destroy the outer plate but the jet of gases and molten metal is dissipated by the air gap, or the metal scab which flies off is kept out by the inner plate. It has also been discovered that some plastics have qualities that defeat chemical-energy rounds but these have the drawback of being virtually useless against high velocity shot. The answer to the double challenge of anti-tank ammunition, therefore, lies in an armour of a composite of materials with the property to resist both types of attack.

Chobham armour

The most highly publicized modern armour that claims to have solved the problems posed by the various anti-tank attacks is the British designed Chobham armour —named after the town where it was invented. The composition of the Chobham armour is a closely guarded secret but it has certain characteristics that make it instantly recognisable on tanks to which it has been fitted—the Leopard 2, XM1 and Shir 2—all of which have flat turret fronts and sides.

Because the Soviet T-64 and T-72 did not have such flat surfaces, it was assumed that their armour construction was of the normal rolled cast type. In a US Army report of 1978, however, it was stated that the T-64 and T-72 have advanced armour of the Chobham type so it is assumed that their protective skin (and presumably that of the modern T-80) has the same capabilities as Chobham armour but the tanks have different roles. Chobham-type armour is probably a sandwich of two plates separated by a highly dispersive, possibly granular, filling. Undoubtedly, it has excellent protective qualities, but it is unlikely to have produced the first invulnerable tank.

The armour on MBTs also provides some protection against nuclear weapons. Although nothing can withstand the full force of a large nuclear explosion, tank armour protects against such blast and heat. Armour also give a lot of protection against the radioactivity released by a hydrogen bomb, particularly when an armoured vehicle is fitted with air filters, and has enough food and water for a lengthy stay. It gives no protection however, against the high-speed neutrons released in the explosion of an enhanced radiation weapon (popularly known as the neutron bomb) which pass through the armour, leaving it intact, and kill the crew. In the last resort, there is no protection against theatre and strategic weapons.

Just as the mobility and armour of tanks have progressed from the earliest models so too has the firepower. When tanks were first introduced, a machine gun or two for disposing of the enemy infantry were considered adequate armament. Occasionally, a heavier gun was used to penetrate the protective shields of field artillery. As it became apparent that a tank is best matched by another tank, more emphasis was placed upon a heavier armament to give an anti-tank capability. Now that the main battle tank has been established as an army's capital armoured fighting vehicle, the type of weapons deployed on each MBT is similar.

Besides matching opposing armoured vehicles, an MBT should also be able to defend itself against lesser enemies, such as determined infantry who could approach and attack it with flame-throwers, rockets or grenades. Although the main armament can be used to fire canisters—spread shots for use against men in the open—the best defence against infantry is provided by machine guns. The general rule for modern MBTs is that one machine gun is mounted coaxially (in line) with the main gun, so that it can be fired from the security of a closed down turret, and another on the turret roof.

Crew cuts

The latest Soviet tanks (T-64, T-72, T-80) have a crew of only three—driver, gunner and commander—so two machine guns are as many as they can handle, but the latest NATO tanks keep to the tried concept of a four-man crew including a gun-loader/radio operator positioned beside the gunner. Interestingly, provision is made for an extra machine gun. On the Leopard 2, a 7.62 mm Rheinmetall MG3 is mounted coaxially and a similar weapon skate-mounted on the left hand side of the turret for the loader's use. At the commander's hatch of the XM1 is a heavier 12.7 mm Browning HB machine gun, which has powered and manual controls for traversing but manual controls only for elevation.

The Russian innovation of cutting tank crews to three has the advantage that there is more room to accommodate armour and machinery. But a disadvantage is that some of this room is taken by an automatic loader, which might not be able to make a quick change in the type of ammunition selected. Also there are a pair of eyes, ears and hands fewer. These disadvantages may become significant as the amount of battlefield information reaching the turret by radio and through various sights and other instruments increases.

Modern electronic equipment is a vital part of the MBT, being used to locate targets and to sight the tank's main gun as well as for communication. The object is always to see and destroy a target before it manages to destroy you. To help the crew locate and destroy the enemy, there is a bewildering assortment of aids. They include Doppler radar to pick out moving men and machines at a distance, seach-lights, heat sensors for detecting objects whose temperatures differ from those of their surroundings, image-intensifiers which use ambient lights to give an observer a bright picture in the dark, laser range finders which can pinpoint an enemy with amazing accuracy, devices that give a warning when they detect enemy

Relatively slow but heavily armoured, Britain's Chieftain Tank Mark V is equipped with a range of technical innovations that make it a formidable weapon. Infra-red detection equipment will seek out any heat—from the engines of enemy vehicles for instance—even in pitch darkness or behind thick camouflage. A laser rangefinder will pinpoint any target with devastating accuracy. And a computer will take in all the information and aim the big 120 mm gun in a fraction of a second.

Sarson/Bryan

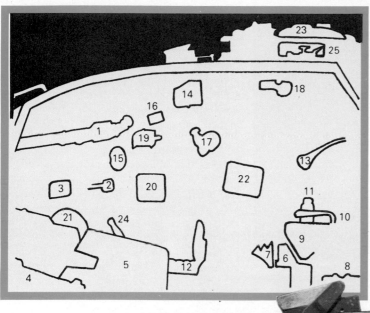

INSIDE THE CHIEFTAIN'S TURRET

1 telescopic sight
2 turret traverse gearbox
3 gyro trim control
4 7.62 mm machine gun
5 120 mm gun breech
6 periscope stowage
7 map stowage
8 cooking vessel
9 commander's seat
10 seat lever handle
11 signal pistol
12 gunner's seat
13 reading light
14 smoke discharger control
15 turret traverse indicator
16 metadyne control
17 commander's turret traverse
18 commander's sight
19 headset box
20 firing control box
21 elevating handwheel
22 rangefinder control
23 cupola and hatchway
24 firing switch
25 periscope wipers

Tanks used to be unable to operate at night and as darkness fell, tank battles would gradually die down. Now, with the aid of powerful searchlights, laser rangefinders and in particular thermal imaging devices *(below)*, tank commanders are hindered neither by fog, smoke or pitch darkness.

Soldier Magazine

Barr & Stroud

Left Thermal image of a Land Rover taken at night over a considerable distance. Like a TV camera, the imager works by a series of line scans that build up into a complete picture. Notice how the wheel arches are brightest, revealing the heat emitted from the engine.

Bar & Stroud

surveillance being made by radar or infra-red scanners, and ensors that register when an enemy shot has passed close by. With all this at his fingertips, a tank commander is hardly handicapped by darkness or smoke, and the probability that he will score a hit with his first round is high.

Not all MBTs, however, are lavishly equipped because of the expense of such sophisticated equipment. Also, too much information can confuse rather than enlighten. As an example of a fairly complete and utterly modern system, one can do no better than look at the excellent American XM1. The sights on this MBT are as comprehensive as can be imagined. The commander has six periscopes which give a full 360° field of view and three times magnification for his machine gun sight. He can also see where the main armament is pointing through an optical extension of the gunner's primary sight which has its own magnification of ten times with a 5.5° field of view or, alternatively, eight times with a 21° field of view. The night vision optics have a 2.6 by 5° field of view at ten times magnification and a 16° field at three times. If the primary sight fails, there is a Kollmorgen 939 auxiliary sight at ten times magnification with an 8° field. Even the loader is not expectd to ride blind and he has a normal non-magnifying periscope, which can be traversed through the full 360°. The night sight itself is a Hughes thermal imaging system and accurate aiming is ensured by a Hughes laser rangefinder linked to a computer.

The gun itself is stabilized so that it points in the same direction irrespective of the vehicle's motion; the gunner simply centres his reticle on the target and uses the laser rangefinder. A muzzle reference system measures the droop of the gun while information from a drift wind sensor and a pendulum static cant sensor are fed into the computer together with the laser findings and the lead angle. The gunner makes certain manual settings, such as the ballistic character of the ammunition about to be used, and the computer then determines and automatically offsets the weapon sight to the angle necessary to obtain a hit. It all seems rather elaborate but similar fire control systems are fitted, or are being fitted, in most NATO frontline MBTs, and the latest Soviet tanks are known to have laser rangefinders and infra-red searchlights for night visibility.

Although a modern MBT has all the latest technological aids to make it quick and deadly accurate, its main armament has not changed vastly since World War 2. It seems strange that, in the electronic age when so many claims are made for guided missiles, all the most modern tanks are armed with guns. This is because guided missiles lack the versatility of the gun and, being slow in flight, they have to rely on HEAT or HESH rounds to pierce armour, whereas a gun can switch from those to a high velocity projectile, whenever necessary.

To impart high velocity to a shot it is necessary to give it a terrific force of propulsion in relation to its diameter. The Leopard 2 has a Rheinmetall 120 mm (4.7 in) smoothbore gun which may well be adopted by the American for the XM1. This uses 7.1 kg (15.6 lb) of propellant to drive 7.1 kg of high velocity projectile but takes a mere 5.4 kg (11.9 lb) of propellant to drive 13.5 kg (30 lb) of HEAT projectile. When it is realized that high velocity shots can be forced up to a speed of 5,760 km/h (3,580 mph), it is evident that a strongly constructed gun barrel to withstand the forces is essential. There has been a certain amount of technological advance in the making of high barrels so that pressures can be attained but they still tend to wear out rapidly. As few as

120 rounds of armour-piercing high velocity shot can be enough to wear a gun barrel out and reduce its accuracy to a quite considerable degree.

Although the gun barrel has a limited life, a gun-armed tank can still fire more rounds than a missile-firing equivalent, before being forced to replenish ammunition. The Israeli experience in war has shown that the more rounds a tank can carry the better. As a result the Merkava can hold as many as 85 rounds, as opposed to a mere 40 on Soviet T-72 or 42 on the Leopard 2. Obviously a surviving Merkava will be able to maintain the battle for a very long time but there are dangers in a large ammunition store—particularly the danger of fire and explosion.

The turret-mounted gun is not a revolutionary new weapons system. Nevertheless, it must be engineered to a high standard of precision. Normally, tank guns weigh more than a tonne and they need a complicated recoil mechanism. All this machinery and the turret have to be finely balanced because they may have to be turned and used by hand if the power assisted mechanism fails or sustains battle damage.

The skills needed to produce a modern MBT gun and fighting platform are at the limits of technology but, in this instance, a comparatively old technology. Indeed, it is striking how much today's MBTs are the product of the refined and developed machinery of yesterday. They are still mobile, armoured gun platforms and their technical characteristics have evolved to their present pitch of excellence by a process of steady improvement. They still move on tracks powered by combustion engines and, if their armoured envelopes are more elaborate than they were half a century ago, that is only to be expected.

Yet it would be a mistake to think that the most modern MBTs are no advance on their predecessors or that the tank concept has had its day. The age old problem has been to balance the conflicting needs of firepower, protection and mobility so that MBT has the best of all worlds. Most experienced soldiers would agree that the latest generation of tanks—spearheaded by the Leopard 2 and the XM1—have come near that elusive goal. The big new engines that give modern MBTs up to 1,120 kW (1,500 hp) have enabled those that weigh about the 50 tonne mark (which can be provided with really strong protection) to have adequate mobility. As for firepower, it seems that the tank gun is still unmatched for versatility and effectiveness as a battlefield weapon and, as long as that remains true, the MBT will be the capital weapon system of land forces in any major offensive.

Below Three modern tank warheads. At the top is the more traditional high velocity armour piercing shot; although effective against even the thickest slab sides, it is easily deflected by sloping armour. However, chemical-energy shells, HEAT (high explosive anti-tank) and HESH (high explosive squash head), are fired at low velocity on curving trajectories so that they hit even sloping armour squarely. Despite their low velocity, their high energy explosions mean that they are very destructive.

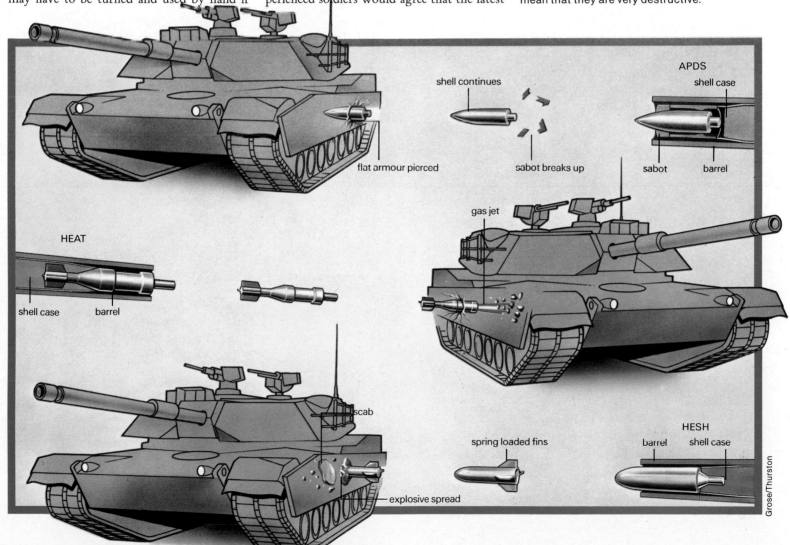

APDS
shell continues
shell case
flat armour pierced
sabot breaks up
sabot
barrel

gas jet

HEAT
shell case
barrel

scab
spring loaded fins
explosive spread

HESH
barrel
shell case

The beaming battlefield

One of the greatest confrontations in the land battles of any future war will occur between invading tanks and defending anti-tank forces. The Soviet Union has long been one of the leading tank-designing nations and its present operational tanks—the T-64 and T-72—are among the best of the current generation of Main Battle Tanks (MBTs). The T-80—entering service in the early 1980s—is expected to be even better.

Russian tanks are not only effective but also plentiful. A recent, unclassified, official estimate totalled more than 20,000 Warsaw Pact MBTs in central Europe, giving the bloc a three-to-one superiority over NATO's tanks—a ratio that military experts consider to be the margin needed for successful attack.

To oppose this massive tank fleet, NATO's generals have a considerable array of direct weapons, ranging from tank guns and small, guided weapons to rocket launchers and mines. But all these weapons are essentially short ranged. Even 120 mm tank guns become less effective beyond 2,000 m (6,600 feet). And bad visibility or covered

This scenario shows a typical NATO unit in a defensive position on a river-line (1). The only bridge (2) has been destroyed. Main positions are situated along the near river bank (3). About 10 km (6 miles) to the rear is a gun position with a battery of 155 mm howitzers. Another unit is equipped with Remote Piloted Vehicles (RPVs) on which are mounted laser target designators (4). In the first engagement, a helicopter (5), equipped with a Target Acquisition and Designation System (TADS) is marking a T-72 tank which has been spotted on the edge of the woods (6). In the second engagement, a forward artillery observer (7) is using a Ground Laser Locator Designator (GLLD) to mark a tank (8) preparing to move forward and ford the river. Another RPV (9) has spotted a column of tanks some distance behind the enemy front-line and is directing a CLGP onto the leading tank (10). To prepare Copperhead to fire, the observer transmits fire mission data to the Fire Direction Centre (11). At the FDC the artillery computer translates the signals into technical data which is transmitted to the Battery Command Post (12). At the gun lines the projectile is removed from its storage container, the settings are put on, the round is loaded and then fired. The gun crew is immediately ready for the next round, while the designator is ready for the next engagement as soon as the previous round has landed. The system is widely regarded as marking a breakthrough in artillery effectiveness.

Jeremy Gower

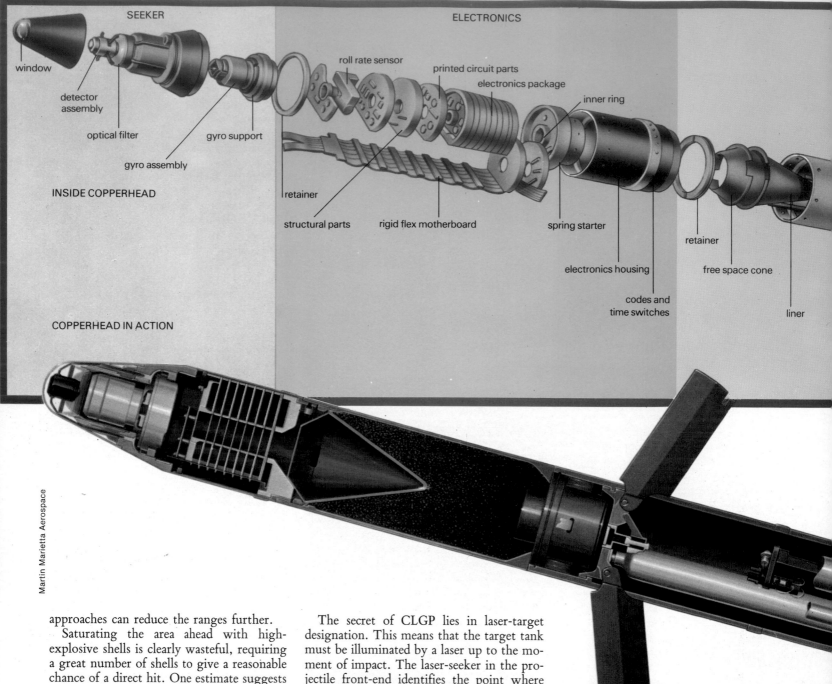

SEEKER

ELECTRONICS

window

detector
assembly

optical filter

gyro assembly

gyro support

roll rate sensor

printed circuit parts

electronics package

inner ring

retainer

structural parts

rigid flex motherboard

spring starter

electronics housing

codes and
time switches

retainer

free space cone

liner

INSIDE COPPERHEAD

COPPERHEAD IN ACTION

Martin Marietta Aerospace

approaches can reduce the ranges further.

Saturating the area ahead with high-explosive shells is clearly wasteful, requiring a great number of shells to give a reasonable chance of a direct hit. One estimate suggests that a minimum of 250 shells would be required per strike. Some means of on-board terminal guidance is therefore needed—together with aerodynamic control surfaces to steer the projectile onto the target. The whole package must fit a shell of calibre small enough to use existing artillery barrels.

The ability to achieve such a system became technically feasible only in the late 1970s, with the development of a small, semi-active laser-seeker that can be fitted into a 155 mm shell—NATO's standard field-artillery calibre.

The Cannon-Launched Guided Projectile (CLGP) is named Project Copperhead after a species of deadly American snake. CLGP are now in production for the US Army as well as European purchasers.

The secret of CLGP lies in laser-target designation. This means that the target tank must be illuminated by a laser up to the moment of impact. The laser-seeker in the projectile front-end identifies the point where the laser beam hits the target and then 'flies' the shell down the beam until impact.

Current laser-target designators are of various types. The simplest is the hand-held model, normally carried by an infantryman or artillery observer. Slightly larger and more powerful is the Ground Laser Locator Designator (GLLD) which is mounted on a tripod. Laser target designators can also be mounted on armoured vehicles, remotely piloted vehicles (RPVs), helicopters or fixed-wing aircraft.

A CLGP consists of three sections: guidance, payload (warhead), and stabilization and control. A projectile is 137.2 cm long and weighs 62 kg. It is aerodynamically controlled by cruciform in-line wings and tail fins that provide roll stabilization and lateral manoeuvrability. The guidance section comprises the seeker and electronic assemblies, which are housed at the front.

The seeker employs folded, body-fixed optics with a spin-stabilized gimballed mirror. This seeker gyro is spun-up mechanically after launch by a steel spring, and is then sustained and torqued electrically. An electronics assembly includes seven annular printed wiring boards, which are supported by concentric aluminium rings.

In the payload section is the high-explosive, shaped-charge warhead in a steel structure with a copper facing. A fuze

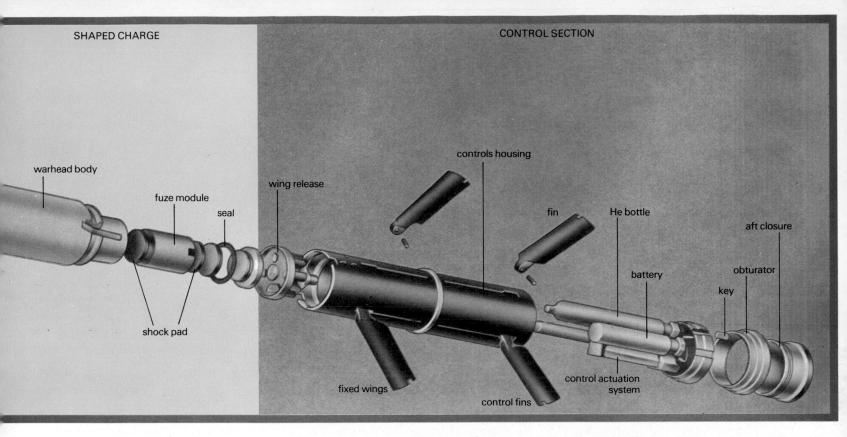

SHAPED CHARGE

CONTROL SECTION

warhead body

fuze module

seal

wing release

shock pad

controls housing

fin

He bottle

battery

key

obturator

aft closure

fixed wings

control fins

control actuation system

Kuo Kang Chen

Above and left Copperhead consists of three separate sections—seeker and electronics, warhead with shaped charge, and stabilization and control. US Army tests show that Copperhead can destroy in two rounds a target requiring up to 1,500 conventional high explosive rounds.
Below Copperhead under test.

module houses the dual-channel 'Safe-and-Arm' Mechanism and the firing circuit train. A fuzing sub-system also includes a direct-impact sensor, mounted in the nose, and six shock-wave sensors are situated in the guidance section.

Shaped-charge warheads are used because the projectiles are designed to hit and penetrate the target's armour. At its front end, the charge has a conical depression. When detonated, this focuses the explosion in a fast, forwards-moving jet—a phenomenon known as the 'Munro Effect', after its discoverer. To enhance its efficiency, this jet is increased by lining the cone with copper.

The stabilization and control section includes the aerodynamic surfaces and their associated actuator mechanisms. The control

actuator itself is a cold gas, three-axis system. Pitch control is provided by the two connected fins in tandem, whereas yaw and roll controls are obtained by independent operation of the other two fins. This section also contains the gas supply, control electronics and the launch-activated thermal battery that supplies all on-board electrical power.

At launch, the slip obturator seals the propelling gases, but because it is slipping it limits the spin to 30 revolutions per second. The acceleration activates the 11-volt battery and, at 800 g, the warhead 'Safe-and-Arm' activates and partly arms the round. As the projectile leaves the muzzle, the four-control fins deploy and maintain a clockwise spin. Also, provided velocity exceeds 214 m/s, the fuse rotor turns into alignment.

Martin Marietta Aerospace

Following launch—during free-flight—the projectile follows a normal ballistic path.

Next, an activation phase occurs at a time set during the pre-launch procedures. Firstly, the 30-volt battery is started, then the gyro is released. The roll-rate sensor and the gas bottle are then activated, unlocking the fins and providing 70 seconds of actuator power—during which the roll rate is reduced to zero. One second after roll control is started, the gyro is spun-up by a steel-spring actuator. After a further two seconds, the wings are extended by a squib piston.

In the mid-course phase, the projectile continues until the laser-seeking finds and identifies the correct laser code.

Finally, in the terminal phase—following acquisition of the target—the gyro slews towards the target and the fuze-arming sequence is completed. Proportional navigation guidance with gravity compensation then controls the projectile down to impact.

Advantages and limitations

Copperhead's merit is that it enables field artillery to attack tanks and other hard targets by accurate and effective fire. But the CLGP concept has limitations. Its operators must find their target and identify it, transmit orders to the artillery, load and then fire. During this sequence, the observer must keep the target under observation.

If the target tank, meanwhile, moves into cover behind a building, into trees or into a dip in the ground, it may elude the missile. Furthermore, because laser illumination is used to signal that a tank is about to come under fire, it will not be long before all MBTs are fitted with laser detectors—and thus, as soon as warning is given, the tank commander will take immediate evasive action. Such a tactic can, of course, be countered by switching the laser designator on at the last moment—when the round is fired from the gun.

Another limitation is the degree of precision required. Because the projectile will hit the precise place at which the laser designator is pointing, the slightest tremor in the point of aim at the critical moment will result in a miss.

There are numerous electronic links in the system, and these, too, are vulnerable to such counter-measures as jamming—which could cause dislocation. Nevertheless, Copperhead represents a significant step forwards in the effectiveness of field artillery, its advantages outweighing its drawbacks.

CLGP could be a very effective weapon at sea—fired from a normal ship's gun, with a laser designator mounted in either an RPV or a helicopter. It could also be used to give a new lease of life to coastal defence guns. Also, because the effect of a shaped charge is directly proportional to the diameter of the cone, the CLGP could lead to a return to larger calibres, such as the old eight inch.

Simplicity is paramount

CLGP, and specifically Project Copperhead, shows how modern technology can be applied to an old problem to produce effective, simple and relatively cheap answers. All too often, designers produce systems of high complexity, with numerous 'extras' and at vast cost. In Copperhead, simplicity is paramount—and NATO's generals welcomed its arrival at their field units.

Below The Ground Laser Locator Designator (GLLD—pronounced 'glid') has been developed by the US Army's Missile Research Command. It enables forward observers to seek out and classify targets for attack by Copperhead.

Hughes

Mines and minefields

Packed in a small, plastic box 1.2 m long and 11 kg in weight is a dose of explosive capable of knocking out a tank and concussing its crew. With the great range of highly complex weaponry available today, the role of the mine is sometimes overlooked. Yet the modern mine is compact, cheap and difficult to clear—whether on land or at sea.

The use of mines at sea goes back to the American War of Independence, and the fundamental principles of these early inventions remain valid. *Moored mines*—designed to explode on contact or when detonated from ashore—were soon evolved.

Although the technology of mines has changed radically over the years, the reasons for laying minefields remain the same. The tactics are basically to close an enemy's ports, restrict his freedom of movement or, if mines are placed closer to home, prevent enemy incursion or protect friendly shipping. Minelaying may be *overt*—publicized to achieve a desired end—or *covert*—aimed at exerting powerful psychological damage by a sudden, unexplained strike.

Moored mines

The 'classic' moored mine is an ovoid, buoyant steel container. This weapon can contain several hundred kilogrammes of explosives and, in its usual form, is detonated by contact. Protruding at intervals are *Herz horns*, soft lead tubes containing a glass phial of an electrolyte (current-conducting liquid). When these tubes are bent, the glass fractures releasing the fluid to complete the battery circuit which fires the mine's detonator. Simple and reliable, this kind of mine can be moored at any depth, making it effective against either submarines or surface ships.

For use against surface ships, a moored mine is attached to a sinker and a small winch of wire cable. The whole assembly sinks to the seabed and after a deliberate delay caused by the dissolving of a soluble plug (to enable the minelayers to get clear), the buoyant mine rises from its sinker with the winch revolving freely to release the cable. At the desired depth below the surface, a hydrostatic device operates in the mine, immediately clamping the cable.

Alternatively, further hydrostatic controls can be packed into an unmoored mine casing to allow the weapon to oscillate in depth as it drifts, or it can be shackled to a length of chain and left to 'creep' along the seabed under the influence of currents. This alternative is particularly useful in rivers.

The basic *minesweeping* (mine-clearing) technique, known as *Oropesa,* deploys a heavy steel cable which is held at an angle to the sweeper and at a set depth by means of a kite and otterboard at the free end. At intervals along the sweepwire, mechanical or explosive cutting devices sever any vertical mine wires intercepted by the sweep, forcing the mines to surface and be disposed of by small-calibre gunfire.

A mine moored in strongly running tidal waters may be vulnerable in one of two ways. Because it is always liable to drag its sinker on an impenetrable seabed, it is in danger of breaking surface and being detected at low water. Alternatively, it may be too deeply submerged. Because of this,

Left Blending well with natural cover, the scatter drop mine is a quick and easy to lay anti-tank measure. The mine can be scattered from helicopters or fast-moving trucks.

Valsella SpA

Sarson/Bryan

ANTI-TANK MINE

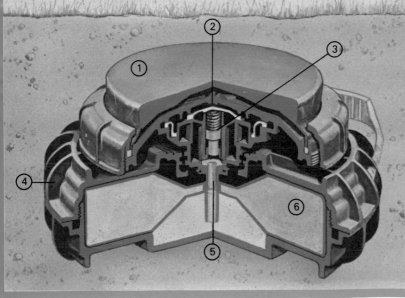

To totally destroy a tank requires a large charge positioned in the right place. However, a smaller charge will quite easily cause the tank to shed a track, thus immobilizing it indefinitely. Most anti-tank mines can be detonated even if they are upside down or partially buried.

KEY
1 Pressure plate
2 Pressure membrane
3 Striker
4 Plastic case
5 Detonator
6 Explosive charge

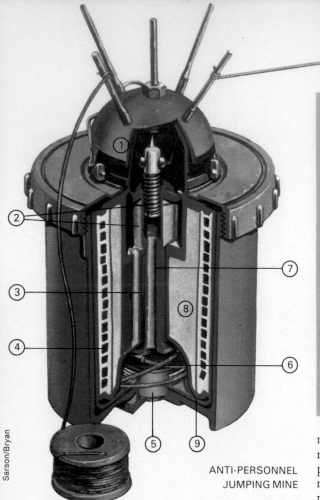

ANTI-PERSONNEL
JUMPING MINE

Sarson/Bryan

1 Fuse cap
2 Strikers
3 Mine fuse
4 Splinters
5 Ejection charge
6 Arming wire
7 Detonator
8 Explosive charge
9 Mine body

Above Jumping anti-personnel mines can seriously wound or kill anyone within a range of 40 m. Such mines are triggered by direct contact or by stumbling across a trip wire.

the mine would be ineffective at high tides.

To overcome these shortcomings, the *influence* mine was developed. It can be placed at depths of 60 m (200 ft) without loss of efficiency. Since no mooring system is required, they can be laid by aircraft or submarine and can destroy their target without direct contact. The best-known are the acoustic and magnetic varieties.

Magnetic mines work on a simple principle. Steel-built ships create their own magnetic fields which is complicated by electrical machinery onboard. As it moves over a magnetic mine, the ship's natural terrestrial magnetic field becomes distorted, creating a local variation which, though small, is sufficient to generate a voltage in a detector. Amplified, this activates a triggering signal.

Each ship has its own magnetic 'signature'. By 'de-gaussing' or neutralizing its magnetic fields, a ship can be afforded a

measure of immunity against magnetic mines. Techniques have been refined to the point where the magnetic properties of each major section of the ship are constantly monitored and the neutralizing current adjusted to suit any combination of machinery.

The most effective method of sweeping magnetic mines is by towing a 'Double L'. Powerful pulses of electric current are passed along a pair of buoyant cables towed from the ship. The current reaches two pairs of bare electrodes, one pair on each cable, and passes through the water between them to produce a magnetic field powerful enough to detonate magnetic mines on the seabed.

Vaisella SpA

Above Illuminating warning mines take the shape of their more lethal counterparts but serve only to warn that a minefield is being penetrated and hence to deter intruders.
Below Hovercraft have carved an important niche for themselves in the realm of minefield clearing. Floating on a cushion of air, they are less likely to trigger magnetic mines and, even if they do, will suffer little damage. Here, a hovercraft employs an Oropesa sweep to bring a field of mines to the surface where they can be disposed of by gunfire.

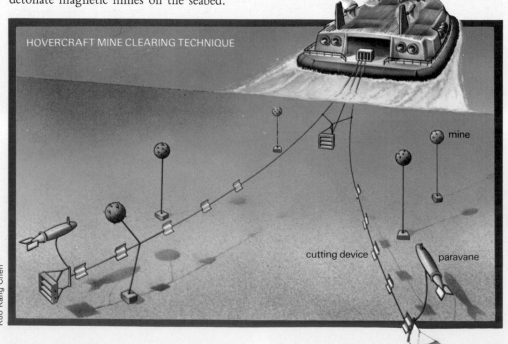

HOVERCRAFT MINE CLEARING TECHNIQUE

mine

cutting device

paravane

depth regulator

Kuo Kang Chen

The main alternatives—aircraft fitted with girdles of cable—have now been replaced by high-speed sleds towed behind large helicopters, a technique used by the Americans to clear Haiphong harbour during the Vietnam war. The other main sort of influence mine—the acoustic type—is activated by the noise of machinery, propellers, and so on generated by the target ship. This mixture of sound frequencies is detected by a sensitive hydrophone and converted into a voltage, which can be amplified to act as a trigger. A filtering and tuning system ensures that the mine is detonated by the sound of a ship only. Each ship has its own characteristic sound 'signature'.

Sweeping acoustic mines—a difficult and dangerous job—is now largely done by towing a drum containing powerful noise generators which can be tuned to simulate most ship-induced noises.

The pressure mine

By contrast, the *pressure* mine, first used in World War 2, still has no effective antidote. This type of mine uses a reduction in vertical pressure generated by a moving ship to distort a flexible diaphragm which then detonates the mine. The difficulty of simulating a ship is aggravated by the fact that many mines combine more than one trigger type. The problem is further complicated by the addition of counters, so the mine will ignore a set number of stimuli and eventually explode in apparently well-swept waters.

Attempts to tackle this particularly effective kind of mine led to the development of a new class of minesweeper, the minehunter. Although it resembles a conventional minesweeper, the minehunter is equipped with sonars which can 'see' both down and sideways, with a discrimination that can detect objects only 20 cm (3 in.) square at a range of several hundred metres (yards). Once located, the object is identified either by divers or by underwater television. Suspicious contacts are buoyed and disposed of by counter-mining with an explosive charge. This is placed alongside, either by a diver or by a remotely controlled vehicle, such as the French PAP (Poisson Auto Propulse) miniature submarine. PAP is equipped with a television camera and transponder which help it to seek out the mine. Once located, explosive charges are locked onto the mine and detonated. The advantage of PAP is that it can operate in deeper and faster waters than can divers.

Modern minehunters, such as the Royal

Navy's HMS *Wilton*, are built of glass-reinforced plastic (GRP) and contain no ferrous fittings. This reduces the ship's 'signature' to a minimum and enables it to tackle all types of mine.

The 'Troika' system, where three crewless minesweepers are 'slaved' from a mother ship, has recently been developed by the West Germans in an attempt to reduce the hazards of mine clearance. Troika sweeps both magnetic and acoustic mines, but pressure mines are detonated only when the slave ship passes over them. Specially reinforced hulls have been designed for the slave ships to help cope with pressure mines.

The main problem encountered by minesweepers and minehunters occurs when they are clearing harbours, which are prime target areas. This is the complication caused by the presence of large amounts of rubbish beneath the waves. Navies can spend a great deal of time and money checking and identifying the various kinds of seabed litter.

One new type of mine is the CAPTOR (captive torpedo), a deep-moored capsule which responds to an impulse from a surface ship or submarine by launching a self-homing torpedo at its target. Also under development is PRAM, a deep-water rocket-propelled mine which takes the principle one stage further.

Hovercraft are proving increasingly valuable in the fight against the mine menace. Various types, including Britain's SR.N4, have undergone trials in a minesweeping capacity and have proved effective. Floating on its air cushion, the hovercraft transmits little noise beneath the waves and has no pressure or magnetic signature worth speaking of. It is, therefore, unlikely to trigger a mine. Moreover, trials have shown that, even if it does, the effect of the explosion is greatly reduced by the air gap, so that the craft is usually only slightly

US Navy Photo/MARS

MOD

Left Loading a dummy mine aboard a patrol aircraft during an exercise. Aerial deployment enables fields of mines to be laid swiftly over vast expanses of sea.
Below HMS *Brereton* leads a trio of minesweepers out of Hong Kong. Worldwide, such craft maintain the safety of shipping lanes.

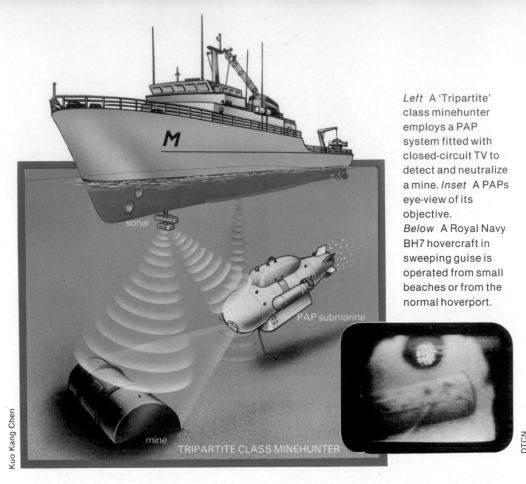

sonar

PAP submarine

mine

TRIPARTITE CLASS MINEHUNTER

DTCN

Left A 'Tripartite' class minehunter employs a PAP system fitted with closed-circuit TV to detect and neutralize a mine. *Inset* A PAPs eye-view of its objective.
Below A Royal Navy BH7 hovercraft in sweeping guise is operated from small beaches or from the normal hoverport.

halt a numerically superior opponent.

Anti-tank mines and anti-personnel types are usually laid together. The latter will discourage and delay clearance teams from their main business and make difficult the recovery of damaged vehicles.

While a small and inexpensive anti-tank mine—typically weighing about 8 kg (18 lb) with a 5 kg (11 lb) charge—will disable an armoured vehicle by blowing off a wheel or track, the crew may remain uninjured and the vehicle subsequently recovered. To increase the destructive effect, the mine needs to be more powerful.

Anti-tank mines

One example is the French HPD which, though weighing only 6 kg (13 lb), contains two charges, one to force it from the ground in which it is buried and a second shaped in order to detonate on contact with the target vehicle and penetrate its belly armour. Penetration through up to 70 mm (2¾ in.) of steel is claimed.

As an alternative, the British have developed the simpler, but larger, bar mine—a small plastic box 1.2 m (4 ft) in length and 110 × 80 mm (4¼ × 3 in.) in cross-section, weighing about 11 kg (24 lb) and containing a charge of 8.4 kg (18½ lb).

damaged, rather than crippled or destroyed. The bigger craft can handle all types of sweep. Being amphibious, they have the advantage of being able to operate from small beaches and from fully equipped harbours.

Mines are as important on land as at sea. Ultimately, land warfare aims at occupying territory, and the mine remains a fearsomely simple means of preventing this. Land mines are of two basic types, each designed to do a specific job. Both the anti-personnel (AP) mine and the anti-tank (AT) mine use a pressure-sensitive trigger, although the AT mine contains a far larger charge and is far less responsive to pressure because of the weight of its target.

Tactical minefields are often not concealed

to any great degree, but have clearly marked 'safe lanes' for use by friendly forces. Should a retreat become necessary, these lanes can easily be blocked off. This arrangement gives great scope for the creation of 'spoof' minefields which impede an advancing enemy in its tracks until the apparently lethal area has been checked.

Covertly laid random fields can also demoralize an enemy completely. A small force can cover a mined area with artillery to

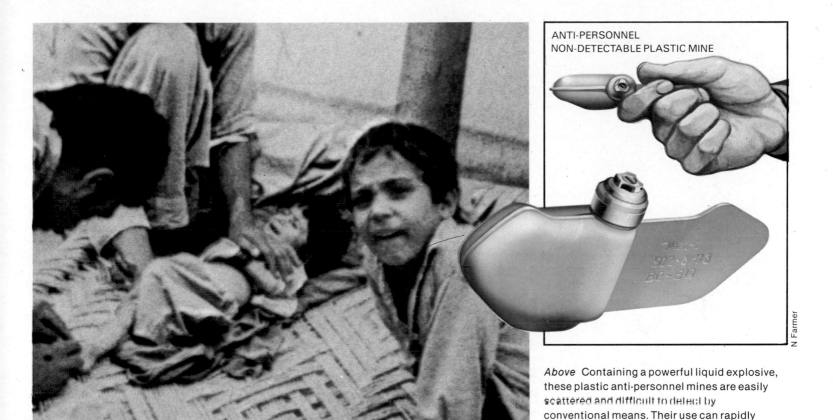

ANTI-PERSONNEL
NON-DETECTABLE PLASTIC MINE

N Farmer

Visnews

Above Containing a powerful liquid explosive, these plastic anti-personnel mines are easily scattered and difficult to detect by conventional means. Their use can rapidly destroy an army's morale. Sadly, they maim indiscriminately and many civilians, often children, suffer as a result *(left)*.

Its sheer length means that fewer need be laid to ensure an incapacitated tank.

Nowadays, large mines are rarely buried manually. Instead, a tractor tows a truck on which the mines are stowed, behind which is a plough which makes a single furrow. A two-man crew can feed over 600 mines per hour down a chute leading into the furrow, which is covered in a single operation. If an armoured personnel carrier is used as a tug, a Ranger anti-personnel mine launcher can be carried on top. This is a simple device consisting of 72 launch tubes each containing 18 mines. The mines, each one a mere 120 mm (4¾ in.) in diameter are capable of immobilizing a foot soldier. A single combination AT/AP mine launcher can thus rapidly lay a very effective combination minefield.

On occasion, a sudden enemy thrust may require even more rapid measures to be taken, and techniques now exist for remote delivery. This may be either from dispensers mounted on helicopters or fired by rocket or artillery. A 155 mm (6 in.) howitzer shell, for instance, can deliver 36 AP or nine AT mines. Such methods are restricted to emergency use, as accurate placement cannot be guaranteed and the weapons cannot be buried or marked. To avoid endangering friends as well as foe, mines fired in this way are self-activating and/or 'self-sterilizing' after set periods.

The modern infantryman must be constantly alert, for small anti-personnel weapons often do not wait to be trodden on. Many types exist which, while weighing only a kilogramme or less, will leap to waist height before exploding into a mass of splinters. Stake mines are a little larger and mounted on short posts. Detonated by trip wires, they will shatter in several hundred pieces, and are lethal at 15 m (50 ft).

Anti-personnel mines

More effective still are the larger mounted mines, such as the American Claymore or the Swedish FFV plate mines. Mounted in prepared positions on low tripods, they are remotely triggered by wire. Exploding, the plate will shatter into several hundred heavy fragments following a low swathing trajectory, extending up to 150 m (490 ft) in the case of the FFV. By placing mines over a wide area an effective ambush can be arranged without a shot being fired.

Though commonly made of metal, mine cases may be made of plastic, wood or even fibreboard, thus minimizing the chance of discovery by radiation loop detectors. New techniques are required to detect minute differences in temperature, earth compaction or dampness. Once detected, mines can be very hazardous to lift by hand as micro-electronics now permit very effective anti-clearance devices to be inbuilt. Rapid clearance, although not wholly effective, can be made by an armoured vehicle with a roller or flail mounted in front. Other methods of clearing a suspected field involve using systems which fire lines of explosive-filled hose across its area or deliver fuel in a fine spray which combines with air to form an explosive mixture instantly detonated over a large area.

The modern mine remains an anonymous, cost-effective weapon. Simple to lay at sea, its effect on shipping is out of all proportion to cost. Laid by ship, submarine or aircraft along the shipping routes of the world, or in inland waters and harbours, they can bring the movement of shipping to a halt, as recent wars such as those in Vietnam and the Middle East have shown. Indeed, the mere threat of mining or the belief that a minefield has been sown may be sufficient to impede shipping and halt a modern army. And with the ever-increasing cost of countermining measures, their use is likely to increase.

The helicopter learns new tricks

On 24 April 1980 helicopters were used by the Americans in the 'mission impossible' hostage rescue bid in Iran. No other craft were thought capable of operating in such hostile conditions, landing without airstrips and carrying the necessary troops and supplies. Although the mission ended in failure, it focused world attention on the development of this multi-purpose flying machine's potential.

There are two species of rotorcraft: the helicopter and the autogyro. Today the helicopter, whose entire flight is achieved by power-driven rotors, is far more widely used; but the autogyro was developed first.

Innovators

Juan de la Cierva, a Spanish aircraft designer, was so bothered by the fact that aeroplanes can stall—suddenly losing the lift of their wings if they fly too slowly, or attempt a sharp turn at too low a speed—that in the early 1920s he perfected a rotorcraft with rotating wings driven by the slipstream, the machine being pulled along by a normal propeller. During World War 2 large numbers of autogyros were used by the Soviet Union, Japan, Britain and the United States for short-range military and naval observation, liaison duties and such odd missions as flying slowly at specific places in the sky to allow ground radar to be calibrated.

The two pioneers who first got helicopters into service were Anton Flettner and Igor Sikorsky. Sikorsky had built a helicopter in Russia in 1910, but it was not until 1939 that he at last got a helicopter into the air in the USA. From this VS-300 prototype stemmed the R-4, flown in prototype form on 14 January 1942. A year later a service-test batch was in operation with the US Army and Navy and the British Royal Navy and RAF even in such tough locations as Alaska and Burma, and aboard ships at sea. But the R-4 played little part in the war. A 74 mph two-seater with a 180 hp engine, it served mainly as a trainer and as a vehicle to explore possibilities. Unlike most previous helicopters it had a single lifting rotor, and a small rotor placed sideways at the tail to cancel out the main rotor's drive torque. From it stemmed the famed family of modern Sikorsky helicopters, nearly all of which use this same configuration.

Anton Flettner built helicopters in Germany from 1932, and by 1939 had settled on the 'eggbeater' configuration in which the engine drives identical left and right rotors which intermesh and turn in opposite directions. Their hubs are close together but tilted to stop the blades from hitting each other. The *Kriegsmarine,* Hitler's navy, was so eager to use helicopters that it placed an order for the FL-265 helicopter in 1938. In 1940 Flettner produced the improved FL-282 *Kolibri* (humming bird) and the navy ordered 30 prototypes and 15 production machines. Although little-known, they were the only

Above and left An international flight of helicopters takes to the air. From the ground up: the Sikorsky Sioux (US), the Aerospatiale–Westland Gazelle (a Franco–British collaboration), the Alouette (France), and the Westland Scout (Britain).

helicopters used operationally in World War 2, at least 20 being delivered by 1943.

Another German company, Focke-Achgelis, made large numbers of the Fa-330 *Bachstelze (water wagtail)*, a simple engineless rotor-kite carried aboard U-boats. The 330 could be quickly unfolded on the deck of the surfaced submarine and towed at a height of about 400 ft, giving the U-boat commander a 'flying crow's nest' with far greater visibility (exceeding 25 miles) than from the U-boat itself. This useful scheme was finally defeated by Allied air power. The Fa-330 could be seen on radar and could lead attacking aircraft straight to the surfaced U-boat. in an emergency the Fa-330 pilot/observer could jettison the rotor and descend by parachute, but with hostile aircraft approaching the U-boat commander could not even wait for him to get back down the conning-tower.

The Germans also pioneered the larger, transport-type helicopter. Professor Heinrich Focke, of the Focke-Wulf company, collaborated with Gerd Achgelis to produce the Fa-61 side-by-side-rotor helicopter, which in February 1938 was flown by Hanna Reitsch from Bremen to Berlin and then demonstrated inside the *Deutschlandhalle* before an amazed audience. From it stemmed

the 1,000 hp Fa-223 *Drache* (kite), which could carry a 1,500 kg (3,300 lb) load. Allied bombing wrecked the factory but a few of these impressive machines reached the Luftwaffe, some serving with Luft-transportstaffel 40. By far the most capable helicopter until the 1950s, the Fa-223 carried out front-line supply and casualty evacuation and also served as an artillery spotter, flying crane, battlefield reconnaissance platform and rescue machine with an electric hoist.

The most capable helicopter of the Korean War of 1950–53 was the Sikorsky S-55, used in many roles by the US Army, Navy Marines Corps and Air Force. Its roles were much the same as had been pioneered by the Fa-223, but there was increasing interest in the helicopter as an offensive vehicle. This has been pioneered in wartime experiments with rotors attached to troop-carrying gliders in Germany, and to jeeps and even Valentine tanks in Britain. The idea was to bring offensive ground forces swiftly and silently straight to their objectives. It finally became a reality in 1951 with experiments by the US Army and Marine Corps, but actual employment in Korea was sporadic and largely experimental. However, they were useful in bringing in such items as mortars, light artillery and ammunition or rations to

troops already engaging the enemy, no matter what the terrain.

The helicopter could be based offshore on a ship, and could bring out casualties on the return trip. A particularly important duty in Korea, often carried out under fire, was rescuing downed aircrew. More than a decade later large 'flying crane' helicopters, notably the Sikorsky CH-54 Tarhe, were to do even better and bring back the downed aircraft itself. It was reckoned that CH-54s in Vietnam recovered 380 aircraft worth $210 million, much more than the cost of the helicopters in the first place.

Kit-bag kites

On a smaller scale such American companies as Bell and Hiller succeeded in building small yet reliable helicopters seating two or three, and carrying two stretcher casualties in pods on the skid landing gear. These soon became regarded by front-line troops as an 'item of kit', issued down to brigade and then regiment level, for every conceivable front-line duty. Although usually unarmed and unarmoured, they proved they could survive quite well in hostile environments, and had counterparts at sea (where an extra task was plane-guard, on station beside an aircraft carrier for instant rescue of a crashed crew). They were also used in the Arctic for supporting remote radar stations on mountains.

One of the original combat roles of the helicopter, for which a special Fa-223 version had been designed but not flown, was in anti-submarine warfare (ASW). Although early Sikorsky machines had been tried in this vital role, it was not until 1955 that purpose-designed ASW helicopters went into production. The aim of the Bell HSL-1 was to find and destroy the new breed of

Left The Westland WG-30 in its tactical military role. It can carry 14 troops fully armed and equipped, or up to 22 as a standard troop transport. The WG-30 has a cruising speed of 150 mph and a range of 375 miles.

Westland Helicopters

Right The instrumentation of the German–Japanese helicopter MBB-BK-117, shown at Farnborough UK in 1980. It is expected to be in service by the autumn of 1981.

Ken Brookes

The Chinook, named after a group of west coast American Indian tribes, is expected to be in service well into the 1990s. It played a major role in the Vietnam war, its heavy payload of up to 11,000 lb being ideal for weapons transport. British Airways plan to use them on their North Sea offshore routes.

nuclear propelled, spindle-hulled vessels which could outrun almost all surface ships. It had rotors at front and rear of a slim fuselage, driven by a 2,400 hp engine, and could lift just over two tons of sonar equipment and ASW weapons.

Although not entirely successful, and used for only a brief period, the HSL-1 pioneered all-weather ASW helicopter operations. It used an autopilot to assist the pilot in the difficult task of hovering at a steady height of about 50 feet above the ocean while 'dunking' a sonobuoy on the end of a cable, to

Left The RAAF has 12 model 165 Chinooks. They have a power increase capacity for single engined flight, and have operated well in the difficult conditions of Papua New Guinea.

listen for submerged submarines. Aeroplanes have to strew sonobuoys in a pattern and listen to their emissions by radio, using each buoy only once. The ASW helicopter needs only one buoy, and this can therefore be larger and more sensitive, and can transmit its signals by cable direct to the crew.

Helicopter design was transformed by the

THE BOEING CHINOOK HC MK 1

1 Heated pitot tubes to measure speed
2 Cover for vibration absorbers
3 IFF aerial
4 Yaw sensing ports
5 Cyclic stick grip with speed trim and winch control switches
6 Rotor hub and oil tank
7 Air inlet to heater and blower
8 Jettisonable two-piece entrance door
9 Hydraulic rescue hoist (600 lb strain)
10 Transformers, rectifiers and generators
11 VHF (AM)/UHF (AM) aerial
12 Troop seats (33 in all)
13 Fixed non-swivelling undercarriage
14 Trailing-edge trim tab
15 Forward drive synchronizing shaft
16 Fire extinguisher (10 each side)
17 Engine intake protective grill
18 Combined gearbox oil tank
19 Oil cooling fan
20 Fully steerable hydraulic undercarriage
21 Lycoming T55-L-11CS/SE engines
22 APU (power for engine starting)
23 Vertical drive shaft to rear rotor

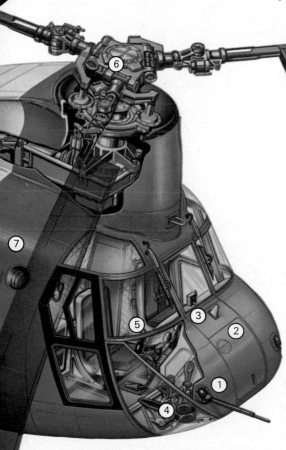

advent of gas-turbine propulsion. This dramatically improved the ratio of engine power to weight; for example, typical helicopter piston engines weighed about 600 g (1.4 lb) per horsepower whereas turboshaft engines reduced this to barely 100 g (0.25 lb). Even more important were two other advantages: the turboshaft engine burns less-flammable kerosene fuel and has a far longer life and better reliability than even the most highly developed piston engines.

Sud-Est Aviation, later merged into Aerospatiale, was the French pioneer of turbine helicopters with its *Alouette II* of 1955. This little machine appeared to be in the same class as the small Bells and Hillers but, with a reliable engine of well over 500 hp, offered much greater performance, especially in hot or high conditions.

Ground attack

Many of the techniques for use of helicopters in really offensive land fighting were worked out by the French Armée de Terre (land army) in the Algerian war of 1956–62. Using such machines as the Bertol H-21 and Sikorsky (Sud-built) S-58, both with 1.525 hp Wright Cyclone engines, the ALAT (Aviation Legere de L'Armée de Terre) fought around the clock, using refined methods first tried during France's war in Indo-China. Algeria was the forcing-ground for helicopter ground attack, using almost every kind of gun, bomb and rocket including the new French-pioneered wire-guided missiles which were ideal for employment from helicopters against armour, or against a particular window in the thick-walled buildings in Algerian villages.

From 1957, the Soviet Union's rapidly growing capability in helicopters was being matched by the growth of airborne assault forces which today exceed anything in NATO, both in size and in equipment. Unlike Western countries, the Warsaw Pact forces have a wide spectrum of weapons from armoured fighting vehicles to anti-tank weapon carriers, and from anti-aircraft missile vehicles to mobile radars. These can be flown in by air, or dropped by parachute, and then driven across every kind of territory including deep rivers. The first helicopter to offer really capable battlefield mobility —including heavy artillery, rockets and armour—was the Mil Mi-6, which appeared in 1957. Several hundred were built, enabling airborne operations to become possible on a scale not seen since World War 2 when the vehicle was the expendable glider. Most Mi-6s can carry 65 troops, 41 casualties or up to 20 tons of cargo, such as Frog missiles on a tracked launcher. Two crane versions, the Mi-10 and Mi-10K, can carry even heavier loads but are used chiefly for civil construction, while the gigantic Mi-12, with side-by-side rotors, has not seen military service.

Similar machines in the United States have remained mere projects, and even the smaller Boeing Vertol HLH (heavy-lift helicopter) for the US Army was cancelled in July because of its high cost. No Western power has helicopters larger than the Sikorsky S-65 or Boeing Vertol CH-47, which are powered by two turboshaft engines of 2,850 to 4,600 hp and are much smaller than even the old Mi-6. One special version of the S-65, the RH-53D, is equipped for minesweeping, and the latest version, the CH-53D, has three engines and will be used for Marine Corps construction and supply, and Navy clearance of crashed aircraft from carrier decks. In the West, development of large new helicopters and jet-lift aircraft such as the V/STOL has lagged behind their Soviet counterparts.

In the field of small and medium helicopters, however, the US Army is probably still marginally ahead. Development of purpose-designed helicopters for direct participation as weapons platforms in land battles was progressing from 1944 onwards, and several new strategies were deployed in Korea, Indo-China and Algeria.

In the mid-1950s US Army Brigadier General Carl Hutton was forcefully advocating an all-helicopter force to deploy firepower quickly over any battlefield, even the scene of a nuclear-weapon exchange. A unit was formed in March 1957 but its title, the Aerial Combat Reconnaissance Platoon, Provisional (Experimental), shows the strength of official reluctance. After 1960 the chief architect of helicopter warfare was

Above Manufacturers are constantly responding to the demands of today's armed forces. This Sikorsky S-70 SH-60B is being developed for use by the US Navy. Flown by a crew of three, it is armed with two Mk-46 torpedoes.

Left A tried and tested warhorse, the Sikorsky S-58 started in service in 1955.

Lt-Gen Hamilton Howze, but he might have made slower progress had not Bell Helicopter, at its own expense, flown a rebuilt Sioux, the Sioux Scout, in 1963.

Unlike the ordinary Bell 47 Sioux, familiar to almost every NATO foot-soldier, the Sioux Scout had a small and streamlined fuselage seating a crew of three in an enclosed cabin. Wings helped it to fly faster, and also carried racks for rockets or other weapons. Under the nose was the first 'tactical armament turret' (TAT), with two rifle-calibre machine guns. It was a major step in the right direction, but the small, low-powered Sioux Scout was the wrong starting point as a machine choice.

By 1964 the US Army had decided what was wanted was a really capable battleship of the sky, with high performance and packed with advanced electronic systems. Bell pro-

duced the AH-1 Huey Cobra, a 'gunship' development of the most numerous of all utility tactical helicopters the UH-1 (previously designated HU-1, hence its popular name 'Huey'). Unlike the Huey, with its broad cabin for up to 16 troops, the Cobra has a slim but tall fuselage seating just two crew in a tandem fighter-like cockpit, with the gunner in the nose and the pilot above and behind. In the air the Cobra is virtually a rotary-winged fighter, with tremendous speed and manoeuvreability.

In the long and tough Vietnam War the Cobra was a prime instrument from 1968 onwards. It often fought as a hunter/killer team with the small and even more nimble Hughes OH-6A Cayuse or Bell OH-58A Kiowa. In 1968–73 US Army helicopters flew more than 600,000 hours in action in south-east Asia, bringing the whole technology and technique of helicopters in land warfare to a new level. It required completely new forms of electronic warfare, sensors, communications, an unprecedented assortment of weapons, and absolute control of helicopters in violent and sometimes seemingly unnatural manoeuvres at very low altitudes. That the helicopters could do this work at all was remarkable.

The Cobra's successor, the Hughes AH-64A, now in production, is due in service in 1981. The Sikorsky SH-60B, due in service in 1983, has a planned role designated as 'light airborne multi-purpose' (LAMP) and a crew of three. The Hughes AH-64A, however, retains the old Cobra crew arrangement for its two pilots.

Its main armament normally includes a 30 mm 'chain gun' able to fire from single shots to 700 rounds per minute, with extemely accurate turret aiming, a range of spin-stabilized rockets, and either TOW or Hellfire guided anti-tank missiles. Both missiles are compact and can be carried in large groups and aimed with deadly accuracy over ranges up to 4 km (2½ miles).

Hellfire and Lynx

With the new Hellfire, the target can be pointed out by a laser held by a ground soldier; the helicopter missile then homes in on the target by itself and the helicopter does not have to remain in the area. With most anti-tank missiles the helicopter crew have to steer it all the way to the target, their commands being transmitted through fine wires unrolled from the speeding missile.

One modern helicopter where the missile sight is relatively high is the versatile Westland Lynx, which in British Army service uses the TOW missile and a sight on the roof of the cabin. This is an example of a multi-role helicopter, fast and tough enough to fly anti-tank and other armed roles, but also able to carry troops and supplies.

The main strike force in the attempt to rescue American hostages in Iran in April

Right The giant Sikorsky CH-54 Skycrane can lift weights of up to 10 tons. Amongst the helicopters of the Western powers only the Chinook and the Sea Stallion are larger.

1980 consisted of Sikorsky S-65A Sea Stallions, heavy assault helicopters. Each was capable of carrying three crew and 55 passengers. Their short range made refuelling problematical, but the main reasons for aborting the mission seem to have been adverse weather and terrain, hydraulic failure in some helicopters, and insufficient back-up craft in the event of such failure.

Beyond the west

The chief Russian tactical helicopter is the Mi-24 (called 'Hind' by NATO). Used in confrontation with guerrillas in Afghanistan, the Mi-24 is heavily equipped with advanced sensors and weapon systems, yet has a cabin for eight troops or several tons of supplies. The Hind A has a 12.7 mm machine gun, four *Swatter* anti-tank guided weapons and 128 57 mm rockets; the Hind D also has a four barrel 23 mm cannon. Formidable as the Mi-24 undoubtedly is, it has been reported that Afghan guerrillas high in the mountains have been able to shoot downwards at the helicopters as they struggle for height in the thin air above this harsh terrain.

Among recent developments in the medium size range is the Bell 214ST (super transport). Deliveries are planned to start in 1982 at a production rate of three per month. The twin turbine helicopter will carry up to 18 passengers and a payload of 3,000 kg (7,000 lb) at a cruising speed of over 150 knots, with a range of 740 km (460 miles). The newly designed rotor blades are of fibreglass, with titanium abrasion strips on the leading edges, to reduce fatigue.

The major influence of helicopters in the conflicts in Vietnam and Afghanistan is proof of their vital role in modern military technology. Ever more sophisticated defence and weapons systems seem certain to keep these versatile craft in the forefront of NATO and Warsaw Pact plans, especially in the fields of ferrying supplies, troop carrying and anti-tank attack.

For troops and pilots operating in battle zones there is the reassuring knowledge that helicopter crashes are not always fatal. One Vietnam veteran, Capt Hugh Mills, was shot down 16 times in 1,019 sorties!

Right In modern warfare the helicopter is indispensable in ferrying arms and supplies when the battlefield terrain is inhospitable. *Below* British Army Lynx helicopters are being fitted with TOW (Tube launched, Optically tracked, Wire guided) missiles capable of speeds in excess of Mach 1.

Future Fighters

General Dynamics/David Baker

McDonnell Douglas/David Baker

A new generation of fighters destined to revolutionize air combat tactics is taking to the air. In the evolution of modern fighter aircraft, each shape and planform has been more efficient than its predecessor. But the early progress in design was only a prelude to modern forms of layout and construction that are radically changing the shape of fighter planes.

The increase in the speed of fighter aircraft has been truly phenomenal. It took 40 years of progress to reach a speed of 900 km/h, yet in just 15 years from 1945 that capability was more than doubled. By the early 1960s, aircraft like the F-106 Delta Dart and the F-4 Phantom were capable of dashing at more than 2,200 km/h to intercept a target several hundred kilometres away.

Close to the speed of sound, the air passing over the aircraft's wing and around its fuselage cannot be compressed any more. As a result, the effect of air flow rapidly increases as the aircraft passes through the sound barrier. (The speed of sound 'Mach 1' is 1,220 km/h.)

Dr Richard Whitcombe, aeronautical engineer at NASA, solved the problem by giving the fuselage an 'area-rule' coke-bottle

shape—pinching it in where the broad chord of the wing joins the fuselage. The configuration made space for the compressed air to release some of its pressure. Dr Whitcomb's design of area-ruling, married to the technology of high-performance engines, has pushed up performance.

By the early 1970s the Russians introduced a growing range of interceptors designed to penetrate European air space and take the fight to the enemy. These penetrators threatened to clear the skies ahead of an advancing Soviet army, protecting it from the ground-attack aircraft that would be an integral part of American defence. Accordingly, the US Air Force sought an aircraft that could go out in all weather—night or day—and turn back the Russian interceptors

that challenged the air space above a battle.

The specification went out in 1965, and the McDonnell Douglas Corporation won the contract to build the F-15 air-superiority fighter. Adopting radical design trends and incorporating revolutionary materials, the Eagle, as it was named, achieved a peak in performance and fire-power. The result was the most lethal interceptor of the 1970s and 1980s with a lifetime that will probably extend into the 21st century.

The Eagle's design rationale is based upon doing the job and surviving to fight another day. It carries two powerful turbofan engines, so that if one is hit or fails the other will provide limited combat capability and get the aircraft home. Several separate control systems are carried, so that if one is hit

The F-16 Fighting Falcon *(above)* with twin external fuel tanks, achieves a degree of excellence that appears insuperable.
Left An F-15 Eagle draws up to the nozzle of a refuelling tanker.

General Dynamics/David Baker

others can take over so the aircraft can be flown and landed safely.

The structure is supported in such a way that if part of the aircraft is destroyed by battle damage the aircraft will not fold up or fall apart before making an emergency landing. And in the cockpit, careful planning of controls and electronics enables the aircraft to be flown through a complete mission by a single crew member.

The Eagle carries a Gatling gun with six barrels. It also has up to eight air-to-air missiles operating through a radar system capable of detecting the enemy more than 200 km away. In every dimension, the Eagle is geared to the task of stopping Russia's Mig-25 Foxbat, a Mach 3 interceptor in the vanguard of Soviet fighter technology. In ef-

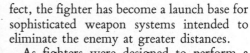

Having a thrust-to-weight ratio better than one, Falcons *(right)* can climb vertically like rockets, and air-to-air missiles *(below inset)* give them a deadly punch. The Eagle is built with ease of maintenance in mind *(below)*.

General Dynamics

fect, the fighter has become a launch base for sophisticated weapon systems intended to eliminate the enemy at greater distances.

As fighters were designed to perform a wider range of roles, their complexity increased and so did their weight. Types like the F-111 and F-15 weighed in at about 22 tonnes, while the Mig-25 tipped the scales at more than 30 tonnes. By comparison, fighters in the early 1960s weighed between 12 and 16 tonnes.

With sophistication and high technology went higher unit costs. High costs reduced the market size and made foreign sales less likely—only the rich countries could afford expensive weapon systems like the F-15 or the F-111. To expand the sales market, reduce unit costs and provide an upturn in the numbers of fighters available to Western air forces, the United States initiated, in 1972, a light weight fighter (LWF) contest.

General Dynamics won the US Air Force LWF contest and by the early 1980s, F-16 Fighting Falcons were rolling from the production line. The F-16 uses the same power plant as that fitted to the Eagle, but carries only one engine. The plane is purpose-built for highly manoeuvrable dog-fighting, where quick turns and snappy flying can readily pull nine *gs* on the stressed airframe and pilot. (One *g* equals the force of gravity at the Earth's surface, so an acceleration of nine *gs* means that a 70 kg pilot would momentarily weight 630 kg.)

The F-16 set a trend that opened unique possibilities for smaller, more responsive aircraft capable of out-turning and out-shooting the heavier interceptors of the Soviet and Warsaw Pact air forces. With a maximum loaded weight of only 11 tonnes (15 tonnes with external fuel tanks), the

General Dynamics/David Baker

McDonnell Douglas

Today's F-16 dog-fighters *(right)*, having met their design specifications, stand ready for battle but a new generation of fighters are already begining to take shape. The development of super-strong and super-light materials have made possible a forward-swept wing design *(left* and *below right)* that promises a host of advantages, including improved manoeuvrability at high speeds.

Falcon has a thrust-to-weight ratio better than one. The aircraft can literally stand on its tail and go up like a rocket. While it is not unusual for twin-engined aircraft to have a thrust greater than the loaded weight, this capability is unusual for a single-engined fighter. It implies a performance equal to that of the most powerful interceptor, but with the manoeuvrability of a lightweight.

In climb, with modest war load, the Falcon can reach a height of 13 km in 60 seconds, while maximum speed is a creditable Mach 1.9. Armed with a Gatling gun and air-to-air missiles, the Falcon can also perform in the ground-attack role.

Dog-fighter

The Eagle can also function as a bomber, but its performance advantage is at its greatest when the fighter-bomber has released its stores. It then reverts to being the highly manoeuvrable dog-fighter its designers intended, capable of responding to threats and fighting back to base. The Eagle was one of the first fighters that could double up as a ground-attack aircraft, but also be light and manoeuvrable enough to tangle with the lightweights and survive.

The F-16 has been used to pioneer new and innovative technologies. For example, it formalized the introduction of novel control modes. It has a seat tilted back 30° for better tolerance at high *g* and a side-stick controller (instead of the familiar control column between the pilot's legs). The Falcon employs 'fly-by-wire' control systems which elec-

General Dynamics/David Baker

Kuo Kang Chen

tronically link the controller to the control surfaces—ailerons, elevators and rudder.

As part of an Advanced Fighter Technology Integration (AFTI) programme, F-16 Falcons introduce the use of digital control systems. These allow the pilot to blend several different motions into the most efficient pattern. In this way, the aircraft can be accelerated up, down, forward, and side-to-side to confuse anti-aircraft systems.

The Soviet position estimators can only calculate an accurate result if the aircraft being followed is moving over a relatively smooth path. However, computers and electronic processing systems enable the AFTI pilot to skid, slew, drift or bob while *simultaneously* making a controlled turn or attitude change.

The F-16 Falcon achieves sophistication of aerodynamic motion that cannot be attained by the conventional Man-machine combination. But, unlike a conventional plane, the F-16 does not utilize push-rods and levers to carry movement from the pilot's control column to the aircraft's control surfaces. Rather, it blends computer software with electronic signals so that the pilot can truly 'fly' the aircraft around mathematically precise patterns.

Support equipment being tested for 1990s fighters includes a special helmet which controls the direction of flight by tracking the position of the pilot's eyes. Combined with a computer control module, this system would free the pilot's hands for other duties, like interrogating radar scanners or controlling the

Nick Farmer

weapons-release system. Other developments include automatic seats responsive to high *g* loads that would automatically tilt the pilot back to a 65° inclination, placing him in a better position to withstand the forces imposed on him by sharp turns.

In several respects, the F-16 represents a starting point for the advanced generation of fighters the US Air Force will seek for the future. With a wing blended into the body, the aircraft's shape provides increased lift, greater internal volume and less structural weight. Stability in flight is improved by the

The Highly Manoeuvrable Aircraft Technology (HiMAT) concept employs a blended wing/body shape and wings that twist and bend in flight to match aerodynamic profile with speed and performance. Launched from a B-52 bomber, a HiMAT model *(above)* has been remotely piloted in fighter development tests before being landed *(far right)* on skids. Weighing just 11 tonnes full-sized, HiMAT in profile *(right)* resembles the Eagle and the Falcon.

NASA/David Baker

Kuo Kang Chen

'forebody strake'—extension of the wing towards the nose each side of the fuselage. With a chin engine inlet (located below the nose), the flow of air into the engine is smoother than on fighters in which two separate intakes are provided—one each side of the fuselage. Also, by having the nose of the aircraft above the inlet duct, the engine gets a better airflow when the fighter pitches upwards and tilts nose high.

A fighter must perform across a very wide spectrum of operating conditions while maintaining high efficiency and optimum manoeuvrability. The F-16 incorporates movable flaps on the leading and trailing edges of each wing to change the camber (or shape of the cross-section) according to the flight conditions so that air resistance and buffetting are substantially reduced.

By the end of this century, fighters will be called upon to perform agile dog-fighting manoeuvres that would tear the wings from any aircraft flying today. They must be able to jiggle out of the path of air-to-air missiles or flip in and out of ravines and fjords to evade detection. Yet, it is hard to see how even blended wing/body shapes like the F-16 could dance such demanding pirouettes.

One suggestion harks back to an old idea that failed to gain acceptance before the age of high-power engines and electronic control systems. By sweeping the wings forwards, instead of back, engineers found that there was far less drag as the aircraft approached the speed of sound. This is because the shock wave—a prime cause of drag close to the speed of sound—is much weaker. However, the forward-swept wing would encounter

great stress because in facing forwards the airstream would try to bend it upwards at the tip, thereby increasing the load.

On a conventional wing, swept rearwards, the tip tends to flex down, decreasing the load on the entire wing. However, this tendancy could also work to the wing's advantage, because with forward sweep the root would begin to stall before the tip. With a modest amount of upward curve, lift would be distributed more evenly, enabling very low handling speeds when necessary, extremely agile performance at around 1,200 km/h, much lower stalling speeds and the possibility for further reductions in the overall size of the aircraft.

Before super-strong composite materials had been developed for aircraft construction, forward-swept wings (FSW) could not be adopted. The additional metal needed to accommodate the dramatically increased flexing loads prohibited use of FSW designs. But now the Grumman Corporation has pioneered the use and application of high-strength composites (made by adding fibres or filaments such as graphite to an epoxy matrix). Super-light as well as super-strong, they suddenly make the FSW an attractive prospect for the future. The US Defence Advanced Research Projects Agency (DARPA) has proved the feasibility of FSW aircraft for lightweight fighter roles in the next decade. Tests indicate that performance can be improved quite significantly by adding a small canard (or forward tail).

When the fighter pulls up in a tight turn the canard is similarly pitched up, deflecting the airstream down towards the main wing and reducing the wing's angle of attack (or inclination to the airstream). This reduces the likelihood of a stall at the wing root. Similarly, when pitching down, the canard directs air up to the top of the wing and again helps prevent root stall.

Digital control

All of these refinements and design trends aim to make the aircraft less inherently stable, and would seem to give the pilot an almost impossible task to achieve steady flight along a fixed path. So computers and electronic digital control systems are used to continuously sense every minute twist and jiggle as the aircraft flies along. They instantaneously command the control surfaces to 'float' along the flight path and smooth out undulations.

For all its promise, however, the FSW may not suit every requirement of a fighter. Thus the advanced HiMAT (Highly Manoeuvrable Aircraft Technology) concept has been proposed. It employs canards and a blended wing/body shape with tip fins and dual rudders. HiMAT is a model of projected fighter concepts—not all of which may actually be employed in the 1990s—aiming for an optimal design. Weighing just 11 tonnes, the proposed fighter would be required to make 8-g sustained turns.

A HiMAT model scaled to 44 per cent of the projected design size has been launched from a B-52 bomber. It is controlled by a ground operator, or from another aircraft flying alongside. With a small jet engine in the tail, HiMAT has proved a valuable research tool for NASA tests aimed at resolving aerodynamic problems connected with future fighter trends. Like the projected FSW, a full-size HiMAT fighter would incorporate aeroelastic tailoring so that wings and canards would twist and bend in flight to the best shape for specific speed and performance. About 30 per cent of the aircraft's construction would be in composites, and fly-by-wire systems would be standard.

By the year 2000, interceptors may be called upon to perform new and demanding tasks outside the capacity of present designs. The trend has already been set by the development of hardware for F-15 Eagles to disable low-orbit satellites—vital components of enemy forces in a future war. Equipped with a two-stage rocket below the fuselage, the F-15 would release the satellite 'killer' at great height, freeing it to zoom towards space and home-in on the satellite.

By the turn of the century, the aero space interceptor may be required to knock out several such satellites with rocket-powered impact heads designed to ram the object or detonate a neutron bomb to burn its electronics. This task would be accomplished by a 'lifting-body' design, an advanced derivative of the blended-wing concept in which the entire shape becomes a lifting surface devoid of separate wings and fuselage. Such super-cruiser interceptors would fly at Mach 3 and travel to the fringe of space, challenging the space-based sensors, which in any future conflict would dictate the capacity of each side to wage war.

Laser guns

At a lower level, heavy fighters like HiMAT would perform like the F-16s of the 1980s but probably carry laser guns to blind bombers, shoot down cruise missiles in flight or attack massed air assaults. They would probably have a composite materials structure and be capable of Mach 2.

At low altitude, and at Mach 0.9-Mach 1.2, will be the FSW dog-fighters, wrestling for air space and dancing a circuitous pattern around the heavier-strike aircraft. Their task will be to clear the opposition to let through deep penetrating, supersonic, nuclear strike aircraft flying at high Mach numbers close to the ground.

Super light, and with a thrust-to-weight ratio better than 1.5:1, FSW dog-fighters will carry their pilots in bubble cockpits for all-round vision and on tilt-back seats to physically protect against high g forces. The FSW fighter promises to be a formidable supersonic integrated Man/machine system, yet be no larger than a WW2 Spitfire.

NASA/David Baker

The A-10 killer plane

Conscious of the threat posed by massive tank forces, the United States Air Force is equipping with a new type of aircraft, the A-10 Thunderbolt II. A radical departure from the sophisticated, expensive, all-purpose miracles of modern aviation on which Western air forces have relied in recent years, the A-10 is by comparison simple, slow and heavy. Yet its rugged workhorse simplicity gives it great advantages over its thoroughbred predecessors. In short, the A-10 is a tank killer. In the role it was designed for, close air support of ground forces against enemy armour, it is powerful enough (experts predict) to neutralize Soviet tank superiority.

With six fighter squadrons now equipped with the A-10, NATO defence chiefs are placing great faith in the aircraft's future. Built by the Fairchild Corporation of America, the A-10 is one example of a return to specialist roles for a variety of aircraft types. No single aircraft is now expected to carry out the three classic air force functions of interdiction (attacking strategic ground targets), air superiority, and close air support. Other aircraft, the F-111 and F-15, will specialize in interdiction and air superiority. The A-10 will operate almost exclusively in support of ground troops, working in close co-operation with them through ground and air liaison officers.

Should Warsaw Pact forces ever invade the West, the A-10's task will be to lie in wait close behind the front lines, and respond rapidly when called upon to deal with armoured formations. To do this job, the A-10 carries one of the most awesome battlefield weapons ever installed in a tactical aircraft. The General Electric GAU-8 Gatling gun has seven rotating barrels, and can deliver 30 mm shells at either 2,100 or 4,200 rounds per minute. With the switch flicked to the fast firing rate, a one-second burst from the trigger on the pilot's control lever will put 70 shells on the target. The gun sight is set at 3,780 m (4,000 yds) in a turn of 1 g; at that range the first projectiles in a one-second burst would be hitting the target before the pilot takes his finger off the trigger. In fact, the pilots of A-10 squadrons are trained to fire half-second bursts. The projectiles are loaded in a continuous belt composed of five armour-piecing followed by one high explosive incendiary bullet.

The GAU-8 gun is 6 m (21 ft) long with seven-foot long barrels. Weighing 900 kg (2,000 lbs), it occupies, with the ammunition drum stored behind it, most of the fuselage of the A-10. Effectively, the aircraft is built around the gun. This type of weapon offers an important training advantage. While air-to-ground missiles are so expensive that the whole United States Air Forces can afford to fire only 200 live examples per year, A-10 ammunition is cheap enough to give pilots ample live ammunition training at sea ranges in the North Sea and Welsh waters, and land ranges in Germany.

In combat, the A-10 carries up to 1,350 rounds, enough for up to twenty firing passes. Although the GAU-8 is its main armament, the A-10 also carries a cluster of other weapons under its wings. The normal armament is six Maverick air-to-ground missiles (three attached to a pylon under each wing). However, the Maverick's future is under review. It is a TV-guided weapon, and requires several seconds of relatively straight and level flight for the pilot to identify the target, lock the missile guidance system on to it, and fire the weapon. This inhibits manoeuvrability which is an essential element in the A-10's defensive ability.

Cheap but tough

The A-10 is in the forefront of a new movement towards cheap, tough, straightforward, easily maintained weapons. At 2.5 million pounds per aircraft, it is also cost effective. It is powered by two General Electric TF 34 GE100 turbofan engines, each giving 4,082 kg of thrust. It has a combat speed of 387 knots (717 km/h). This is relatively modest, yet it can survive where its supersonic rivals might be brought down.

Part of the key to A-10's survivability lies in the simplicity of its design. Its two engines are mounted separately, and externally. If one is knocked out, the aircraft will still fly on the other. It has two tail fins. If one is damaged, the pilot can manoeuvre on the other. All its hydraulic systems and basic

Above right Four A-10s flying in line ahead. These single-seater, twin-turbofan aircraft were designed for close air support missions, with the emphasis on the anti-armour role. Standard armament is a GAU-8 30 mm cannon, but the underwing pylons can also carry a variety of other weapons, including Maverick AGM air-to-ground missiles, seen *(right)* being fired during evaluation tests.

electronics can run from either engine, or from an auxiliary power unit in the fuselage. If the hydraulic controls are cut, the pilot has old-fashioned cables linking his stick to the control surfaces.

One of the few predictable features of the modern battlefield is that there will be dense anti-aircraft fire rising from enemy ground units, especially the Soviet ZSU-23 four-barrelled radar-directed gun. The A-10s are expected to take their share of punishment, and still get home. The pilot's cockpit, and all the vital controls, are protected by several centimetres of armour plating.

Heat-seeking, surface-to-air missiles are also likely to feature in modern all-out war. A-10 pilots feel that their aircraft is less vulnerable than most. The wings obscure the heat from the engine to any missile fired from the ground in front of the aircraft. The tail assembly obscures the exhaust heat fired from behind. A missile fired from the side stands a chance, but will hit only one engine. The engines themselves aid in deceiving heat-seeking SAMs. They are the modern type of high bypass ratio engine that uses a central small burner not only to provide thrust but to drive the large turbofan.

Quiet and manoeuvrable

The fans provide a pocket of cool air that envelopes the hot gases from the burned jet fuel. The cool air not only disperses the heat quickly: it also obscures the hot part of the exhaust, except when a missile is directly behind. With its high turn rates and exceptional manoeuvrability, the A-10 needs to present this target to a missile or missile-launching site for only the briefest moment. This design of engine is quiet running, so there may be little engine noise warning to ground gunners before they themselves come under fire from the A-10.

Ground gunners will be lucky to get a clear sight of the A-10. It is not designed for a fast approach and getaway, but to fly low and slow among the tree-covered slopes and rolling contours of the German border areas. The defending gunner may have little time to fix range and direction as the A-10 slips out from behind a hill at treetop height, fires a burst, and banks hard to disappear rapidly back into cover.

If the enemy's own aircraft should succeed in establishing air superiority and go on the hunt for the A-10s, the aircraft has one asset that few other aircraft can match—manoeuvrability. Although at 16.25 m it is as long as a World War 2 Wellington bomber, it has a turning circle of only 1,220

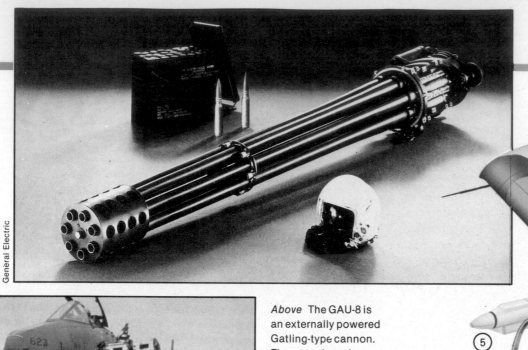

Above The GAU-8 is an externally powered Gatling-type cannon. The seven barrels rotate at speed to deliver up to 70 shots per second. With an innovative automatic ammunition loading system *(left)*, the rearming turnaround time has been effectively reduced from 3 hrs to 15 mins.

m and can out-turn almost any pursuer. Supersonic F-14s flying combat-training against A-10s have found the A-10 turn agilely to meet them head on. Then a burst from the GAU-8 is enough to destroy a thin-skinned F-14, or Mig-25 fighter.

The A-10 pilot also carries, in 16 wingtip cusps and on the undercarriage fairings, Tracor flares that he can release to thwart incoming missiles, and chaff dispensers to baffle enemy observers if his receiver warning set tells him he is under radar surveillance.

If all else fails, the pilot has a new design of ejection seat, the Aces II, which reportedly has proved efficient and reliable in trials. Supersonic fighters almost invariably have to make a sighting pass followed by an attacking pass over a target. Because of its slower speed, manoeuvrability, and powerful gun, the A-10 has no need to pass over the target. If the pilot is forced to eject, he is likely to land in friendly territory, where he can be picked up by ground troops or helicopter.

Operationally, the aircraft represents a new departure. The 108 A-10s to enter service will be flown by the 81st Tactical Fighter Wing, based at the neighbouring airfields of RAF Woodbridge and RAF Bentwaters in eastern England. From there, the six squadrons of 18 pilots each fly short missions to one of four Forward Operating Locations (FOLs) in Germany—at Leipheim, Sembach, Noervenich and Ahlhorn.

Bigger than any other fighter wing in the West, the 81st works on the simple principle of forward basing and rearward maintenance. Maintenance for the aircraft takes place in bomb-proof shelters in East Anglia. The Forward Operating Locations in Germany are austere, simple, and lightly manned with only 50 permanent personnel at each. They exist only to turn the planes round and get them back into the air.

Three times per week, or daily in the event of war, a C-10 Hercules transport flies necessary spares, and perhaps maintenance engineers, out to the Forward Operating Locations. The men solve all the problems on the spot, and have been remarkably successful.

The eighteen aircraft of one squadron have

FAIRCHILD A-10 THUNDERBOLT II

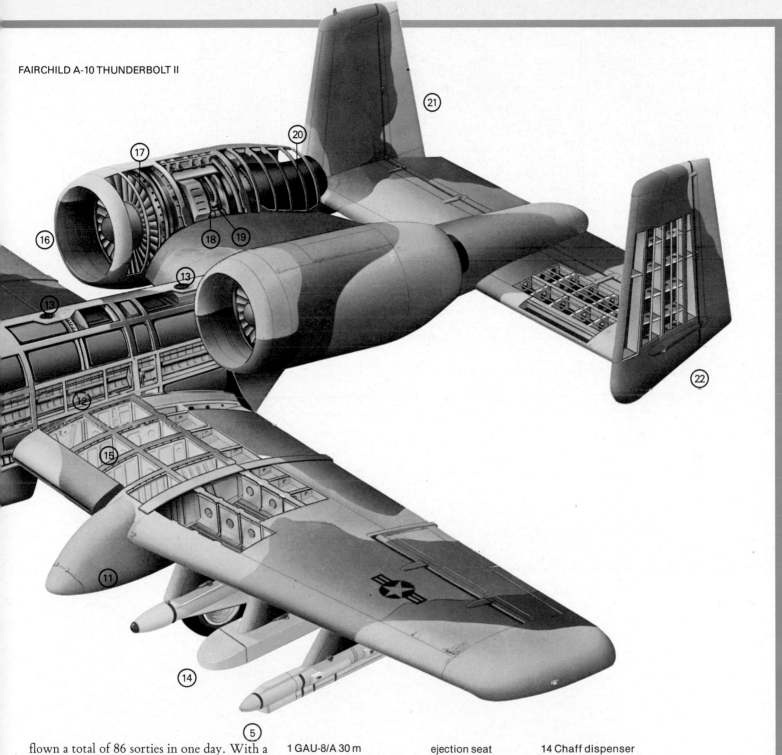

Steve Cross/Pilot Press

flown a total of 86 sorties in one day. With a combat radius of 250 nautical miles, the A-10 can fly to the border area, operate for 1 hour 45 minutes, and return to its Forward Operating Location. Alternatively, it can go to temporary sites located only 40 km (25 miles) or five minutes' flying time behind the fighting line and be back in the air attacking enemy tanks within about an hour.

The only limit on the number of sorties the aircraft can fly is the endurance of the

1 GAU-8/A 30 m Gatling rotary gun
2 Electrical system relay switches
3 Pilot's head-up display screen
4 Ammunition feed
5 Electronic counter-measures pod
6 Ammunition drum
7 McDonnell Douglas

ejection seat
8 IFF aerial
9 Avionics
10 UHF/TAKAN aerial
11 Port Mainwheel housing
12 Longitudinal control and service duct
13 Gravity fuel filler caps

14 Chaff dispenser
15 Port wing integral fuel tank
16 Starboard intake
17 Engine fan blades
18 Oil tank
19 General Electric TF 34 GE100 turbofan
20 Engine exhaust
21 Starboard rudder
22 Port tailfin

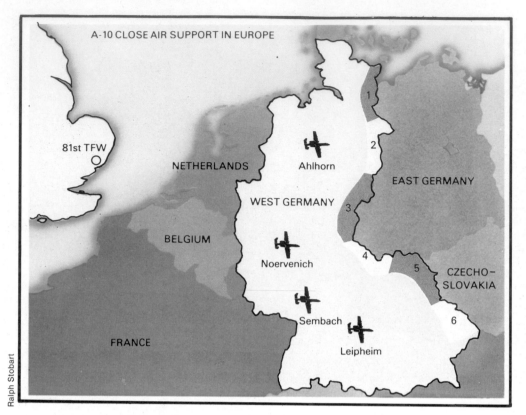

A-10 CLOSE AIR SUPPORT IN EUROPE

81st TFW

NETHERLANDS

BELGIUM

FRANCE

WEST GERMANY

Ahlhorn

Noervenich

Sembach

Leipheim

EAST GERMANY

CZECHO-SLOVAKIA

Ralph Stobart

Left The USAF 81st Tactical Fighter Wing is based in East Anglia, UK. Six operating squadrons of A-10s will become so familiar with zones approximately 30 km deep along the eastern border of West Germany that they can work quickly and efficiently without reference to maps and with a minimum of radio traffic. *Below* A-10s on the line at Myrtle Beach Air Force Base in South Carolina being refuelled.

pilot himself. The 81st TFW pilots are determined and eager, but after three sorties in one day they begin to lose their edge.

Should the FOLs come under attack, there are several alternatives. The A-10 needs only about 1,200 m to take off and land, and the USAF are developing a programme to locate refuelling and rearming facilities in hundreds of German airfields, both military and civilian. All the A-10 needs is a tanker to fuel it and a trailer to arm it. The pilot is trained to carry out his own refuelling if he has to. Fire retardant foam both outside and inside the tanks cuts down the risk of irreparable fire damage, but if one fuel tank takes a hit, it can be isolated at a bank of switches, and the pilot and ground crew can fill the others

U.S. Air Force Photo

to get the A-10 back up in the air again.

If the aircraft has taken punishment, it may be possible to repair it on the spot. Bullet holes in the structure are not a significant problem. The A-10 flies subsonic, so a smooth skin is not important and the panels are simply and cheaply rivetted on from the outside. The panels are mostly single curvature, so a competent mechanic can bend a patch from a piece of sheet metal and rivet it in place. All the inspection panels are designed to be easily accessible from the ground.

Motorway landing

With such sparse servicing facilities needed to keep this aircraft flying, USAF planners are even working on a scheme to use short stretches of *Autobahn* as runway. The pilot can roll his aircraft under a flyover, go on ground-alert with his communications systems driven by the auxiliary power unit, and be ready to respond to requests for air support at the front within minutes.

Ways of deploying the new aircraft to best effect are still being evaluated. The most promising idea is to use it as part of a Joint Air Attack Team (JAAT), composed of two A-10s and two AH-1 Cobra helicopters. The basic operation is for the helicopters, called in by the ground and air liaison officers, to locate and identify targets, and for their team leader (the Forward Air Controller) to call

Under the port wing this A-10, one of the pre-production models, carries a Hobos television-guided bomb, while under the starboard wing is fitted a laser-guided bomb.

up the A-10s' fire power if he needs it.

The helicopters, lurking among the treetops, will attack missile launching sites with their 'stand-off' weapons, normally TOWs (tube-launched, optically tracked, wire-guided missiles). The A-10s will signal their approach one minute before arriving over the target area, at which time the helicopters will lift out of the treetops and hover, with their noses pointing towards the target at a distance of 3,000 m.

The A-10 pilots can spot the target by smoke or flames, or any other visual means, and they also carry among their few elaborate aids a Pave Penny laser spot tracker. When an observer on the ground or in a helicopter directs a laser beam at an identified target, the beam is dispersed by vegetation, but not by a solid object like a tank. The A-10s underwing 'tracker' picks up the beam's reflection, and gives the pilot a red diamond shape to aim for on his head-up display.

At 1,200 m the pilot is at his optimum range and opens fire. He can turn away and come in for a second pass, without having to fly over the target area. One pilot has

described it as like throwing stones at a rattlesnake. 'You can stay out of danger until you have destroyed it'. After circling twice, the A-10 will have lost momentum, and needs to pull back from the battlefield to prepare for another run. On its withdrawal, it will have support from the helicopters, before they descend into the trees.

Simple to fly

The aircraft is virtually without complicated flying aids. There is a radar-controlled landing device, but no computerized navigation system. A head-up display gives basic flying information—air speed, altitude and attitude of flight. Beyond that, the pilot is in total control of his own aircraft. Pilots in the FOLs have been training in the skies where they will fight, if called to. They know the terrain and can find their way back from anywhere in a radius of 160 km (100 miles), just on their knowledge of the landmarks. For flying outside that range, they carry maps.

Pilots who fly the A-10 take an individual approach to their new tank-killer. They are on their own in a simple, reliable aircraft. They know they have to rely on their eyes, on their training, on their understanding of the aircraft and on combat courage. The emphasis is back where it was in air warfare 25 years ago: on pure airmanship.

The bombers return

Streaking through the air at the speed of sound, just a few hundred metres above ground, a delta-winged bomber weaves an undulating path across ridged terrain as it seeks out a target. Fast from behind, a supersonic fighter dives to attack the intruder but in a single burst from the bomber's tail is vapourized by a flash of laser light. It might not happen tomorrow but, if aircraft engineers have their way, the new shapes developed for high-speed intrusion combined with the product of exotic experiments, might produce such a weapon for the 1990s—and all wrapped up in a package that is almost completely invisible to radar.

A widely held belief during the 1960s that manned combat aircraft were rapidly becoming obsolete was not borne out by events of the 1970s. In fact, the 1980s brought a new awakening to the reality that far from being on the brink of obsolescence, fast attack and strike aircraft were needed as much as ever. But it will be the 1990s before the major new developments in heavy-bomber technology catch up with production orders and service operations.

The idea of an intercontinental heavy bomber goes back only to the end of World War 2, yet the growth in bomber capability has been phenomenal. Each B-52 now carries more explosive power than all the bombs dropped in World War 2, including the atomic weapons dropped on Hiroshima and Nagasaki. The first intercontinental aircraft did not fly until August 1946 but, when it did, the Convair B-36 ushered in the age of global bombing.

The B-36 was developed as the prime US nuclear delivery vehicle, capable of carrying 38 tonnes of bombs in two massive bays. Some B-36 flights lasted more than 40 hours, and the 28-man crew operated the aircraft in shifts. For protection, the B-36 carried up to 16 cannons in eight remotely controlled turrets, retracted behind closed doors that smoothed the airflow across the fuselage.

The aircraft bridged eras in aviation history, effectively providing a transition from piston to jet engine power. When

Crown Copyright/MOD

designed, the B-36 carried six 28-cylinder radial engines driving pusher propellers mounted along the trailing edge of the wing, thus improving air flow across the front of the aircraft's wing.

A standard piston-engined B-36 could reach a height of 12,000 m (39,000 ft), but with four General Electric jet engines in two separate pods—one under each wing outboard of the piston engines—the aircraft could reach much higher altitudes. Some crews claimed to have flown at 18,300 m—well above the operating ceiling of the most advanced jet fighters of the day. The enormous range was made possible through expedient use of fuel, shutting down the four jets and two of the six piston engines.

The Peacemaker, as the B-36 was known, was the long arm of US nuclear deterrent at a time when the Soviets had no means of delivering the few atomic bombs they possessed. The main advantages of the B-36 were that it extended conventional heavy-bomber capacity (one Peacemaker could carry the bomb load of a whole squadron of Lancasters), and it had trans-continental range.

When the Peacemaker entered service with the Strategic Air Command in 1948, it represented a completely new way of carrying the fight to the enemy—the perfect tool for early concepts of nuclear deterrence by which the heartland of a nation's territory was accessible to atomic bombers. But the B-36 was slow and needed too many people to keep it flying.

By 1958, the massive B-36 was withdrawn from service, its place taken by a completely new all-jet bomber—the B-52. Also known as the Stratofortress, the B-52 will probably be in service at the turn of the century.

The Stratofortress took several years to reach prototype stage. Conceived in 1945 as a long-range replacement for the Peacemaker (then only just emerging from the factory as a prototype), Boeing submitted a straight-wing, four-engined design.

Subsequent work by German aircraft designers conclusively demonstrated the advantages of the swept-wing for high-speed flight. The new bomber was expected to fly close to the speed of sound, so the swept-wing was incorporated in a new design offered by Boeing as Model 464-49.

An eight-jet, swept-wing aircraft with a range of 13,000 km was just what the Air Force wanted. Refinements were made, but it was essentially Model 464-49 that was eventually built.

Following successful test flights with the B-52 Stratofortress in 1952, the Air Force ordered full-scale production and the Strategic Air Command had its replacement for the lumbering ten-engined Peacemaker. From 1966, only the B-52 was left in service as a strategic intercontinental bomber. The

David Baker

Photri

Far left A Soviet TU-20 Bear long-range bomber, one of about 45 used for shipping surveillance.
Centre Equipped with 1,200 km range Hound Dog missiles, the B52G changed the B52's emphasis from a bomber to missile launcher *(above)*, incorporating many technical changes.

search for a successor has been impeded by a series of policy changes and funding cuts that resulted in two significant new aircraft being cancelled soon after the prototypes flew.

The need to keep a manned bomber is fundamental to the Western system of deterrence, based on the idea that no single technical breakthrough by a potential enemy should disable the complete US strategic strike force. From this idea came the so-called 'triad' in which bombers, land-based Intercontinental Ballistic Missiles (ICBMs) and ballistic missiles in submarines ensure a three-part response.

The need to expand the missile capability made less visible the campaign for a B-52 replacement. And that fact, probably more than any other, has kept the B-52 as the prime intercontinental bomber since the mid-1950s. Nevertheless, even before the

first B-52 joined Air Force units, a specification had been written for its successor, seeking a faster aircraft able to penetrate hostile air space and get through the fighters and surface-to-air missiles thrown up to shoot it down. Boeing and the then North American Aviation responded with proposals that verged on science fiction.

North American's design was about twice the weight of the B-52, and would have flown three times faster at nearly two and a half times the speed of sound. The North American and Boeing designs were rejected, but their new proposals were no less spectacular, though far less futuristic. In 1957, North American Aviation was ordered to build their prototype, an aircraft called the Valkyrie, or XB-70. While still under construction in 1961, US Defence Secretary Robert McNamara cancelled the programme

because, it was claimed, bombers were too vulnerable and missiles would fight the next war that broke out.

Despite the negative attitude, North American Aviation engineers rolled the first XB-70 out of the hangar doors in May, 1964, knowing that their product would never fly as a supersonic replacement for the B-52. It was about 25 per cent heavier and about 20 per cent longer, with a delta-shaped wing little more than one-half the span of the Boeing B-52. But it was powerful, carrying six massive engines in a huge pod beneath the fuselage. The combined thrust was more than four times that of early B-52 aircraft, providing a maximum speed of 3,700 km/h, or three times the speed of sound.

Although the XB-70 promised a range of nearly 7,000 km (4,300 miles), it could carry

Right The swing wing Rockwell B-1 bomber. It has twice the bomb-carrying capacity over the same range as the B-52 and is designed for prolonged periods of low-level sonic flying. With the development of the Cruise missile, however, the B-1 has remained largely at prototype stage, developing gradually.
Inset An FB-111, a B-1 and a B-52 fly together.

THE ROCKWELL B-1

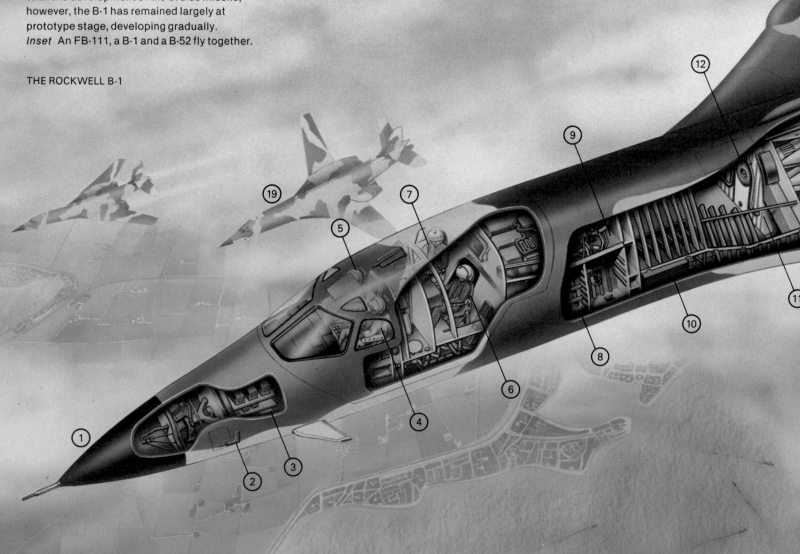

Kuo Kang Chen

1 Radome
2 Dynamic pressure
   sensor
3 Forward
   electronics bay
4 Pilot
5 Co-pilot
6 Defensive systems
   operator
7 Offensive systems
   operator
8 Fuel-cooled heat
   exchanger
9 Weapons avionics
refrigeration unit
10 Side looking
   radome
11 Weapons bay
12 Weapons actuator
13 SRAM missiles
14 Swing wing
15 Undercarriage
16 General Electric
   F101, turbo-fan
17 Aerials
18 Rudder
19 Subsonic position
   of wings

David Baker

David Baker

only a small bomb load. Yet, because of the superior performance, it needed at least as much fuel as the B-52 which, with modifications introduced in the 1960s, was capable of lifting four times the XB-70s projected bomb load, or more than 36 tonnes. No sooner had the XB-70 been relegated to an expensive flying laboratory than work began to define a more practical successor to the B-52.

In studies that began during 1962, the search was for an efficient, manned, supersonic, penetrating bomber, or Advanced Manned Strategic Aircraft (AMSA) as the resulting analyses were known. The need for speed had reduced payload significantly and there was justification for dropping the XB-70. Engineers were asked to design an aircraft capable of flying twice as fast as the B-52 at high altitude and faster than the B-52

close to the ground, while carrying more bombs than the B-52 across similar distances.

By 1970, North American Rockwell (now the North American Aircraft Division of Rockwell International) was awarded a contract to build the B-1. It was smaller than the Stratofortress and had a variable-geometry or 'swing' wing, but its four powerful turbojet engines were in a class of their own.

Significant improvements had been made to the diminishing fleet of B-52s, and the latest G and H models of the massive aircraft had eight engines totalling 29.2 tonnes of thrust; the four engines of the B-1 would generate 54.4 tonnes of thrust. At high altitude, it would fly at Mach 1.6 (1.6 times the speed of sound) and go nearly supersonic at an altitude of less than 60 m. Moreover, it could carry 84 bombs each weighing 227 kg.

Top One of the many stages in the US bomber programme, the XB-70. Four times more powerful than the early B-52s, the XB-70 could not match them for load capacity yet required the same amount of fuel. Modified in the early 1960s, the B-52 yet further proved its worth by being able to carry four times the XB-70's bomb load, relegating its rival to the role of an expensive flying laboratory.

Above A pack of Short Range Attack Missiles (SRAMs) inside the belly of a B-52.

Opposite page Refuelling a B-52 in mid-air from a C-135 tanker.

Right The concept of Stealth. A combination of three separate technologies in order to achieve discreet penetration of an enemy's defences, Stealth bombers were first tested in Nevada during 1976. The chances of detection are dramatically reduced by taking all the following precautions: shaping the aircraft to minimize radar reflection, using materials that will absorb as many radar signals as possible and utilizing electronic shielding equipment.

Nick Farmer

Photri

promising was a concept called Stealth, an appropriate name for an aircraft design capable of penetrating hostile airspace virtually unseen.

In 1976, a top-secret airfield in Nevada, about 50 km north-west of Las Vegas, began to test strange-looking shapes constructed as demonstration replicas of what a Stealth bomber could do. Essentially, Stealth is a combination of three separate technologies in an attempt to achieve discreet penetration of enemy defences.

One technology concerns carefully shaping the aircraft's contours, minimizing radar reflection from jet engine intakes and turbo fan blades. The second concerns a material applied to the aircraft's skin to absorb radar signals. The composite, fibrous material has been developed for the space programme and shows remarkable promise for Stealth designs. The third, more conventional technology involves electronic equipment to partly shield the aircraft from detection.

Stealth is not a single aircraft, but rather a completely new concept involving every design step and each assembly phase. Combined, the new technologies could produce by the early 1990s a bomber almost transparent to radar and capable of hugging the ground at high speed.

There are inevitable gaps in Stealth's ability to evade detection, so some form of defence is essential. The sting in the tail is still the most promising means of shooting down an attacking fighter. The existing force of B-52 bombers has been extensively modernized, and each aircraft has advanced electronics combined with a single cannon in the rear.

Future designs

Future designs could be equipped with high-energy weapons, such as lasers. By placing a laser in the tail and power generation equipment in the rear fuselage, it would be possible to give a Stealth-type aircraft defence over great distances.

By the early 1980s, revised priorities in the US economy made the prospect of a replacement for the ageing B-52 better than ever. Stealth would be the preferable option for the future, but an interim solution might be a B-1 or an FB-111 derivative.

Development of the winged, pilotless, flying bomb—the cruise missile—will also continue. For that, a large aircraft capable of flying at subsonic speed to a release point some distance outside enemy territory will be favoured. Paradoxically, the best platform might be a military version of the jumbo jet.

In a nuclear role, the B-1 could take up to 24 Short Range Attack Missiles (SRAMs) in its large bomb bay, compared with 20 in the B-52s. The SRAM has an explosive yield of the Hiroshima bomb. It is nearly 4.3 m long and has a rocket motor that starts up when the missile is dropped, carrying the single warhead to a target up to 161 m away.

The first B-1 flew in 1974, and proved the confidence of its designers from the outset. But it was not only the B-1 that Strategic Air Command sought to have in the inventory. Developments in propulsion, electronics and miniaturization brought to reality a concept almost as old as the flying machine—the cruise missile, which was not just a flying bomb but an intelligent, thinking device able to twist and turn, fly a circuitous route to its target and achieve unparalleled accuracy.

Built by Boeing, the Air-Launched Cruise Missile (ALCM) was just what the Air Force wanted for use with the ageing B-52s and

the new B-1. The older aircraft would not survive long if forced to penetrate enemy defences, but with a 'stand-off' capability, whereby it could stand outside enemy airspace and release up to 20 small cruise missiles, its service life would be extended.

Each ALCM can fly 2,400 km (1,500 miles) to its target, allowing the launch aircraft to remain over friendly territory, and serve merely as a launch platform and not an intruding bomber. But to fly undetected under radar to hit enemy defences and knock out the many anti-aircraft posts requires a fast penetrator. For that task, the B-1 would have proved indispensible.

But in 1977, in an attempt to extract concessions from the Soviet Union, President Jimmy Carter cancelled the B-1 outright and retained the ALCM. He did, however, agree to keep the B-1 flying as a test laboratory for new electronic equipment.

Nevertheless, Air Force studies continued on future bomber technology. The most

The rise of the all-purpose warplane

If Soviet tanks were to surge into Western Europe on a stormy winter's night, most of the thousands of aircraft in NATO's armoury would be powerless to stop them. Some would be out of range. Some would be grounded by the bad visibility. So only a hundred or so F-111 swing wing attack bombers of the US Air Force would be available to stem the attack.

But from late 1980 onwards Europe's air defences are due for a dramatic boost — 809 new aircraft which can outperform the F-111, yet are much smaller and lighter. This extra muscle will be provided by the Panavia Tornado, the unique warplane whose design and construction has been probably the biggest technical programme ever carried out jointly by several countries, in armaments or in anything else.

There have been occasions in the past when the aircraft designs of one country have been wholly or partly built elsewhere. Tornado is unique because it was created, starting with a clean sheet of paper, by three countries working as partners.

It is also unique in satisfying an amazingly varied demand. What the three countries originally wanted seemed to

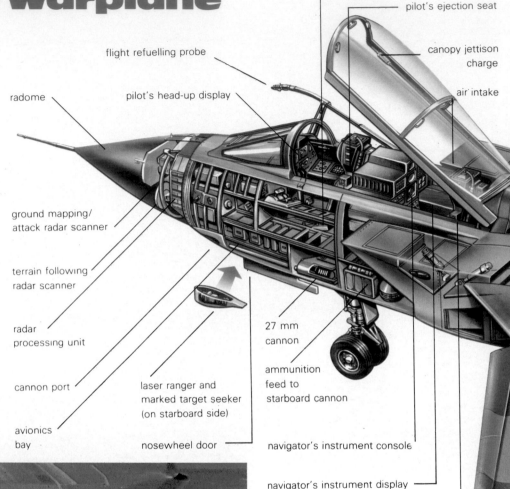

autopilot control panel

pilot's ejection seat

flight refuelling probe

canopy jettison charge

air intake

radome

pilot's head-up display

ground mapping/ attack radar scanner

terrain following radar scanner

radar processing unit

27 mm cannon

ammunition feed to starboard cannon

cannon port

laser ranger and marked target seeker (on starboard side)

avionics bay

nosewheel door

navigator's instrument console

navigator's instrument display

navigator's ejection seat

full-span leading-edge slats, extended

port navigation light

wingtip antenna

Above Crammed with weaponry, electronics and fuel tanks, Tornado is exceptionally compact for a long-range reconnaissance aircraft. Its wing span (extended) is 13 m, and its length only 16.7 m. It is also extremely versatile: the aircraft it will replace range from the Italians' and Germans' Starfighters (wing area 20 sq m) to the RAF's Vulcan bombers (370 sq m).

Left Tornado prototype 03, in a tight turn over the German countryside, gives a fair imitation of a fighter in action.

Panavia

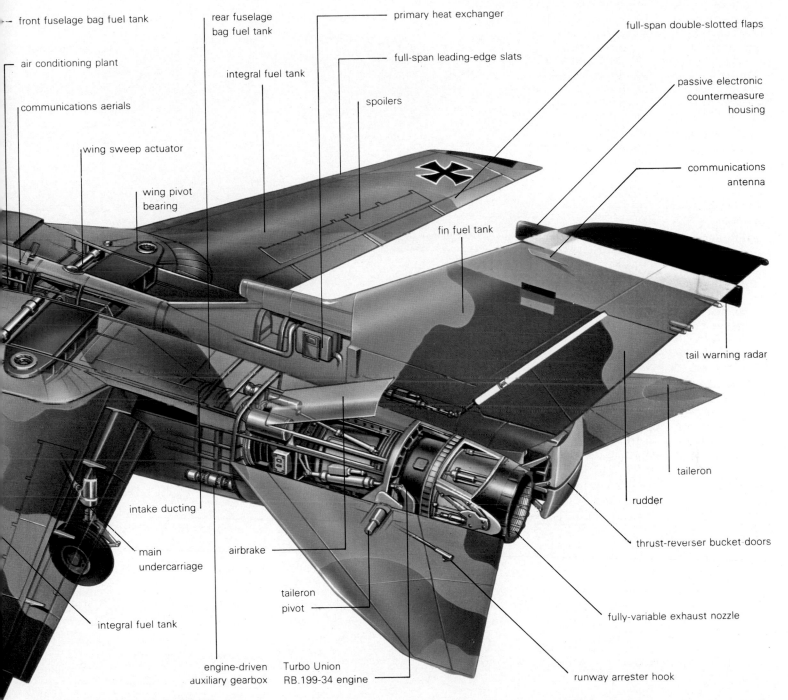

front fuselage bag fuel tank

rear fuselage bag fuel tank

primary heat exchanger

full-span double-slotted flaps

air conditioning plant

integral fuel tank

full-span leading-edge slats

passive electronic countermeasure housing

communications aerials

spoilers

communications antenna

wing sweep actuator

fin fuel tank

wing pivot bearing

tail warning radar

taileron

rudder

intake ducting

thrust-reverser bucket-doors

main undercarriage

airbrake

taileron pivot

fully-variable exhaust nozzle

integral fuel tank

engine-driven auxiliary gearbox

Turbo Union RB.199-34 engine

runway arrester hook

Grose Thurston

require a whole range of radically different aircraft. Britain's RAF wanted a short-range strike aircraft or 'tank buster', plus a long-range, low-altitude bomber which could make pinpoint attacks without ever seeing the target. Germany's Luftwaffe wanted a strike aircraft designed to take off from short, bomb-damaged runways, and also needed an interceptor fighter. The German naval air force wanted a low-flying anti-ship weapon and reconnaissance plane. Italy's AMI (air force) wanted similar

aircraft to Germany's, but with different detailed specifications.

A brilliantly advanced engine designed by Rolls-Royce was selected, and an international company called Turbo-Union was set up by Rolls-Royce, MTU of West Germany and Fiat of Italy to develop and produce it as the RB.199. And, as an agreed aircraft design emerged, a similar three-nation company was formed by BAC, MBB and Italy's Aeritalia. It was called Panavia.

Panavia's engineers and military advisers

were determined not to be sidetracked by the political arguments which had plagued some international projects, notably those involving France. It was not important, they decided, in which country the first flight took place, nor what the nationality of the pilot was. Work on the airframe was to go 42.5% to Britain, 42.5% to Germany and 15% to Italy — roughly corresponding to the number of aircraft each country planned to buy — irrespective of which country most needed the jobs it would create. And ancillary equipment was to be the best available, whatever its source. When the billion-dollar contract for the internal radar went to a US company, Texas Instruments, for example, some politicians and newspapers were furious. But Panavia stuck to its decision.

At first the four customers' demands seemed hard to reconcile. The RAF alone demanded 'wet wings' — that is, wings sealed to become fuel tanks to give extra range. It also insisted on a two-seat aircraft, whereas the Italian AMI was equally adamant about a single-seater. So two versions were planned, the two-seat version being brought into aerodynamic balance by adjusting the angle of the swing wings.

By 1970, however, everyone agreed that extra range was a good thing, and wet wings became standard. Then all four customers decided that, since there was to be a two-seat version anyway, they might as well have some as training aircraft. And by 1974, having further studied the problems of aerial combat, all four agreed that a two-man crew was better for the fighting version also. So in that year a decision was taken to standardize on a tandem-seat model, just 500 mm (20 in.) longer than the original. The new, standard aircraft was named Tornado.

In the meantime the first prototype, called the P-01, had been rolled out for testing. Considering its multi-role function, most observers were surprised at how compact it was. Its overall length of 16.77 m (55 ft) compared with 19.21 m (63 ft) for the Phantom, almost 19.52 m (64 ft) for the F-15 and 18.91 m (62 ft) for the US Navy's F-14 — and all of these were fighters. Similarly, the F-14 has swing wings like Tornado's, but its wingspan varies from more than 19.52 m (64 ft) extended to 11.59 m (38 ft) at maximum swept-back angle, whereas the corresponding Tornado figures are 13.88 m (45 ft 6 in) and 8.54 m (28 ft).

Tornado is packed with equipment intended to give it a greater capability and efficiency than any rival. Some of these, such as the F-111, were designed in the days of the 'tripwire' policy, in which any aggression against NATO was to be met by instant nuclear retaliation. Tornado was designed for the later policy of 'flexible response', in which nuclear warfare is deferred for as long as possible and — with luck — any aggression is halted by some means without it. This produced profound changes in the demands likely to be made on

Right Armed with four Kormoran anti-submarine missiles, prototype 03 blasts off from the test runway. So that it can operate from bomb-damaged airfields, Tornado is designed to take off from runways only one-third normal length. The wing-tip pods in the illustration contain electronic countermeasure equipment.

tactical aircraft. Instead of carrying one or two nuclear bombs and dropping them on or close to a particular target, the attack aircraft has to carry dozens of bombs and drop them day or night, in any weather, on tanks, guns or troop formations.

Big bomb load

So Tornado has weapon racks all the way along the broad flat underbelly of the fuselage, on each side, as well as pivoting pylons both inboard and outboard along the wings. This means that a bomb load as heavy as 4,500 kg (10,000 lb) can be carried under the fuselage while leaving the wings free for enormous drop tanks, self-defence missiles and electronic countermeasure (ECM) pods to confuse hostile defences.

For pinpoint delivery, no aircraft has ever had better equipment. The back-seat crew member has three electronic displays with which he can first plan the whole mission and then study its progress as he sits and manages the entire operation. His biggest electronic item is the main radar, which can operate in no fewer than 14 different ways for contrasting purposes concerned with navigation, weapon delivery, air combat and testing. He also has a TFR (terrain-following radar) with which Tornado can

roar along just off the ground faster than any other aircraft ever built, automatically avoiding trees, radio masts, hills and other obstructions, and thus confusing defending radar, fighters and missiles.

Under the nose is a laser for exact measurement of the range to a surface target, or for automatic guidance of the aircraft or missiles to a target picked out by a 'friendly' laser on the ground. And the diversity of weapons that can be carried exceeds that of any other aircraft in history: at least 48 different kinds of bomb, missile, pod, tank or dispenser.

For close-range operations against an invading army, Tornado might well need to operate from an airfield blasted by air or missile attack. So it is designed to take off at maximum weight in about 900 m (3,000 ft), one-third or less of the length of its normal runways. On landing, too, it can stop better than any other aircraft except for a jump-jet such as Harrier. It has spread-out wings, leading-edge slats, full-span trailing-edge flaps, large rear-fuselage airbrakes 'like barn doors', automatic reversers on the two engines to make them pull instead of push, and powerful anti-skid brakes. Just in case all this is not enough, there is an arrester hook for catching a runway wire.

And although it has small, fuel-efficient engines tailored to long range and economy, its performance as a fighter compares quite well with aircraft designed for that role.

For long-range sorties, all four air services have done more than follow the RAF's lead by using wet wings. They have now decided that, as the aircraft is so amazingly capable, they might as well add two 1,500 litre (330 gallon) drop tanks and provision for an in-flight refuelling probe. Total fuel capacity is thus almost three times that of the original aircraft as envisaged in 1969.

Although it is ten years since the design was settled, Tornado is still regarded as completely up to date and ideal for the rest of this century, if not longer. Now Panavia and the NATO air forces are considering the next project: a small, but equally versatile, tactical fighter.

Below An array of the weaponry Tornado can carry on its diverse range of missions. Perhaps its most extraordinary load is the German MW-1 dispenser, which looks rather like a freight container strapped to the belly. It is designed to shower the ground with precision-aimed mines and bomblets to destroy advancing tanks.

Return of the flying bomb

Like a small, pilotless jet aircraft, to-day's cruise missile flies to its target, following a predetermined path imprinted in its computer memory. Armed with a nuclear warhead, it flies low to avoid enemy radar and automatically avoids any obstacles in its path. Incredibly, it can hit its target with an accuracy of a few metres after a journey of more than a thousand miles.

The concept of the cruise missile is not new. When in June 1977 President Carter announced his decision to cancel the B-1 strategic bomber and to accelerate production of a supposedly new weapon called a cruise missile, he was in fact proposing a weapon as old as the bomber.

The earliest cruise missiles were produced in World War 1. Britain, Germany and the United States were the leaders, but France and Italy also produced various types. Most were designed for land take-off and thereafter to fly under autopilot or radio control for either a set time or a set distance (logged by a small free-wheeling air propeller which unwound from a screw thread). At

Left A Tomahawk cruise missile is launched for the first time from its 'transporter erector launcher' (TEL) at the Utah test and training range. A typical ground installation will have four TELs, each capable of firing four cruise missiles.

Below A Tomahawk releases explosive charges over a target runway during a test run.

General Dynamics Corporation

the right moment it would be commanded to dive on its target.

Between the wars the RAF pioneered cruise missile missions over quite long ranges, using live warheads. The RAF Lynx did most of its flying in Iraq, carrying a 113 kg (250 lb) bomb distances up to 225 km (140 miles) with great accuracy. Several other groups, particularly the US Navy, tested cruise missiles in the 1930s.

By far the most important programme at this time, however, was one that stemmed from a new kind of engine developed by German aerodynamicist Paul Schmidt from 1928. Called a *pulsejet,* or *resonating-duct engine,* it had no moving parts except a grille containing sprung-steel flap valves. In operation these vibrated 47 times a second, alternately admitting fresh air (which mixed with fuel behind the valves) and being blown shut by the resulting explosion. This, the first production jet engine, was such a simple idea that in June 1942 the Nazi leaders gave the go-ahead for a missile powered by the Schmidt duct, and on 13 June 1944 the first V-1 flying bombs fell on London. By April 1945 no fewer than 29,000 of these missiles had been produced, but the defences had gained the upper hand and Hitler's idea of destroying the British capital was thwarted.

Mace and Regulus

After World War 2 most countries built cruise missiles. The US Air Force received more than 1,000 Martin TN-61 Matadors and then switched in 1959 to the much more deadly TM-76 Mace, later called MGM-13A in its hardened-shelter version and CGM-13B in the type fired from mobile launchers. Carrying various warheads, including thermonuclear ones, the Mace had either terrain-comparison or inertial guidance, as do today's cruise missiles. The US Navy's chief cruise missiles were RGM-6 Regulus I and RGM 15 Regulus II, both of which were lethal long-range weapons fired from submarines. The reason for their withdrawal was simply a belief that ballistic missiles made them obsolete.

Largest of all the cruise missiles were monsters used by the US Air Force Strategic Air Command. First to be developed was SM-62 Snark, a tailless missile by Northrop Corporation, with a powerful turbojet for cruise propulsion after a launch from a mobile trailer under the thrust of two large rockets. Snark cruised at Mach 0.9, rather

faster than jet bombers. Near the target, the massive 5-megaton warhead was separated to fall at transonic speed. Snark had astro (star-guided) inertial navigation, and could make various manoeuvres to avoid defences, even after an 11-hour mission over more than 9,700 km (6,000 miles).

An even bigger programme was SM-64 Navaho, but this was cancelled in July 1957 just as it was about to go into production. A real leviathan, weighing 132 tonnes, it rode up vertically on the back of a giant rocket with three of the most powerful engines then constructed, and then pitched forward into cruise at Mach 3.25, or 3,460 km/h (2,150 mph), faster than any combat aircraft. It was powered by two immense ramjets that formed the main part of the fuselage. Again the reason for rejecting it was that cruise missiles were thought to be obsolete.

Stand-off bomb

Among the dozens of other cruise missiles of the 1950s and 1960s were some that were designed not to destroy targets but merely to carry *electronic countermeasures* (ECM) which would confuse and dilute the defences during an attack by manned bombers. France, Sweden, Norway, Italy and Britain all made cruise missiles, the British term for an air-launched example being 'stand-off bomb' because it allowed the carrier aircraft to stand off from the target at a distance where, it was thought, it would be less likely to be shot down.

By far the most numerous type was a Russian missile, called SS-N-2 Styx by NATO, which was relatively simple and could be carried in groups of four aboard small high-speed patrol boats. Western navies appeared not to notice the hundreds of these missiles spreading across the navies of the Soviet Union and its clients, and made no attempt to counter or copy it, until 21 October 1967. Then Egyptian missile boats sank the Israeli destroyer *Eilat* without even leaving Port Said harbour. Instantly the naval world was in turmoil, and within weeks this rather primitive cruise missile was being copied, not least by Israel.

The largest air-launched cruise missiles have also been Russian. The first of the Soviet Union's many bomber-launched missiles entered service in 1957. Called AS-1 Kennel by NATO, it resembled a miniature jet fighter and carried various types of warhead but was used mainly in an anti-ship

Above The anti-ship version of the BGM-109 Tomahawk cruise missile with the land attack warhead (11) below. A submarine launch begins with the flooding of the torpedo tube. The tube is then opened and the cruise missile ejected from its protective capsule. The booster rocket (8) fires about 10 m (30 ft) out and its four jet tabs (9) vector the thrust to steer Tomahawk out of the sea. The missile surfaces at an angle of 55° with a speed of 88 km/h (55 mph) and the tail fins (7) spring out to roll the missile the right way up. After 6-7 seconds the booster burns up and is jettisoned while the wings (4) extend. The air scoop (10) deploys and the turbofan engine (6), spun up to 20,000 rpm by hot gas from a starter cartridge, begins the cruise flight at about 300 m (1,000 ft). The ship attack missile has a 450 kg (1,000 lb) high explosive warhead (3) and 550 kg (1,125 lb) of fuel (5). The land attack version has a 120 kg (270 lb) nuclear warhead (13) and 175 kg (385 lb) of extra fuel (12). Both versions have an airspeed indicator (1) linked into a highly sophisticated terrain-following navigation system (2 and 14) which guides the missiles along a pre-set path to the target.

role. Its carrier was the TU-16 'Badger' bomber, which was also developed in a different version tailored to the much larger AS-2 'Kipper' missile. The bomber carried two AS-1s, one under each wing, or a single AS-2 under the fuselage. Armed with a nuclear warhead, AS-2 could fly more than 210 km (130 miles). The next missile, AS-3 'Kangaroo', is still the biggest air-launched weapon. First seen by the West in 1961, it was carried by the monster Tu-20 (Tu-95) 'Bear' turboprop bomber, and was more than 15 m (49 ft) long. It carried a thermonuclear weapon to distances exceeding 650 km (400 miles). Later came AS-4 'Kitchen', a supersonic missile for the supersonic Tu-22 'Blinder' bomber, the neat rocket-propelled As-5 'Kelt' carried under the wings of later Tu-16s, and the pinpoint supersonic AS-6 'Kingfish' carried under the wings of both the Tu-16 and the Tu-22M 'Backfire' swing-wing bomber.

What President Carter presented as a new kind of weapon originated in January 1963 when the US Air Force decided to study a 'subsonic-cruise armed decoy' (SCAD). To

be carried by the B-52 bomber, this was a smaller cruise missile than previous designs, and was powered by a Williams WR 19 turbofan engine little larger or heavier than a wide-carriage typewriter. SCAD was intended to be released by the B-52 when approaching its target, partly to confuse the defences by looking, on radar screens, like a B-52 and partly by itself carrying a nuclear warhead. To fit inside the B-52, SCAD had to incorporate new design principles, with wings, tail and air-inlet duct all being extended after it was dropped from the bomber's weapon bay. Boeing had already developed an extemely fast missile called SRAM (short-range attack missile) which the B-52 carries in multiple. SCAD was intended to replace SRAM on a one-for-one basis, with longer range, up to 1,200 km (750 miles), but flying at only the same speed, around 800 km/h (500 mph), as the B-52.

After considering all the possibilities for future bomber attacks, in 1972 the US Air Force recast the SCAD as the ALCM (air-launched cruise missile), retaining the same

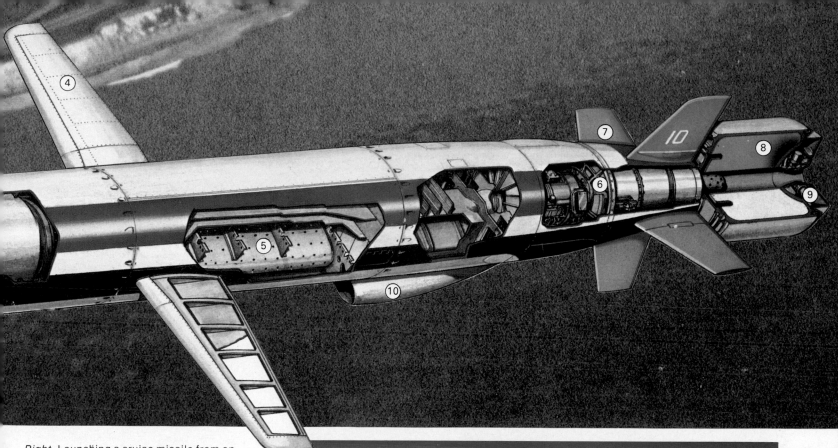

Right Launching a cruise missile from an armoured box launcher installed on the deck of USS *Merrill.* The box launcher is designed for use on certain US Navy surface ships, and can carry up to four cruise missiles which would primarily be used against enemy ships.

designation of AGM-86A. Boeing Aerospace had already been awarded the AGM-86A contract, and switched to the modified weapon. Instead of being a bomb-carrying decoy, ALCM was simply a missile, similar in size and again interchangeable with the SRAM. Williams produced an improved engine, the F107 turbofan rated at 280 kg (600 lb) thrust and with reduced fuel consumption. The engine is started by an explosive squib fired through the turbine as the missile is dropped. The air inlet duct extends, the wings swing out from under the rear fuselage, and the tail unfolds.

AGM-86A first flew from a B-52 at White Sands Missile Range, New Mexico, on 5 March 1976. Subsequent testing solved various problems, but the original expectation was that AGM-86A would go into production, each B-52 carrying up to 12 missiles externally and 8 more on a rotary launcher inside the aft weapon bay. Externally carried ALCMs were to have an underbelly fuel tank added to extend the range. But the programme was made more complex by the development of a longer ALCM version, AGM-86B, and the construction of a completely different cruise missile by a rival contractor, General Dynamics. The latter did

Salamander Books

Above The sea-launched version of the Tomahawk is so compact that it can be fired like a torpedo from a submarine. It reaches the surface at about 88 km/h (55 mph).

not originally intend to build an air-launched weapon but a sea-launched cruise missile (SLCM), under a US Navy contract awarded in January 1974.

General Dynamics named its missile Tomahawk, and although able to carry the W-80 thermonuclear warhead (the same as both versions of ALCM and also the SRAM), it was made so compact that it could be fired like a torpedo from the tubes of a submarine. In this form, designated BGM-109, Tomahawk is sealed in a steel capsule. After being fired underwater, the capsule is discarded, the boost motor behind the tail fired and the missile programmed into a steep climb at about 90 km/h (56 mph). To transform the compact underwater projectile into an efficient aerodynamic vehicle, the boost motor drops off, and wings and tail extend as BGM-109 climbs to cruise level. One difference between this and AGM-86B is that, while the latter's turbofan air inlet is on top, Tomahawk's inlet extends from below.

Tomahawk quickly became an exceptionally versatile missile. In addition to the strategic land attack version, with Litton in-

ertial and Tercom (terrain comparison) guidance and the W-80 warhead, it was built in an anti-ship version with a modified Harpoon (another missile), large conventional warhead and active radar guidance to home in on hostile ships. A third and quite different model is AGM-109, the air-launched model, also called TALCM (Tomahawk ALCM). Without either a surrounding capsule or a boost motor, it has been carried by US anti-submarine aircraft which do not normally have much capability against surface targets, at least not against distant cities. Yet another version, perfected in 1977, is GLCM, the ground-launched cruise missile.

Designated BGM-109B, this emerged without much fuss and was an obvious and predictable development. If the USA has a missile able to penetrate hostile defences with a city-destroying warhead, which can also be fired from a torpedo tube, obviously the next move is to take some out of the submarine and put them on land. All existing strategic missiles and aircraft in the West, except those in submarines, are launched from vulnerable fixed bases which automatically become targets for Soviet counter-force missiles. For example, there are just two bases in Europe for the F-111 swing-wing bomber: RAF Upper Heyford, Oxfordshire, and RAF Lakenheath, Suffolk. But a mobile cruise missile changes the picture completely.

In any time of crisis it would be dispersed to some distant location. No Soviet missile could be targeted on it, because its location would not be known.

In mid-1980 the British government announced that from 1983 a force of 160 GLCMs would be based in Britain. Like all GLCMs they will be deployed aboard launchers able to drive unobtrusively along public highways, especially by night, to preselected but secret firing locations. There will be just ten launchers in Britain. By 1988 it is intended to deploy 464 GLCMs in Europe, the others being in Italy (7 launchers, 112 missiles), Germany (6 launchers, 96 missiles), and Belgium and the Netherlands (each 3 launchers and 48 missiles). GLCM had a successful first firing from its launcher in May 1980.

Meanwhile a stern battle had been going on between AGM-86B and its unexpected rival AGM-109 Tomahawk, to decide the ALCM for Strategic Air Command's B-52G and B-52H bomber force. Whereas AGM-86A, the original Boeing ALCM, had

Right The Boeing air-launched cruise missile (ALCM). As the missile is dropped from the B-52 *(inset)*, it is started by a cartridge fired through the turbine. Then the air inlet duct extends, the wings swing out from the fuselage and the tail unfolds.

Right A Tomahawk ALCM under construction. This version of the Tomahawk has no boost rocket or surrounding capsule. It has been carried by US anti-submarine aircraft.

been designed to be interchangeable with SRAM (and at 4 m, 14 ft, was precisely the same length), the AGM-86B is no less than 5.8 m (19 ft 6 in.) long. The rival AGM-109 has a length of 5.5 m (18 ft 3 in.). The reason for the greater length of AGM-86B is that the fuel tankage making up most of the length of the fuselage was almost doubled in capacity, extending the range from 1,200 km (750 miles) to rather more than twice this distance. But it has had the effect of making the missile no longer interchangeable with SRAM, and incompatible with the B-52 rotary launcher. Costly modifications to the B-52 carrier aircraft are therefore needed. (For more than three years the USAF studied the possibility of buying a fleet of existing transport aircraft, such as 747s or DC-10s, to pack with cruise missiles, but dropped the idea.)

After prolonged and not partularly successful testing and evaluation throughout 1979, the Secretary of the US Air Force, Dr Hans Mark, announced in March 1980 that SAC would adopt the Boeing missile, AGM-86B. It is now expected that the US Air Force will buy as many as 3,400 AGM-86B missiles costing well over $1 million each, for deployment from December 1982.

Used entirely in the strategic deterrent role, with nuclear (probably W-80) warheads, these sleek pearl-grey missiles will navigate for most of their flight by the well-proven inertial method. When they cross the coastline of a hostile country their Tercom system will start operating. Its purpose is to measure the exact distance vertically to the ground beneath. As the missile flies straight at quite low level, to keep as far as possible below enemy radar, the result of the measures is an exact plot of the enemy country's terrain profile, showing every hill, valley and even buildings. Tercom compares the results with measures stored in its memory and keeps adjusting the track of the

missile until they exactly correspond. The guidance becomes accurate to within centimetres; unlike most long-range systems it becomes more accurate the further it flies.

Some critics have questioned AGM-86B's ability to penetrate modern defences. They point out that it is only a modernized version of the old V-1, and much slower than many cruise missiles of the past. It is no longer thought possible to escape detection by modern mountain-top radar, and SAMs (surface-to-air missiles) could destroy an AGM-86B or a Tomahawk GLCM in a split second. The official answer is that this has not been overlooked, and that after careful tests of these new cruise missiles against simulated defences the plan is going ahead.

SAM, the portable peril

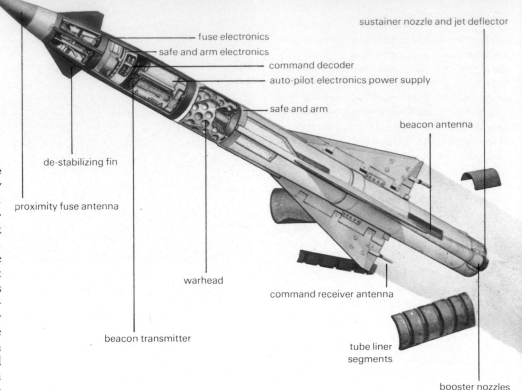

fuse electronics
safe and arm electronics
command decoder
auto-pilot electronics power supply
safe and arm
sustainer nozzle and jet deflector
beacon antenna
de-stabilizing fin
proximity fuse antenna
warhead
beacon transmitter
command receiver antenna
tube liner segments
booster nozzles

The surface-to-air missile, perhaps more than any other piece of current military hardware, epitomizes modern warfare. Lightweight, it is packed with sophisticated technology enabling it to seek out and destroy the most evasive target. Since 1939 war in the air has played a large part in every high-intensity conflict. Specialist ground-attacking aircraft and helicopters have proved to be so deadly that it is very difficult for land forces to operate effectively without some counter to them. Ideally, the best defence is to have combat aircraft which can both win the struggle in the skies and protect ground troops. But due to the high cost of aircraft, surface-to-air weapons have been evolved to give soldiers the chance to strike back against aircraft.

The most established form of anti-aircraft defence is the AA gun, but until the 1960s, mobile self-propelled AA guns were optically controlled and could only engage aircraft in clear weather. In recent years more modern batteries of guns with all-weather capability have been deployed but these are nearly always used in conjunction with surface-to-air missiles (SAMs) and some countries—notably the United Kingdom—rely entirely on SAMs. The Soviet Union, probably the world leader in anti-aircraft systems, still uses guns but only in conjunction with a great variety of SAMs which are integrated into a highly effective system.

All true surface-to-air missiles have some means of guidance. Guidance systems are enormously varied. Some require operator guidance from launch to target, others have a largely automatic, computer-controlled guidance system. Homing devices sensitive to an aircraft characteristic are also common.

Heat seekers

Methods of missile propulsion are equally diverse. Nearly all SAMs have a solid-propellant booster to provide a high initial launch speed but some—such as the Soviet SA-2—have two stages, the first using solid and the second liquid fuel. Warheads, on the other hand, are fairly uniform in that they are explosive although they differ in that some are triggered automatically and others are

detonated by operator guidance.

Among the latest developments in SAMs has been the design of missiles as 'single man systems'. For the first time, infantrymen (and terrorists) wage personal war on aircraft. Naturally, these weapons are among the simplest and most compact in use.

The best known of these missiles, and among the longest established, is the Soviet SA-7 (NATO designation GRAIL). Weighing a little over nine kilograms (20 lbs) with a length of 1.29 metres (4 ft 2 in) it has a range of 9–10 kilometres (6 miles) and is widely used to intercept low-flying aircraft, particularly helicopters. It has simple optical sighting and tracking—the operator simply points it at the target—and relies on its heat-seeking, infa-red warhead to direct it into the heat emissions of its target's exhaust. Once pointed at the target, with the heat seeker activated, an indicator light shows when the target has been 'acquired' by the homing device. The operator can then fire. This means of guidance was, at first, countered by flares dropped from the aircraft to confuse the heat-seeking equipment. But the latest models of the SA-7 are reported to have filters to combat this tactic.

Among the more recent NATO manportable devices is the United Kingdom's Blowpipe System. It has the advantage that the missile itself comes sealed inside its launching canister and thus needs no removal for arming or assembly. The operator simply

clips his aiming unit to it, puts it to his shoulder, sights and fires. The tail and wing assembly are folded in the canister but spring-loaded to open out as they emerge. A first-stage motor lifts the missile out of its case and coasts it a short distance toward the target. Then, at a safe distance from the operator, the second stage motor fires and boosts the missile to supersonic speed. The operator has a monocular sight and aims the missile at his target by straightforward up/down left/right movements. These movements transmit radio commands to the missile's control surfaces. Movements of the cruciform tail-fin wings alter the missile's flight direction. When Blowpipe is near enough to its target it is detonated by a proximity fuse.

Blowpipe can be linked to an identification radar which transmits and recieves pulses to 'interrogate' the target and identify it as friend or foe. Such a radar system is known as IFF (Identification Friend or Foe) and is designed to save friendly aircraft from the attentions of enthusiastic infantrymen who have made an inncorrect visual identification. Other man-portable NATO weapons differ from Blowpipe chiefly in their homing systems. The US Stinger has an infra-red seeking head and a new improved version of it seeks both infra-red and ultra-violet images, giving it a better performance and making it more difficult to counter. (Ways to jam or negate guidance systems are as actively developed as any other technical branch of to-

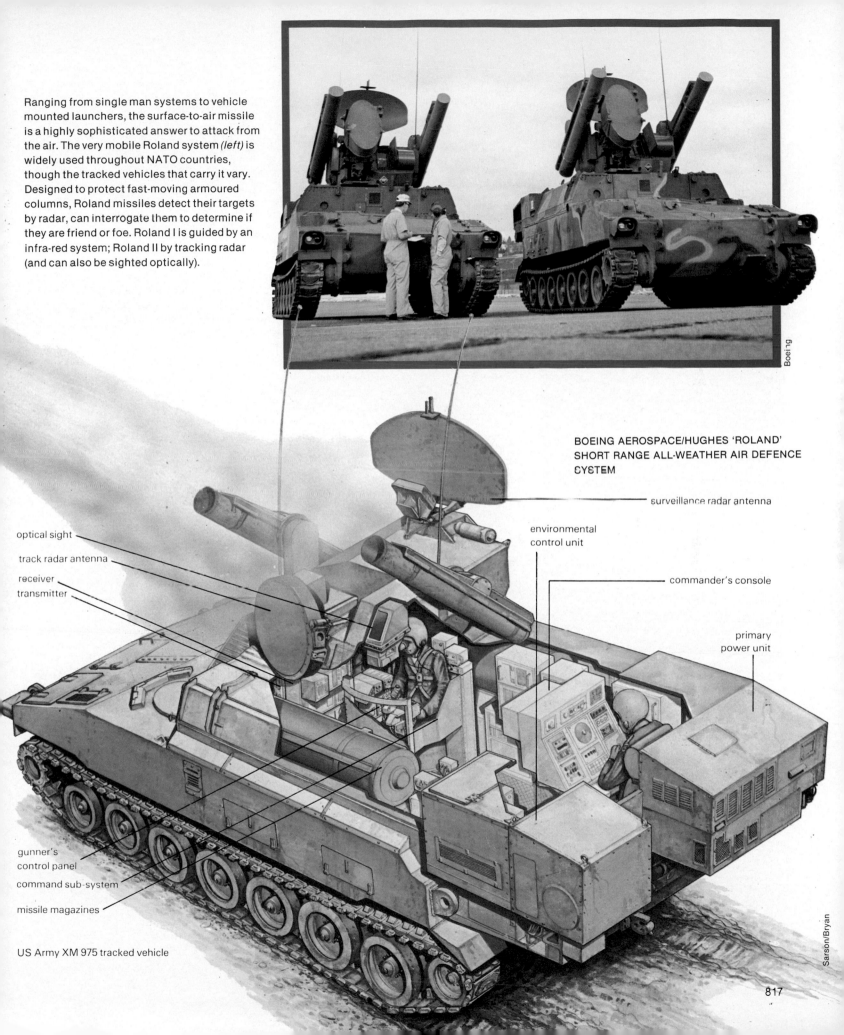

Ranging from single man systems to vehicle mounted launchers, the surface-to-air missile is a highly sophisticated answer to attack from the air. The very mobile Roland system *(left)* is widely used throughout NATO countries, though the tracked vehicles that carry it vary. Designed to protect fast-moving armoured columns, Roland missiles detect their targets by radar, can interrogate them to determine if they are friend or foe. Roland I is guided by an infra-red system; Roland II by tracking radar (and can also be sighted optically).

Boeing

BOEING AEROSPACE/HUGHES 'ROLAND'
SHORT RANGE ALL-WEATHER AIR DEFENCE
SYSTEM

surveillance radar antenna

environmental
control unit

commander's console

optical sight

track radar antenna

receiver

transmitter

primary
power unit

gunner's
control panel

command sub-system

missile magazines

US Army XM 975 tracked vehicle

Sarson/Bryan

British Aerospace

day's complex warfare.)

Another portable SAM system, the Swedish RBS 70 is considerably more cumbersome. This missile is used from a stand and rides a laser beam to its target. The laser beam is directed by the operator and the missile carries a receiver which keeps it on course. The advantage of 'laser beam-riding' is that it is hard to jam. The RBS 70 is, however, halfway between a true man-portable SAM and the next stage up—the portable integrated SAM battery.

As the RBS 70 uses precision target information from search radar, a number of them can be grouped around a radar. The G/H band 'Giraffe' radar, developed for the RBS 70, is fully mobile so a battery of RBS 70 firing systems would not be confined to static

position. The RBS 70 thus provides Sweden with a flexible SAM system.

The Soviet Union has a wide variety of SAM systems which can be considered land-mobile. The SA-8 (NATO code name GECKO) being the nearest equivalent to the numerous smaller NATO systems. The SA-8 is often compared with the internationally constructed Roland SAM system. Roland is highly mobile, designed specifically to protect armoured columns, and similar fast-moving formations, from low-level air attack. The whole system can be carried in a tracked vehicle. The tank chassis of the Main Battle Tank of each country to adopt Roland is generally used. The French fit Roland systems on the AMX 30R chassis, the Germans use the Marder SPZ, while the Americans rely on the M-109 tracked vehicle.

There are two types of Roland. Roland I has a clear-weather capability. Roland II has all-weather capability. Each vehicle carries two missile launchers, reloaded from magazines within the vehicle.

Rapier (below) is effective against low level strikes at tree-top height, can engage subsonic or supersonic aircraft at an altitude of several thousand metres. Flying at supersonic speed itself, Rapier is highly accurate, mobile and can be used independently or as part of an integrated system. It is an ideal defence against surprise attack.

Typical of most SAMs, both Roland I and Roland II employ radar for target acquisition and IFF, but the tracking systems vary between the two types. Target detection is made by radar carried on the launch vehicle. The radar uses the Doppler Principle, whereby the speed and direction of objects is detected by a shift through the frequencies of the electromagnetic spectrum. The interrogatory IFF system varies from country to country.

Roland I has an infra-red guidance system. The operator aims at his target with his optical sight, while tracer flares from the missile's tail are monitored by a precision *goniometer*, an instrument for measuring angles. If the missile departs from the line of sight, an angular error signal is produced. Correction signals are sent by radio to steering vanes in the rocket motor.

Roland II is guided by tracking radar. The relation between missile and radar beam is established by continuous wave transmission from a radar beacon on the missile. Roland II can also be sighted optically in clear-weather conditions. This makes a kill possible even in the face of radar jamming techniques.

Tracked Rapier

The Roland SAM is closely rivalled by the British Rapier system, now evolved into the Tracked Rapier. A complete SAM system, mounted in a tracked and armoured vehicle, the Tracked Rapier has a very similar guidance principle to Roland I. It differs by employing a television camera linked to a computer in place of Roland's precision goniometer and correction command system. Tracked Rapier also has the optional addition of the Blindfire radar equipment. This gives it the all-weather, day and night capability of Roland II, together with both radar and optical tracking systems. Rapier has proved particularly deadly with a very high single shot kill probability.

Roland and Rapier are not the only land-mobile SAM systems but they are alone in being contained in a single vehicle. There are other mobile rapid-reaction missile systems, but they are often divided into two or even three sections—launcher, command centre and radar unit. Some of the most successful SAMs of this sort have a duel land and sea role. Typical are the Soviet X SA-3 (NATO code name GOA), the British Tigercat/Seacat and the French Shahine.

The Shahine, a much improved version of the earlier Crotale, is mounted on an AMX 30 tank chassis. Target acquisition and IFF radar is carried in a separate vehicle and can work with three or four firing vehicles, each

On trial in Australia *(right)* and the Hebrides *(below),* the British built Rapier system has proved highly effective. *Below* The operator controls the firing from inside the cab. The microwave command link antenna is raised to obtain an unobstructed view. Symbols on the side of the Tracked Rapier vehicle denote the number of successful firings made.

British Aerospace

British Aerospace

carrying six ready-to-fire missiles. The firing vehicles use radar for guiding up to two missiles simultaneously. The radar tracks both missile and target—as a result, one blip on the screen can be steered into the other missile through commands from a digital radio link. In the event of radar jamming, a television system is fitted to do the necessary tracking. Shahine is mobile enough to do the same sort of job as Roland or Tracked Rapier but it is more elaborate as it requires more

than one vehicle to function.

The man-portable and other highly mobile SAM systems have been designed chiefly as an antidote to low-flying aircraft. But the more elaborate systems—particularly in their naval versions—must have an additional anti-missile capability. Because of the increasing numbers of surface-to-surface and air-to-surface missiles coming into service it is evident that major targets, such as warships, are going to have to fight off missiles at close

Marconi Space and Defence Systems

a radio frequency signal to a receiver in the missile for self-guidance onto target.

Unlike the HAWK's two-stage solid-propellant motor, Sea Dart has an initial solid fuel booster—but then switches to a liquid-fuelled Rolls-Royce Odin Ramjet engine. The rapid launch rate of Sea Dart means that it can deal simultaneously with many targets. A lightweight version can be fitted to ships down to the 300 tonne size.

From the less complex man-portable SAMs to an area system like Sea Dart there is a wide variety of homing or guidance methods. But increasing sophistication has meant more reliance upon radar and radio signals. Other aiming methods (such as those which use laser beams) are limited by the fact that their target has to be in sight. The favourite method of all-weather targeting has to be through radar. There are four radar aiming systems generally in use.

Guidance systems

The first is *radar command guidance*—as used by the French system: Shahine. Two radars are employed—one to track the target and one to track the missile—and data are fed through a computer which sends out commands to correct the missile's course.

The second is *beam-riding guidance:* in which a radar beam is locked on to the target and the missiles rides down it. The third is *radar illumination,* as used by the US HAWK and the British Sea Dart. The target is illuminated by radar, so that radar signals are reflected back from it and the missile homes in on the source of the reflected signals.

The fourth system is *active homing*—in which the missile both transmits and receives radar signals independently of any control (this system entails large and complex missiles). All these means of guidance suffer from the fact that there are some well-known ways of confusing radar.

In an attempt to avoid such drawbacks the Americans have designed the Patriot (XMIM-104) tactical air defence system which is due to replace HAWK and the Nike-Hercules systems. Patriot is a land-mobile system, but the radar set, engagement control station, electric power plant and between five and eight launching stations put it

range rather than aircraft (which are able to stand-off as they launch weapons).

As most missiles are guided, or have homing devices, the first line of defence is to counter this guidance by electronic counter measures. But, in the last resort, the capacity to destroy an incoming missile is a necessity.

The first known kill of one supersonic missile by another was made by the excellent US HAWK SAM on an Honest John Battlefield missile in January 1960. The HAWK (Homing-All-the-Way-Killer) is, in fact, a land-based missile system which tracks its targets with illuminating radar. The reflections from this provide a reference point for a radar receiver in the missile itself. Information obtained from the received radar energy directs the missile to home, intercept and detonate. Although it was designed as a low altitude SAM, HAWK has a proven ability to

intercept successfully at up to 11,000 metres.

HAWK's success in intercepting another supersonic missile was encouraging to the specialists. As NATO navies became increasingly concerned with defending themselves against missiles, demand for SAMs grew.

Perhaps the most impressive sea-going defence SAM system is the British Sea Dart. Sea Dart can intercept missiles and aircraft at very high and low altitudes. Most Sea Dart performance data are secret but the missile is claimed to be superior to most other radar guided systems in its air defence role (and comparable to many when used as a surface-to-surface weapon). Like the HAWK, Sea Dart has its targets designated to tracking, illuminating radar by a main acquisition radar. Just like the HAWK the Sea Dart's tracking radar illuminates the target with radar signals which are reflected off the target and provide

firmly among the larger SAM systems. This has given the designers the necessary elbow-room to elaborate and build in electronic counter measures (ECCM) which means that the missile still performs well against electronic counter-measures (ECM).

The real heavyweights of the SAM division are the anti-ballistic (ABM) missiles. These systems are not portable because they are assigned to protect fixed targets, such as major cities, and the missiles involved are very large. Up until 1976 the US maintained the Safeguard ABM system with such SAMs as the Nike Hercules (MIM-14B) and the Russians had a comparable system around Moscow with the ABM-1B (NATO code named GALOSH). GALOSH is about 20 metres long (66 ft) and is believed to be directed by Radar Command Guidance and to have a range of 300 kilometres (160 miles). It is armed with a multi-megaton nuclear warhead which is designed to be exploded outside the Earth's atmosphere in the path of incoming Inter-Continental Ballistic Missiles. However it now seems that the US has lost faith in SAM as an ABM system and the Russians may be doing the same. The US has an active programme developing new means to counter ICBMs. With physicists working on laser killers and enhanced particle beams it may be that SAMs have had their day as ABMs. But even if this should turn out to be so, the SAM certainly has its place as a first line conventional air or missile threat. This is particularly true on the battlefield, in the hands of the common soldier, who, until recently, had little defence against attack from the air.

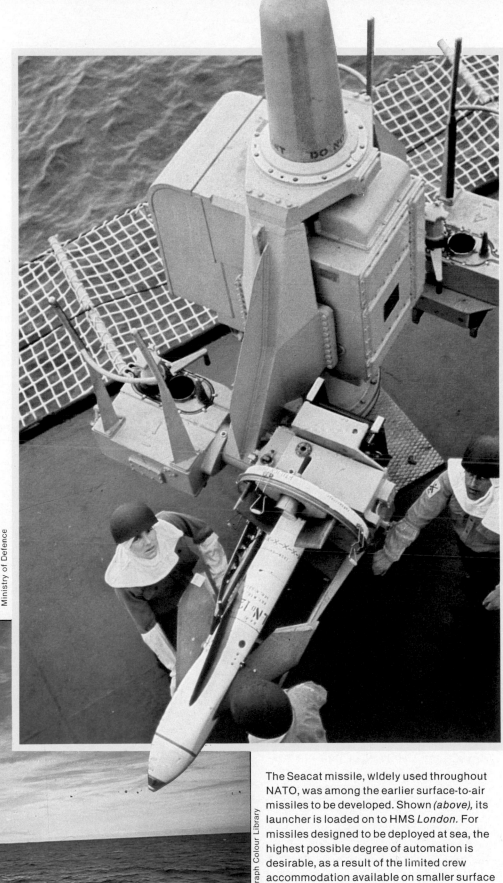

Ministry of Defence

Daily Telegraph Colour Library

The Seacat missile, widely used throughout NATO, was among the earlier surface-to-air missiles to be developed. Shown *(above)*, its launcher is loaded on to HMS *London.* For missiles designed to be deployed at sea, the highest possible degree of automation is desirable, as a result of the limited crew accommodation available on smaller surface vessels. Automatic action also speeds response to a large-scale attack, enabling a high rate of return fire against the aggressor.

How MIRVS find their mark

In the early 1960s the USSR carried out a series of high-altitude nuclear tests, and they were promptly followed by Premier Kruschev's famous boast about the accuracy of Soviet missiles. He claimed that they were able to hit 'a fly in outer space'.

To the Americans it was becoming apparent that the arms race was taking a sinister new turn. The Soviet Union was deploying anti-ballistic missiles, or ABMs, in strategic places, designed to destroy US missiles before they could reach targets in the USSR.

For a time it seemed likely that the Americans would be forced into an expensive programme of providing their own ABM cover. Then a way to get the US missiles through the Russian ABM barrier was suggested. Based on the concept of fooling the enemy's ABM-steering radar guidance systems, ideas such as using clouds of chaff —radar-reflecting metal strips—and dummy warheads, both distributed from the warhead vehicle on approach to its target, were considered. It was not long before the economy of using dummy warheads instead of clusters of the real thing was questioned, and so a simple *Multiple Re-entry Vehicle,* abbreviated to *MRV,* was born.

Simple MRV

Initially, the MRV assemblies possessed multiple warheads that fell to the target in an organized pattern, a technique that was the logical direct descendant of single warhead delivery. This made the last stage of a ballistic missile—the business end of it—an increasingly complicated piece of hardware. Preliminary stages of a rocket are simply a mechanism for carrying the last stage up and out of the Earth's atmosphere on a predetermined course and speed. Once into the near-vacuum of space, the final stage coasts on and upward to its highest point, about 800 miles up with an Inter-Continental Ballistic Missile, before the pull of Earth's gravity drags it back on a shallow arc into the atmosphere again. On re-entry, increasing air resistance of the atmosphere will pull the missile more and more sharply into a vertical line of descent until—if all the calculations were correct—it falls upon and devastates

the target at which it has been aimed.

Towards the end of the 1960s it became apparent that the US had the technological capacity to do more with multiple warheads than fling them in clusters at their targets. A programme of building and deploying defence and communications satellites had shown great economy in launching a number of satellites for the expenditure of a single booster rocket. The secret of this was the use of a highly flexible post-boost rocket vehicle powered by *hypergolic* propellants, substances which ignite on contact, and incorporating restart and guidance methods. The ability to transport a load in space with this vehicle made it possible for satellites to be distributed in various suitable orbits.

It became obvious that, instead of

Above The sharp contrast between cause and effect. The three 'dunces caps' on a 'plate' are 200 kilotonne nuclear warheads capable of enormous devastation. The 'plate' is their compact bus vehicle. This assembly is of Mk12 warheads for the US Minuteman III, and beside them is the missile's protective nose cone.

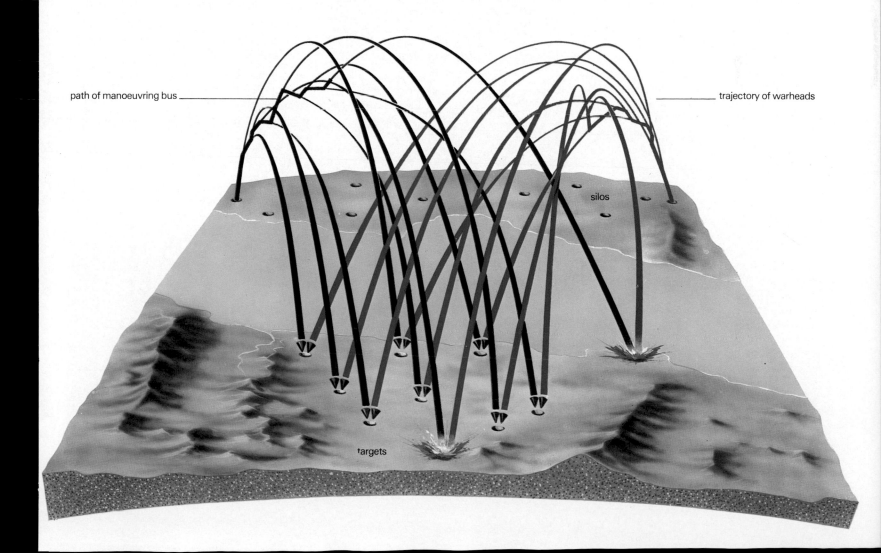

SINGLE WARHEAD
DELIVERY

ICBM trajectories

silos

targets

MIRV WARHEAD DELIVERY

Above The old technique of sending one ICBM against one of the enemy's missile silos. This tactic, in addition to being expensive, was threatened by ABM protection.

Below The confusing trajectories of MIRVs, directed to targets by the manoeuvring bus vehicle. Fewer launch vehicles are needed and defence is made more difficult.

distributing satellites, a vehicle of this nature could distribute warheads—to fall on Earth.

By 1968 all the essential engineering problems for this type of MIRV, using *Multiple Independently-targeted Re-entry Vehicles,* had been solved in the course of this satellite-launching programme.

As the Soviets were adamant in continuing to develop their ABM system, the first US Minuteman III missiles were 'MIRVed' in 1970. The merit of MIRVs was that they were a comparatively cheap and 'technically sweet' way of swamping any ABM system that could possibly be devised. It cost much less to add warheads to a nuclear armoury than to build an equivalent number of ABMs. MIRV delivery could also be designed to confuse defence controls.

The task of aiming the warheads at individual targets is performed by the missile's last stage. It become known as the *bus* because it carried the warheads as passengers to be dropped off at intervals. Once launched, the bus has enough momentum imparted to it by the booster stages of the missile to

path of manoeuvring bus

trajectory of warheads

silos

targets

Grose Thurston

reach a target, but it also possesses an inertial guidance system and small rocket motors which can modify its velocity and attitude. The inertial guidance system is self-contained within the missile, and it usually relies on accelerometers and gyroscopes to detect any deviations from a programmed flight path, automatically correcting them. It is also possible for a missile to continuously monitor its position by making a reference to some fixed object such as a star or a communications satellite. Since all these guidance methods have no need of ground control, they cannot be jammed.

The bus

As the bus coasts through space, it shares its momentum with its deadly cargo. Warheads are ejected, usually by a coiled spring, and they race away to their target with the initial speed of the bus vehicle. After each release, the bus will manoeuvre in order to place successive warheads on different trajectories, so ensuring they hit different targets.

The sequence of delivery can be varied in a number of ways to confuse enemy defences. After releasing one warhead, the following one may be positioned uprange by use of a rocket motor to increase bus speed or, alternatively, it may be positioned downrange by decreasing bus speed, again by use of the rocket guidance motors.

Further confusion may be caused by sending two warheads to the same target, but on different trajectories and at different times. There is, however, a limited area one MIRV group of warheads can cover.

The distance separating targets reached by the contents of a single bus is typically of the order of hundreds of miles. The critical factor which determines this is the total thrust available to the bus's propulsion system. As the payload carried by a bus is limited by the power of the booster rocket that puts it into orbit, it is obvious that some compromise will have to be made between the amount of fuel its guidance motors can have for their task, the extent and complexity of the bus guidance systems and the total weight of the warhead load—a measure of the vehicle's effectiveness as a weapon.

If there are a few hundred miles between the targets allotted to a bus, this distance forms a small percentage of the total range of the missile, so that most of the momentum will have been provided by the original booster rocket. Of course, it would be possible in principle for a very powerful rocket to orbit a bus which had enough power to

criss-cross continents and spread its destructive load anywhere on Earth, but missiles available to the world's nuclear powers currently do not possess the thrust necessary to achieve this and carry a useful warhead load.

One of the reasons for wanting to limit the weight of a bus, in order to increase its warhead payload, is that a MIRV missile will always deliver less total explosive force than a single warhead missile.

Fusing

This situation is caused by the need to fuse and detonate each of the warheads. The fusing mechanism may itself become complicated if the warhead is to explode at a predetermined height. Each warhead also needs an effective heat shield to withstand high-speed entry into the Earth's atmosphere. These factors tend to emphasise the need to reduce the weight, and therefore complexity, of the bus in order that the maximum warhead load may be carried. In fact, the more warheads a bus vehicle must carry, the less will be their total explosive force, and this represents yet another compromise that must be taken into account in the development of MIRV warhead delivery systems.

US Air Force Photo

The tall Minuteman III launch vehicle *(right)* carries Mk12 MIRV warheads. It is protected from enemy attack by a strong underground concrete silo, into which it must be lowered *(above)*. These silos can survive nearby nuclear blast, but not a direct hit. Unlike the successor to the Minuteman—known as Missile X, or MX— an enemy will know in advance, from information gathered by spy satellites and ground sources, the precise location of the silos. This makes them vulnerable to attack from accurate or very powerful enemy missiles.

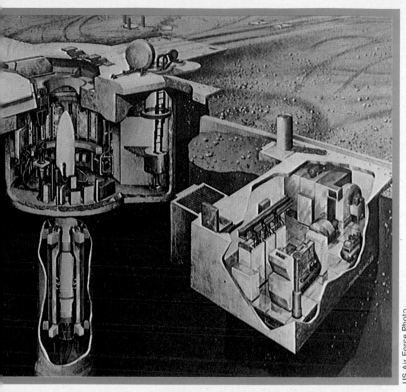

Left The silo complex of a Minuteman III missile and *(below)* a launch tube with a missile in place. The probability of survival of such silos is considered good, since a nuclear warhead of no less than 750 kilotonnes would need to be detonated within approximately 183 m (550 ft) in order to destroy them. Huge sliding concrete doors over the silo mouth contribute to this protection from a nuclear blast.

US Air Force Photo

Photri

gyroscopes that direct long-range ballistic missiles are very precisely engineered, but they are not perfect and, in addition to this, a warhead may be pulled off course by the gravitational effects of the Sun and Moon or by weather conditions and air density over the target. The result is that US land-launched missiles, for example, have the capability of landing 50 per cent of their warheads within 150 m of the target, whilst submarine-launched missiles can bring 50 per cent down within 400 m of the target. This is obviously quite accurate enough to destroy cities and similar 'soft' targets, but it is not good enough to cope with 'hardened' military targets such as the reinforced concrete and steel silos in which ballistic missiles are kept ready for launching.

It can be theoretically calculated that, if the Minuteman III system is 80 per cent reliable—an optimistic estimate—and the reliable missiles deliver one new Mark 12A warhead to each silo, 50 per cent landing within 150 m of a silo as predicted, only one-third of the silos aimed at will be destroyed. If two Mark 12As are aimed at each silo the figure rises to 55 per cent. Under these circumstances the MIRV can be considered a good but not completely effective anti-silo weapon. The most important rule in attacking hardened targets is that accuracy—up to a certain point—is more lethal than high explosive yields.

Accuracy

During the 1970s, advances in micro-miniaturizing electronic components held out several possibilities for refining the accuracy of MIRVs. By this time, two of the three stages of a ballistic missile's flight used some sort of guidance system, but the third did not. The path of the booster rocket out of the atmosphere was programmed and controlled, as was the movement of the bus in space, but once the MIRVs had been detached from the bus they could not alter course. If the original computations had been right, and there had been no build-up of mechanical error, they would hit their target, but any deviation would be uncorrected.

There were certain ways in which the performance of the bus and the booster rocket could be improved, but controlling the re-entry phase of the warhead was also an important prospective development. With 'large array' microelectronics and ultra-sensitive radiation sensors, it had become possible to construct terminal guidance systems for the MIRVs themselves.

Details of terminal guidance systems have

However, certain improvements in miniaturization and engineering can produce dramatic improvements in the yield of a MIRV. A good example of this is the Mark 12 re-entry vehicle, used on US Minuteman III missiles, which carries three warheads of 200 kilotonne force, equivalent to 200,000 tonnes of conventional explosive. This is being succeeded by the Mark 12A, which is externally identical but which, it is claimed, has at least double the yield per warhead, due mainly to miniaturization of the arming and fusing components. The US is also developing MIRVs shielded by an especially light metal known as oralloy.

The fact is, however, that whatever

engineering skills are employed, multiple warheads will have less yield than a single warhead. Paradoxically, this does not mean that they will be any less destructive. The destructive force of an explosion relies on the amount of 'blast over-pressure' which will cause the specified degree of damage, and in effect two separately aimed one-kilotonne devices will produce more blast than a single two-kilotonne device. So what MIRVs lose in total yield they gain in increased destructive effect by separate targeting.

All the benefits of separate targeting would be thrown away, however, unless each warhead could be delivered with the necessary accuracy. The accelerometers and

not been made public but it has long been evident that a technique known as 'terrain matching' is probably being employed. Large array microelectronics have enabled millions of logic circuits to be crammed onto a square centimetre of silicon. By placing logic circuits and memory banks in a terminal guidance system, it can be programmed to retain a complete map of the target area and steer a warhead in by comparing this stored image against what it sees of the target on approach, through electronic sensors.

Digital maps

The most typical way in which the guidance system can be taught to distinguish a target area and make sense of information about it is to construct a *digital map* of the terrain. In a digital map, the area is divided into squares and each square is given a different numerical value, in accordance with certain characteristics which are recognized by the sensors or 'eyes' of the guidance mechanism. As sensory signals can be provided by a variety of devices—including electromagnetic radiation sensors, radar, microwave radiometers, infra-red detectors and laser altimeters—there is a choice in the type of terrain characteristic which can be used to provide the numerical values on a digital map. The favourite variants of ground location may be altitude above sea level or terrain radiation.

If altitude above sea level is chosen as the key to the map, this can be very accurately gauged by a laser altimeter which provides a resolution better than 10 cm vertically, and 20 cm horizontally from a height of 2,000 m (6,500 ft). It allots digital values to the squares of the map, according to their height above sea level, performing this task at high speed as the MIRV hurtles down towards its target. The stream of data produced is fed to the computer and its memory bank, where it is built into a map of the target area which is then compared with the stored map. Any positional error of the warhead can then be determined and corrected by the guidance system, which may consist of moveable fins or an offset centre-of-gravity.

Altitude is only one of a number of ways in which terrain varies. The Earth's surface receives radiation from space and reflects it back. Different materials or structures on the ground will reflect radio waves differently. The change in reflectivity at a given wavelength caused by a field, forest or road can also be measured and used to construct a digital map of the area.

In case even these sophisticated guidance systems prove inadequate, the US maintains a research programme aimed, amongst other things, at realizing the possibility of producing warheads that will manoeuvre in space, and on re-entry, cause even greater confusion to any defence forces. It currently seems unlikely that these refinements will be required however.

Pre-emptive strike

Once MIRVs began to be introduced on any scale, each side developed the capability of aiming two or three warheads at its rival's silos while maintaining a formidable force in reserve. The advantage had shifted to the aggressor so decisively that it would be tempting to make a pre-emptive strike in a time of tension. As the accuracy of warheads increases, the advantage to the aggressor becomes yet greater.

But, as the USSR followed the US lead in developing MIRV missiles it became apparent that they had a potential advantage. A far larger proportion of their strategic missile force was deployed in the more accurate land-launched mode, and many of these missiles were so much larger and more powerful than US equivalents that they could carry greater numbers of high-yield MIRVs. Although the Soviet Union has been appreciably behind America in MIRV development, it was obvious that ultimately the Soviet Union might gain a decisive advantage unless the United States developed its strategic forces further.

MX solution

Once again, American ingenuity has proved equal to the challenge—but at some cost. A new missile system—known as the MX—is under development and should come into operation by 1985. The MX solution to the MIRV threat will consist of distributing a few missiles at random among a very large number of silos. Any large increase in Soviet warheads or in their accuracy can be met with a similar US increase of empty silos. As the Soviets can never be sure where American missiles are hidden, they would be forced to target every silo in a pre-emptive attack and the Americans can always make sure that there are too many silos for them to do this with any degree of real confidence.

It only remains for the Soviets to protect themselves with an equivalent of the MX system and, again, the precarious balance of ability between the two super-powers—before Kruschev's proud boasts—will have been achieved.

Left Adjusting a 'floated ball' missile guidance system component. Inside, there are inertial instruments that are protected from vibration, temperature changes and magnetic fields by floating the ball.

Northrop Corp.

Below Thermal infra-red imaging by electronic sensors can clearly differentiate the features of a city. This sort of information may be gathered by a warhead on its descent in order to achieve target recognition.

Science Photo Library/Daedalus Enterprises Inc.

The MX missile system

The next decade will see a remarkable new direction in American nuclear strategy as construction begins of an entirely new land-based missile system. Instead of sitting in one spot, special launchers will race the powerful MX missiles between thousands of dummy silos to dodge any pre-emptive enemy missile strike.

American military strategists have put the case for an advanced system with some urgency because they believe that their existing fields of Titan and Minuteman ICBMs could be knocked out in a surprise attack by Soviet missiles. The root cause of this worry is that although both the Soviet Union and the US have limited the number of their intercontinental ballistic launcher vehicles by Strategic Arms Limitation Treaty (SALT), many of these vehicles are capable of carrying a number of independently-targeted warheads. Because of the increasing accuracy of these warheads, it is becoming theoretically possible for the Soviet Union to destroy the US land-based missiles in their silos by expending a mere fraction of their launcher vehicles. Once they had achieved this they would then be free to use their remaining strategic missiles to blackmail the US into surrender—or so the theory runs.

Act of suicide

In cold fact it would be an act of suicide to attempt a sneak attack on the Titan and Minuteman silos. Even if the attack were successful the Americans would be left with all their submarine-launched ICBMs and a number of their air-launched ICBMs intact. American strategic nuclear power is fairly evenly divided between the three methods, and even after the loss of Titan and Minuteman they would still have colossal retaliatory power. Although the submarines are not invulnerable, the Soviets have never succeeded in tracking one throughout its patrol and the USAF would almost certainly get one or two of its strategic bombers airborne before the strike arrived. Besides, the

Photri

Above The Minuteman silo—a reinforced concrete and steel cylinder designed to withstand a blast of several thousand pounds per square inch. To destroy the Minuteman a 750 kiloton nuclear warhead would have to detonate about 550 ft from the silo.
Right Minuteman after launch—the MX Missile can carry three times its payload.

Soviet Union could never totally rely on its ability to take out the land-based silos.

To be certain of destroying Minuteman, the Russians would have to detonate a 750-kilotonne warhead at ground level within 170 m (550 ft) of the reinforced concrete and steel silos that house the missiles. It is hard to know the exact characteristics of Soviet warheads, but experts believe that those with a 750-kilotonne yield will have attained an accuracy no better than 22 km (12 nautical miles) in *Circular Error Probable* (CEP) by 1985 (the CEP is the radius of a circle around a target within which half the warheads aimed at that target can be expected to fall). It is also extremely unlikely that the total reliability of launcher vehicle and independently-targeted warhead will exceed 80 per cent so that—in the worst case from the US point of view—an expenditure of two warheads per silo would destroy no more than 87 per cent of them. As the US expects to have 1,004 land-based launcher vehicles in position in 1985, this means that

well over a hundred of them would survive.

A retaliatory force of 100 missiles—many armed with multiple warheads—would probably be sufficient to destroy the Soviet Union unaided. And if they arrived at their targets after a massive submarine and air-launched counter-strike they could be little more than 'rubble-bouncers', hitting targets already destroyed.

Still effective

So the US deterrent should still be effective despite any loss of credibility that their land-based ICBMs may have suffered from the proliferation of accurate warheads. Nevertheless, the Americans seem determined to pay the price of producing a less vulnerable land-based system for good strategic reasons. First of all, they are unwilling to abandon any of the three parts of the deterrent—the so-called 'Triad' of land, sea and air launch capability—in case a sudden technological break-through (perhaps a new method of pinpointing submarines) should

increase the vulnerability of one of the Triad.

They might also be tempted to launch the land-based missiles as soon as warning of a Soviet launch was received to get them safely on their way to Russia before the incoming Soviet warheads could reach their targets. There have been enough false alerts in the past to suggest that this policy may involve a high risk of accidental nuclear engagement. And it is obviously preferable to have a system able to survive an actual, incontravertible Soviet strike before missiles are committed beyond recall. An added consideration is the greater accuracy and control possible with land-based missiles. So there are a number of good reasons for building a land-based nuclear deterrent which will be less vulnerable to a pre-emptive strike.

Of the numerous ways open to the US to increase the survival rate of their ICBMs, the authorities have chosen the MX system. This is designed to overload and confuse Soviet ability to locate missiles by constantly moving them between otherwise empty silos.

At any particular time, the Soviets will have no idea which silos contain missiles and which are empty. Each missile will move randomly between the silos on a special vehicle called a Transporter/Erector/Launcher

Popperfoto

Above An explosion of a Titan II missile silo on 19 September, 1980 blew the top off the installation, jarred the countryside for miles, and injured 22 Air Force Personnel.
Centre The mobile basing scheme currently favoured for the MX would require building some 200 'closed loop' roads, each with a system of spur roads leading to 23 horizontal protective shelters.
Right A test proves an MX can burst through over 10 ft of dirt and concrete.

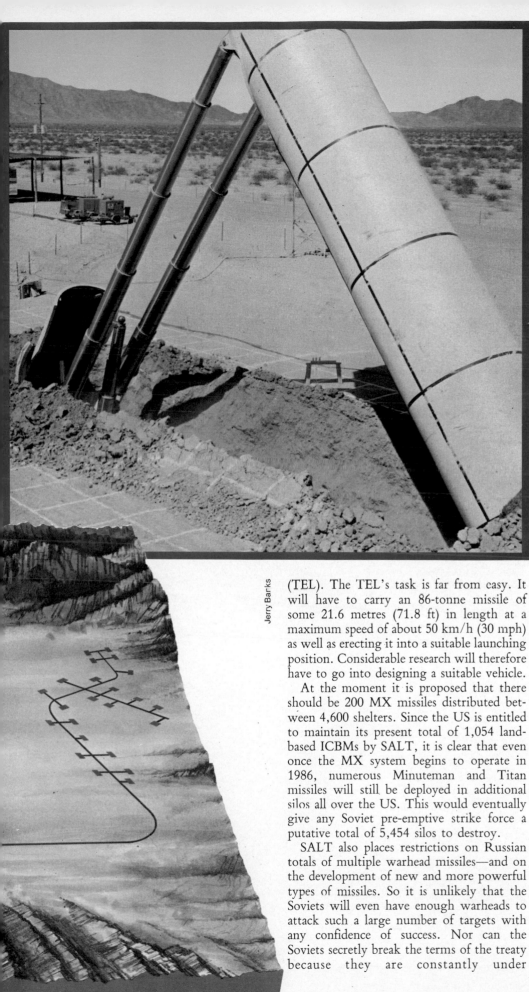

Jerry Barks

Photri

surveillance by US spy satellites, infra-red sensors, photography techniques and various types of radar.

Thus the MX system gives the US a degree of invulnerability for the land-based leg of its strategic Triad. But each MX silo must be far enough from its neighbour to be separately targeted, and yet near enough for every TEL vehicle to have a choice of shelter to run to during the thirty minutes that elapse between detection of a Soviet launch and arrival of the warheads. To achieve these objectives the MX will be provided with an arrangement of roads between silos that is comparable to a racetrack. Every individual missile will be placed on a loop road which will have a number of shelters located on spurs which lead off it. Each shelter will be at least 1.6 km (1 mile) from the next so that the enemy could never hope to destroy more than one silo with a single warhead. Yet they will be so accessible that the TEL could reach any of them in the vital half hour.

While Minuteman III can only carry three independently targeted warheads, the MX will be able to carry ten, each with a yield of over 400 kilotonnes. So, although only 200 missiles will be built, the MX systems will obviously have an importance and power out of proportion to their numbers.

Accuracy

Until recently, the immobile missile, tied to a single silo would have had a considerable advantage in accuracy over one that might be launched from a number of points anything up to six or seven miles apart. This was because the guidance system of ICBMs depended upon accelerometers of great accuracy that could measure the distance travelled by a missile in any direction. The distance between the launcher silo and the point at which the fast warhead vehicle should re-enter the Earth's atmosphere was computed with extraordinary care. It became possible for the accelerometers to detect any discrepancies in the programmed path of flight and to trigger off course corrections. However, any change in the point of launch meant that the flight distance had to be re-computed and a new programme assigned to the missile control mechanism. This takes time and effectively ruled out the employment of really accurate mobile missiles. Recent developments in inertial guidance systems, however, have enabled missiles to take a highly accurate check on their positions in mid-flight by reference to constantly positioned stars or space satellites.

The MX missile itself will have an Ad-

(TEL). The TEL's task is far from easy. It will have to carry an 86-tonne missile of some 21.6 metres (71.8 ft) in length at a maximum speed of about 50 km/h (30 mph) as well as erecting it into a suitable launching position. Considerable research will therefore have to go into designing a suitable vehicle.

At the moment it is proposed that there should be 200 MX missiles distributed between 4,600 shelters. Since the US is entitled to maintain its present total of 1,054 land-based ICBMs by SALT, it is clear that even once the MX system begins to operate in 1986, numerous Minuteman and Titan missiles will still be deployed in additional silos all over the US. This would eventually give any Soviet pre-emptive strike force a putative total of 5,454 silos to destroy.

SALT also places restrictions on Russian totals of multiple warhead missiles—and on the development of new and more powerful types of missiles. So it is unlikely that the Soviets will even have enough warheads to attack such a large number of targets with any confidence of success. Nor can the Soviets secretly break the terms of the treaty because they are constantly under

vanced Inertial Referenced Sphere and will not need re-programming after each move. The system continues to navigate during the moves and is self-aligning as well. It is also supposed to have a greater resistance to shock and vibration than the Minuteman navigation system and an increased ability to survive the radiation effects of a nuclear explosion. These refinements make it suitable not only for its mobile role, but emphasize the fact that it could be expected to defy a pre-emptive strike.

Indeed the MX will incorporate every technical advance that could increase its ruggedness, reliability and carrying power. The propulsion cases will be made of Kevlar composite material which is said to be both lighter and stronger than the steel, titanium and glass used in earlier missiles. In the first two of the four stages, MX will be powered by the same propellant as Minuteman, but in the third stage the powerful 'Class 7' propellant developed for use in the Trident will come in. By incorporating every improvement known to ICBM science in this way, the US hope to produce a weapon which is a generation's advance on the old system.

Mutual Suspicion

There is a drawback, however. For although this powerful concealed missile is clearly good for US strategy, it clashes with the principles of SALT. One of the purposes of the Treaty is to lessen mutual suspicion, and it therefore contains certain provisions that are designed to enable each party to verify that the other is respecting the Treaty restrictions. In theory, the MX system enables the US to gain a significant predominance of nuclear weaponry by cheating the Treaty terms in secretly constructing many more than 200 missiles and hiding them away in the warren of 4,400 empty shelters. In practice it is extremely unlikely that such large-scale deception could go undetected by the Soviet Union's National Technical means of verification (spy satellites and radar) or by traditional methods of cloak and dagger spies. Nevertheless, the Soviets might find it difficult or embarrassing to actually produce proof of American cheating. So great care has been taken to evolve means of checking on the MX programme to ensure that it is conducted with scrupulous honesty.

In the first place the US has agreed to assemble each missile and its launch equipment in the open. Every assembly area will be located at some distance from the group of 'race-tracks' which it is designed to serve.

At no little cost, special railway tracks will be laid to connect each assembly site with every shelter complex in its group and the assembled missile will be loaded onto a train for its journey to the TEL vehicle—presumably the only time each stretch of track will be used. As the assembly process and the journey to the silo will be carried out under the full glare of Soviet space satellites there is effectively no chance of the US evading the Treaty terms.

Besides being complicated the MX system is bound to be expensive. So far, projected funding for the programme between fiscal year 1979 and 1982 is $4,550.6 million—and that is for engineering development alone. The final cost is beyond accurate estimate.

It seems that the MX system is bound to have some effect upon the Soviet Union. History has shown that every improvement or innovation in nuclear weapons delivery systems (multiple independently-targeted warheads or cruise missile development, for example) has been pioneered by the US but then quickly emulated by the Soviet Union. In the case of the MX there is more than ever a reason for the USSR to construct something similar—the fact is that the Soviets have no equivalent of the Strategic Triad and have to rely heavily on land-based missiles for their deterrent. By 1985 it is estimated that the USSR will deploy an imposing 6,200 deliverable warheads on its land-based ICBMs—but only 1,200 submarine-launched warheads and 1,000 air-launched. In contrast the US will have 2,000 land-based, 6,272 submarine-launched and 4,560 air-launched. In the face of recent US official announcements that Soviet military targets are being given priority over civilian ones, it would not be surprising if the USSR began to doubt the effectiveness of its land-based deterrent. Of course, the Soviet Union may not slavishly copy the whole MX system complete with silo complexes and 'race-tracks'. But it may well take some steps to prevent America acquiring the ability to destroy three quarters of its nuclear armoury in a pre-emptive strike.

As a system, the MX project has its critics. Among them are the ranchers of the remote areas where the massive construction will disrupt lives. Others fear that Russia will develop the technology needed to guess the real from the dummy rockets.

Right The MX launch mode utilizes a cantilever—formerly called 'plow-out' launch mode—in place of breakout mode (in which the missile and protective structure broke through the top of the shelter prior to launching). MX will now slide out of its tunnel until clear of the shelter doors. It is then cantilevered to an upright position. The bottom of the shelter is 2.1–2.8 m above the road level so the launcher need not plough through post-attack debris. MX has three solid rocket fuel stages and a liquid fuel post-boost for a total throw of 3,607 kg.

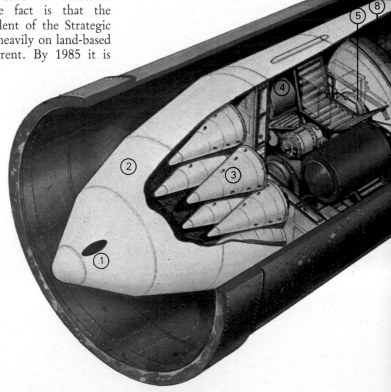

1 Shroud Eject Engines
2 Shroud
3 Re-entry vehicles
4 Guidance
5 Post booster
6 Stage 3
7 Stage 2
8 Extendable Nozzle Exit Cones
9 Stage 1

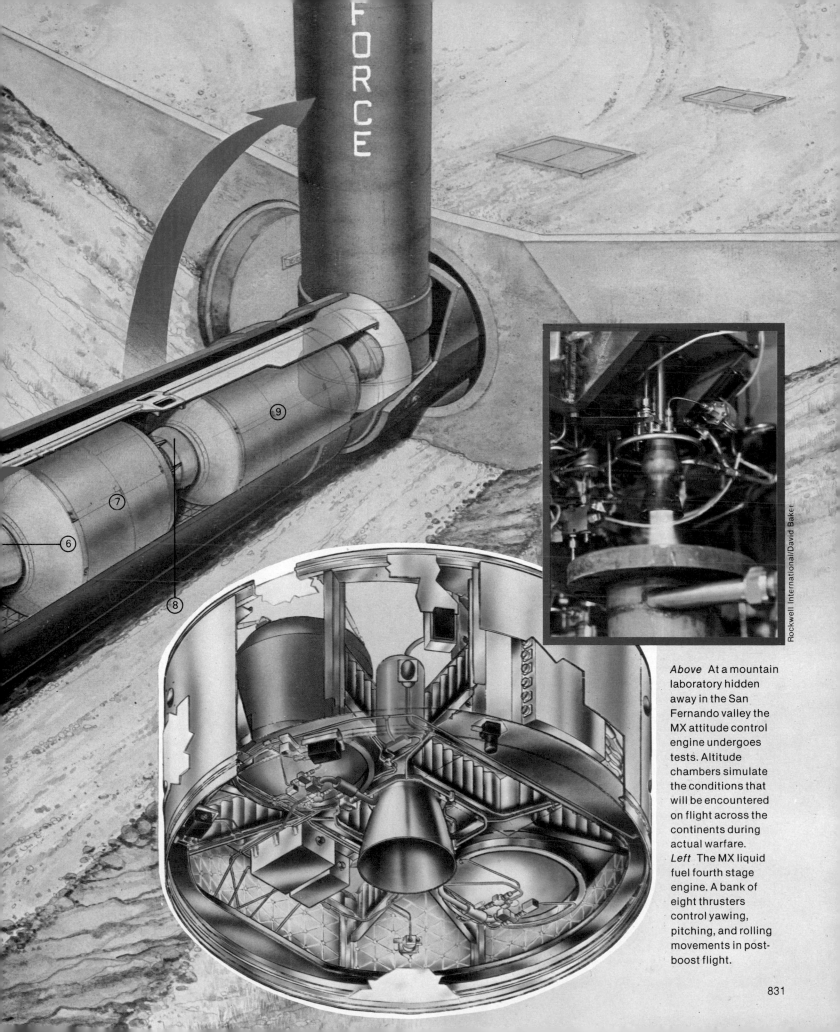

FORCE

⑨

⑦

⑥

⑧

Rockwell International/David Baker

Above At a mountain laboratory hidden away in the San Fernando valley the MX attitude control engine undergoes tests. Altitude chambers simulate the conditions that will be encountered on flight across the continents during actual warfare.
Left The MX liquid fuel fourth stage engine. A bank of eight thrusters control yawing, pitching, and rolling movements in post-boost flight.

Chemical warfare: death drop by drop

After more than fifty years in the shadows, a range of weapons—so deadly that even Hitler shunned them—is back in production. Known as CBW, chemical and biological warfare includes poisons lethal in quantities of one pinhead-sized droplet on exposed skin. Again and again, nations have agreed to ban the use of such weapons, most notably in a famous Geneva Protocol of 1925. But those who signed this and later resolutions reserved the right to experiment and stockpile CBW weapons.

Both Russia and the US promise only that they will not be the *first* to use the formidable armoury now available.

In the trenches of World War 1, the dense, yellowish-green clouds of chlorine gas that rolled across the lines choked 5,000 men to death on the first occasion of its use; 15,000 more were seriously injured for life

that day. By the war's end, 1,296,853 casualties had resulted from gas poisoning. No more barbarous form of warfare could be imagined, it seemed at the time.

But that was before the advent of nuclear weapons—costly, destructive of property as well as life. There are those who now argue that chemical warfare is comparatively 'humane': especially in variants that stun, not kill, the target into submission—or change his mental state to terror or carefree intoxication.

Others see CW as a practical—if sinister—solution to their war aims. Small nations know that the cost of producing chemical weapons is relatively cheap and easily mastered by talented chemists and engineers. Possession of such an arsenal can win respect from surrounding 'enemy' states.

Aggressors with an eye to future wealth know that chemical weapons destroy people

but not valuable industrial plant, mineral resources, and agricultural land and crops. Alternatively, certain chemical agents can be used to *destroy* a nation's food crops, causing famine, and defeat.

Commanders contending with guerrillas may regard chemical agents as ideal 'localized' weapons: the enemy can be flushed out of underground bunkers and from mountainous inaccessible areas.

Terrorists, with minimum funds and limited laboratory facilities, could manufacture small quantities of chemical agents and use them against any opponent with superior forces. Curiously, their task has been made easier by the fact that the details of some of the most lethal agents have been patented —and the patents can be freely inspected.

One such agent is codenamed VX. Some ten million pounds (4.5 million kilos) of it were produced in the US between 1961 and

1968. Less than half a mg is a lethal dose. Much of this stockpile, it is claimed, is intact and based in Western Europe.

Recent indications are that the arsenal of CW weaponry is being enhanced in both East and West. On a 1978 estimate, the US stockpile stood at some 40 million pounds (18 million kilos)—with the compound known as Sarin composing 75% of the total.

In 1980, the US House of Representatives amended a military construction bill to include $3.2 million for the building of a nerve gas plant in Pine Bluff, Arkansas. And in the same year, Britain's research institute into CBW at Porton Down, Wiltshire, announced extensions to include a new testing range.

Reasons for such developments stress fears that the Warsaw Pact bloc now has one in three of its missiles armed with CBW warheads—and that Western research is purely defensive. But it is also true that Western military research has cracked one of the main drawbacks to CW from the

handler's point of view—the perils that ensue if a container of lethal agents is broken prematurely.

A new programme for so-called *binary* weapons has been planned. In these, two canisters of relatively non-toxic chemical compounds combine to form the final lethal mixture, only after the missile is fired. The spin imparted to the shell on firing aids the mixing process. The binary principle can be adapted for use in aircraft spray tanks and rockets—but is suitable for bombs.

Whatever the means of delivery to the general target, the effectivity of any CW agent is dependent upon its chemical and physiological properties, and also upon the meteorological conditions existing at the time of its dispersal.

There are *non-persistent agents:* those which remain effective for only relatively short periods (from minutes to a couple of hours).

There are *persistent agents:* those which remain a threat for much longer periods of time. Unlike non-persistent ones, these

Daily Telegraph Colour Library

Top and above Samples of the deadly rain of chemical warfare. Over Vietnam, defoliant agents were sprayed to strip trees (exposing troop movements) and to ruin crops. Though most nations have long held back from full-scale chemical warfare, some now argue it is more humane than nuclear strikes.
Right An accident at a chemical plant at Seveso, Italy, in 1976 let loose a cloud of dioxyn. Skin burns and long-term health disorders resulted. The entire region needed decontamination.

Frank Spooner Pictures

agents are frequently in liquid or powdered form and their slow evaporation (into gas) ensures lethality for days to weeks under certain meteorological conditions.

Weather is a key factor in delivery. Strong winds will rapidly disperse non-persistent agents in open country areas. But winds will build up very high concentrations inside woods, buildings, bunkers, and dugouts. High air temperatures enable certain chemical agents, such as a pool of liquid gas on the ground, to give off heavy vapours over long periods. Low temperatures tend to delay this process and thus extend the persistence of the threat. Rain tends to wash away liquid or powdered agents, removing the hazard. If there is a 'temperature inversion'—when a sandwich of warm air occurs between colder ground and upper atmosphere—then the vaporized agent will persist for much longer periods. Snow and ice will trap and freeze many 'liquid' or powder agents, thus considerably extending their persistency. For the best results from a chemical attack, weather conditions should be stable and wind speeds down to around 3–6 miles per hour (or 4.8–9.6 kilometres per hour).

Chemical agents are usually classified into groups which characterize their effect upon people or plants as the chart on page 440 shows. So far as nerve gases are concerned, the principal varieties are organo-phosphates.

Battlefield protection against chemical agents will be essential for troops in the future, even though the necessary all-enveloping suits restrict free movement. Filters on the front of the face masks seen here purify incoming air whilst waste air is vented at the sides.

Haywood Art Group

Ministry of Defence

Ministry of Defence

Far left The make-up of a chemical grenade. Release of the starting lever (1) occurs when the grenade is thrown, causing the primer mechanism (2) to send primer (3) down to break a glass phial containing the delayed start mixture (4). This provides a time interval before the grenade explodes, triggered by the ignition mixture (5). Fuel (6) carries the active agent (7) from the canister by way of the outlet port (8) to contaminate the surrounding area.
Left Decontamination of a casualty occurs before removal from the 'noddy' suit.

Insecticides in use today, such as Malathion, Parathion, and Dimethoate are in the same family. Their destructiveness in the human body is a result of their action on an essential body enzyme, acetylcholinesterase, whose role is vital to processes in the nervous system and in blood cells.

There are other more exotic CW agents—among them a class known as 'psychotogens'. Their official patents define them as 'compounds or compositions which induce, supplement, or amplify in humans or animals a state comparable or similar to the manifestations observed in a diseased mind'.

Conceivably, therefore, the battlefield of the future might be peopled by lunatics, though how useful that might be to an opposing army remains in debate.

Defence against such bizarre weapons is problematic—but increasingly developed. The human body has its own defence system: and the 'success' rate at which an age can overcome this has led to a formal classification system.

Human body functions include a detoxification process which can handle low levels of poison. But this natural mechanism can be overwhelmed if the amount of poison entering the body exceeds the rate of detoxification. For example, the lethal dosage for a particular war gas might be 2 milligrams to one cubic metre of air inhaled for one minute. If the victim breathes this level into his lungs for that period of time he will become ill and die. The inhalation of fractional amounts of this dosage may, however, have no ill effect because the rate of detoxification exceeds the input level of the toxic substance. Conversely, there will be a midpoint level, where the detoxification processes may be overwhelmed and the victim could stand a 50 per cent chance of becoming incapacitated and possibly dying. In the world of chemical warfare, that situation is known as an LD-50 dosage.

A substance or gas (or dose of radiation) that can kill 50 per cent or 100 per cent of those exposed to a specific dosage rate, is thus expressed as the LD-50 or LD-100 dose. For example, the LD-50 for VX is 15 milligrams for liquid on the skin contamination. (Fifty per cent of those receiving this dosage will die, the remaining 50 per cent will recover after serious illness.) Ricin, the toxin used recently to kill a Bulgarian dissident working in London for the BBC Overseas Service, is reported to have an LD-50 'injected' dosage of just 0.5 milligrams. Incapacitating doses are express-

CHEMICAL AGENTS WHICH MAY BE USED IN FUTURE WARFARE

Name	Type	Effect	First Aid/Medical
Tabun (GA) Sarin (GB) Soman (GD) VX	Nerve Agents In gas or liquid form.	Breathing difficulty, runny nose blurred vision, nausea, sweating, vomiting, giddy, muscular spasms, paralysis, death.	Oxime and Atropine injections. Forced respiration. Washing contaminated parts with water and soap.
Distilled Mustard or Nitrogen Mustard	Blister Agents In gas or liquid form.	Eye and skin irritation. Blisters, external (and internal if inhaled or swallowed) Bronchopneumonic Effects, following initial irritation may be delayed up to 48 hours! Can prove fatal immediately or after years of illness.	Wash off contamination. Use mydriatics, antibiotics. Treat blisters like burns
Phosgene	Choking Agent Gas.	Damages lungs. Victim coughs and drowns in his own fluid.	Give fresh air, oxygen.
CN DM CS BZ	Incapacitating Agents Gases.	Irritates eyes and skin. Breathing difficulty. Nausea and vomiting. For BZ, flushed skin, irregular heartbeat, high pulse-rates, hallucinations, maniacal behaviour.	Fresh air and oxygen. Use restraint for BZ victims if violent.
Botulin (X & A) Ricin Saxitoxin (TZ) Entero-toxin (B) Tetrodotoxin	Toxin Agents In powder or liquid forms.	Blurred vision, tingling limbs, headaches, numbness, fatigue, cramps, breathing difficulty, dizziness, vomiting, paralysis, death.	These agents are derived from various poisonous plants, fungi, bacteria, and animals. Unless the serum or antidote can be found survival is not expected.
LSD and other mind affecting drugs.	Psycho Agents In gas, liquid, or powder.	High pulse rate, flushed skin, hallucinations, incapacity to think clearly, open to suggestion. Stupors. Unconsciousness?	Counter-acting drugs can be used if a particular agent is identified.

DEFOLIANTS AND HERBICIDES USED AGAINST PLANTS

Name	Effect
2, 4. D Dichlorophenoxyacetic acid 2, 4, 5. – T Trichlorophenenoxyamatic acid Cocodylic acid	Specially developed to defoliate jungles and tall grasses. Kills animals. Stunts the growth of young plants, alters hormonal balance in crops. Dioxin, a contaminant of 2, 4, 5 – T is extremely toxic to humans and particularly to expectant mothers.

Left Large Civil Defence shelters require comprehensive air filtering to remove nuclear fall-out, chemical and biological warfare agents. This shock-proof air-handling unit houses large cylindrical gas filters which cleanse incoming air of contamination from radio-active particles, and certain bacteria. Activated carbon within the filters is capable of removing all known war gases and the units remain effective for at least 30 years.

1 Contaminated air drawn in.
2 First particulate filter collects dust.
3 Activated charcoal granules absorb gas and toxic vapours.
4 Final particulate element.
5 Non-return valve.
6 Cool input air prevents misting.
7 Air drawn, via non-return valves, into nose piece.
8 Air inhaled.
9 Exhaled air leaves through the non-return valve.

⑨

Complete family protection against chemical agents is provided by a decontaminable over-suit, inner NBC garment and respirator.

Civil Defence Supply, Lincoln, England

pressed similarly as—IcD-50 and IcD-100.

Protection against agents as efficient as this has become a prime consideration of all nations, but the greatest research going on in this area remains the province of the Soviets, Chinese, Americans, and the NATO and Warsaw Pact countries. In addition, there is evidence that South Africa, Israel, and Arab nations are deeply involved in small but sophisticated chemical weapons pro-grammes.

Modern chemical warfare demands com-plete, whole-body protection. Military and civil defence masks have been designed to ex-clude nuclear fallout dusts, chemical agents, and biological aerosols. The mask consists of a close-fitting helmet and hood. Vision is provided by a plastic visor or goggles set into the material of the mask. Exhaled air passes out through a non-return valve and inhaled air passes through a filter container. The filter, usually cylindrical, is filled with various gauzes and fine meshes of metal. Sandwiched between these are several chemical compounds, usually in granular form. One of the most important chemical substances included is 'activated charcoal'. It consists of minute particles of specially prepared charcoal, each grain pocked with thousands of microscopic tunnels. The total surface area of the tiny granule—including the tunnel walls is phenomenal: one gram

(1/28th ounce) of these granules has a surface area of about 334 square metres (3,600 square feet). These surfaces trap gases as they pass through them. The process is called ad-sorption. It greatly reduces the concentration of gas penetrating the filter. Other layers of neutralizing chemicals take care of what re-mains. The lifespan of a filter depends on the concentration and types of gases (or dusts) it encounters. The used filter simply unscrews from the face-mask and the replacement is then screwed back into position. Another important feature concerning the provision for replacement filters is flexibility. Although scientists have tried to anticipate all of the likely war agents which might be used in some future conflict, it is feasible that some new gas might be utilized—one which penetrates the existing filter design and causes heavy casualties. The screwthread ar-rangement built into existing masks greatly simplifies the time and effort needed to adapt existing equipment—by quickly designing a new filter and screwing it into place.

Various forms of protective clothing have evolved over the past 30 years but scientists are still contending with the problems of severe heat exhaustion experienced by wearers performing heavy work loads. For instance, a soldier dressed in a complete NBC suit (NBC designates nuclear and bio-chemical) must be expected to run across

country and climb hills. The best 'noddy' suits (as they are affectionately called) are manufactured from plastic materials coated with various neutralizing chemicals and sandwiched with activated charcoal. Boots and gloves are likewise treated. The entire suit can consist of an under-suit and outer coveralls. This protection permits operations in nerve-agent battlefield environments.

Decontamination techniques are also fully developed. The majority of chemical agents used in modern warfare (including nerve agents) can be greatly weakened by various alkaline chemicals, bleaches, etc, mixed with soapy water. Steam hoses will clean up most agents, particularly blister agents like mustard gas (liquid). Pads of alkaline-soaked material can be carried to wipe off splashes of liquid agent when it contaminates clothing.

All shelters for military and civilian per-sonnel should ideally be well below ground level and their air ventilation systems must be fitted with giant versions of the kind of filters used by respirators.

One day, perhaps not so far off, chemical warfare and other kinds of weaponry will in-duce soldiers to wear closed-air system suits, like astronauts. Or perhaps the battlefield will become such a hazardous place that only closed atmosphere vehicles, hovercrafts, and robot flying machines directed from remote bunkers will fight the wars of tomorrow.

The battle of the bugs

In every major war this century, a more deadly weapon than bomb or bullet has stalked the battlefields. Infectious diseases claimed more casualties than armed conflict—in World Wars 1 and 2, in Korea and in Vietnam.

These outbreaks were 'natural'—the result of mass movements of humanity in close company and insanitary conditions. But if natural disease is so effective in removing or incapacitating troops, why not introduce disease deliberately and overcome the enemy without firing a shot?

How far research has trodden down the path to this form of warfare remains largely secret. But biological warfare—the deliberate breeding and dissemination of virulent organisms—is under active research in many nations. Much of the research is 'defensive', it is claimed. If protective counter-measures against BW can be developed, strategists argue, aggressors will abandon it.

In the course of microbiological research, the specialist institutes involved have hit on many beneficial ways to combat known diseases to Man, animals and crops. But the development of highly potent germ warfare agents, even only with defence in mind, remains a hazardous military activity.

In theory, BW could be used to exterminate an enemy population entirely. It could be aimed against his livestock or his agriculture. Man has survived major outbreaks of plague and pestilence before. But after a massive and intensive onslaught of deliberate infection it is possible that parts of the planet would never recover. A Scottish island, Gruinard, on which experiments

with the disease Anthrax were made in the 1940s, remains out of bounds.

Biological poisons exist as living organisms or as toxins derived from germ cultures and living creatures. Poisons derived from snakes, spiders, fish, and many plants can be processed into liquid or powder form and stored inside pressurized aerosol containers. Possible methods of delivery also include shells, bombs, missile warheads, and 'poisoned' bullets.

Spraying from aircraft could disperse several tons of germ-laden material into the atmosphere over an enemy target. But from the moment the aerosol is sprayed into the air, various physical and environmental factors come into play. Meteorological conditions may shorten the lifespan of the microbes and hence their ability to cause extensive harm. But under stable weather conditions, with low wind speeds, and dispersed from a height of 152 metres (500 ft), the individual particles of the aerosol pouring from the nozzles of the disseminating aircraft will slowly approach their target.

Low altitude attack

The ideal size of aerosol particle (for penetration into the lungs) is between one and five microns (1 micron = 1/1,000 of a millimetre). This size of particle will, in theory, descend at around five feet (1.5 metres) per hour. In reality, air turbulence causes this descent rate to vary from place to place. A disadvantage to BW, as a result, is that too much aerosol may land in one area, too little elsewhere. High wind speeds and other atmospheric instabilities will intensify this scattering effect.

Hans Reinhard/Bruce Coleman Ltd.

Dry air is a hazard to effective delivery. The longer it takes for a moisture-laden particle to fall to Earth, the greater the amount of evaporation it will experience. More and more salts cluster to the microbe cells, drawing still more water from them and causing greater dehydration to occur. Some microbes survive drying, but many do not. In addition, the ultra-violet light contained in sunlight will destroy microbes in time. As a result, it is likely that a BW attack would come at low altitude, above the selected target and during the hours of darkness.

There are several other methods available for delivering biological agents to a target. Certain species of insect attack Man, animals, and crops. They are thus valuable 'carriers' of disease organism. The normal

Oxford Scientific Films

Left The attractive black berries of Deadly Nightshade—a member of the potato family—contain a potent poison that affects the body's central nervous system. It may be extracted and used on the battleground.
Right A lethal dose for a nation. This quantity of Botulinus toxin—just seven grams—could kill the entire population of France. It is one of the most powerful poisons known and is secreted by spores common in garden and farm soils. *Bottom right* Diseases that can kill cattle and other farm animals may be of great long-term effectiveness in biological warfare, destroying the enemy's food supply. *Below left* Disease carriers. Ticks and mosquitoes feed on human blood by puncturing the victim's skin and inserting a proboscis. *Bottom far left* An Argasid tick, bloated after feeding on human blood. *Bottom left* A female mosquito feeding and excreting on a human arm, its proboscis in the flesh. In biological warfare, these vectors could be enlisted to spread the disease deliberately among the civilian populations.

Daily Telegraph Colour Library

Topham

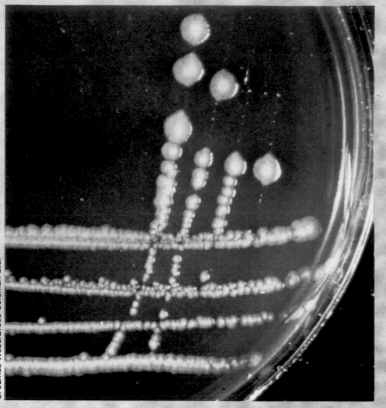

C. James Webb/Bruce Coleman Ltd.

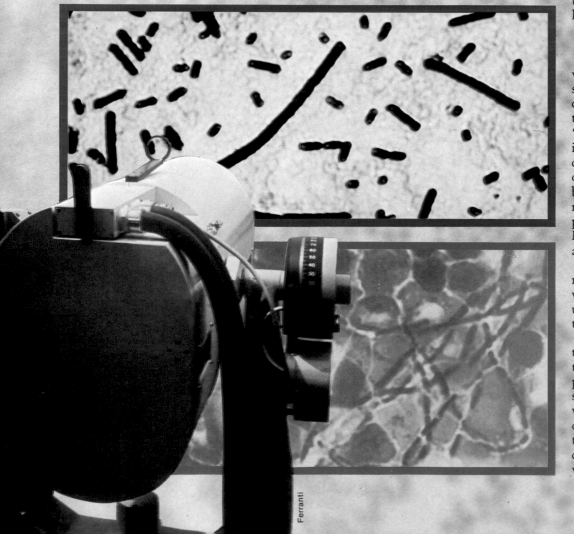

Ferranti

Left Typhoid colonies growing on a culture plate and *(below)* Typhoid bacilli responsible for food poisoning. Typhus can be spread by aerosol dust, which is inhaled, or by lice when they bite. *Bottom* Anthrax in pleural fluid that surrounds the lungs. Although an animal disease, Anthrax spores are potent against humans, causing fever, breathing difficulty, even death. The virulent spores may last for years in the soil. *Bottom left* Laser beam monitoring equipment can detect particle clouds.

lifespan of certain insects offers the BW aggressor an attractive method of circulating a disease for several weeks or months, as opposed to an aerosol operation which might have a lifespan potential of a few hours. (Some types of mosquito live up to two months; some fleas survive seven months.) In addition many insects possess the ability to pass the disease organisms to their young, hence extending the persistency of the threat indefinitely. Such *vectors*, or carriers, make ideal 'seek out and destroy' weapons; once they have been infected and released into enemy territory over 80 viral diseases can be transmitted to Man by this method.

Mosquitoes, for example, carry yellow fever, dengue fever and varieties of encephalitis. Flies transmit typhoid, cholera, and forms of dysentry. Tsetse flies bite their victims and transmit 'sleeping sickness'. Fleas pass on typhus and various forms of plague. Lice carry epidemic typhus and trench fever. Ticks and mites transmit Rocky Mountain fever, tularemia and Colorado fever.

In the case of the last three vectors, an aggressor might choose to speed them on their way by carriage on their usual hosts—rats or other small animals.

Biological cocktail

Insecticides and pesticides are counter-weapons. And even after infection medical science can fight back. But the task will be complex. Aggressors may disseminate a mixture of disease—in a kind of exotic biological 'cocktail'. This would greatly complicate the identification of the diseases appearing in the community, and slow down the application of counter-measures. However, the chaos brought about by such an attack could return on the aggressor unless his own population has been effectively immunized. Much research is being directed to perfecting a 'blanket' immunity injection, as a result.

Because biological warfare may pose as many problems for the aggressor as for his victim, and because many hazards remain unmapped, military researchers have tended to concentrate their investigations.

Of special interest is the work on toxin extraction. Toxins are biological materials extracted from a wide variety of creatures and plants. Bacterial and viral materials store such poisons. The toxin, separate from its vector, can be delivered to the target without risk of spreading contagions beyond those who inhale or swallow it. Toxin prodution is where 'chemical' and 'biological' warfare techniques join forces and the

Above Deadly bite: venom being 'milked' from a Rhinoceros-Horned viper. Biological toxins from animals and fungi can be used effectively against Man. Fleas, lice and ticks *(right)* can carry a wide variety of deadly diseases on to humans. After infection they may be delivered on, for example, the brown rat. This system can extend the life span of biological agents from a period no greater than a few hours to many weeks.

methods of disseminating the poisons are the same in both instances.

A more speculative form of biochemical research involves attempts to isolate the biochemical substances which trigger the emotion of 'fear' in animals. In the course of such experiments, animals (usually rats) have been subjected to a series of 'frightening' situations. The terrifying events, brought about by loud noises and sudden electric shocks, trigger the production of sudden complex substances within the brain cells—a kind of 'chemical' memory, locked inside the animal's chemical conditioning. (The process might be compared with biochemical production of *adrenalin* after physical and psychological stress.) The next step is to transfer the 'chemical' memory bank to another creature which has not been subjected to the fear-making experiment. If this is achieved, the 'novice' behaves in exactly the same way as his terrified companion when small amounts of chemical are injected into the bloodstream. He suffers the illusion that he is receiving severe electric shocks.

The implications of such research are im-

PARASITIC DISEASE CARRIERS

Harvest mite

Plague flea

Fever tick

Brown rat – parasite carrier

Lyn Cawley

Norman Myers/Bruce Coleman Ltd.

Ian Beames/Ardea

DISEASE	INFECTIVITY	TRANSMISSION MAN/MAN	INCUBATION	SYMPTOMS	DURATION OF ILLNESS	THERAPY
ANTHRAX (Pulmonary)	Moderately High	Minor	1/5 days less if inhaled	High fever. Breathing difficulties. Collapse. Probability of boils forming on hands, feet, and body.	3 – 5 days	Antibiotics: Penicillin, Aureomycin, Terramycin & Chloromycetin. With sulfadiazine and immune serium
BRUCELLOSIS	High	None	1/3 weeks less if inhaled	Irregular fevers over prolonged periods. Sweating, chills, painful joints/muscles. Great fatigue. Illness can last years.	weeks to months	Antibiotics: particularly Dihydrostreptomycin and the Tetracyclines. Vaccines not yet perfected.
CHOLERA	Low	High	1/5 days	Nausea, vomiting, diarrhoea, loss of body fluids. Toxemia. Collapse.	One to several weeks	Use intravenous hypotonic saline drip to replace lost body fluids. Also treat with tetracycline and vaccine.
PLAGUE (pneumonic) (bubonic)	High	High	2/5 days less if inhaled	High fever. High pulse. Red eyes. Coated tongue. Swollen glands. Pneumonia. Extreme fatigue. Black spots on skin.	up to 3 days	Vaccines and Streptomycin and Sulpha drugs.
TULAREMIA	High	Minor	1/10 days	Chills and shivering. Fever, fatigue. Pneumonia. Enlargement of lymph glands. Ulceration possible.	few days to several weeks	Antibiotics: Streptomycin, Aureomycin & Chloromytin. Vaccine available.
'Q' FEVER (Nine Mile or Queensland Fever)	High	Practically nil.	3/21 days	Acute fever. Chills & headache. Fatigue and sweating.	1 – 3 weeks	Tetracycline antibiotics. Vaccine available.
PSITTACOSIS	High	Medium to High	4/15 days less if inhaled	Acute fever. Coughing. Bronchitis. Muscular ache. Disorientation.	1 to several weeks	Antibiotics: Chloramphenicol, Aureomycin, Terramycin. No vaccine available.
ROCKY MOUNTAIN SPOTTED FEVER (San Paulo fever)	High	Nil	3/10 days less if inhaled	High fever with joint and muscle pains. Rash spreads along arms and legs, covers body by day 4. Aversion to light.	2 weeks to several months	Antibiotics: Chloro-tetracycline, Chloramphenicol, and Oxytetracycline. Vaccine exists.
EPIDEMIC TYPHUS	High	Nil	6/15 days less if inhaled	High fever with severe headache. General pains. Delirium, fatigue, rash over body, feeble pulse.	few weeks to months	Antibioti: see Psittacosis.
DENGUE FEVER	High	Nil	4/10 days less if inhaled.	Sudden chills and fever. Intense headache, backache, pain in eyes, joints & muscles. Fatigue, rashes. High fever lasts days.	few days to weeks	Vaccine developed but no specific treatment available – just supportive.
Eastern enquine ENCEPHALITIS	High	Nil	2/15 days	Inflammation of the brain tissue. Headache, fever, dizziness, drowsy, stupor, tremor, convulsions, paralysis.	1 – 3 weeks	Some vaccines have been made but no proper therapy exists.

portant. Man is a biochemical creature, his entire life-support system based on a complex web of chemical compounds, each controlling or triggering behaviour and normal functions. If scientists succeed in isolating the chemical compounds which trigger 'fear', 'suggestibility', or 'crazed anger', military and civilian behaviour could be altered drastically.

Protection against biological weapons can be achieved by a number of methods. The first, and most natural, form of protection has been practised for many years—immunization. Immunity to a particular illness may be lifelong or limited to a much shorter period of time. But genetic engineering may create new forms of virulent bacteria against which no immunity is possible.

Protection can also follow the same techniques used by the survivors of a nuclear or chemical war—by donning protective masks and head-to-toe clothing; by sanitization of the environment.

But biological warfare might be used successfully not against Man but against his crops and farm animals.

Zoonoses (the name given to the diseases of animals), include anthrax, rabies, brucellosis, and psittacosis. Foot-and-mouth disease, rinder-pest, and swine fever are equally severe in their effects. An attack on farm animals would deny food to the enemy, hopefully forcing him to surrender with a minimum of loss to the attacker.

Against food crops, various fungal, bacterial and viral organisms could be deployed, again with a view to creating famine. Such warfare would be a modern-day version of a siege—the will to fight on is overcome by hunger and instinctive self-preservation.

Thus, though as yet untried, biological warfare offers a range of effective methods of attack. BW also offers the least expensive method of destroying the enemy. A one-megaton nuclear weapon would cause extremely heavy casualties over an area of several score square miles, but it must be expensively delivered. To kill an equal number of people by Anthrax spores would require a mere eight grams of aerosol spray.

DEATH RATE	OTHER REMARKS
Nearly 100%	Spores only destroyed by long-term boiling, steaming, fire, and intensified disinfectants.
High for inhalation 6% for oral	Destroyed by boiling food and contaminated clothing. Disinfectants.
up 80%	Killed by fire and boiling, high pressure steaming, disinfection.
between 80 – 100%	Killed by burning. Destroy all vermin and insect vectors. Boil or steam food for at least 30 minutes. Use fire and disinfectants.
up to 60%	Destroy by fire, steam, boiling. Disinfectants kill.
very low 1%	Boil all milk and water. Boil food. Use fire and disinfectants for decontamination.
40% fairly high	Burn or boil contaminated materials for destruction of disease. Kill vermin carriers.
up to 80%	Use fire to destroy or boil for long periods.
up to 70%	Do not eat any contaminated food. Destroy carriers and bodies by fire.
very low 1%	Kill insect vectors with insecticides and fire. Spray areas continuously for some weeks/months following attack.
high 60%	Difficult to kill – fire and high pressure steam hoses are best method for decontamination.

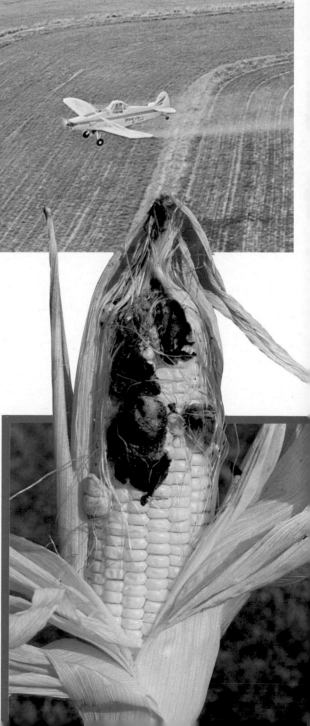

David Moore/Colorific!

Various fungal, bacterial and viral organisms can be introdced by spraying *(above)* to cause serious damage or total destruction of crops. Mildew and Rust Fungus can be used against wheat, barley, rye and oats.
Below left A field of barley infected with rust fungus and *(below right)* Ergot fungus on corn. These biological agents could be fought by use of further toxic spraying.

Oxford Scientific Films

How big is a nuclear bang?

A ten-megaton atomic bomb exploding with a force equivalent to ten million tons of TNT produces a blaze of light brighter than the sun that would blind people 200 or 300 miles away. Within forty seconds it would grow into a blazing fireball three miles across. People out in the open would be incinerated up to twenty-two miles away. It would produce a crater up to a mile wide and over 200 feet deep: millions of tons of debris would be sucked up into the mushroom cloud, then deposited as deadly radioactive fallout over more than 7,000 square miles. Blast, fire storm, and radiation could kill more than one million people.

The explosive force of 1,000,000 tons of TNT is called a megaton. During all of World War 2, a total of three megatons were detonated. Today, some hydrogen bombs have an explosive power of 65 megatons. A nuclear attack on Britain would involve (it is envisaged) 200 weapons in the one-megaton range—equivalent to about 13,000 bombs of the type dropped on Hiroshima. The Hiroshima bomb slew some 200,000 people. Globally, the world's nuclear arsenals contain the equivalent of more than three tons of TNT for every man, woman, and child.

Nuclear bombs derive their devastating power by releasing the energy that binds together the inner core, or nucleus of the atom. Energy is released when a heavy nucleus breaks into two (a process called fission) or when two light nuclei join (fusion). Heavy atoms tend to be naturally unstable because their nuclei contain so many constituent protons and neutrons that the nuclear force is not strong enough to stop them spontaneously disintegrating. Substances which are unstable in this way are radioactive. The rate of decay and energy release of a radioactive atom may be extremely low—sometimes it takes millions of years—but if such an atom is given an extra jolt, for example, by being hit with a neutron, it will break up into two parts, thereby releasing its nuclear binding energy

Right A 10 MT bomb has a fireball 6.9 km (4.3 miles) wide which climbs at 250 mph (400 km/h). If the fireball touches the surface of the Earth vast tonnages of dirt, dust, and debris are swept into the sky to form the mushroom cloud. Deadly fallout later rains down over more than 7,000 square miles.

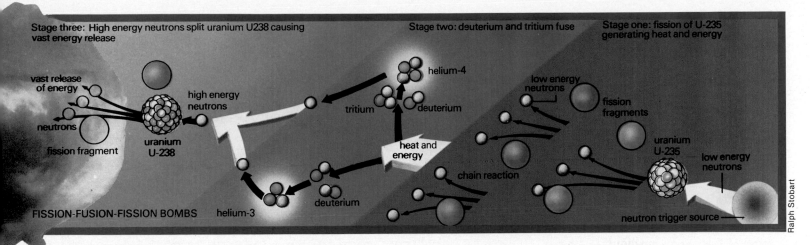

Stage three: High energy neutrons split uranium U238 causing vast energy release

Stage two: deuterium and tritium fuse

Stage one: fission of U-235 generating heat and energy

vast release of energy

high energy neutrons

helium-4

low energy neutrons

fission fragments

neutrons

tritium

deuterium

fission fragment

uranium U-238

heat and energy

chain reaction

uranium U-235

FISSION-FUSION-FISSION BOMBS

helium-3

deuterium

low energy neutrons

neutron trigger source

Ralph Stobart

instantaneously. This is the process that takes place in an atomic explosion.

Uranium and plutonium were the chemicals used in the early bombs. The first step in the manufacture of an atomic bomb based on uranium is to enrich the U-235 component of natural uranium from 0.7% to at least 70%. When this is done, a self-sustaining nuclear chain reaction can occur in the uranium. The secret of U-235 fission is that having initiated it with one bombarding neutron, *two neutrons* are ejected which collide with two other atoms which in turn eject *four additional neutrons*. They ideally, col-

lide with four new atoms. These splitting atoms give off *eight neutrons* which seek out eight more atoms—and then there are *sixteen neutrons* on the loose. And so on. Thus we have a very rapid chain-reaction of events which flashes through the entire lump of uranium in 1/100,000,000th of a second.

But this chain reaction continues only if there is sufficient uranium present, and if the fissioning mass of uranium is held inside a very heavy material long enough for the energy output to build up to an astronomical figure. Otherwise the early phases of the chain reaction will blow unfissioned pieces of

uranium apart and thus halt the explosive process. This is called a 'fizz out'.

The amount needed for a self-sustaining reaction to take place is known as the critical mass (about 15 kg in the case of U-235). An A-bomb contains two hemispheres of radioactive material which are each smaller

Below 6 August 1945 Hiroshima—devastation after the bomb. In the first few seconds vast amounts of heat, blast, and radiation annihilated 78,000 men, women and children. Within 30 minutes a terrible firestorm engulfed the city. The final toll was some 200,000.

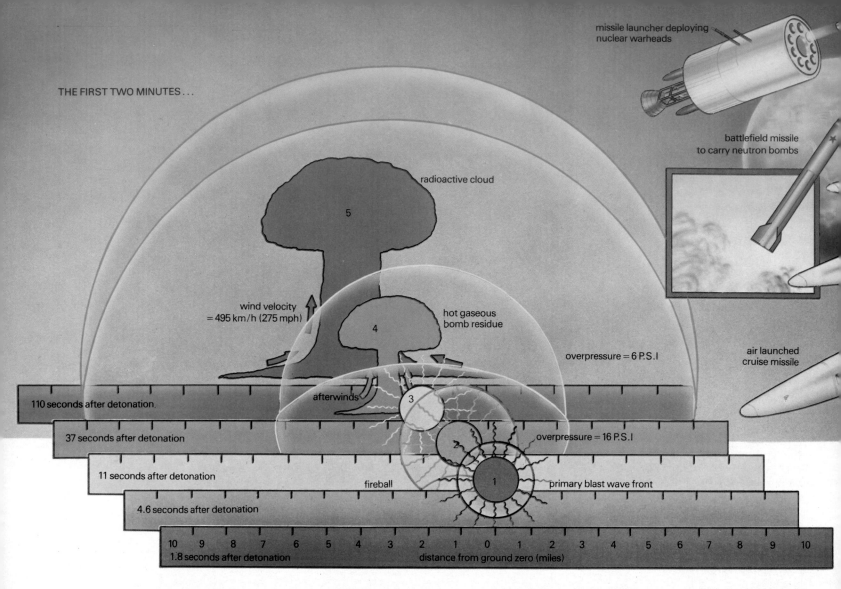

missile launcher deploying
nuclear warheads

battlefield missile
to carry neutron bombs

radioactive cloud

5

wind velocity
= 495 km/h (275 mph)

hot gaseous
bomb residue

4

overpressure = 6 P.S.I

air launched
cruise missile

afterwinds

3

110 seconds after detonation

37 seconds after detonation

overpressure = 16 P.S.I

2

11 seconds after detonation

fireball

1

primary blast wave front

4.6 seconds after detonation

10 9 8 7 6 5 4 3 2 1 0 1 2 3 4 5 6 7 8 9 10

1.8 seconds after detonation

distance from ground zero (miles)

The effects of a one-megaton explosion at a height of 6,500 ft over New York. About half the energy goes to heat and radiation, the rest creates a supersonic blast wave. It exerts an overpressure that squeezes and then explodes structures and human tissue.

than the critical mass and are kept apart when the bomb is assembled. The bomb is enclosed in a heavy metal tamper which holds the hemisphere in position. The atomic explosion takes place when the two sub-critical masses are slammed together by a conventional chemical explosive charge.

Fission is also used for nuclear power generation. In this case a slow, controlled, release of energy is needed and the amount of uranium is regulated so that the self-sustaining chain reaction does not 'take off'. This is, in effect, a controlled nuclear explosion, with the heat being removed to generate electricity in the same way as in coal-fired and oil-fired power stations. However, the *waste by-products* of nuclear power generation are more dangerous than those which are produced by burning oil

or coal, because they are radioactive.

Plutonium on the other hand is a *useful* by-product of nuclear power generation and is made from the bombardment of the U-238. It happens like this: when neutrons from the fission of Uranium-235 hit atoms of U-238, they are absorbed by the nucleus to produce another isotope—Uranium-239—which then undergoes its own radioactive decay process and eventually becomes plutonium (Pu-239). For every gramme of uranium used in this type of nuclear power plant, about a gramme of plutonium is produced—and this plutonium is itself radioactive and can be used as further nuclear fuel.

Breeder reactor

So nuclear reactors can be built to produce electricity *and* to 'breed' new fissionable material. Future nuclear reactions may then use this plutonium to produce further nuclear energy. But, unfortunately, plutonium can also be used to make atomic bombs. Indeed, the first atomic reactors were intended to produce plutonium; they generated electricity only as a by-product.

The very first nuclear reactor started production in 1942 and within a very short space of time other nuclear reactors supplied the plutonium used for the Alamogordo and Nagasaki explosions. The Hiroshima bomb was the only one to use uranium.

Today's world is dotted with peaceful nuclear power programmes, and because of the relative simplicity of plutonium 'breeding', the United Nations Organisation is constantly alert to the possibility of non-nuclear weapon states developing their own arsenal of atomic bombs from the illicit production of plutonium hidden beneath a veil of secrecy and 'innocent' peaceful nuclear power operations. Notwithstanding these frightening possibilities, the nations which already possess a nuclear weapons arsenal have embarked upon a 'plutonium breeder' reactor programme. The justification for such action rests on the fact that uranium ore supplies are not infinite. More and more countries are turning to nuclear power as an answer to the dwindling stocks of oil. To the nuclear engineer and scientist it makes sound sense to 'breed' plutonium from the

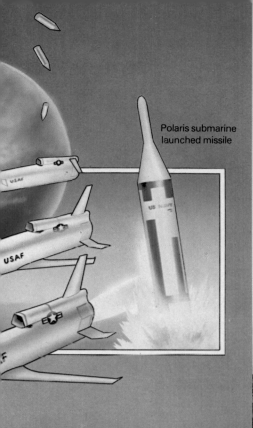

Polaris submarine launched missile

abundant and 'unwanted' Uranium-238. Meanwhile, the international concern over illicit nuclear weapons development includes the spectre of terrorists gaining possession of some bomb-grade fissile material and manufacturing a crude atomic weapon.

Hydrogen bombs are even more powerful than A bombs. They obtain their energy by the same process that occurs in the centre of stars. Just as in the Sun, the nuclei of atoms of hydrogen are forced together to form a new heavier element, helium. This chemical alchemy releases a vast amount of energy, but will only occur when the temperature of the hydrogen is raised to between one and ten million degrees Centigrade. This fiery temperature can only be generated in one way—by the explosion of an atomic bomb.

So a thermonuclear bomb is a fusion bomb encompassing a fission device. When the fissile material's chain reaction reaches its peak output and localized temperature soars to ten million degrees, the hydrogen atoms fuse together and liberate their vast energy.

And whereas the size of a fission explosion is limited by the critical mass, there is no problem of this in a thermonuclear device —the doomsday machine can be as big and destructive as desired.

The first detonation was conducted in the spring of 1951 at Eniwetok, an atoll in the Marshall Islands in the Pacific. There was some thermonuclear reaction, but most of the energy still came from fission. The second Eniwetok test, on 1 November 1952, was more successful and released energy equivalent to about ten million tons (ten megatons) of TNT. It was reported to have blown an island off the face of the sea.

Even more powerful

The fission-fusion bomb is an even more powerful version of the thermonuclear bomb. First tested at Bikini Atoll on 1 March 1954, it comprises an ordinary fusion bomb, triggered by a fission bomb, and encased in ordinary uranium.

Although Uranium-238 does not fission when bombarded by low-energy neutrons, this is not the case when high-energy neutrons are involved. And since a thermonuclear 'fission-fusion' hydrogen bomb generates enormous quantities of high-energy neutrons, it became the next logical step in the technology of destruction to manufacture the bomb tamper from heavy U-238 metal. Once the atomic bomb trigger detonates and liberates its huge amount of thermal energy, setting 'fire' to the hydrogen fusion process, the generation of very high energy neutrons also occurs. These energetic particles bombard the U-238 tamper and trigger a second fission reaction.

Because ordinary uranium is used, and there is no need to go through the expensive process of enriching it, the 'FFF' bomb achieves even more destructive power for relatively little extra cost. The detonation of a single 65-megaton weapon of this nature over any of the world's cities would be a disaster unprecedented in history.

The detonation of a nuclear weapon raises the temperature of the bomb parts and the

Colorific!

Left Within hours, radiation illness causes sickness and vomiting. Then diarrhoea, weakness, and depression. Hair falls out. Bleeding starts from mouth, nose and bowels.

HEAT/RANGE EFFECTS FOR 100 MT BLAST

Heat effect zones for
a 100 MT atomic blast
(from centre).
(1) Metals vaporize.
(2) Metals melt.
(3) Plastics ignite.
(4) Wood burns.
(5) 3rd degree
burns—skin chars.
Paper, fabrics ignite.
(6) 2nd degree
burns—blistering.
(7) 1st degree
burns—skin
scorched.
(8) Effect of open
oven door.
(9) Global fallout.

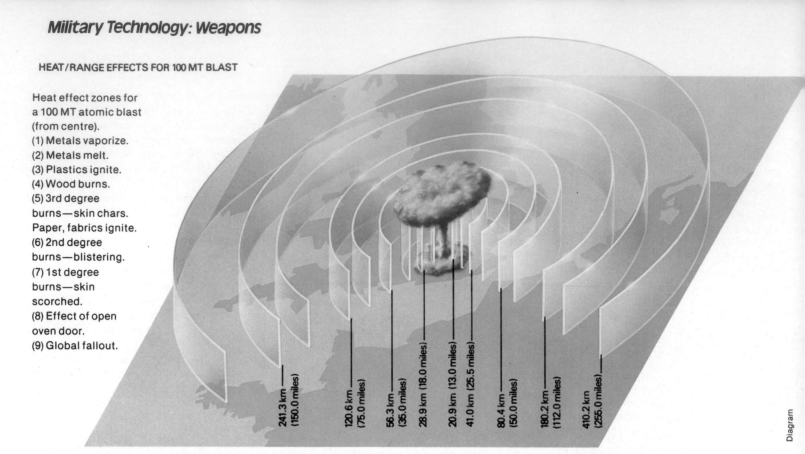

241.3 km (150.0 miles)   120.6 km (75.0 miles)   56.3 km (35.0 miles)   28.9 km (18.0 miles)   20.9 km (13.0 miles)   41.0 km (25.5 miles)   80.4 km (50.0 miles)   180.2 km (112.0 miles)   410.2 km (255.0 miles)

Diagram

surrounding air to a temperature of about 10,000,000°. Enormous pressures are generated with this tremendous heat and within microseconds (one microsecond = one millionth of a second) all of the bomb parts and nearby air expands into an intensely hot incandescent sphere called a fireball. In about the same time that it takes for a clap of thunder to follow a flash of lightning—the blast follows the brilliant flash of a nuclear explosion. Since the blast wave travels at a speed just slightly above that of sound, moving outwards from the explosion at 1,150 feet or 350 metres per second, it will take approximately ten seconds to reach a person standing two miles away from the burst.

The mechanical motions of a nuclear explosion are analogous to those of a tidal wave. The shock front is literally a wall of compressed air. As it passes, structures are exposed to a nearly instantaneous rise in the local atmospheric pressure, and crushed.

Extremely high speed winds having gusts of more than 670 mph (1,072 km/h) frequently attend these pressure phases. The drag forces of these winds cause additional damage to buildings as well as hurling dust, wreckage and people through the air, inflicting further damage and fatal injuries.

The enormous number of fires that would be started over hundreds of square miles would not remain isolated. A 'firestorm' would be produced on a scale bigger than the enormous fire raids of World War 2 at Hamburg, Dresden, Tokyo and other cities.

Then, the fires from thousands of incendiary bombs joined together to form huge pillars of fire which sucked in winds of up to 150 miles per hour (strong enough to uproot trees). People caught in the streets were burned to death, and others in fire-proof shelters suffocated, because the air that came in from the street was denuded of oxygen and scorchingly hot. Something similar, only worse, can be expected following a nuclear attack. The fire storm might well destroy everything within twenty miles.

Radiation

People who survived the heat and blast would still have to face the third destroyer: radiation. Most of the initial radiation emitted by a nuclear detonation comes from countless billions of neutron particles generated by the rapid fission chain-reaction. This swarm-like beehive of nuclear particles generates a wide spectrum of radiations. Ultra-violet light, X-rays, heat radiation, and gamma radiation. Both neutron and gamma radiations can penetrate thousands of feet of air, through tens of feet of water and soil and feet of other solid material.

The almost instant release of neutron and gamma radiation by a nuclear explosion poses one of the greatest threats to living organisms—neutrons move through the atmosphere at speeds of 100,000 miles per second (1.6×10^8 metres per second), and gammas traverse at the speed of light —186,000 miles per second (3.0×10^8

metres per second). Consequently there is no time to take evasive action—when you 'see' the blinding flash you will already be bathed in these potentially lethal radiations.

If, at the moment of explosion, the fireball reaches the Earth's surface, thousands of tons of earth and rubble are sucked up by the strong winds that rush towards the centre of the fireball as it climbs at an initial rate of 250 mph (11 metres per second). The fireball is an atomic cauldron that vaporizes the debris and thoroughly mixes it with highly radioactive bomb parts. Their nuclear fallout is a long-term hazard. The fine dust from the mushroom cloud can stay aloft for two or more years and return to earth thousands of miles from the point of the explosion.

The three most dangerous longer-lived radioactive substances in fallout dust are strontium-90, caesium-137 and carbon-14. These three are easily absorbed in our bodies, so that when they decay over a matter of years *all* their emitted radiation affects living matter. The effect is to change the chemical composition of living cells—a highly toxic form of internal poisoning.

The special danger of strontium-90 is that it is chemically similar to calcium and so is absorbed into our bones and blood. Growing children who drink milk from cows which have grazed on fallout-laden grass are especially vulnerable. So people born after 1955, when many atmospheric nuclear tests were conducted, have more strontium-90 in their bones than older people. By irradiation of bone marrow, or bone, strontium-90 may cause leukaemia or bone tumours.

Caesium-14 can replace ordinary carbon anywhere in the body, and decays to nitrogen-14. So carbon-14, like caesium-137, can cause abnormal children to be born. Even today, childen of Hiroshima and Nagasaki are more likely to be born dead or deformed than elsewhere—the effects of radiation are transmitted to the unborn generations.

Even more exotic nuclear weapons are on the drawing board. Back in 1960, Herman Kahn (the Futurologist), suggested that it may soon be possible to manufacture a tiny atomic bomb no larger than a rifle bullet or cannon shell. Just as plutonium can be manufactured in the 'fires' of a nuclear reactor (by transmuting U-238), so neutron bombardment can produce even heavier artificial elements of fissionable material; the 'critical mass' is considerably smaller. And if this element is machined into a bullet and fired at a solid structure, the impact will compress the californium into a 'super-critical' mass and explode with a force of several tons of TNT. The foot soldier of the future would be a formidable opponent armed with such a weapon.

Although by no means new in concept (its origins go back to the 1950s), the *'neutron bomb'* has now entered the vocabulary of 'doomsday language'. These weapons are small fission-fusion devices—miniaturized H-bombs. When a neutron bomb is exploded several hundred feet above its target, the effects of the blast and heat will be minimized, whilst the 'prompt' emission of neutron particles and gamma radiation will be enhanced. As the world's newspapers put it—'the neutron bomb is designed to destroy people, not buildings'.

The importance of the neutron bomb will undoubtedly increase with time, as more and more nations gain the ability to manufacture nuclear weapons. It was earmarked originally for battlefield use, against tanks and large numbers of troops. Its unique 'killing' capacity lies with the fact that neutron irradiation has a lower threshold of lethality.

But if the world's nuclear arsenals are put to use there will be global effects even more serious than the short or long-term radioactive hazards. It has been calculated that a massive 10,000 MT nuclear exchange would release enormous quantities of nitric oxides (NO). The artificially generated NO, between ten and 50 times that normally encountered within the Earth's atmosphere, would seriously affect the level of ozone existing in the upper atmosphere.

Devastated planet Earth

If the predictions are correct we can expect the entire atmospheric machinery to go out of alignment. Atmospheric circulation will change, bringing with it an alteration in solar heating. This will cause a reduction in the average global surface temperature between 0.2 and 0.5 degrees Centigrade. Many experts believe that similar fluctuations in the past were responsible for the start of the Ice Ages.

In addition, ozone layers protect all surface life from the harsh ultra-violet rays of the Sun. Too much UV combined with the immense quantities of radiation released in the explosions would result in cancer and sterility, and an increased incidence of spontaneous abortion. After a nuclear war, deformed, weak, and short-lived children would be born to inherit a devastated, intensely radioactive planet Earth.

Nevertheless, a developing market for shelters is being exploited and offered for sale to the public. And, in Britain, official leaflets are available advising the public on techniques such as extinguishing curtains with the aid of bucket and water after a nuclear explosion.

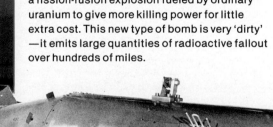

Below 'Little Boy'—a replica of the atomic bomb dropped on Hiroshima.
Left First tested at Bikini Atoll on 1 March 1954 a fission-fusion explosion fueled by ordinary uranium to give more killing power for little extra cost. This new type of bomb is very 'dirty' —it emits large quantities of radioactive fallout over hundreds of miles.

War in space

Every day, vital security and defence information travels back and forth between ground stations and orbiting satellites. If ever an enemy were to seek a quick means of knocking out potentially hostile forces, he could do no better than blind the sensors that, during conflict, would direct weapons to within a few metres of their targets on the ground, at sea and in the air.

By the end of the 1980s, both US and Soviet defence systems will be tied to the success or failure of space sensors linked with navigation satellites. By the 1990s, space war could be a reality.

The move towards conflict in orbit has been both rapid and unexpected. Made feasible by technology impossible to understand when Neil Armstrong put Earthman's footprints on the Moon in 1969, beam weapons and laser gunships are a product of research in high-energy physics. Yet, the race for space weapons emerged only in the second half of the 1970s, when Soviet killersats (killer satellites) were observed from the US on tests aimed at proving their feasibility.

As early as 1956, more than a year before Russia put the world's first artificial satellite in orbit or the first intercontinental ballistic missiles had flown, engineers in the USA studied possible methods for knocking out an incoming warhead before it reached its target. In 1963, a Thor booster successfully flew to within 'killing' range of a previously launched target rocket, demonstrating the ability of a ground-fired anti-satellite (ASAT) device to knock out hardware already in orbit. Both projects proved prohibitively expensive.

Killersat tests

In the Soviet Union, however, where costly military projects are easier to finance, work went ahead on a killersat programme. In October 1967, just 10 years after Sputnik 1 (the first satellite), Cosmos 185 was placed in an almost-circular orbit about the Earth, followed by several other vehicles of a similar type at the start of an active killersat test programme. Just how active that test series was expected to be emerged twelve months later when Cosmos 249 shot past its target (Cosmos 248) and blew up on command from the ground. Preserved for later tests, Cosmos 248 would have been destroyed had the interceptor detonated at close range.

Western observers noted a comment in an East German newspaper that it was now possible to destroy unfriendly satellites 'with the help of weapon systems which the Soviet Army has at its disposal'. Such confidence was boldly premature, for the Russians had many tests still to perform with several different systems.

In 1978, Soviet killersat tests proved conclusively that Russia would soon have an operational ASAT system. In further demonstrations, they made interceptions regularly within 1 km (0.6 miles) of the target, a distance at which delicate apparatus aboard sensitive satellites could be destroyed by flying shrapnel.

American response

At the same time, the United States, too, felt compelled to renew serious studies of a comparable system. The studies resulted in an ASAT weapon based on the marriage of a small two-stage rocket and the McDonnell Douglas F-15 Eagle fighter aircraft.

The two-stage rocket was packaged in a single container to which was attached an impact head (developed by the Vought Corporation). The barrel-shaped Vought device (0.3 m long and 0.37 m in diameter) was to be released from the F-15 and propelled by rockets to impact at high speed with an enemy satellite, guided to its target by an active radar seeker and radiation sensors.

This 'hot-metal kill' mode was deemed suitable for satellites in low orbit but, like the Soviet killersats, would be unable to knock out the high flyers. Nevertheless, it is a unique response to Soviet initiatives and could be fully operational by 1985.

But low-flying satellites, such as reconnaissance vehicles and electronic ferrets designed to gather radio and communication signals from a potential enemy, are but a few of the several hundred pieces of operational military hardware orbiting the planet—highly vulnerable in time of war. Whereas spy satellites operate from only a few hundred kilometres above Earth, navigation, communication and early-warning satellites occupy orbital lanes far above the

Artist's impression of war in space. In the foreground, a Soviet killersat fires a laser beam to destroy a US military satellite. Soviet killersat technology was well advanced in 1978, research having started in 1967. In 1976, work on a US killersat programme was put in hand to counter Soviet developments.

Tony Roberts/Young Artists

atmosphere. If they are to be eliminated in the opening stages of a war, a more-advanced system is called for.

The Russians are known to have developed high-energy physics to a level where laser and particle-beam weapons are feasible. A laser beam would take little more than one-tenth of a second to travel the distance separating Earth from a *stationary orbit* satellite 36,000 km (22,400 miles) above the clouds. The important military satellites are at that height because a satellite there takes 24 hours to orbit the Earth once—the time the Earth takes to spin once on its axis. Hence, a satellite in stationary orbit will appear to remain over the same spot on Earth.

But a laser beam projected from the ground is severely distorted and de-energized as it passes through the atmosphere. The problem is solved if pulsed lasers are used. Pulsed lasers switch on and off at high speeds, discharging bullets of light with a significant drop in interference. Yet, compared with a continuous-wave laser operating in a vacuum, even a pulsed laser in air is much less efficient. Because of this, high-altitude satellites are best countered by a laser weapon based in space. Towards this objective, the Russians have been experimenting with several laser-weapon components on Salyut space stations and are believed to be planning the operational deployment of laser gunships before the end of the 1980s.

The United States is vigorously researching the best way of deploying a similar system. And it is in this application that the laser is most efficient. The weapon is good at disabling electronic equipment rather than killing people or destroying industrial targets. The destructive effect of a laser weapon held on target for several seconds is equivalent to only a few kilogrammes of high explosive—far too low an equivalent yield for major destruction but just right for knocking out a vulnerable satellite or the nose cone of a sophisticated missile.

Lasers can be most effective only if they are placed in space to patrol orbital lanes populated by defence satellites. But even lasers firing small bullets of energized light are vulnerable to attack, and as insurance against the possibility of an effective Soviet anti-ASAT device, strategists in the USA have pressed for a high-altitude ASAT based on conventional technology.

Taking the existing F-15-launched ASAT and expanding its size and capability, engineers have proposed the same kind of hot-metal kill head but fixed to a submarine-launched Trident C4 missile. In the event of

McDonnell Douglas/David Baker

McDonnell Douglas F-15 Eagle fighter aircraft which is armed with a compact two-stage rocket to which is attached an impact head. This forms the US weapon to destroy enemy warheads before they reach their target.

war, existing launch sites, such as Cape Canaveral, would not long survive, and by using an underwater launch pad the ability to keep on knocking out enemy satellites would deny information essential to their military needs.

Whereas the F-15 would release an impact head to destroy satellites a few hundred kilometres above Earth, the Trident missile would launch an impact head on course for its target 36,000 km away. But would there be a need to keep on knocking out each other's satellites? Once destroyed, would the comparatively few orbital eyes be effectively put out? Not so, for both America and Russia can hide satellites in space during peacetime, switching them on only when the prime satellites are knocked out.

The USA and Russia constantly track, with great precision, the more than 4,500 pieces of 'junk' in various paths around the

Earth. It would be almost impossible to hide in space a piece of metal larger than a football, but simple to make a stand-by satellite look like useless debris on a radar monitor. Both countries might have to counter an unknown number of passive, replacement satellites waiting to be switched on for communications, observation, navigation or monitoring purposes.

Hence, the need to find a system that can operate in the extreme environment of a nuclear war. The F-15, or any other aircraft of its class, is incapable of lifting the larger weight of a stationary-orbit ASAT device, so the submarine serves as a survivable base with the capacity to support several weapons of this kind.

Lasers in space will become an essential part of any orbital defence system. Just as unwanted debris could hide a replacement satellite, so could fragments and discarded

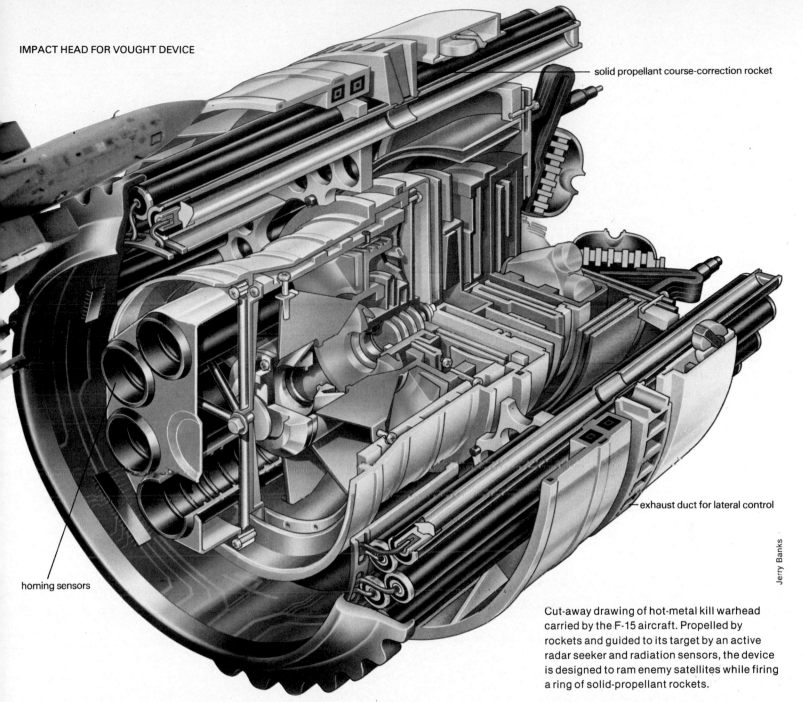

solid propellant course-correction rocket

exhaust duct for lateral control

homing sensors

Jerry Banks

Cut-away drawing of hot-metal kill warhead carried by the F-15 aircraft. Propelled by rockets and guided to its target by an active radar seeker and radiation sensors, the device is designed to ram enemy satellites while firing a ring of solid-propellant rockets.

rocket stages veil the presence of mines in space. Not the passive type waiting for a chance collision; rather a self-propelled warhead ready to move rapidly and impact a military satellite nearby. Rapid response would be vital, and only a laser could hit the mine before it struck its target and blew up.

Yet, most scientists working on high-energy physics in the United States believe such capabilities to be only a start. Particle beam experiments show an improvement in capability similar to that provided by the hydrogen bomb over the atom bomb—increased destructive potential and fewer means to defeat the weapon. A particle-beam device would generate a neutron beam in space with a radiation equivalent to many

thousand neutron bombs; they would minimize the blast effect while maximizing radiation.

The major US particle beam project is at the Los Alamos Laboratory, New Mexico. Called White Horse, the work is aimed at developing a space-based weapon ultimately capable of a power output equal to 100 million electron volts. Lasers are effective satellite weapons at a power output of 5 million electron volts.

Both devices are classed as direct-energy weapons. Although possible methods to defend satellites against lasers are known, nobody knows how to stop a particle beam from reaching its target.

The spur to US particle-beam research

came in the late 1970s when intelligence information about Soviet developments reached the defence establishment. In 1976, US reconnaissance satellites photographed a large research base at Sary Shagan, site of Russia's work on anti-ballistic missiles. And then, at Azgir, near the Caspian Sea, US early-warning satellites detected tests with enormous power levels generated by fusion-pulsed nuclear explosions—proof that the Russians were using a hydrogen bomb to generate large quantities of electrical energy.

In 1978, satellites picked up contamination in space coming from the Semipalatinsk area where explosive generation of electrical energy was being researched. By 1980, disbelieving critics conceded defeat and the

OFFENSIVE SPACE SYSTEMS OF THE 1990s IN SIMULATED COMBAT

US laser battlesat

US battlesat

US navsat

Soviet killersat

Soviet reconsat destroyed
by ground-based laser

US battlesat

early-warning satellite

air-launched
anti-satellite weapon
aimed at Soviet metsat

silo-launched anti-satellite missile
attacking Soviet navsat

US reconsat

early-warning satellite

Soviet killersat

Soviet communications satellite

Jeremy Gower

United States embarked upon particle-beam research programmes of its own, at Los Alamos and the Lawrence Livermore Laboratory. In that year, US satellites mapped the construction of a directed energy weapon at Sary Shagan capable of firing pulsed-laser or particle-beam energy at military reconnaissance satellites, blinding their camera lenses.

The real application of particle-beam weapons, however, is in anti-ballistic missile (ABM) systems. At present, and until the mid-1990s, the possibility is remote that either America or Russia will develop a satisfactory means of screening their countries from missile attack. Re-entering the atmosphere at more than 24,000 km/h (15,000 mph) nuclear warheads are almost impossible to stop. It is that fact alone that preserves the balance of nuclear deterrence, the threat of mutual annihilation.

Breaking the stalemate

But the belief that one side or the other could break away from this stalemate spurs both sides to seek the means by which each can effectively destroy the other's warheads before they explode over selected targets. As both major countries have found to their cost, the price for a conventional ABM system is prohibitively high. There is little assurance that all the many mechanical components would work with sufficient reliability to warrant the investment.

But this view changes with the technology of particle-beam weapons. Before the end of this century, it is probable that both Russia and America will have placed in space large particle-beam generators capable of burning holes in nuclear warheads. If an effective screen can be put up to missiles from another country, it will insulate the population of the threatened state from nuclear attack by ballistic rockets.

An important argument against the probability of anyone starting a nuclear war is the sound logic that so much destruction would ensue that no-one on Earth would escape its effect. Countries only marginally concerned in the conflict would be reduced to rubble, and prime contestants would be almost annihilated. By using particle-beam weapons from Earth orbit, however, people would be killed but buildings, factories, power stations and warehouses would all remain standing, totally unharmed.

It would serve the interests of an aggressive country to strike first, at the speed of light, with particle-beam generators based in space. A nuclear-tipped missile takes near-

ly 30 minutes to reach its target. A particle beam would take only one-hundredth of one second to propagate death for millions on the ground. Because of the ultimate threat to balanced power, many scientists are concerned about the application of particle-beam weapons, for the consequences of developing a system to make nuclear weapons obsolete go beyond the obvious advantages.

However, during the 1980s, the search for an ABM is concentrated upon several satellites placed in orbit, each with a capacity to switch quickly from one target to another. Several warheads could be knocked out by a single weapon.

Where the USA plans to use Trident missile-launched impact heads for satellite targets in stationary Earth orbit, the Russians will employ powerful lasers from the ground to knock them out. Called battle stations, the US ABM devices planned for the late 1980s will comprise several separate weapons in orbits 1,750 km (1,090 miles) high. Capable of handling up to 3,000 separate nuclear warheads, the three-tier defence structure would also employ homing interceptors launched from the ground and low-altitude missiles despatched from underground shelters.

At most, 30 battle stations are deemed necessary for this level of screening. The ability to launch, service and tend these large vehicles in space is made possible because of the re-usable Shuttle. The US Defense Department plans to fly more than 25 per cent of all Shuttle missions expected by the mid-1990s, and new capabilities afforded by this revolutionary transport system will significantly expand the potential uses for military Man in space.

Even the Shuttle would be unable to operate during a nuclear war, however. Its exposed launch site would be prey to incoming warheads. Accordingly, the United States expects during the 1980s to provide satellite launch facilities from underground or undersea locations, preserving the ability to quickly replace satellites knocked out by killersats, lasers or particle-beam weapons and maintain its defence systems.

From late in the 1980s, the outcome of major wars in space will depend not on direct conflict among teams of orbital soldiers but on the success or failure of eliminating the opponent's sensors—essential to tactical and strategic decisions. The threat beyond that will be the particle-beam weapons which, from the mid-1990s, could imperil the safety and freedom of millions back on Earth.

On the watch for war

The science of radar and the development of military early warning systems, in which gigantic strides had been made during World War 2, were relatively neglected after the war ended. There was little new development in these areas during the Korean conflict between 1950 and 1953. However, all this changed towards the end of the 1950s with the introduction of ground-to-air nuclear missiles.

As both the US and the USSR made strenuous efforts to keep track of each other's new developments during the 'Cold War', early warning systems were devised to provide notice of attack by enemy aircraft or missiles. These consist of chains of powerful radar stations located so that their arc of coverage overlapped with that of their neighbours on either side, forming an 'electronic fence' whose outer limit is a series of arcs. One of these, running across Canada and the USA, is known as the DEW (Distant Early Warning) Line.

However it soon became apparent that the DEW Line was not able to give information on approaching missiles fast enough. This led to the development of another radar chain known as BMEWS (Ballistic Missile Early Warning System), designed to detect missiles launched from the Soviet mainland. It consists of three radar stations, at Fylingdales in Yorkshire, Thule in Greenland, and Clear, Alaska. The Thule radar was the largest and most powerful in the world when it went into service. It can detect an approaching missile at a range of 4,800 km (3,000 miles). Nevertheless, even BMEWS' radar can give only about 30 minutes' warning of an approaching ICBM fired from Russia. Radar stations in California and Florida keep watch over the Pacific and Atlantic respectively, to detect missiles launched from submarines.

As well as developing ground-based early warning systems, both the superpowers use a wide range of specially equipped planes, ships and submarines to gather 'Elint'—electronic intelligence—in an effort to discover details of the enemy radar systems and thus develop countermeasures.

The best known of the weapons in this bizarre electronic cold war was the American U-2 spy plane developed by Lockheed. Operating from bases in Turkey, Iran, Pakistan, Norway and Japan, U-2 aircraft flew at high altitudes over the Soviet Union, China and other Communist bloc countries between 1956 and 1960. They produced quantities of remarkably detailed photographs taken through long-range cameras, and also recorded a wide range of electronic signals, including radar.

The U-2 was followed by more advanced spy planes, such as the SR-71 'Black Bird', which was not only able to carry out simple

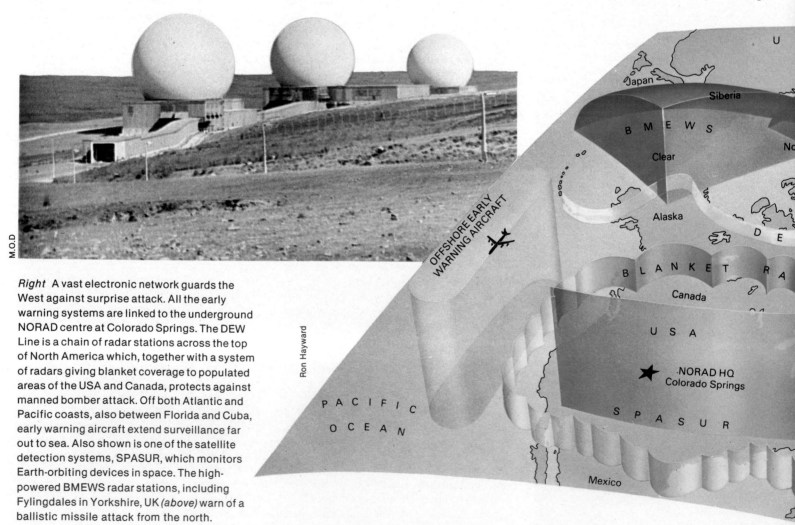

M.O.D

Ron Hayward

Right A vast electronic network guards the West against surprise attack. All the early warning systems are linked to the underground NORAD centre at Colorado Springs. The DEW Line is a chain of radar stations across the top of North America which, together with a system of radars giving blanket coverage to populated areas of the USA and Canada, protects against manned bomber attack. Off both Atlantic and Pacific coasts, also between Florida and Cuba, early warning aircraft extend surveillance far out to sea. Also shown is one of the satellite detection systems, SPASUR, which monitors Earth-orbiting devices in space. The high-powered BMEWS radar stations, including Fylingdales in Yorkshire, UK *(above)* warn of a ballistic missile attack from the north.

Above The sleek, pencil-slim contours of this USAF Lockheed SR-71 Black Bird surveillance aircraft conceal complex electronic equipment. Capable of speeds over 2,000 mph, Black Bird is the fastest jet aircraft in the world.

battlefield surveillance but was also equipped with sophisticated systems capable of specialized surveillance of up to 155,400 sq km (60,000 sq miles) of territory in one hour. However, one problem with using aircraft for spying is that the vibrations from their engines upset the delicate cameras. A more serious problem, as U-2 pilot Gary Powers found to his cost when his spy plane was shot down by a Soviet anti-aircraft missile in 1961, is that aircraft are vulnerable to enemy surveillance and attack.

For these reasons, both the Americans and the Russians decided to use satellites for gathering information. There have now been several generations of spy satellites since the first SAMOS (Satellite and Missile Observation System) vehicle was launched by the Americans in 1961. Since then, the Americans have also sent into orbit a number of devices, including one known as LASP (Low Altitude Surveillance Platform), which is 15 m (50 ft) long and weighs 10 tonnes.

While SAMOS performs the role of an 'eye' in the sky, with its electronic, photographic and infra-red cameras, LASP acts as an 'ear', orbiting 177 km (110 miles) above the Earth's surface and relaying Elint and communications intelligence (Comint) to listening posts on Earth.

Big Bird

The Americans then went on to design an even larger and more sophisticated spy satellite, an improved version of SAMOS, known unofficially as 'Big Bird'. The first one was sent up in June 1971. Launched by a powerful Titan 3D booster rocket, it weighs over 10 tonnes and is 15 m (50 ft) long and 3 m (10 ft) in diameter. Orbiting at an altitude of over 160 km (100 miles), Big Bird satellites can photograph with complete clarity objects less than 30 cm (1 ft) square.

Combining the technical advances of several earlier satellites, Big Bird can eject its capsules of exposed film, which parachute down to be retrieved in mid-air by a device

Below Bristling with surveillance equipment, a Soviet Tu-95 'Bear' long-range reconnaissance aircraft observing a NATO naval exercise is intercepted by a British Phantom jet fighter.

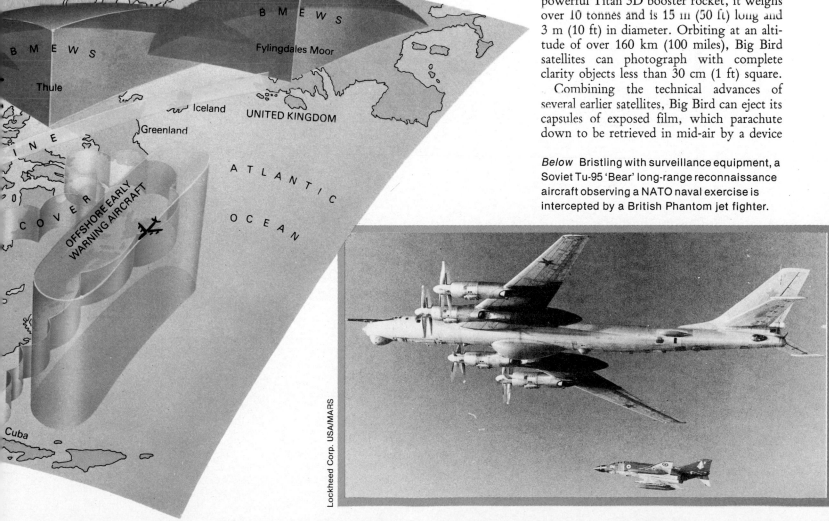

Lockheed Corp. USA/MARS

attached to aircraft over the Pacific; it can transmit high-quality pictures by radio; and its infra-red heat sensing equipment enables it to 'see' through ice and snow to locate underground weapons or submerged nuclear submarines.

The Russians, too, have their own watchdog system of spy satellites circling the globe. They have put hundreds of Cosmos satellites into orbit, each one built up from standardized sections and sub-assemblies on automated production lines.

Their orbits can be changed during flight by signals from ground control to direct them over particular objectives.

Soviet spy trawlers and snooper aircraft scored a victory in this technological war of wits by effectively neutralizing the £40 million missile and bomber early warning station built and operated by the Americans at Orford Ness in Suffolk. This complex, forming part of a comprehensive ground radar and satellite surveillance system which the West had used to eavesdrop on high-frequency radio communications deep inside the Soviet Union, was closed down in 1973, after only two years of operation. The Americans, who believed that the huge array of 189 masts on the 40-hectare (100-acre) complex would enable them to 'look right down Russia's throat', found the view so distorted by Soviet electronic counter-measures that they had to abandon the operation. The station's transmissions were probed extensively by Soviet spy ships and planes operating just outside British territorial waters, and its sophisticated 'back scatter' over-the-horizon radar system (known as 'Cobra Mist'), which could detect extremely weak and distant signals, proved to be highly susceptible to Soviet jamming and spoofing.

Submarine detection

The build-up of nuclear submarine fleets, capable of firing their missiles from anywhere in the oceans, has led to the development of increasingly effective submarine detection systems. Eight NATO countries collaborated to produce an underwater sonar system in the Atlantic Ocean, codenamed AFAR (Azores Fixed Acoustic Range). A number of 30-m (100-ft) high towers were lowered on to the seabed 600 m (2,000 ft) or more below the surface. The towers form a triangle with sides 32 km (20 miles) long. Despite the fact that AFAR has been kept under surveillance by Soviet spy trawlers, the system has proved its usefulness and has been duplicated in other ocean areas to monitor Soviet submarine movements.

A similar arrangement, for keeping track of Soviet submarines passing through St George's Channel between Wales and Eire, was revealed in 1975 when an American ship accidentally fouled cables which were part of the detection system. Denmark has installed a similar unit to spot Soviet submarines entering and leaving the Baltic.

Towards the end of 1974, Soviet underwater 'spy bins', which listen for nuclear submarines, were discovered close to British waters. Two similar devices were found near the NATO base at Keflavik in Iceland early in 1975. The canisters, weighing just under a tonne, contain 32 *hydrophones* (underwater microphones) which listen for submarine engines and propellers.

Submarine 'signatures'

Each submarine possesses a unique sound 'signature' by which it can be identified. A number of Soviet spy bins have been washed up around the coasts of Scotland and Northern Ireland.

Aircraft are also used to detect and track nuclear submarines. The aircraft that is probably best equipped for this task is the British Nimrod. Based on the De Havilland Comet airliner, Nimrod is powered by four Rolls-Royce Spey jet engines, giving it a top speed of almost 965 km per hour (600 mph), and enabling it to reach patrol zones far out in the North Sea or Atlantic. Once it has reached the target area, Nimrod can shut down two of its engines so that it can patrol for long periods without refuelling. The plane is crammed with complex equipment for submarine detection. A recently developed radar system called Searchwater allows the crew to detect and identify submarines by examining their periscopes and exhaust pipes; it can do this at great range, even during periods of high seas.

The Nimrod crew also use a computer linked to a new acoustic system to work out the submarine's exact position by monitoring and analysing information from *sonobuoys* (devices for detecting underwater sounds and transmitting them back to the plane by radio), which the plane drops beneath the waves. Navigational precision is ensured by a tactical control system which co-ordinates information from all the detecting devices. When Nimrod has found its prey, the crew can attack the submarine with torpedoes launched from its weapons bay.

Towards the end of 1967, the Americans were disturbed to learn that the USSR appeared to be developing a Fractional Orbital Bombardment System (FOBS) that could be

1 Bailout jettison mechanism
2 Data processor functional group
3 Computer operator console
4 Special purpose console
5 Radar maintenance station
6 Radar receiver and signal processor
7 Communications equipment
8 Surveillance radar antenna
9 Antenna ancillary equipment
10 Identification Friend or Foe antenna
11 Spare survival equipment
12 Rest area
13 Auxiliary power unit
14 Radar transmitter
15 Navigation and identification
16 Display consoles
17 Power supply and distribution
18 Bailout chute
19 Flight essential avionics
20 Communications equipment
21 Communications console

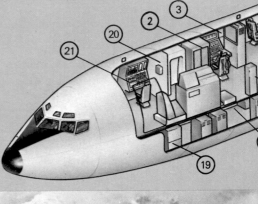

Steve Cross

launched into orbit some 160 km (100 miles) above the Earth. Before completion of its first orbit, its retro-rockets would be fired, causing the weapon to slow down and fall on a predetermined target. This weapon was expected to be more expensive and less accurate than conventional intercontinental ballistic missiles. Nevertheless, from the Soviet point of view, it had the advantage that it could be used to strike with less warning. The FOBS could be orbited so that it approached its target from the south, travelling in the opposite direction to that safeguarded by the American DEW Line, which was designed to detect missiles or aircraft approaching over the North Pole. At the end of 1970 the United States attempted to counter the threat from the FOBs by launching IMEWS (Integrated Missile Early Warning Satellite). The efficiency of IMEWS was tested by ar-

Left The Boeing E-3A AWACS aircraft forms a fully integrated surveillance, command and control system. The huge 9 m diameter radome on top of the fuselage rotates constantly to survey the scene up to 400 km away.

ranging for it to monitor many of America's own space missile launches. The IMEWS satellites are placed in *geostationary* orbits—orbits allowing them to remain in a constant position above the Earth's surface. They use infra-red sensors (IRS) to detect the exhaust heat from enemy missiles. These are coupled with a visible light sensor (VLS) rather like a TV camera, which enables observers on the ground to see the rocket plume of a missile as it is launched and rises above the Earth's atmosphere. The design is intended to prevent false warnings, each sensor system acting as a check on the other in the detection and identification processes.

If an IMEWS satellite detects a missile launch through its infra-red sensors, it then relays its visible light sensor (VLS) data to the nearest ground station, either on Guam, in the Pacific, or at a point some 480 km (300 miles) north of Adelaide, Australia. From the ground station, the electronic and telemetry data are relayed by communications satellites to NORAD, the North American Air Defence Command headquarters at Colorado Springs, Colorado.

NORAD is the heart of the West's early warning systems, the place where the bewildering multitude of radar 'blips', electronic data and other information from various sensors on land, or aboard surface ships, submarines and satellites is brought together. The headquarters is underground, built into Cheyenne Mountain, and is designed to withstand a nuclear attack.

This is the nerve centre of the West's defence system. NORAD passes on its information to 40 US agencies and to the US president. In the NORAD command centre, there is a huge illuminated map of North America and the Atlantic which shows the whereabouts of any Soviet missile-carrying submarines within firing range of the United States. These submarines have been detected by sonars (like the AFAR system), by satellites and by other submarines. On the map, the position of each Russian submarine is marked by an illuminated 'X'. The map also shows Soviet aircraft when they follow flight paths near the United States.

A second screen shows the 'decay track' of each worn-out or decayed Soviet defence satellite when it begins to fall back to Earth. The screen also indicates the satellite's expected impact point on Earth, calculated by the centre's computers.

Signals from NORAD's satellites, along with those from other American defence and reconnaissance satellites, can be collected by ground stations on America's Pacific coast, and at New Boston, New Hampshire; Vandenburg Air Force Base, California; Oahu; Hawaii; Kodiak Island, Alaska; the Seychelle Islands, in the Indian Ocean; and on at least six shipboard stations, as well as by the ground stations in Guam and Australia.

The latest American reconnaissance satellites can be equipped with a whole range of sensors able to cover a broad spectrum of signals: infra-red, gamma rays, X-rays and neutron emissions, as well as various types of radio and radar waves.

The use of reconnaissance satellites as part of an early warning system which is not susceptible to electronic countermeasures has made it necessary to devise means of destroying the reconnaissance vehicle itself.

The Russians began experimenting in 1967 with 'hunt and kill' satellites and over the following four years conducted 16 tests in space using 'killer' satellites to search out

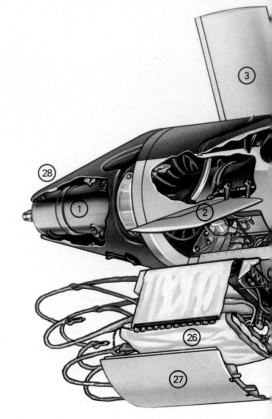

and destroy target satellites.

Soviet tests with killer satellites were subsequently resumed in February 1975 and continued into 1977. At least six tests were carried out during this second phase.

This resumption of 'hunt and kill' tests after a five-year lapse prompted the United States to draw up its own plans to counter the Soviet threat. In November, 1976, Dr Malcolm Currie, director of defence research and engineering in the United States, warned the Russians that development of a 'war-fighting' satellite was 'a dangerous road'. Unfortunately, however, it is a road upon which both sides seem to have embarked.

The US Defence Department began development of satellites that can sound an alarm if they are approached, then set off a second alarm if they come under attack, and finally fire at an enemy satellite if it comes too close. Contracts for the study of all three defence methods were given to manufacturers in 1976 and no doubt by now some form of defensive mechanism is being built into America's early warning and recon-

Richard Cooke

Left The Advanced Early Warning version of the RAF Nimrod uses 'look down' radar to spot low-flying bombers or missiles. It can carry on surveillance and direct fighters even when the enemy is trying to jam its radar.

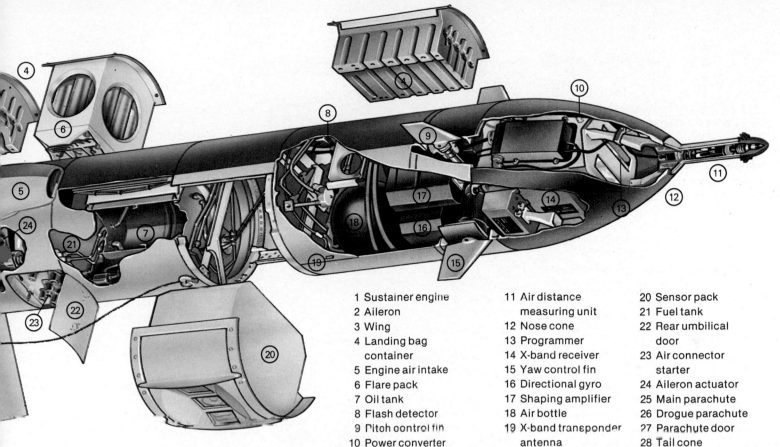

1 Sustainer engine	11 Air distance	20 Sensor pack
2 Aileron	measuring unit	21 Fuel tank
3 Wing	12 Nose cone	22 Rear umbilical
4 Landing bag	13 Programmer	door
container	14 X-band receiver	23 Air connector
5 Engine air intake	15 Yaw control fin	starter
6 Flare pack	16 Directional gyro	24 Aileron actuator
7 Oil tank	17 Shaping amplifier	25 Main parachute
8 Flash detector	18 Air bottle	26 Drogue parachute
9 Pitch control fin	19 X-band transponder	27 Parachute door
10 Power converter	antenna	28 Tail cone

Will Stephens

Above Small pilotless 'drones' such as this 8 ft long Canadian type serve as short-range battlefield early warning systems. Fired from an army truck, they give information on troop and vehicle movement by day or night and return by homing in on a radar beacon.
Right Pilotless electronic countermeasures aircraft, such as this recoverable US Ryan AQM-34V, can confuse enemy early warning systems by jamming and other methods.

USAF/MARS

naissance satellites. The Defence Department also began a study to provide more manoeuvreability in space for its satellites so they can take evasive action if threatened.

The Americans are putting great store by the Airborne Warning and Control System (AWACS) they have developed and which they have been trying to sell the NATO allies. This was designed to meet the need to improve the air-warning situation dramatically in view of a growing Soviet aircraft capability. Each AWACS aircraft is a souped-up Boeing 707 jumbo jet, crammed with various sophisticated electronic devices ranging from several types of radar, sensors, computers and communication equipment. The whole system comprises some 80,000 separate electronic components. At its heart is an IBM computer, believed to be the largest airborne computer. At a cost of over

$120 million, the AWACS Boeing is the most expensive aircraft ever produced for the USAF. The initial order was for 31.

The plane is effectively a command post on a flying platform with special all-altitude surveillance systems enabling it to keep track of air, ground and sea activity over an extremely wide area. A demonstration carried out over the Atlantic showed that the AWACS aircraft could simultaneously keep an electronic watch on more than 600 civil and military aircraft and a large number of ships, despite the efforts of two aircraft which were using a variety of ECM equipment in an attempt to jam its capabilities. A prototype which underwent European trials in 1975 could direct simulated battlefield

operations as it flew over the North Sea, while at the same time keeping track of air traffic as far distant as Moscow. It was able to pick up planes flying below normal radar by means of its special 'look-down' radar system and feed this information direct to the radar screens of ships and command stations operating below it, as well as providing target co-ordinates for ground-based missile batteries. All-in-all, AWAC has proved itself the heart of an integrated command and surveillance system.

Despite the successful European trials of the AWACS, Britain decided not to buy the American version but to adapt a new generation of Nimrod aircraft as its own AWACS system.

The art of camouflage

Camouflage is justifiably described by the military as an art. But where once it was an art that needed to deceive only the naked eye or binoculars, modern practitioners must contend with the likelihood of their cover being 'blown' from a bewildering variety of sources and angles. Photography and television, night observation equipment and thermal imaging from high-flying aircraft and even satellites can now be brought to bear on the detection of an enemy position.

The two world wars established various techniques for camouflaging equipment and installations which have remained largely unaltered to the present day. For instance, the theory behind the dazzle painting of warships is that a submarine is unable to calculate the speed and range of the ship because of its apparent odd shape. This technique was extended to the painting of false bow waves, and attempts were even made at making the ship appear foreshortened by painting her bow and stern a light grey and the rest of the hull a darker colour.

On land, buildings such as power stations and airfields were painted to blend with the local terrain. Nets were strung from the roofs to the ground to conceal any give-away shadows, and trees planted to fringe the buildings. Sometimes a 'wood' was continued over the buildings by constructing a false roof which was then covered with dummy trees. The Germans painted runways and buildings to make them look as if they were bomb-damaged and cratered, so that air reconnaissance photographs would mislead Allied intelligence officers.

Patterns of deception

With large targets, such as intelligence centres and munitions factories, the patterns painted are very large. They do not need to be detailed, because at close range it is impossible to mistake the building. By contrast, a soldier uses camouflage in an attempt to stay concealed as long as possible and to avoid detection even at short ranges. His camouflage often incorporates very small patterns. With his face painted with brown or green stripes, his uniform covered by a mottled smock and his helmet garnished with leaves and twigs, he can easily be mistaken for a bush.

Surprisingly, aircraft can be effectively camouflaged even when they are in flight.

Richard Cooke

Above This soldier uses simple methods to achieve effective camouflage in the snow.
Below Using camouflage netting covered with leaves and twigs, this tank crew have broken up the angular outline of their vehicle so that it is difficult to spot at long range. Such techniques have changed little in the last few hundred years of armed conflict.

(On the ground, provided there has been advance warning of an attack, it is a simple matter to move them out of sight and under cover behind blast walls and nets.) In the air, they can still be effectively hidden if they are painted with colours resembling the local terrain, and fly low. At night bombers can be painted with black undersides so that AA gunners and searchlights have a slightly harder task locating them—though with radar-directed guns the task has become much easier.

Since World War 2, major ground installations have become harder to conceal since aerial photo-reconnaissance, from planes and satellites, has greatly improved. However sophisticated the information, though, the intelligence expert must have some indication that there is something to look for. Unless an important military installation is made to look unusual or surrounded with security, it is likely to be accepted by the local inhabitants and be regarded merely as an underground sewage works or part of the nation's TV network.

Once the intelligence expert has spotted something unusual in a photograph, he will concentrate on that area. So one uncamouflaged vehicle, or even a single soldier, can make the rest of the camouflage valueless under the detailed scrutiny of a skilled photo-interpreter.

An additional technique is to examine matched pairs of pictures through a stereoscopic viewer so that they appear as a single, three-dimensional image. Odd shapes stand out from the jumbled texture of scrub or

Spectrum

Left Looking remarkably like a tree-stump, this sniper has made himself difficult to spot even though he is in open country.

Below Soldiers on exercises in Norway camouflage a tank by covering it with white netting. White sheets are sometimes used.

Bottom The outline of this tank has been broken up by painting it with black and white patches, so that it merges into the background. This treatment also makes it difficult to determine its range and speed.

woodland and even single figures show up on low-level photographs.

Flying over enemy territory can be a dangerous practice, and though the loss of a reconnaissance aircraft may indicate the presence of an enemy, it is an unacceptably high price to pay. *Remotely piloted vehicles,* or *drones,* are used to fly over an area, photograph it and return to a collection point. Even more sophisticated machines can be used to hover over the target with TV cameras giving 'real time' information. TV can even be used at night for air or ground surveillance using 'low light' equipment, which gives the enemy no indication that he is being observed.

Night into day

Developments in electronics have changed night into day on the modern battlefield. The first night-vision devices were *infra-red* (IR) sights. During World War 2, both the Allies and the Germans developed equipment that could be fitted to a rifle—the Germans used a device known as 'Vampire', while the US Army developed 'Sniperscope'.

Both worked on the principle of an IR filter over a white light source. The filter converted the beam of ordinary white light into a beam invisible to the naked eye. Only when an observer had a suitable filter fitted to his goggles or to his telescopic sight did the picture resolve itself. The early equipment was bulky, normally consisting of a power supply in a haversack, a source of illumination, and a sight with a large lens to gather all the available light from the scene. IR searchlights and driving lights were fitted to vehicles and tanks, and the crews issued with special goggles.

When it was first introduced, IR equipment appeared to be a formidable weapon, but it has subsequently proved relatively simple to evade. When an enemy IR light is trained on a patrol, the men can be alerted by a passive warning device which operates a buzzer in an earpiece. Alternatively, they may discover that they are being observed through an IR viewer because, even though the beam of light it produces is normally invisible, it can, of course, be detected by a member of the patrol equipped with an IR sight. Through such a sight, an enemy soldier using an IR viewer becomes as obvious as a man holding a torch.

Another drawback with IR and other night-viewing aids is that they give the operator 'tunnel vision'. He observes the terrain down a narrow band, and when relying on such devices to navigate a vehicle, he has the confusing experience of not knowing if a fold in the ground is a ditch, or vice versa.

Infra-red viewers are now used mainly by Warsaw Pact armies only, but there are other IR devices which are in wider service. *Infra-red photography* is extremely useful in aerial photo-reconnaissance. It works on the principle that vegetation reflects green light and infra-red light and absorbs all other wavelengths, principally because it contains the green colouring matter, chlorophyll. This is not detectable in pictures taken on ordinary film, but does show up if *false colour film* is used, loaded into a special camera. Any living vegetation appears red or pink, while men and equipment stand out as a contrasting blue or grey colour.

Escaping infra-red

To avoid detection by infra-red photography, troops are clothed in uniforms which have been treated with chlorophyll. Chlorophyll is also added to the paint used on vehicles and to the camouflage netting used to break up the angular shape of trucks and tanks. Although the treatment remains effective with the nets and paint, laundering of combat clothing eventually washes out the chlorophyll.

Another useful device is the *passive night sensor.* It uses visible and near infra-red wavelengths of light to produce an image on a screen or a gunsight. The systrem used by the US infantry is known as Starlightscope. Passive night sensors are not only lighter and more compact than IR viewers, but they also

Right Mimicking the natural camouflage of a carpet shark, this US Navy research submarine reproduces the patchy surface of the fish and its mouth lobes, which help break up its shape when it is viewed against the background of the sea floor.

Crown Copyright (MOD/RAF/MARS)

Left Hiding this RAF Harrier jump-jet deep within a wood not only makes the plane harder to spot by conventional photography, but also helps defeat the prying eye of the infra-red camera, since the trees obscure its 'heat picture'.
Below A Harrier hides beneath a roof of camouflage netting.

Richard Cooke

Photri

Photri

Richard Cooke

Above The camouflage of these two RAF Harriers is so effective that it may take you several seconds to find them. The vertical take-off Harrier does not need a runway—it can use confined spaces such as fields, forest clearings or roads, where it can be effectively hidden. Its small size and extreme manoeuvreability make it a difficult target to pinpoint.

Left The dark and light bands on these Harriers help to break up their shape when seen from above as they fly over a background of snow-covered forests.

have a greater range. The development of these new aids to nightsighting has been concentrated on three areas: gunsights, observation devices and navigational aids.

An example of a passive night-sensor gunsight is the HV5X80AT developed in the Netherlands. It will fit onto the NATO self-loading rifle and has a ×5 magnification, allowing a soldier to identify men up to 500 m (1,600 ft) away and vehicles up to a distance of 1,500 m (nearly a mile).

A typical observation device is mounted on a tripod for use in a static position. It weighs about 12.8 kg (28 lbs), and while the electronics are identical to that of the smaller device, it has a bigger lens which not only allows it to gather more light but gives a longer range. With a ×7 magnification it

can detect infantry at 1,000 m (3,200 ft) and tanks at 3,000 m (almost 2 miles).

Detection ranges with all night-viewing equipment depend not only on the conditions at night—they cannot 'see' so far if it is very overcast or raining—but also on the background against which the soldier is viewing the object, and the amount of training he has had in using the equipment.

Goggles and armoured fighting vehicle sights make it possible for helicopter crews and tank drivers to operate at night. Compared with the other passive night-sensor devices, this equipment produces less eye-strain because both eyes are used.

Like infra-red viewers, passive night sights have made it much more difficult for men and machines to rely on the cover of darkness

865

for protection. However, like the IR devices, they have their limitations. They are defeated by fog, smoke and heavy rain, and the observer may be dazzled by flares, street lights and fires.

The most effective countermeasure against detection by passive night sights is for the soldiers to use the terrain as if they were operating in broad daylight, avoiding the temptation of taking short cuts by crossing open ground, and by limiting their movements as much as possible.

Just as the armed forces have come to terms with the problems of active infra-red and passive night sights, technology has presented them with a new and more daunting challenge.

Devices called *thermal imagers,* which make use of *far infra-red* wavelengths, can measure the temperature of an object and build up a 'heat picture' of it on a screen. For example, a light truck will show up as a warm engine and transmission, with cooler bodywork —and, as the crew emerges, the thermal imager reveals their warm faces and hands contrasting with their cooler clothing and boots. Attempts to protect a vehicle from detection produce a heat pattern which is obvious due to its lack of heat texture. Dubbed a 'black hole', it is as eye-catching to a trained observer as is an uncamouflaged vehicle.

Thermal imagers can even be used to identify aircraft or vehicles after they have left the target site, by revealing the heat pattern which they leave behind them. The ground below the object will be cooler (since it has been in shadow) and it will show the clear outline of an aircraft or tank. If an aircraft has recently taken off, it will leave a mixture of residual heat patterns—cool ground and a warm area where the engines have been running. A fuelled aircraft leaves a different pattern from that of one with empty tanks.

Thermal imaging is less of a challenge to the infantryman, but it is a real threat to units with large concentrations of vehicles. Engineer bridging units or tank squadrons use large vehicles which not only have a distinctive shape, but also move *en masse.*

The best way of avoiding detection by a thermal imager is for the vehicles to move into deep cover, such as into a wood with dense undergrowth and a high canopy of trees. Alternatively, they may enter a village

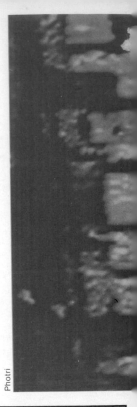

Advances in infra-red techniques make it harder to escape detection. Aerial infra-red photograph *(right)* shows houses, even at night, as red areas, representing heat lost through their roofs.
Bottom left Examining high-altitude infra-red pictures.
Below This infra-red picture, taken with a thermal imager, shows the images of Phantom and Canberra jets that just left the airfield.

or town. The latter move is less acceptable in peacetime exercises, but during wartime, built-up areas offer a multitude of advantages. Not only does such an area have a variety of thermal textures; it also contains garages, barns or other buildings that can be used to hide major heat sources such as generators. Unlike a remote wood, villages contain people—even in war—so movement will be less likely to attract attention.

Foiling radar

Compared with the thermal imager, radar has a long history of use as a formidable tool for penetrating camouflage and other forms of deception. Various ways of responding to this challenge have been devised. To evade detection by radar, aircraft carry an Electronic Countermeasures (ECM) pod containing sophisticated equipment to scan radar frequencies and jam them if necessary, after giving the pilot warning that the radar beam has locked onto his aircraft.

A much older but still effective countermeasure is to scatter a cloud of metal foil, known as 'chaff' or 'window', into the air to produce an echo resembling an aircraft on the enemy aircraft radar screen. Ships as well as aircraft use chaff.

On the ground, battlefield radar is used for locating the position of artillery. Some devices are small enough to be carried by the individual soldier. He can use his radar to discover the range, bearing and elevation of the enemy, but it can only 'see' in straight lines and has the additional disadvantage of being susceptible to jamming.

An example of a portable radar pack is the French Rasura (DR-PT-2A) battlefield surveillance radar. This equipment can detect vehicles at 10 km (6 miles), a walking man at 7 km (4.4 miles) and a crawling man at 2 km (1.3 miles). The operator can scan the terrain either manually or automatically in 4° arcs, and receives visual or audible warning of a radar contact.

Faced with such a challenge, a platoon of soldiers or a tank unit must rely on using hills, hollows in the ground and so on as a screen to deflect the radar signals.

The effectiveness of the radar ultimately depends on the soldier using it. He is quite

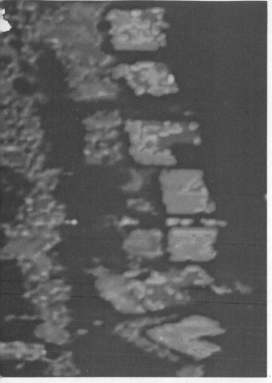

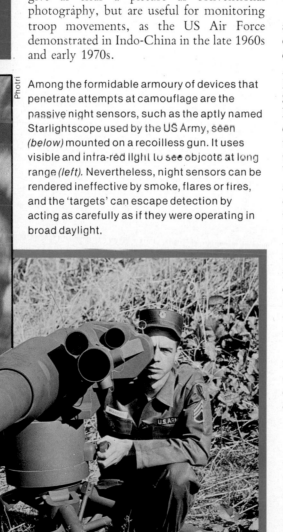

likely to ignore a brief signal resulting from the rapid movements of a patrol breaking cover, but lengthier manoeuvres are easier for him to spot.

Radar can also be used against ground targets by reconnaissance aircraft. Sideways Looking Reconnaissance Radar (SLAR) can be used to look into enemy territory, or cover the side of a mountain range. It operates in two modes: a mapping picture or Moving Target Indication (MTI) picture.

The mapping picture has various peacetime uses, but MTI is a strictly military device. It gives a printout of the terrain on which each moving object is revealed as a white dot. Mapping radar and MTI do not give as clear a picture as conventional photography, but are useful for monitoring troop movements, as the US Air Force demonstrated in Indo-China in the late 1960s and early 1970s.

Among the formidable armoury of devices that penetrate attempts at camouflage are the passive night sensors, such as the aptly named Starlightscope used by the US Army, seen *(below)* mounted on a recoilless gun. It uses visible and infra-red light to see objects at long range *(left)*. Nevertheless, night sensors can be rendered ineffective by smoke, flares or fires, and the 'targets' can escape detection by acting as carefully as if they were operating in broad daylight.

One of the developments of the Vietnam War was the Unattended Ground Sensor, or UGS. Small seismic intruder alarms were already in use by many armies for guarding against ambush, but the US Air Force and Army produced a whole new range of equipment. The PSID, Patrol Seismic Intrusion Detector, and smaller version known as MINISID, warned of an enemy's approach by means of a *geophone*, which detects vibrations in the ground. They could be fitted with other sensors, such as MAGID, for detecting metal, and AAU, Add-on Audio Unit, which was activated by two or three seismic pulses and recorded and transmitted speech. It could be fitted with a switch to operate only at night.

The whole system was linked to a monitor and event recorder. With this equipment an enemy's movements could be monitored in comfort from a command post. During US Air Force operations against the Ho Chi Minh Trail, UGS units known as AD-SIDs—Air Delivered Seismic Intrusion Devices—were dropped from the air. They consisted of a spiked dart-like device with a tail that was made to resemble a jungle plant, to help them escape detection by the enemy. When they hit the ground the spike stuck in and the UGS began to transmit. There was one report of a recorded conversation between two Viet Cong who had discovered the UGS and were discussing what to do with it. If they had decided to tamper with the sensor, they would have activated a built-in device which would have burnt out the circuits using the power from its battery. At the end of its transmitting life, the ADSID would self-destruct, using the residual power from the battery.

Methods of detection on the battlefield may have changed greatly since the early days of modern warfare, but many of the countermeasures are still based on the principles used in the past.

Life or death

Even satellite cameras can be rendered useless by thick cloud. If a soldier takes care when moving himself or his equipment he can avoid alerting his counterpart on the other side who evaluates the information he receives from his array of surveillance equipment. The greatest threat results from an observer with an alert eye linked to a quick brain. His second glance at the radar screen or infra-red image in front of him, for instance, may literally mean the difference between life and death for the man within the area under observation.

The unexploded bomb

The problems of bomb disposal did not end with World War 2. If anything, they have increased. Thousands of anti-aircraft shells are still being discovered, as builders dig down to uncover them. Mines, demolition charges, stores of bombs and explosives hidden during the war and then forgotten emerge regularly. Decades after the war, bomb disposal squads in major European cities are dealing with over a thousand explosive items every year. An international rise in terrorism has escalated the bomb disposal task still further.

Cold nerve is still the basic requirement for bomb disposal men. But the growing threat of terrorist bombs has brought science and technology into play, and today the bomber is faced with a vast array of counter-techniques and devices. Many of these are not publicly revealed since to do so would give the bomber an advantage. But even the few techniques made public indicate the weight of scientific effort in this field.

Normally, the first task of bomb disposal teams involved with World War 2 relics is to dig for the bomb, since the terminal velocity of the bomb was generally sufficient to drive it deep into the soil. Often, the digging takes the disposal men below the water table, and pumps and 'dewatering kits' have to be brought in to control the water.

The World War 2 German bomb fuse is electrical—as it left the aircraft a charge of electricity was sent into the fuse, and stored in a capacitor. When the bomb landed, the impact would close a switch, allowing the current to flow into a detonator and fire the bomb. Occasionally, the bombs failed to work because switches failed to operate properly. But disturbing the bomb—turning it to get at the fuse, or unscrewing the fuse itself, could cause the switches to close. So the first piece of technical equipment to enter the bomb disposal armoury was an electric lead which clamped on the fuse and safely discharged the fuse capacitor—making the fuse completely inert, and safe to remove.

Clockwork complications

A slight complication was the fact that impact with the ground frequently stopped the bomb's delay clock. The clock might then resume action of its own accord, or might be jolted into action by the activity of the disposal squad.

The solution to this was to listen to the bomb with a stethoscope as soon as its surface was exposed. This was done by one member of the squad, while the rest worked to disarm it. If the clock activated, instant evacuation of the site took place, until the bomb stopped ticking, or exploded. But if all seemed safe, the fuse could be attacked by various methods. For example, an extremely powerful magnet, clamped on to the bomb, would stop most clocks by jamming the mechanism together. Carbon dioxide, packed around the bomb, could freeze the mechanism solid. Or a hole could be drilled in the fuse, and a quick-setting plastic com-

Above An explosives detector such as this distinguishes between explosive and non-explosive vapours.
Below The electronic stethoscope detects activated clockwork fuse mechanisms.

pound injected to solidify around the clock mechanism.

As the bomb disposal men became more proficient, and rendered more bombs harmless, this battle of wits turned into a straight duel between the bomb designers on the one side and the bomb disposers on the other. The next step in the race was the incorporation of booby traps to prevent the fuse being removed without activating it.

Some booby traps were simple mechanical triggers under the fuse. The trigger would then fire a hidden detonator as soon as the fuse was removed. Others were more sophisticated devices, such as photo-electric cells which reacted to the admission of light when the fuse was withdrawn. Photo-cells were sometimes discovered by luck when the device failed to operate as intended.

But each new development meant that a new method of disabling the bomb had to be carefully worked out. In addition to aircraft bombs there were other explosive devices which had to be dealt with—among them washed up sea mines and torpedoes, land mines in combat zones, booby-traps and demolition charges in captured towns. All these gave scope for the bomber's ingenuity and had to be countered by the disposal men's astuteness.

Today, the bomb disposal business has entered a new phase as the terrorist and urban guerrilla have appeared on the scene. Indeed, so many and varied are the devices that have appeared that a new term has been coined—'explosive ordnance disposal'. The battle is now between the terrorist and the authorities, with the highly trained explosive ordnance disposal (EOD) man playing a key role. It is not unusual for a bomb to be planted as bait, and the area surrounded with mines and perhaps snipers, with the intention of putting the EOD disposal team out of action. The bomb itself may be fitted with devices to prevent it being moved or handled, and it may have sensitive electrical contact switches which react to vibration, and will fire if the bomb is moved or tipped.

At first, the amateur terrorist bombers made many mistakes. They assembled bombs incorrectly so the devices either failed to go off or went off too soon. More than one mysterious explosion in a car or building was due to the premature functioning of a badly-designed bomb. Devices which failed to work allowed the disposal squads to analyse

Daily Telegraph Colour Library

SAS Group of Companies

Above An EOD (explosive ordnance disposal) suit weighs over 22 kg (48 lb). It can be fitted with a cooling system, enabling a bomb disposal expert to work in comfort.
Left Tracker dogs are trained to sniff out hidden explosives, a feature of recent troubles in Northern Ireland.
Far left Many modern bomb disposal methods were originated during World War 2.

their intended method of operation, so that the squads gradually became familiar with these unorthodox weapons and some of the more common techniques of the bombers.

But, as had happened during World War 2, as soon as the disposal squads began to gain the upper hand, the battle changed its nature. While retaining its political character, the battle became a personal duel between the opposing technicians. Bombs were planted not simply for the damage they could do, but with the intention of killing the men who made them safe.

As terrorists in the UK started to incorporate anti-lift, anti-open, anti-disturbance switches into the bombs it became obvious that a remote approach to disposal had to be devised and developed—'Wheelbarrow' was one answer.

Remote control

'Wheelbarrow' is the name of a small, remotely-controlled tracked, vehicle which can be adapted to carry a number of devices. It has an articulated arm carrying a closed-circuit television camera and a floodlight. This allows the device to be driven up to the bomb, and then to transmit a picture back to the operator. With the picture displayed on a screen, several experts can look at the bomb at the same time, compare opinions and debate the very best way of dealing with the device. If necessary, Wheelbarrow can lift the device, turn it about so that the camera can examine it more closely, or even move it to some other location. If any of this movement sets off the bomb, then all that is damaged is the machine, and not the highly trained specialist operating it.

Alternatively, a 'disrupter' can be carried on the Wheelbarrow arm and fired at the bomb. It can take the form of a special shotgun charge, or simply a 'slug' of water discharged at high velocity by a small explosive shot. Either of these projectiles, moving at high speed, will rip into the bomb and sever electrical connections. They smash switches, and break up the circuits and mechanisms so quickly that the bomb has no time to function. Other tasks within the machine's scope include opening cupboards and doors—and even nailing them open if required—breaking windows in order to examine the interior of a car or building, cutting open suspect packages, and attaching a hook to tow away suspected car bombs.

At present, the British Army is testing a much improved model of Wheelbarrow known as 'Marauder'. Marauder has longer articulated tracks to give better travel over rough surfaces. The machine can even climb up and down stairs.

During a three-year period ending in December 1975, Wheelbarrow was used on over 5,000 bomb incidents in the UK. By the end of 1978 this figure had been doubled, and if one takes into account the bomb incidents in 32 other countries where Wheelbarrow is operative the figure could well be in excess of 13,000. In some cases Wheelbarrow has been totally destroyed. For security reasons, the exact number destroyed cannot be stated—but when one considers the damage done to over 30 Wheelbarrows,

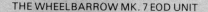

THE WHEELBARROW MK. 7 EOD UNIT

1 Operator's helmet with shatterproof visor
2 Remote control handset
3 100 m (300 ft) control cable
4 Caterpillar tracks for rough terrain
5 Hydraulic tilting mechanism
6 100 m (300 ft) detonating cable
7 Closed-circuit television camera
8 Remote-controlled trigger mechanism
9 Multi-shot automatic rifle
10 Aiming sight, lined up with camera lens

Over 300 Wheelbarrow explosive ordnance vehicles are in service throughout the world. The Mk-7 has been developed during extensive service in Northern Ireland and over 30 other countries. The machine can be adapted to deliver X-ray equipment and recover the exposed film. It can transport foam generators, and also serves as a remote-controlled weapon if necessary. Towing hooks and a window breaker can also be attached.

Daily Telegraph Colour Library

Wheelbarrow in action. *Above left* Closed-circuit TV camera on the extension boom reveals the nature of the bomb. An appropriate detonating charge is attached.
Above centre and right A controlled explosion.

it can be safely claimed that at least 30 EOD men's lives have been saved.

Some bombs, however, cannot be dealt with by remote control, or by removal in a safe container. Sooner or later, somebody has to approach the device and dismantle it. In the past, this was simply a matter of the operator emptying his pockets of any metal (since some bombs used magnetism to trigger them) and walking up to the bomb. Nowadays, he is more likely to be protected by an 'EOD Suit'. Made of ballistic nylon it is designed to offer as much protection as

Sarson/Bryan

possible, should the bomb detonate, and still allow the operator some measure of mobility. The suit is capable of stopping, or slowing down, the fragments of metal from the bomb, and will also deflect most of the blast away from the wearer's body. His head is protected by a helmet with shatter-proof visor, and sonic valves protect the wearer's ears from the effects of blast. The suit also contains a compact communications system so that the EOD man can report his findings, and request advice or special equipment. Boots and armoured leggings protect the EOD man's legs, and a steel plate apron protects his abdomen.

As a rule, the first step in examining a suspected bomb is to X-ray it. This reveals the internal mechanism, the arrangement of component parts, and the layout of electrical wiring. The first X-ray equipment used for this task was extremely cumbersome, requiring a truck to move it into position. But the miniaturization made possible by modern electronics has led to portable equipment.

The operator can also use an explosives detector to check that the device really is a bomb, and not a time-wasting hoax. The explosives detector 'sniffs' the air around the bomb, and detects traces of explosive vapour as low as one part in several million parts of air. Having determined that the device really is a bomb, and having 'seen inside' (as far as the X-ray equipment permits), the operator now has to decide how to disarm the bomb.

Pressure switches

In the simplest case he can slice open the outer container and cut the wires leading to the bomb detonator—the key component. But this remedy can only be used comparatively rarely. The designer of the bomb will, as a rule, have taken this remedy into consideration, and incorporated some form of trap to prevent it. Such traps could take the form of pressure- and attitude-sensitive switches which would close the firing circuit at the least disturbance.

If the bomb relies upon an electric battery to provide power for the detonator (most bombs incorporate a battery), then freezing the bomb will render the battery inert. The bomb can then be safely dismantled. Alternatively, a disruptor aimed at some sensitive part of the bomb discovered by X-ray examination can be fired at short range. Or, the container can be pumped full of plastic foam, short-circuiting the electric circuits and rendering any mechanical devices inert.

All these, and several more techniques which are not made public, are available to the EOD disposal team.

One certain thing is that no disposal technique is ever forgotten or discarded. As bombs get more complex, there is always the chance that an astute designer may take a step backwards to some old technique in the hope of catching the EOD man out. It was recently admitted, by a British Army spokesman, that freezing bombs was a technique which has not been used for several years. But, he stressed, it was still taught and practised in case it should be required for some particular problem.

The battle will continue. So long as some one is prepared to construct and plant a bomb, there will be an explosive ordnance disposal operator to take on the task of rendering it safe. And he will have an ever-increasing array of techniques available to help him do so.

Two more applications of the versatile Wheelbarrow. The handling grab *(left)* is suitable for moving improvised explosive devices (IEDs). The telescopic boom *(right)* enables the closed-circuit camera to examine the top of a petrol tanker.

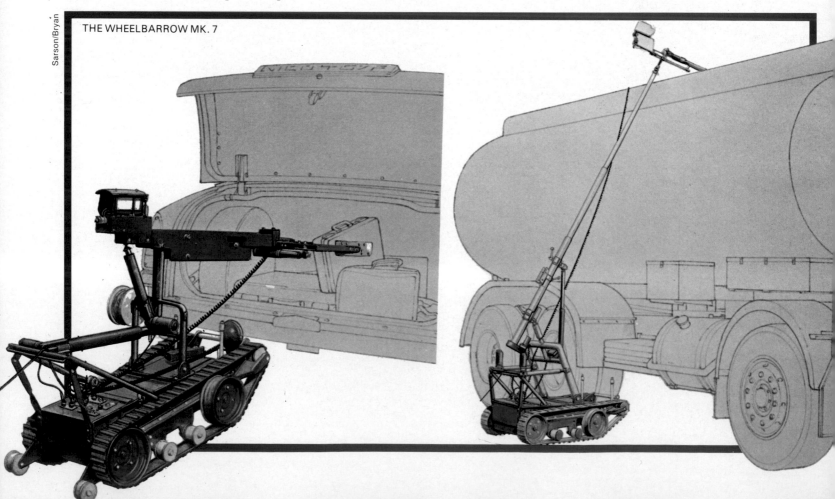

Sarson/Bryan

THE WHEELBARROW MK. 7

Interrogation techniques

Interrogation, both civil and military, has become a complex and ruthless contest between captive and captor. Faced with up to six interrogators backed by electronic research and recording equipment, the victim is in an unequal struggle. His body may be left unscarred. But his mind—the prime target—will be picked clean.

As a technology, the machinery involved in extracting information from an unwilling captive reached its peak in the torture chambers of Europe's dark past. But the rack and thumbscrew were less effective than today's techniques.

Two-edged weapon

However, pain in any form is a two-edged weapon in information extraction. Its infliction may bring quick results—but a victim pushed to the extremes of pain may babble anything he feels his questioners wish to hear. Torture can also harden a few rare individuals. They may resist until death—or prove poor exhibits at a subsequent trial. Interrogators frequently question the wrong man, and inflicting torture on an innocent person can help the opposition acquire new recruits or give enemies excellent material for propaganda.

In information extraction speed remains a priority. Picked up on the eve of a suspected crime, a gang member must be pursuaded to tell what he knows before his colleagues can disperse. Members of illicit or resistance organizations must reveal their colleagues promptly—for most underground movements have a system that enables active members to change hiding places within 24 hours of one of their number being captured. A city curfew is an advantage to the authorities for that reason. The hours of darkness give them extra time to break the suspect and thus the next link in the chain of command that binds the unit together.

Isolating a suspect is the first line of attack among today's practitioners of organized interrogation. The prisoner must be prised away, mentally, from the group whose ideals claim his loyalty. Even his captors will intrude as little as possible, at first, on the prisoner's solitude. On a battlefield, he will be gagged and blindfolded, led by a rope attached to handcuffs, and prodded forward by a boot or rifle rather than by commands. He will not know if he is alone or with a group of fellow prisoners. With little to orientate him to the world outside, fears for his own future will begin to take precedence.

Sensory deprivation

In civil custody, the same isolation is used as a tool throughout many Western countries. Police forces can deny access to solicitors or friends on the grounds that information may be passed to the suspect's

Stop and search: October 1979 at Les Fourons, a French-speaking community within Flemish Belgium, 30 were wounded and 150 arrested. Only the intervention of the police and a search for offensive weapons prevented the rightists, led by Erikson, from creating even more bloodshed than actually took place.

PNS/GAMMA/Frank Spooner Pictures

Mike Sheil/Daily Telegraph Colour Library

Gilbert Uzan/GAMMA/Frank Spooner Pictures

Jeremy Gower

associates in crime. Techniques of sensory deprivation can aid the process of isolation.

Removing all sensation

Methods have been perfected in many police and army forces. Hooded, or crowned by an upturned bucket, the simple lack of light and vision can swiftly break a prisoner's grasp on normal realities. The use of 'white noise'—a recording of sounds across the spectrum not unlike the hiss of escaping steam—blots out auditory contact with the world. Allegedly, drugs were used by Syrian captors of Israeli soldiers during a recent conflict. The chemicals had the temporary effect of removing all sensation of sight, smell, hearing and touch—but left the brain active.

Such techniques, the equivalent of a

Above left A British soldier checks shoppers for concealed weapons in Northern Ireland.
Above right 25 January 1978, after the kidnapping of one of Europe's most powerful tycoons, police make identity checks in an effort to find the kidnappers.

lightless medieval dungeon, can be modified to speed the process of disorientation. Time can be stretched by alternating periods of light and darkness irregularly. Meals can be produced at odd intervals so that a prisoner loses count of the days of his captivity. Even before a formal interrogation begins, he has lost contact with important areas of reality.

Confusion and uncertainty are increased if his captors treat him with absolute 'correctness'. Many experts now regard such an ap-

proach as more effective than abuse or hostility to a suspect—which gives him a focus for his aggression and a recognizable opponent. The captors, instead, will reveal no emotion. They will not talk amongst themselves. They will restrict conversation with the prisoner to monosyllabic commands and orders.

Since Man is a social animal, the surge of relief encountered by a prisoner when he is eventually led into a room and confronted by an apparently friendly interrogator can overwhelm his determination to keep silent.

Alternating severity and amiability has long been a tool in the interrogator's basic approach. The 'soft man, hard man' routine remains crudely effective. After bouts of abuse or violence, the hard man is replaced

The interrogator will ask general questions which both parties know are safe to answer. By answering them, the prisoner has made the first move towards cooperation.

The prisoner will be pushed a long way towards cooperation by a successful interrogator's skill at implying that he already knows all the answers and only requires a few simple details clarified—to save both his and the captive's time. If the interrogator knows names of the captive's close associates, he will often use them casually in conversations and suggest that they were most helpful in earlier interviews.

Persuading a man to break faith with his group—whether a crime gang or a national army—is the most subtle part of the interrogator's art. He can attempt to convince the prisoner that his group has rejected him, or that at least they too have cooperated, thus exonerating him from silence. At his most effective, the interrogator uses a mixture of suggestion and deprivation to persuade the captive to identify with the new group that the interrogator represents.

Brainwashing, the term given broadly to the phenomenon of switched allegiance, is the crude label for a complex process. American servicemen, captured in Korea, astounded the West by espousing their captors' cause. Soviet party bosses appeared at show trials to denounce themselves and their colleagues as conspirators. More recently, hostages in bank raids have emerged expressing sympathy for their tormentors. The reasons for such 'conversions' are comparable to some dramatic religious conversions. All involve isolating the new recruit in

Above Computer used by the police to call up and store information. The computer stores fingerprint indexes, criminal records, and data banks of creditworthiness, enabling the interrogators to build up a detailed picture of the suspect and thus guide them in their line of questioning. It also means that the smallest scrap of information can lead to a wealth of knowledge which daunts the suspect and diminishes his chances of successfully lying to his interrogators.

by a more civilized interrogator who may apologize for his colleague and ply the captor with cigarettes and drink. Despite awareness of the game he has been caught up in, the prisoner finds it extremely difficult not to relax and lowers his guard.

The most effective interrogations establish a friendly relationship by opening a conversation, rather than by conducting a question-and-answer session. From there on, each exchange will build upon the rapport achieved.

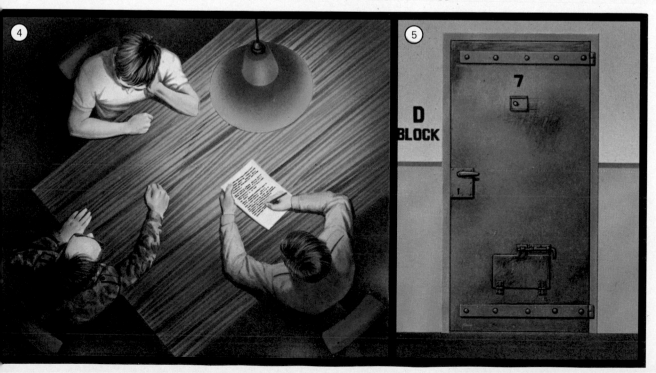

Left The five 's':
1 Stop and search: check for weapons under clothing.
2 Segregation: helps to break down the suspect's will and allows statements to be checked.
3 Silencing: the bag disorientates and isolates a subject.
4 Speed of interrogation: initial 'safe' questions throw a suspect off guard.
5 Safeguard: locked doors bar escape and crush the suspect's will.

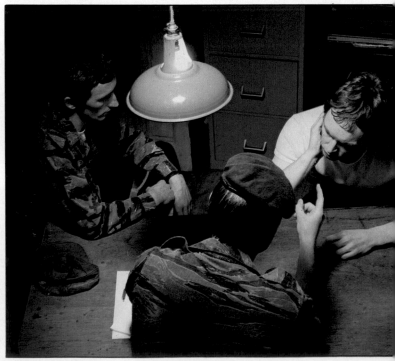

Right A harrowing scene from the John Schlesinger film *Marathon Man*. The hero is tortured by his captors in an attempt to make him reveal information to them.

a highly charged, emotional atmosphere—away from normal influence and under constant pressure from the leaders of the group.

Computer interrogation

Though the subject of an interrogation is isolated from the world outside, his questioners are not. In today's electronic world, they can move swiftly to pursue scraps of information revealed by a suspect and then confront him with a daunting display of knowledge. Police computers, for example, may have access to an entire nation's car ownership data. Fingerprint indexes, criminal records and data banks of creditworthiness can now be sifted at high speed, enabling interrogators to build a background of information about their suspect and his activities and thus diminish his chances of successful lying.

Other widely publicized technical methods of testing the veracity of a subject—truth drugs and lie detectors—have a mixed success rate. Various drugs—forms of sedation that reduce inhibitions—increase the subject's suggestibility but they may also put him in a mood to invent tales in order to please his questioner.

Lie detectors, or polygraphs, are more reliable and have been used as evidence in legal proceedings in the United States. These devices monitor the heart rate, breathing and perspiration of the prisoner under interrogation. Questions that cause special anxiety induce increased production of adrenalin and consequent disturbance to normal body rhythms. However, the stress and tension of being wired up to a lie detector may be sufficient to provoke an abnormal response. And the client's knowledge that some questions are crucial, even if truthfully answered, may cause him to react strongly. On the other hand, people suffering from certain pathological mental states are able to lie with equanimity about crimes they have committed or taken part in.

Without doubt, the most potentially effective aid to the interrogator—particularly in police work—is the tape or video recording. 'Confessions' recorded only in writing as statements signed by a suspect can be challenged in court by defence counsel, who may claim they were extracted under duress

876

or they were fabricated by the questioner.

In theory, a video recording should demonstrate whether or nor an interrogation has been fairly conducted. But a nervous or mentally feeble suspect can still be pursuaded to confess to crimes of which he or she is innocent. In such cases, a video recording could give spurious authenticity to a confession apparently given voluntarily but in reality the result of fatigue and stress.

In many legal systems, the 'right to silence' means that a civil suspect who refuses to answer any questions during interrogation by police cannot, as a result, be asssumed in court to have implicated himself just because of his silence. In practice, few people are psychologically capable of resisting pressures to answer questions.

Tape recordings of a confession can be used as evidence against a suspect. But this is a system open to abuse. Sound tapes can be easily edited to change or delete words and thereby alter or even reverse the original sense of a suspect's statement.

One national agency with a particular concern for aspects of police custody believes that 75 per cent of cases known to it would not have reached a guilty verdict in court if the suspect had refused to make any statement to the police during interrogation. In bringing criminals to book, therefore, interrogation remains a powerful tool of the state. The risk that, in the process, the innocent will also be indicted is one that remains a matter of constant vigilance throughout the free societies of the world.

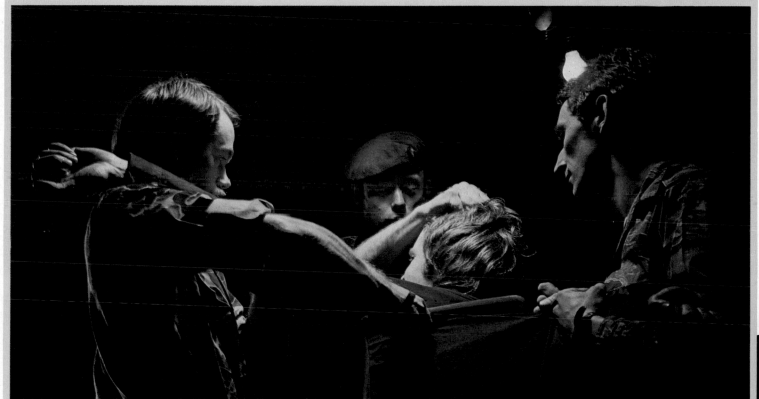

Theo Bergstrom/M.C.

One of the oldest techniques of conducting an interrogation remains the most effective. The 'hard man, soft man' routine can wear down most people's will to resist parting with information. Alternately, the interrogators cajole and then assault the victim either verbally or physically.
Far left The subject's guard is lowered, usually after a period of isolation, by an amicable approach. Cigarettes and refreshment are given and apologies made for the more brutal behaviour of the interrogator's colleagues. Then the pressure steps up as the 'hard man' returns *(left),* culminating in violence *(above).* Lack of sleep strongly reduces the will to resist. *Right* Wartime agent Violette Szabo (played by Virginia McKenna) is shown here in a still from the film *Carve her Name with Pride.*

Kobal Collection

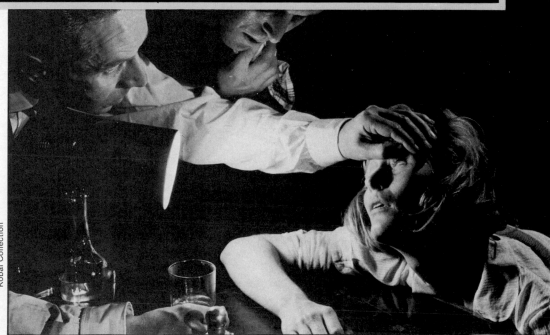

Riot equipment

Shooting demonstrators is a speedy way of ending a riot. But to do so is normally unacceptable, creates martyrs, and will inflame an angry gathering. A more civilized answer to civil disturbance is to use riot equipment to control, rather than crush, a crowd.

Since the 1960s and 1970s many governments have had to confront the problem of sustained riots in city areas. However, the army, National Guard, and police can no longer consider the sort of tactics that were developed by the British Army during their long history of dealing with colonial troubles. Among these was the calculated use of guns against a rioting march. The crowd would be confronted by a line of soldiers with banners and loud speakers warning them that if they crossed a white tape on the ground, they would be shot. A sniper was in position, and when the first man crossed the tape he was shot.

But using fire arms in this way escalates the level of violence. Although a single shot fired at one man may seem discriminate, it can pass through him and wound an innocent member of the public—and since the round has been distorted, the wounds from ricochets are more severe than those from a medium range direct hit.

A modern police force or army in the

Above The black clad French CRS face taunts and stones from Parisian demonstrators.
Right French anti-nuclear demonstration against the Super Phoenix reactor at Creys-Malville, in August 1977.
Far right Israeli police use pepper gas against rock throwing Arab protesters during a demonstration in Jerusalem.

Rex Features

Western world must bear in mind that riots and demonstrations are covered by TV and press reporters. In this respect, authorities in the West are at a disadvantage compared to the Soviet Union—they cannot ban the press or arrest them. The Soviets, however, having denied reporters access to the victims, or the scenes of riots, have put down insurrections with a startling degree of violence.

High pressure

One of the first riot control techniques to be developed as an alternative to the use of fire arms was the use of high pressure water. For example, Civil Rights marchers in the southern states of the USA were drenched by the local fire service. High pressure hoses can be used to knock a man over, and bowl him along the street. However, firemen have no protection if the crowd decides to retaliate by throwing rocks. The Water Cannon—an armoured truck, with high pressure hoses armed from turrets—has been used in European riots. The water can be used with vegetable dye, which turns rioters green or blue, thus making subsequent follow up and arrest easier for the security forces. CS smoke, a powerful irritant, can be added to the water to make it additionally effective. However, water cannons have a major disadvantage—they can run out of water before the riot ends. They also tend to be regarded as an extreme measure, and may indicate to the crowd that the police have lost control.

The British army came to Northern Ireland in 1969 with three weapons in their riot control armoury—batons, shields and CS smoke. The shields were metal (in Cyprus the British Army had used wicker shields, but they were not strong enough to withstand a steady battering of bricks).

From police experience of the Hong Kong riots in the 1960s, came the idea of the baton round. In the Far East this had taken the form of a wooden rod fired from a riot gun. It hit the ground just short of the rioters, and then bounced into their ranks, at about knee height. In Northern Ireland, and subsequently in other countries, it has been refined. The rubber bullet version introduced in 1970, is 152 mm (nearly 6 inches) long, 38 mm (1½ inches) in diameter, and weights 141 g (5 ozs). It is superior to the wooden round since it does not splinter. Later, an improved plastic round (L3A1/L5A2) was introduced which is harder.

Baton rounds are fired from weapons like the Schermuly 1.5 inch Multi-Purpose gun, or the Smith and Wesson Grenade Launcher. These single-shot weapons are like short-shot guns, and can fire a variety of ammunition. Besides despatching rubber bullets, they can be used to fire rounds of CS smoke. The rounds have a range of 100 m and a burning time of ten to 25 seconds. The launcher is, therefore, an effective weapon that can flush terrorists out from a building. (It was used in SAS operation against the Iranian Embassy in London in 1980.) But unless a number of CS rounds are fired, the launcher is not effective for controlling crowds in the open.

CS dangers

CS is a non-lethal irritant. It can, however, cause harm to people with respiratory problems—if those people are confined in a small area and exposed to a very heavy gas concentration. In the open, it makes its victims feel as if they have been attacked with pepper—the eyes and mouth sting, and the nose runs. Exposure to the thick clouds of the smoke leads to coughing, or even vomiting. The smoke also irritates the skin (particularly of individuals who have been sweating).

CS has its limitations. For example, security forces in the USA found that rioters are quite prepared to grab the grenade after it has landed, and then throw it back. In Northern Ireland rioters discovered that they could defeat CS by creating a mask from a wet hankerchief wrapped around the nose and mouth. And because the CS smoke

Frank Spooner Pictures

forms a visible grey cloud it can be avoided.

One answer to the unplanned return of grenades is a version designed to split up like a high explosive. However, for riot control the 'split up' grenade's body has to be non-metallic, or the explosion would cause permanent injury. The L13A1/L16A1 grenade has a rubber body which ruptures when a small ignition charge explodes. This charge also spreads 23 CS pellets over an area of 25 to 35 metres. The pellets only burn for eight to nine seconds, but the CS cloud will hang for a longer period if the air is moist and still.

The problem with hand thrown grenades is that their range is limited by the ability of the particular individual to pitch them at the crowd. Launchers allow security forces to fire CS canisters at greater distance over the heads of the crowd. The launchers can be fitted to rifles, or grouped in a battery on riot control vehicles. The Paris riots of 1968 saw the special riot squad (the CRS) firing smoke grenades directly at the crowd, and then following up with rifle butts and batons.

CS can also be sprayed from hand-held aerosols. This form is particularly potent since the victim receives a concentrated dose of CS at a range of 1.5 metres. It will subdue the most violent man. The British Army uses a can designated SPAD (Self Protection Aid Device), while the US police forces carry a Mace spray as part of their individual kit. Some 400,000 Mace weapons are in service with 4,000 Police Departments in the USA.

Pepper fog

At the other end of the scale is the Smith and Wesson Pepper Fog Tear Smoke Generator. This man-portable machine will pump out CS or CN smoke to cover thousands of cubic feet, for up to 45 minutes of continuous operation. The Israelis use a projector which has a range of 15 m in still air, and is slung from the shoulder.

Besides CS smoke and rubber bullets, there are some more exotic weapons in the armoury of the security forces. The 'bean bag' is a round fired from a riot gun. It consists of a bag delivering the impact of a boxer's punch, but once the energy of the bag is expended the bag is hard to throw back. The Ring Airfoil Grenade (RAG) is another idea from the USA. It consists of a rubber ring about the size of a large napkin ring which can be aimed with accuracy up to 50 metres. It is fired from a projector on an M-16 rifle, and the ring can be designed to take a small amount of CS powder.

Riot control can be more effective if the police are less obvious. For example, the

The illustration below shows the Swiss idea of Soviet riot control methods with tanks. Demonstrators are 'herded' out of the square by sealing one side of the area, and driving a line of tanks towards the other 'outlet' side. The outlet is deliberately kept open by the occupying troops. After the square has been cleared the enemy will push forward for some distance down the outlet streets and then close them off to stop people returning.

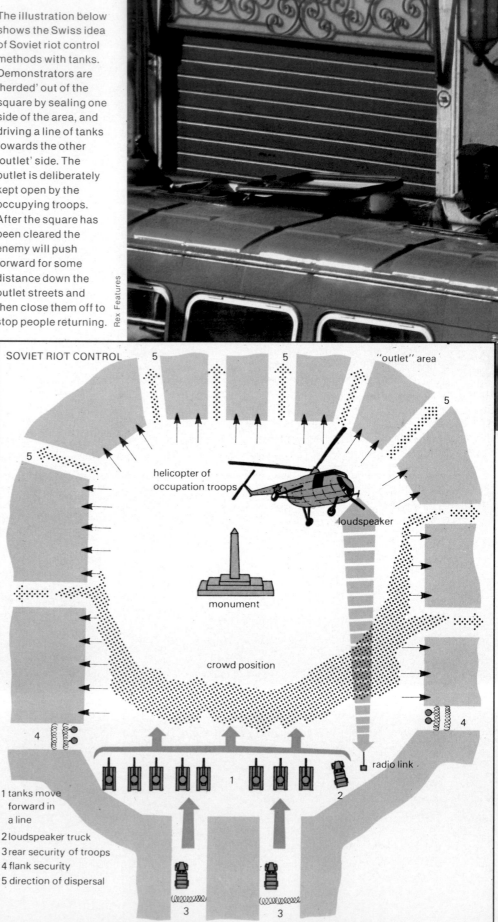

Rex Features

Sarson/Bryan

SOVIET RIOT CONTROL

"outlet" area

helicopter of occupation troops

loudspeaker

monument

crowd position

radio link

1 tanks move forward in a line
2 loudspeaker truck
3 rear security of troops
4 flank security
5 direction of dispersal

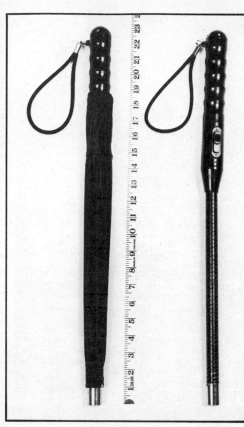

Above left Italian anti-riot vehicle.
Above right The shockstick is *not* a rolled umbrella—it emits 6,000–7,000 volts in a hail of flashing blue sparks.
Below View out of the 'Pig'.

Shorland armoured vehicles (developed in Northern Ireland) have become a successful export from the UK. They are less conspicuous than other armoured vehicles, and being based on the conventional Land Rover, they are easy to drive. Ventilation allows the crew and passengers to stay safely enclosed, even in hot weather.

Since they move faster, wheeled vehicles are more effective in urban areas than tracked vehicles, such as tanks or Armoured Personnel Carriers. In the United Kingdom there is a strong feeling that the use of tanks or heavy armoured vehicles is politically unacceptable for security forces. This is in marked contrast to the Soviet bloc, where a tank at every street corner is an accepted tactic. Certainly tanks are intimidating, by their sheer bulk and noise. But as Russian commanders discovered in Budapest in 1956, battle tanks are unwieldly in town streets.

Armour at the barricades?

Specialized armoured vehicles normally have a boat shaped hull, searchlights, loud speakers, armoured glass for vision ports, and the facility to mount a machine gun. Some have 'dozer blades' at the front for clearing the barricades. They serve both as a troop carrier and as an offensive vehicle. Police forces have also used armoured vehicles for operations involving armed criminals, and this helps to justify the high cost of these specialized vehicles.

France and Switzerland have produced a variety of vehicles. But the British have either adapted the Land Rover, or used standard military machines such as the Alvis Saladin armoured car, the Alvis Saracen armoured personnel carrier (APC), and the Daimler Ferret Scout Car. The one-ton Humber APC, known universally as the 'Pig', has been used by every regiment that has served in Northern Ireland.

Other vehicles can be used as human bulldozers. Fitted to a wheeled digger, these 'crowd pushers' are used to move immobile crowds. Another crowd pusher is the security screen fitted to a 4-ton truck. It consists of a screen over 3 metres (10 ft) tall, and 6.1 metres (20 ft) long, which is used to block a road. It can also be used to screen the view into, or out of, a building during a siege.

In South Africa the police operate from 'Hippos'—these are 4-ton trucks which have the cab and seats mounted above the chassis. The Hippo is, therefore, mine-proof and puts its occupants nearly 3 metres above the ground where they can see clearly.

A South African vehicle has been produced

by a fire-fighting equipment firm. Based on a 1½ ton chassis, it includes a public address system with a tape deck. (This allows the crew to give advice or warnings without any danger of mistakes, since the message is always pre-recorded.) The vehicle can also broadcast soothing or cheering music.

One of the most significant materials used for protecting troops and police is makrolon polycarbonate. This is a clear, tough plastic material, which was first developed for astronauts' visors. When fitted as a protective helmet, it will withstand rocks thrown at short range, and though it may crack after a sustained assault, it does not shatter. After visors, two types of shield were introduced in Northern Ireland, and those have now become more widely available to police forces. The small shield is useful for mobile groups, while the one and a half metre (five foot) shield gives excellent protection to men who have to form a cordon separating rival groups or stone throwing youths.

The larger shields allow troops to see the most active stone throwers, and enables the 'snatch squad' to emerge from cover and make a quick arrest. The snatch squad normally wear flak jackets and carry a baton,

respirator, and a belt with a first-aid field dressing and a water bottle.

The flak jacket is now a vital part of most internal security operations where a crowd is likely to be used as cover for a sniper. The normal jacket has about 16 plies of textile sewn together with plastic armour sheeting so that it is flexible.

Metal plates

Some jackets can be fitted with metal plates which give greater protection against high velocity rounds. But these are normally only worn by static units. Most protective clothing will only stop pistol rounds or fragments from home-made bombs. However, some US and British clothing will stop an AK-47 7.62 armour-piercing round at one metre range.

Human inventiveness has reached some extremes in riot control, and a substance known as 'Instant Banana Peel' is one of the more bizarre measures on the market. In the

form of a spray, it makes the ground very slippery. But it may hinder the police as much as the rioters. Foam is another hindrance—on the theory that no one would want to push through a thick mass of bubbles. However, after the riot, foam involves a tedious street cleansing job.

Devices designed to control crowds without violence are of little concern to the less sophisticated police and security forces of the world. They still favour the use of fire arms in riots—even though firing over the crowds is potentially dangerous, since the rounds can hit people in buildings or pass beyond the area of disturbance.

Even where the tactics used are simple, riot control equipment plays an important part. Research has even reached areas like gloves and footwear with padded gloves designed to give protection to the knuckles. In the USA, a private firm produces 'sap gloves' which have powdered lead sewn into the lining—it gives a policeman a discreet pair of 'knuckle dusters'. For long periods on foot, in cities, lightweight boots can be issued, and crepe soled shoes can be used during operations like sieges.

The ideal way to end a riot is to disperse it

RIOT CONTROL HARDWARE

1. The .38 Special Mighty Midget Grenade Launcher. The launcher permits the firing of the Mighty Midget Grenade—a pocket-sized version of the military grenade.
2. Tru-Flite TM 37 mm Penetrating Projectile. For use in shoulder type gas guns only, this fin-stabilized, hard rubber projectile is designed for situations that demand accuracy and penetrating ability. They will penetrate 16 mm plywood at 100 m (330 ft).
3. Rubber Ball Grenade. This innovative Smith & Wesson grenade virtually eliminates the possibility of throwback and minimizes chance of injury. It can be hand thrown or shotgun launched.
4. The Mace spray can is designed to orient in the hand like a handgun. The device can be reloaded with replacement cartridges. They are carried on a belt holster.
5. Military Type Continuous Discharge Grenade. A crowd control grenade that will emit CS, CN, or smoke. Designed to be hand thrown, the grenade has a military type safety pin.

Sarson/Bryan

Top 15,000 riot policemen at Narita, Japan. Airport police prepare to confront demonstrators—hundreds were injured.
Above Football riot—Glasgow 1980.
Right Bullet-proof vested riot policeman.

peacefully, and the use of a sedative gas would enable a police force to calm a hostile crowd. A novel by Aldous Huxley, *Brave New World*, predicted the scene as 'soma' gas was pumped over a minor disturbance and the rioters were told by a firm but friendly voice that they had made a mistake and should return to their work. The use of taped music and announcements has already brought this fantasy a step closer.

Authorities will soon have to face the moral dilemma of whether to quell violence with violence—and in so doing at least respecting the character of their opponents—or to 'play God' by manipulating the rioters' brains with chemicals and psychological suggestion.

The deadly frontier

Along the length of the East–West German border runs a minefield to deter would-be escapees to the West. Concealed in a narrow band no more than 20 metres wide lie 17 million mines— one for every man, woman and child in East Germany. In the technology of warfare, this frontier is one of the most macabre monuments to Man's efforts to keep people in their place.

A meeting in September 1944 between the US, Britain and Russia arrived at an agreement on the partition of Germany. The participants decided that the boundary dividing east from west should run from the bay of Lübeck in the north to the Czechoslovakian border in the south.

The Germans were the beaten enemy and the feelings of the local population were not heeded. The arbitrary line cut through villages, farms, 27 federal roads, 140 secondary roads, rivers and across the war-torn tracks of 32 railway lines. Towns were simply bisected, geographical boundaries were ignored, natural features bypassed. Whole communities, small families, perhaps no more than yards apart, suddenly found themselves citizens of different states.

Alert squads

At the time of the decision to create the border, Berlin was controlled by four occupying powers. The city and its inhabitants were isolated in an East German state.

The border, then, was a political fabrication. In October 1949 the Soviet occupation forces announced that a 'Peoples' Council', established during a Soviet-controlled 'Peoples' Congress' set up in December 1947, had been given the power to create the German Democratic Republic. The people to run this 'democracy' were selected by single-list elections, which meant in effect the voter had a choice of one.

A 50,000-man force of *Bereitschaften* ('Alert Squads') was raised by the Soviet controllers. They were armed with Russian machine guns (MGs), Kalashnikov rifles, small-arms, anti-tank and anti-aircraft (AT and AA) artillery, mortars and tanks. In Western eyes this force was illegal, being

Right, above and below From these watchtowers and using cameras and high-power binoculars, the East German security forces maintain a constant surveillance on all movements on both sides of the border.

manned to a great extent by Germans, who were forbidden under the terms of the surrender to carry arms.

When the Berlin Wall went up in August 1961 the Warsaw Pact countries called on the East German authorities to establish the border as a permanent zone. Immediately, the 1,346 km (836 mile) border between the two halves of Germany began to be strengthened, and this process has continued to the present day.

The border is designed not as a defensive position against aggression, but as a means of preventing the East German population from leaving. Thus, it becomes progressively

Ullstein

more menacing from the inside outwards.

The first obstacle the would-be East German escapee finds is a rear no-go area some 30 km (18 miles) deep which can only be entered by those with passes. It is occupied by the security forces of the GDR, the *Grenzpolizei* (border guards), known as 'Grepos', who patrol a concrete road which is constantly floodlit.

Minefields

In this area are the control centres, underground bunkers in case of war, and shelters for off-duty border guards. The wall and fences as well as the open ground carry trip wires connected to alarm systems in the control centres. Other sensors—infra-red, acoustic and image-intensifying—make unobserved movement at night impossible.

Beyond the no-go area is a high, metal mesh fence along which runs a 0.6 m (2 ft) high rail. Along the rail, freely tethered, run dogs, each with a limited area of its own. There are about a thousand dogs, mainly large alsatians on 250 runs. This part of the border also includes a trench 2.7 m (9 ft) deep, to halt any wheeled vehicle that might have crashed its way so far without being hit by rifle and MG fire.

Once over the fence and past the dogs, there lies a 10–20 m (33–65 ft) wide area which at first sight looks like bare earth. But it is sown with some 17 million PM/P70 anti-personnel mines which, on impact, detonate at a pressure of 6 kilos (13 lb).

Beyond the minefield lies another wire mesh fence 2.5 m (7 ft) high. Mounted all along it are cone-shaped SM70 spring guns. There are estimated to be 20,000 of these lethal weapons along the border.

The placings of the SM70, 2 m (6 ft 6 in) apart, and staggered at leg, chest and head heights, are so arranged that each one is covered by the range of those around it. The small trigger wires need only to be nudged 2 cm for the device to fire. As a warning goes to the nearest control post, the cone-shaped gun spits out over 100 steel dice, each with an edge of 4 mm (0.16 in), and weighing 0.5 gr (0.018 oz). These square projectiles have a dum-dum effect and a lethal range of 3.5 m (12 ft). The first border area to carry the SM70 was a 200 km (124 mile) stretch near Dannenburg.

Western knowledge of this weapon, apparently so unapproachable, is due to a West German, 30-year-old Michael Garten-

Left Berlin's Checkpoint Charlie on Friedrichstr. as seen from the east. The chicane prevents a vehicle from crashing through at speed to the West. *Below* The Berlin Wall at Wedding, an area to the east of West Berlin. The smooth surface of the cylinders has no hand-hold.

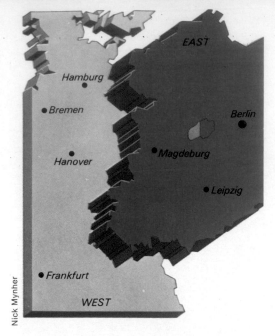

Above The East German border runs from the Bay of Lubeck in the north to Czechoslovakia.
Right Two Grepos in their watchtower, seen from the eastern sector of Berlin.
Below Border guards, well-armed, ever-suspicious and alert, ceaselessly check the length of the frontier for any weak spots.

schlaeger. On the night of 30 March 1976, he crawled across the 20 m (65 ft) strip in front of the spring-gun-mounted fence, supported by two friends, who kept him covered with guns.

He got near enough to cut the cable which leads to the trigger device and sends a signal to the nearest guard post. Gartenschlaeger then separated the four wires inside. When the cone-shaped weapon was brought back to be stripped down it was found to weigh 2,995 g (6.6 lb) and was fitted with a threaded, electrically detonated cap filled with a mixture of 20 mg (0.0007 oz) of potassium chloride and an unidentified copper binding agent.

Under the metal cap is a cup of brown primer composed of 300 mg (0.012 oz) of Blutron and lead acid; and a rose-brown secondary charge of 550 mg of nitropenta.

Shrapnel

An explosive relay charge of 8.8 g (0.31 oz) of nitropenta circles the detonator. The main charge is 102.4 g (3.6 oz) of TNT (trinitrotoluene). When the trigger wire is cut or pulled, the striker pin in the

mechanism makes contact with two circuits. One raises the alarm in the nearest guard post while the other causes the SM70 to fire. A laboratory study, including X-rays, of the SM70 produced a sequence with which to disarm the weapon: 1, Cut the two fuse cables running from the base of the cone. 2, Cut open the mantling round the cable between the support and the push-pull fuse and separate each cable in any order.

The first SM70 to be retrieved and studied was marked '06-10-73' and 'Briselang'. This was probably the date of manufacture and presumably identification of the makers, the Briselang factory, part of the VEB rubber organization in East Berlin. It is believed that the SM70 is based on a device used to prevent escape from German concentration camps in World War 2.

As a measure to prevent further removals, a refinement of the SM70 placing was made in late 1976. In addition to the sets of three guns at leg, stomach and head height, a fourth was added, positioned to fire its dice-shaped shrapnel at an angle of 45 degrees. Metal deflector plates were added to the fences, ensuring that anyone in the field of fire received the ricocheting dice at head height as well.

In front of the mesh fence and the SM70s, there is another control strip and in front of that the 1.8 m (6 ft) striped border posts and smaller, 50 cm (1 ft 7 in) frontier stones.

Wooden and concrete towers stand at intervals along the full length of the border. They are manned by members of the East German border force, who keep the whole area under constant observation at all times.

On the west side are narrow, strictly controlled lanes. The instruments of 'defence', the SM70s, anti-personnel mines, dogs, trip wires can be seen—all of them rigorously

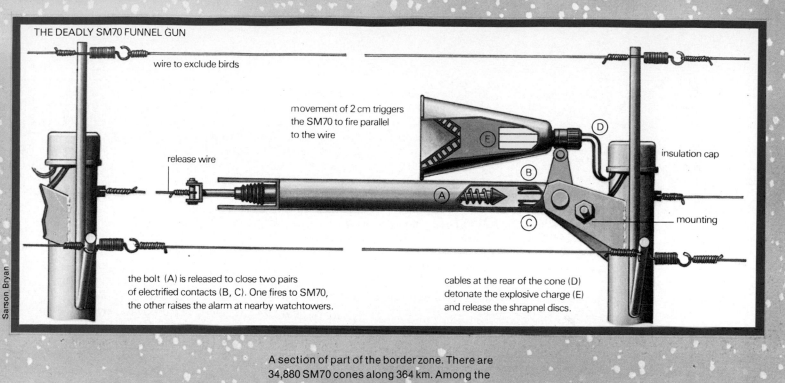

THE DEADLY SM70 FUNNEL GUN

wire to exclude birds

movement of 2 cm triggers the SM70 to fire parallel to the wire

release wire

insulation cap

mounting

the bolt (A) is released to close two pairs of electrified contacts (B, C). One fires to SM70, the other raises the alarm at nearby watchtowers.

cables at the rear of the cone (D) detonate the explosive charge (E) and release the shrapnel discs.

Sarson Bryan

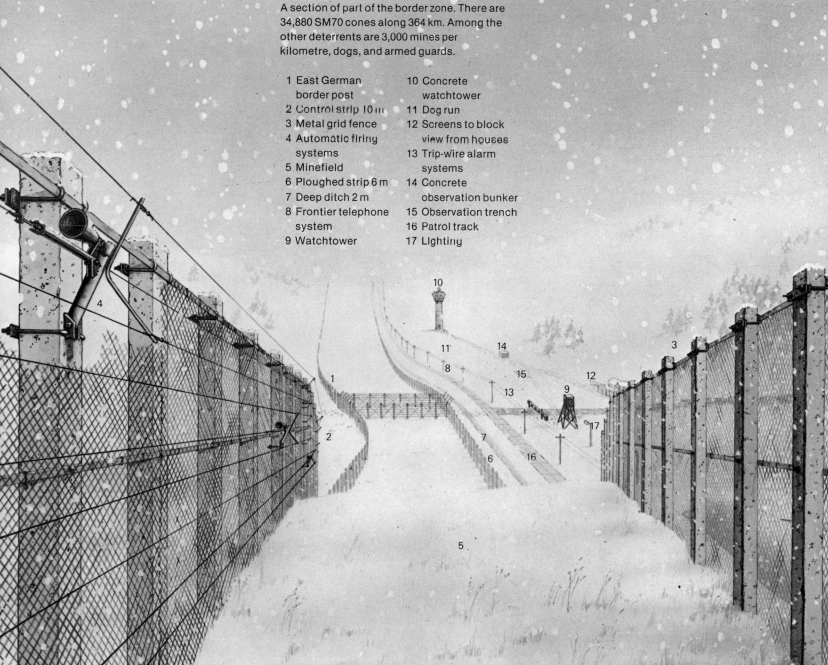

A section of part of the border zone. There are 34,880 SM70 cones along 364 km. Among the other deterrents are 3,000 mines per kilometre, dogs, and armed guards.

1 East German border post
2 Control strip 10 m
3 Metal grid fence
4 Automatic firing systems
5 Minefield
6 Ploughed strip 6 m
7 Deep ditch 2 m
8 Frontier telephone system
9 Watchtower
10 Concrete watchtower
11 Dog run
12 Screens to block view from houses
13 Trip-wire alarm systems
14 Concrete observation bunker
15 Observation trench
16 Patrol track
17 Lighting

controlled by attentive East German guards.

The guards constantly monitor activity taking place on the Western side of the border. Personnel in the vicinity are tracked through binoculars until they move on to the next sector, where other soldiers take up the task, having been alerted by telephone. Should visitors to the border show more than casual interest or be seen to use a camera, there is a quickening of activity. More telephone calls are made, high-powered glasses are trained across the border and long-distance telephoto-lensed cameras record any activity.

In the Bay of Lübeck, the frontier extends out to sea. In one instance the West German naval frontier vessel *Duderstadt* intercepted an East German Condor-class minelayer a mere 8 k (5 miles) off Fenmarn Island in West German waters.

Checkpoint Charlie

To enter East Germany by land there are eight rail crossings and nine roads. The main road entry is at Helmstedt on the autobahn to Berlin. The last recorded death of a would-be refugee at Checkpoint Charlie took place in 1974. An East German border guard made a run for it using one of his officers as a shield, but he was then shot by an East German sniper.

In November 1980 an East German guard, Cpl Ulrich Steinhauer, was shot by an escaping 19-year-old soldier as he fled across the border to West Berlin.

In the centre of the German Democratic Republic lies Berlin, the former capital of all Germany. After 1945 many attempts were made to make the whole of Berlin free for all Germans, notwithstanding the presence of US, British, French and Russian troops. But the Russian insistence on control of the Eastern sector remained relentless.

The Western powers were accused by the Soviet military governor Marshal Vassily Sokolovsky of using West Berlin to infiltrate spies into Eastern Germany, and of subversive activities, including the luring of East German citizens by espionage organizations running 'slave traffic'. The West was also accused of using the air corridors to fly out kidnapped East German children.

In the great Berlin airlift, during the 15 months from June 1948 to September 1949, 2,325,809 tons of coal, food and supplies were flown into the Western sector of the city. It took 227,804 flights by the RAF and USAF using three 20-mile air corridors. It cost the lives of 49 RAF and USAF and ten civilians. The Germans lost nine men.

But the Soviet blockade soon began to recoil on itself. With the passage of all goods from West to East stopped, the East German economy began to fail. The result was the lifting of the blockade.

The next move in the edgy 'peace' came quickly. On the morning of 13 August 1961 Berliners found that overnight the East Germans had sealed off their sector from the West. From barbed wire and white lines painted on the street, the political barrier soon became a physical one.

The wall was 4.5 m (15 ft) high and 160 km (99.5 miles) long, topped by a large-diameter smooth glazed pipe. There is no hand-hold there for anyone who can climb to the top without getting shot. Deaths on the wall from 1961 to 1980 total 170.

Though the frontier ended effective escapes, between 1969 and 1976 a 'trading' in political prisoners was carried on. By agreement, the Federal Republic 'bought' 6,000 political prisoners for £50 million as part of a deal called by Bonn the *Freikaufaktion* or 'Freedom Purchase Plan'. The last batch of 200 East Germans crossed the border near Geisson in late 1975.

The security of the border zone keeps thousands of members of the German Democratic Republic forces tied down. The costs are astronomical, and form a large proportion of the GDR budget. Those members of the *Bereitschaften* who were responsible for laying the first minefields were moved from their sectors immediately their task was done. The replacement forces were not informed of the exact positioning of the mines. But this did not stop some 2,700 East German border guards escaping between 1945 and 1976. And from 1949 to 1980 three million civilians have fled from the East.

From 1977 to 1980, 13,000 East German citizens were allowed through to the West, although it has become increasingly difficult to cross the border zone clandestinely. In order to counter escapes, refinements are continually being made. To test these, it is reported, GDR athletes try to get past obstacles, over fences, and avoid guards on mock-ups of the border zone.

On either side of this most sophisticated barrier ever to divide two nations, lie the arsenals of East and West. It was political, rather than military, expediency which caused the wall to be built in the first place, and it will be politicians who decide its future.

For nearly 200 East German citizens, wreaths and flowers mark the place where their last desperate bid for freedom failed.

Who dares, wins

The elite, crack force consisting of several highly trained and motivated men operating in conditions of absurd risk behind enemy lines, has primed the imagination of children all round the world. It is the stuff of wartime heroic fantasy, of the big screen and the thriller. But, ironically, such troops of men, deployed with the right combination of timing, imagination and intelligence, have become increasingly important in contemporary military and political struggles.

In recent years, for example, the role of the Special Air Service as a force specifically designed to deal with hard core civil and military problems, has become widely recognized as a necessary safeguard in Britain during an era of mounting terrorist activity. The British people seem thankful for this 'last resort', and all over the world specialist commando forces similar to, and even modelled on, the SAS are being established to cope with the disease of terrorism.

But when a nation state has to deal with fanatics who are often prepared to die for the 'cause' in a situation that is far from the battlefield and usually involves the taking of civilian hostages, then the working principles of the commando unit of necessity differ fundamentally from those that hold sway in conventional warfare. To some extent the fanaticism of the terrorist has to be equalled by the imposition of a fanatical discipline and training on the part of the commando group, together with a dedication unimaginable to the average mildly patriotic citizen. And the group often has to adopt the very same guerrilla techniques that the terrorists use.

Early models

The analysis of tactical principles of guerrilla warfare described in Marighella's 'Minimanual of the Urban Guerrilla' followed the teaching of Che Guevara, Mao Tse Tung and Régis Debray. It is considered highly relevant to the training, organization and equipment of recently established commando groups. The German anti-terrorist unit of the Federal Border Guard uses this literature as its principal training manual.

Along with the evolution of new types of tactics, recruitment of commando personnel has become highly selective, since there is far more value placed upon each individual in a special fighting force. The man selected will often have to mingle unnoticed with civilians or work in total unison with a partner. He may be required to speak more than one language fluently—even to pass as a native in a foreign country—and he may have to be something of a diplomat in certain situations. He will certainly be required to be adept in nefarious methods of combat as well as an expert in the use of conventional small arms. He is usually trained intensively in a small group before being deployed. And along with the more traditional units within his commando—such as the leadership, communications experts, technicians, and the supply and services section, there is often a psychiatrist watching him and his colleagues very closely. Response to the high stress levels associated with training and fieldwork must be carefully monitored.

Elite commando units, such as these US troops in full battle kit, are in constant readiness to combat terrorism in any part of the world. The leading man is armed with an M 60 machine gun and the one behind with an M 16 rifle.

Ian Wright/Spooner/Gamma

Royal Marines

Commandos of the Royal Marines undergoing fighting and survival training. Rock climbing *(above)* promotes discipline and confidence. The Rigid Raider Squadron ferries commando units to where the action is *(top right)*. *Right* A prototype fighting vehicle being put through its paces at sea.

In 1972 in Munich the Black September Commandos—a fanatical Palestinian group—managed to horrify the whole world by seizing 11 members of the Israeli Olympic team. At the time the Federal Republic of Germany did not have a specialist commando group of the SAS type and was forced to use its police force to deal with the emergency. The results were disastrous. All 11 Israelis were killed in the struggle to release them and three of the terrorists were captured only to be released later as a result of another terrorist coup, this time in the form of a hijacking. Munich was certainly a watershed in the evaluation of many nations' security and defence organizations. In Germany in particular, the infamous Baader-Meinhof gang, the Red Army Faction and the phenomenal increase in bombings and armed robberies combined to demand a major rethink on the part of the Federal Republic government's Defence Committee.

The answer to much head-scratching and soul-searching came with a special police unit called the Federal Border Guard (BGS) and known simply as Group 9. It was based on no apparent model and only resembled a police unit insofar as its activities were to be confined to a civilian environment.

Essential to the success of any such commando group is its adaptability in terms of operational structure and of the tactical concepts it employs in preventing or combating terrorist activities. This means that—especially in the early stages—its organization must be continually refined.

At present Group 9 consists of the traditional command element, three operational units, a communications and documentations unit, a technical unit and three technical squads, a training unit, an aviation element and support/supply services.

The members of the leading section side-step Command HQ conventions, and lead from the front. This section is also responsible for collecting intelligence data by liaising with intelligence services and reaching command decisions based on the collated information. Liaison officers from the German Intelligence agencies, the Federal Bureau for the Protection of the Constitution, responsible for collecting information abroad and Military Intelligence, all assist the unit. During an operation reports are relayed to the leaders by means of a 'hot line'. The communications and documentation supplementary unit has the role of providing a link-up with other military services and police units which may be deployed, and is responsible for the utilization of long-range or unconventional means of communication such as tracking or monitoring devices. In conditions of insurrection, Group 9 has a large measure of autonomous control.

Special tactical teams

Each of the three operational units has its own leading section and between five and eight special tactical teams. These squads have replaced the former conventional squad and the emphasis within them is on tighter leadership, greater manoeuvrability and better communications. Everybody in the unit has his own mini-radio, which allows a more versatile positioning of fire-power at short notice. Everybody within the unit is cross-trained so that each member can operate in any position in the team, including that of leader. Within the unit, the special tactical squad is the smallest tactical element, working in groups of between two and 13 members dressed in combat dress, civilian clothing or disguise.

Group 9's wide range of modern weaponry extends from the .357 calibre revolver to various types of submachine gun and special items of unconventional equipment. Their use as sniper teams with silencers and specially adapted infra-red aiming devices is a secondary role for each of the squads and explosives experts are readily available should their services be needed.

Sophisticated equipment

On large scale operations the German Group uses troop-lift and reconnaissance helicopters and fast specially equipped cars and armoured vehicles. The M13 and the M130 light armoured vehicle are typical of the models used. One of Group 9's hallmarks is the co-ordination of airborne and motorized forces in the accomplishment of a joint mission. The basic training of each individual lasts seven months. The first three and a half are designed to create a motivated

Richard Cooke

Philippe Letelier/Gamma/VSD/Frank Spooner Pictures

NATO exercises in Norway. Royal Marines Commando *(above)* practising snow-survival drill. *Left* Norwegian troops, seen here firing a bazooka, are training US Marines in the techniques of polar combat.

unit member and the emphasis during this time is on psychological fitness. Specialist advanced training occurs in the second half of the course when the individual is trained together with the special tactical squad and learns the essentials of thoroughly interdependent teamwork. Each individual also learns the use not only of his own range of firearms but also of as many types of firearm used by terrorists as can be provided. Preparation for specific situations such as hijacks requires a knowledge of airport equipment—even learning to drive the airport's catering vehicles—or becoming part of the ten-man team trained as cabin personnel.

Group 9 is representative of a movement that has spread through virtually every country in Western Europe. In Holland there are essentially two different commando units designed to combat terrorist activity. Members are selected with great care, and each applicant undergoes severe medical and psycho-technical selection tests. The Dutch government treats terrorist activity as a criminal act rather than a political one and for this reason keeps the combative force under the aegis of the police. The reason for having a part-time force is due to the fact that work solely in anti-terrorist activity can create undue stress in individual members. Conventional police or Service activity alleviates this stress.

The second security and defence force of this type in Holland is the established single company of the Royal Netherlands Marine Corps. This company operates as a permanent emergency stand-by. It comprises 113 men including 14 staff and three platoons of 33 men. Close combat is not in fact the main task of this unit, which is essentially a normal rifle company contingent, but each platoon takes it in turns to be in a state of alert so that a specially organized, equipped and armed contingent under the direction of a company commander can be called into action at a moment's notice.

For anti-terrorist activities each platoon is divided into groups of five with one group commander and two teams of two men. All are armed with machine pistols, for their light weight and rapid fire capability, and a

A commando recruit has to develop battle fitness and skills to make him a worthy member of a troop of some 25 men in the field. He learns to shoot and handle his gun, the basics of drill, and a whole range of more specialized activities from camouflage to unarmed combat. A Royal Marine recruit takes a fall during an unarmed combat training session *(left)*, while US Marines take part in a strenuous 'Tarzan' assault course *(below)*.

Royal Marines

revolver. To join this unit each member must have finished his basic training, be in excellent physical condition and must—like the members of the Special Police Unit—pass a psycho-diagnostic test lasting one day.

His individual training includes target practice with several weapons including the 9 mm calibre Israeli machine pistol, the American Police Special Lawman Mark .38 in. calibre revolver. He will also get used to the larger machine guns for weight of fire during an operation and the equipment he uses will have infra-red and image-intensifying devices attached. Like the German commando group, the leading members of each section carry mini-radios and all members are trained in street fighting, fighting in buildings and co-operating with motorized ground and air equipment. The final theoretical back-up is delivered in the form of lectures from unit officers, senior police officers and psychiatrists. In recent years, the units have been called to the French Embassy in Holland, to Scheveningan prison and to the Indonesian Consulate.

The SAS came under the glare of the public eye when it ended the siege of the Iranian Embassy in London in the spring of 1980. But it benefits greatly from being able to refine its capabilities ever further in complete secrecy. One reason for its success has been the intensity of its training over a relatively long period. Again, stress was placed on the calibre of the individual recruit—always from the Army—with the greatest emphasis on the applicant's ability to endure situations of great discomfort for a long time. The ability to make on-the-spot decisions—which is effectively the ability of leadership and usually the prerogative of the officer—is necessary for each member of the group and this kind of alertness is used to engender a teamwork that is almost acrobatic in its timing.

Royal Marines Commandos

The Royal Marines comprise about 7,000 men led by a nucleus of about 700 officers. As the Navy's soldiers the Royal Marines can be likened to an amphibious infantry division. In addition they are Britain's commandos acting as a spearhead force working in the air, at sea and on land, and no longer simply as an invasion force securing enemy shores. Together with the Royal Netherlands Marine Corps, the Royal Marines' primary role is to protect the northern flank of NATO. Consequently Commandos 42 and 45 are specifically trained,

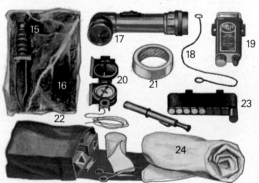

Clothing and personal equipment: (1) Poncho liner, also used as lightweight sleeping bag. (2) Half towel. (3) Spare shirt. (4) Washing and shaving kit. (5) Camouflage stick containing insect repellent. (6) Socks.

Domestic equipment: (7) Entrenching tool. (8) Sandbag, for storing refuse or carrying extra loads. (9) Stainless steel canteen, also used for cooking. (10) Water bottle. (11) Spoon and can opener. (12) Brown foil bag, containing various accessories to accompany meals, e.g. sugar, salt. (13) Cans containing a range of staple food. (14) Solid fuel tablet.

Survival aids: (15) Aviator's knife with whetstone. (16) Aerial map. (17) Flashlight. (18) Wire saw. (19) Strobe light for use as distress signal at night. (20) Compass. (21) Emergency tape. (22) Aviator's first-aid kit includes dressings, aspirin and morphine. (23) Flare kit with various colours for signalling and illumination. (24) Water filter bag, for use with sterilizing tablets.

Unofficial field dress worn by US soldiers belonging to the Special Forces in the Vietnam War. Both clothing and equipment were chosen for combat and serviceability. The uniform itself was of Taiwan 'tiger stripes' and the headband might be made from parachute 'silk'. Patrols often wore the issue jungle hat instead. The webb gear carried magazines for the M 16 rifle, with grenades attached to the outside of the pouches. Canvas rucksack contains the complete survival kit.

Nick Farmer

organized and equipped to operate in mountainous and arctic conditions. This means that they have to be able to fight and survive in temperatures as low as −40°C. In Arctic Norway survival and success go together.

The main large weapons used by the Royal Marines are the 81 mm mortar and the modern Milan anti-tank gun. Mobility in the field is mostly afforded by Gazelle and Lynx helicopters. On the water a variety of craft are used, ranging from the larger vessel assault ships such as HMS *Hermes* and HMS *Intrepid* and the landing craft which take tanks, lorries and troops to shore, to the smaller, more versatile rigid raiders which can take a full contingent of eight men to shore at a speed of 27 knots.

Special Boats Squadron

Within the elite spearhead commandos, there is an even more highly trained nucleus in the form of the Special Boats Squadron. This can be considered the marine equivalent of the SAS. Its activities are altogether more clandestine than the rest of the commando group, with each member usually known by his first name by officers and men alike, so that in situations of subterfuge his rank will not be betrayed. Each man is a skilled parachutist, canoeist, frogman, diver and long-distance swimmer and is trained for such reconnaissance and forward operations sites, using explosives and employing various ing beaching sites and helicopter landing sites, using explosives and employing various deadly methods of hand-to-hand combat.

All officers in the commandos (not merely the SBS section) are trained in modern riot techniques for duty in Northern Ireland as well as in Nuclear, Biological and Chemical (NBC) warfare. For this they have to wear heavy and extremely uncomfortable lead-lined suits and masks for up to eight hours at a time on certain tactical exercises. The suits, designed for protection in the event of germ, gas or radioactive pollution, are so cumbersome as to severely restrict fighting ability but do ensure a few hours survival.

The border line between the military- and the terrorist-orientated commando group is increasingly difficult to draw. Since 1967 the Israelis have developed the identifiable commando unit and successfully employed it all along their border in deep-raid incursions into Arab territory. The fighting is a combination of guerrilla tactics and standard military practice. But it was their extraordinary deliverance of Israeli hostages held by Palestinians at Uganda's Entebbe airport in 1976 which showed the ultimate commando technique. In a world of infra-red, image-intensification, psychological understanding and complex training methods, intelligence and guts can get the results.

HECKLER AND KOCK MP5 A2
SUB-MACHINE GUN

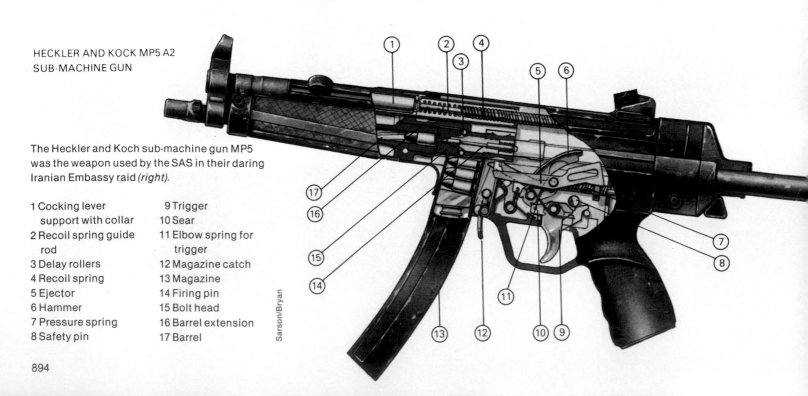

The Heckler and Koch sub-machine gun MP5 was the weapon used by the SAS in their daring Iranian Embassy raid *(right)*.

1 Cocking lever support with collar	9 Trigger
2 Recoil spring guide rod	10 Sear
	11 Elbow spring for trigger
3 Delay rollers	12 Magazine catch
4 Recoil spring	13 Magazine
5 Ejector	14 Firing pin
6 Hammer	15 Bolt head
7 Pressure spring	16 Barrel extension
8 Safety pin	17 Barrel

Sarson/Bryan

The rescue of the hostages held by terrorists at the Iranian Embassy, London, was carried out by a 12-man team from the Special Air Services, resulting in this normally secret force facing a glare of publicity. The rescue was launched from the roof (1). The first man of the leading pair accidently broke a window (2) but continued to the ground. Another pair descended to the first floor balcony (4),

followed by a third pair, one of whom became entangled in his rope (5). SAS commandos then tossed concussion grenades into the building and stormed the main stairwell (6), killing a gunman (7). Meanwhile, the pair on the floor above, having entered through the balcony, were attacked by a gunman but were saved by a hostage (8). The gunman on the landing was then shot (9) while a hostage ran into the front

room (10) and the SAS broke into the front of the building using explosives (11). The terrorists (12) then began to shoot the hostages but were interrupted by CS gas bombs thrown through the windows (13). With hostages screaming, (14), the gunmen were disarmed and their weapons thrown out of the windows (15). In the confusion, two other terrorists were shot (16). After the action, the SAS returned to obscurity.

Jeremy Gower

END OF THE EMBASSY SIEGE: MAY 1980

ELECTRONICS IN ACTION
PART SIX

INTRODUCTION TO PART SIX

Only two centuries have passed since the death of Benjamin Franklin, whose epitaph was written by Minister Turgot:

He snatched the lightning from the skies
and the sceptre from tyrants.

The latter act does not concern us here, but the former does. It is only a moment in the Earth's life since the first man lived who understood something about electricity. Nor did Franklin, enormously clever though he was, understand very much. He almost electrocuted himself in his first experiments and never got very far beyond believing that electricity was a plaything.

Today we know it isn't. In fact, electricity probably makes the most difference between Franklin's world and our own. As few as a hundred years ago there were no electric lights — that difference alone would make the "old days" almost unendurable to anyone living today. Nor would it be much easier to live without electric motors and electric heat and the millions of machines, both large and little, that electricity makes run.

Of all these machines the most important, the most fraught with peril as well as promise, are what we call electronic machines — computers, mostly, but there are others, too. It is hard to realize that computers have only been a part of our lives for a single generation.

Was life really possible without them? (It was.) Could we ever go back to living without them? (Probably not.) The question is whether computers have improved our life on Earth. But how silly it is to try to answer that now, after only one generation! Imagine the world two generations hence, say in 2050. There will be more computers than people. There will be nothing they can't do. But will they be docile servants? Or will they rebel against their masters? There are arguments on both sides, although so far the machines have not spoken.

For the moment, electronics gives us great gifts. Lights in the dark. Electron microscopes. Advanced stereos. Brilliant new recording devices. Memories that never forget, and machines that talk. Electronic companions, and robots to do our dirty work. So far, all quite docile — not a sign of rebellion. We can breath easier, or can we?

Chips for everything

The microprocessor is perhaps the most glamorous product of the silicon revolution. It puts into a tiny sliver, or chip, of silicon the computing power that in 1950 weighed 40 tonnes. And it can be made in its millions, at a unit cost of a few dollars.

Add to the microprocessor additional chips to handle memory, input and output devices such as a keyboard and a visual display unit (VDU), and you have a computer. At least one company, Intel, has managed to put both memory and microprocessor functions on to a single chip 5 mm (0.2 in) square—the microcomputer on a chip.

But what *is* a chip? What can it do and how is it made?

A chip is a small thin rectangle of the element silicon measuring about 6 mm (0.25 in) from end to end. One surface has an intricate pattern whose broad outlines are just visible. The pattern extends into the chip as a complex sandwich of layers of pure and impure (doped) silicon, silicon oxide and aluminium. The pattern is an electronic circuit containing (in mid-1980) up to 250,000 electronic devices— transistors, resistors, capacitors and conductors. The whole chip can be designed to be part of a computer, a computer in its own right or almost any other circuit.

Programs

Microprocessors perform tasks according to coded instructions fed to them. A list of instructions is called a program, and for many applications the programs needed by the microprocessor to perform its various tasks are stored on a memory chip. To change the function of the microprocessor, all that is needed is to change the program. This means that a relatively small number of standard microprocessors can be used for all kinds of applications, from calculators to controlling washing machines.

The basic building block of the chip is the transistor, strictly speaking the *field effect transistor* or FET. A field effect transistor has three electrical connections, called the *source*, the *drain* and the *gate*. Current can flow from source to drain only when a suitable voltage is applied to the gate electrode. So the FET

Left In reality only some five millimetres across, this 'chip' contains hundreds of thousands of electronic components intricately connected together.
Inset Very fine gold wires are attached to a chip before it is encased in plastic. The chip is too small and fragile to be directly connected to an external circuit.

works rather like a push-button switch, with voltage at the gate being equivalent to a finger on the button. By linking a number of these FET switches together, they can be made to carry out the simple logical operations that are the basis of modern computers.

A field effect transistor is made of two different types of silicon, *n-type* and *p-type*. N-type is silicon 'doped' with an impurity (such as phosphorus) to produce an excess of electrons in the material. P-type silicon, on the other hand, has a shortage of electrons and is made by doping silicon with a different impurity (such as boron). Two regions of p-type silicon in contact with a region of n-type silicon go to make up a typical FET. The two p-type regions form the source and drain, and an electrode lying over the n-type material between source and drain forms the gate. When manufacturing chips, which may contain thousands of FETs, the base material (called a 'wafer') is made of one material, for instance p-type, and islands of the opposite type are created where necessary by doping. Areas which do not need to be doped are carefully masked.

Production of a new chip begins with a product specification and a design team. The team includes not only electronics specialists who will design the circuits to perform the required tasks, but also production experts.

When the circuits have been designed, they are tested by computer simulation to see if they are likely to produce the right results. Once a satisfactory simulation has been achieved, the circuits are laid out in a separate graphical computer, and different parts of the various circuit devices are assigned to particular layers of the silicon sandwich. This computer also produces a large-

scale layout for checking, and coded magnetic tape to control the mask-making equipment. The masks are used to create the patterns in the different layers of the sandwich. Masks were once made by photographically reducing in several steps a very large layout of the required pattern, but nowadays an electron beam (e-beam) machine, controlled by the magnetic tape, writes the pattern directly on to the mask.

As already mentioned, chips are constructed on wafers of silicon. To prepare the wafers, pure silicon is melted in a furnace and

Ferranti

Above This picture gives some idea of the size of a typical silicon chip.
Below When designing a chip, giant plans of the circuits are prepared in order to make the various circuit element visible.

Fairchild

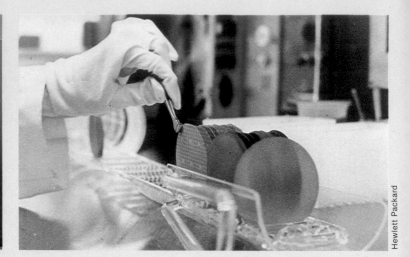

Stages in the manufacture of silicon chips. An electron beam machine *(above left),* controlled by magnetic tape, writes a circuit pattern on to a mask. And silicon with the 'dice' clearly visible are loaded *(above right)* into a carrier ready for a doping operation in which semiconducting regions will be formed. An oven *(left)* for treating silicon wafers.

grow a precise depth of oxide. The oxide layer is now coated with a chemical called the *photo-resist,* which hardens when exposed to ultra-violet light.

Ultra-violet light is shone on to the wafer through the appropriate mask—a single mask is used for each wafer, so it carries the same pattern repeated several hundred times over. The mask pattern is applied either by *contact printing* or by *projection alignment.* In the former, the mask lies directly on the wafer, in the latter an optical system projects the pattern on to the wafer. Whichever system is used, where the light falls the photo-resist hardens. The unhardened areas of photo-resist are now dissolved away by means of a chemical developer.

The next step is *acid etching.* The exposed regions of oxide are dissolved down to the surface of the wafer—only those areas coated with photo-resist remain unscathed. Hydrofluoric acid is the usual etching agent. Finally, the remaining photo-resist is removed with developer.

This leaves a pattern of oxide areas which screens the wafer where doping is not required. The wafer is now ready for doping. To apply the dopant, the wafer is heated in an oven in an atmosphere containing phosphorus or boron—depending on whether n-type or p-type regions are being created. The wafer remains in the oven until enough dopant atoms have penetrated the exposed areas of silicon.

The steps described produce one layer of the circuit. They are repeated, with variations, to produce all the other layers—a typical chip has 11 layers involving 60 or so process steps. Finally, a layer of metal (such as aluminium) is deposited, masked and etched to provide a network of connections linking the various n-type and p-type regions to complete the circuit. This layer of metal also includes contact pads which allow the chip to be connected to the outside world. Normally the chip is coated with a final layer of oxide for protection.

'Clean rooms'

Because the dice are so tiny and so intricate, the slightest contamination in any of the process steps is liable to ruin the circuit. So all operations are carried out in 'clean rooms' whose environment is very closely monitored, with air filtered and temperature controlled within precise limits. Operators wear protective clothing to prevent them from contaminating the atmosphere.

Once the protective oxide layer has been deposited, the dice on the wafer are tested. A set of probes makes contact with the pads of each dice in turn, and a computer runs through a series of tests. Those dice that fail the test are automatically marked with ink and are discarded when the dice are separated. Originally, separation was done by snapping the wafer along the crystal cleavage line, but modern techniques use either diamond saws or laser cutting. The successful dice are glued to a frame, and fine gold wires are welded between the contact pads and connector pins attached to the frame. Finally, the whole package is coated in plastic or ceramic and again tested.

A new product may only have a success rate of 5%, so the great majority of chips made are duds and have to be thrown away.

the required amount of 'dopant' is added. A large single crystal is then grown from the molten silicon. The crystal is cylindrical, and a flat is ground on one side, aligned with the crystal structure, to provide an accurate registration during processing. The crystal is then sawn into circular wafers of 100 or 120 mm (3.9 or 4.7 in) diameter. Once the surface of the wafer has been polished to a high degree of smoothness, it is ready to receive the layers of circuits. Each wafer will produce as many as 500 chips. While on the wafer, before being packaged, chips are often called 'dice'.

A typical layer of the chip circuit might be applied in the following way. First, a silicon dioxide layer is grown by passing steam or an oxygen-rich gas over the surface of the wafer in a furnace. By controlling the temperature and amount of gas, it is possible to

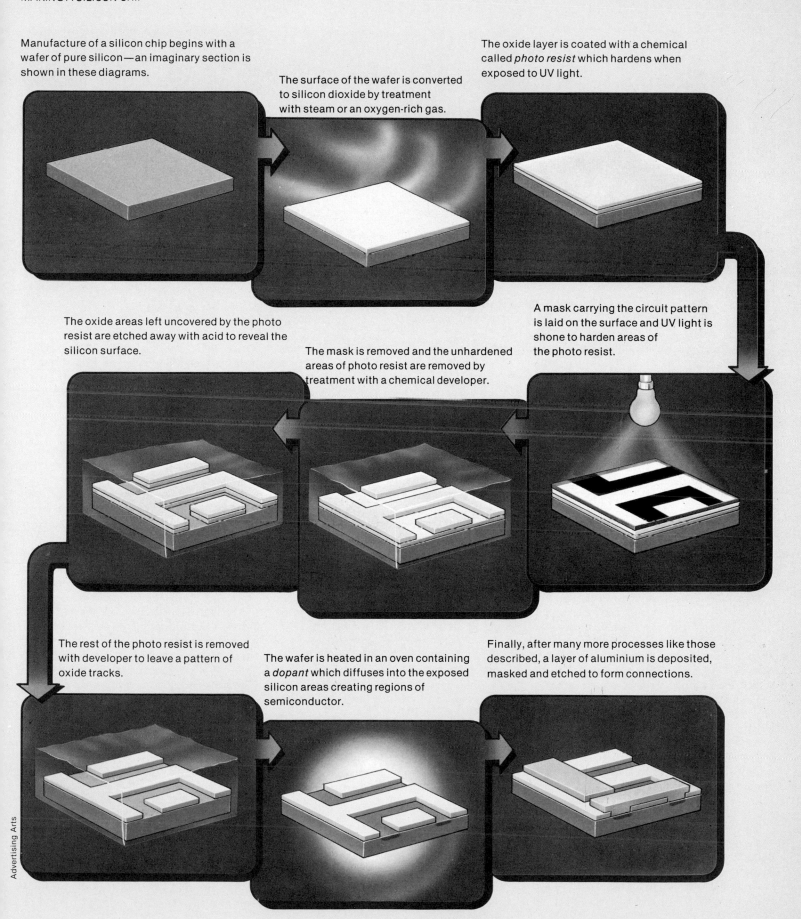

Manufacture of a silicon chip begins with a wafer of pure silicon—an imaginary section is shown in these diagrams.

The surface of the wafer is converted to silicon dioxide by treatment with steam or an oxygen-rich gas.

The oxide layer is coated with a chemical called *photo resist* which hardens when exposed to UV light.

The oxide areas left uncovered by the photo resist are etched away with acid to reveal the silicon surface.

The mask is removed and the unhardened areas of photo resist are removed by treatment with a chemical developer.

A mask carrying the circuit pattern is laid on the surface and UV light is shone to harden areas of the photo resist.

The rest of the photo resist is removed with developer to leave a pattern of oxide tracks.

The wafer is heated in an oven containing a *dopant* which diffuses into the exposed silicon areas creating regions of semiconductor.

Finally, after many more processes like those described, a layer of aluminium is deposited, masked and etched to form connections.

Advertising Arts

And a successful mature product may still have a pass rate of no more than 25%.

Improvements are made by redesigning troublesome areas of the circuit, by modifying the process, and, most important, by shrinking. Shrinking reduces the size of each dice, putting more dice on to a wafer. Apart from an obvious immediate increase in the number of chips produced for the same amount of processing, this also increases yield, since the same number of random faults (particles of dust, surface flaws and so on) will affect a lower proportion of dice. By decreasing the length of the connectors, shrinking also increases the speed at which the chips operate, an important consideration in computer circuits processing thousands of calculations per second.

'Moore's Law'

Before making any forecasts for the future, it is worth looking back to the mid-1960s when chips were in their infancy. Gordon Moore, founder of Intel, then predicted that the number of components that could be squeezed on to a single chip would double every year. Few people believed him, but so far 'Moore's Law' has proved more or less correct. In 1965 the number of components on a single chip was about 32, by 1980 the figure was approaching 1,000,000, in the laboratory.

It seems likely that chips will continue to follow Moore's Law. This will prove easier for memory chips, which have regular patterns of elements, than for microprocessors which are much more complex. As more and more elements are crammed on to the chip, production methods will have to change. For example, it will soon be impossible to use ultra-violet light in the masking steps—the wavelength is too long for the most intricate circuits. However, the problem can almost certainly be overcome by using X-rays, which have a much shorter wavelength.

Looking further ahead, scientists are examining possible alternatives for silicon as the fundamental material for chips. *Gallium arsenide* is one such contender—its advocates believe it will lead to much faster circuits with lower power requirements. Low power consumption is important because it means that the circuit elements will not dissipate so much heat and can therefore be physically smaller. And the smaller the elements the more can be put on the chip. Unfortunately gallium arsenide circuits are not as simple to manufacture as their silicon counterparts.

Cookers, washing machines and sewing machines are already being controlled by

Hewlett Packard

Above Evaporating a layer of aluminium on to wafers carrying chip 'dice'. The layer will be masked and etched to provide a network of connections and contact pads on each chip.

microprocessors. They are much more reliable and much more flexible than the mechanical devices they replace. A cooker, for instance, can turn on at a particular time, cook for a predetermined period and temperature, increase the temperature for a second set period and then cool to a 'keep warm' temperature until the food is removed. Programming these operations could be done by a sequence of key strokes (giving times and temperatures) or by summoning a program from memory. This is a trivial operation for a microprocessor, requiring little computing power. Extending memory and processing power only slightly could allow recipes to be recalled with appropriate cooking times and quantities. And the cooker could soon be linked to a household viewdata set, allowing further recipes to be called up from a central computer.

Experimental house

In the US an experimental house has been built with microprocessors controlling heating, ventilation, alarm systems and, when necessary, electrical equipment. One of the chief advantages of such a system is that it cuts energy loss to a minimum.

But it is at work that the chip will have its biggest impact. The most obvious influence will be in industries whose products are superseded by silicon products. For example, the clockwork-driven watch has virtually disappeared; so has the mechanical adding machine. Industrial robots will be another area of influence. Already Fiat are producing a car largely made by robots, and specialist robots (welding robots, paint spraying robots and so on) are certain to become more

and more commonplace. This sort of job displacement by new technology is by no means new; it has been happening since flint knappers were displaced by bronze casters in prehistoric times.

White-collar workers will be affected most of all. Word processors, typewriters that store and correct text, are beginning to make a real impact. If these are linked to video screens and a central electronic memory, paper copies and multiple files begin to disappear. Inevitably this means less work for typists and filing clerks.

Moving money from one bank account to another electronically is already done, and computer-based stock control systems and computer-controlled warehouses are in existence. A little imagination in coupling together these systems through an electronic communications network, and one can envisage orders being placed by a stock control computer, supplied by a warehouse computer and paid by an accounts computer instructing a bank computer. This sort of arrangement will provide management with much more detailed and accurate information, but at the same time will lead to a reduction in the number of people employed in routine accounting jobs.

Just what will be the long-term impact of the chip is a matter of debate, but if past experience of new technologies is anything to go by, the silicon revolution will create at least as many jobs as it destroys.

Quartz: a cut ahead

The word 'quartz' has become synonymous with accuracy in electronic watches, hi-fi turntables and tuners. But what has a clear, crystalline rock to do with electronic accuracy? Remarkably, that accuracy does depend on tiny, carefully cut slivers of the mineral. The electronics industry has come to depend on the properties of this rock and there is every sign that reliance will increase.
In its natural state, quartz is a cloudy or clear mineral, composed of silicon dioxide which has crystallized into hexagonal crystals. The main supplier of natural quartz is Brazil and the material exported varies from comparatively small pieces to large blocks weighing up to 90 kg (200 lbs). It was widely used up to the beginning of the 1970s in a variety of optical instruments required to transmit ultra-violet light (which ordinary light blocks).

Synthetic quartz

But natural quartz is expensive, and sold in irregular lumps. Their useful yield can vary considerably, but is in the region of 50% to 60%, according to the defects in a rock and the way it is to be cut. For some purposes, considerable waste is produced by the need to cut along particular optical axes. Thus a search developed for ways to produce faultless synthetic quartz in quantity.

A breakthrough in this quest came in the 1960s, pioneered mainly by Japan and the USA. Coincidentally, a boom in electronics technology produced a massive requirement for the material. These developments led a humble mineral to a new pinnacle of esteem.

The process of synthetic quartz production varies among manufacturers, but typically, the crystals are grown in large steel autoclaves—thick-walled containers—at a temperature of 350°C to 400°C, and at a pressure of 2,000 atmospheres (30,000 psi or 2,100 kg/cm^2). Inside is a saturated solution of silicon dioxide. The growth process starts with 'seed' crystals, sometimes cut from natural quartz, but predominantly selected from synthetic quartz. These are planted in the autoclaves. From a seed measuring 2.3 cm by 3.8 cm, a crystal weighing about ½ kg (1.11 lbs) could be grown in three weeks. Over a period of six weeks, the weight would increase by a further kilogram.

Without the cost of quarrying or transportation, the yield from this process exceeds 90%, compared with the 60% yield

generally obtained from natural quartz.

For optical use, synthetic quartz offers superior transmission properties to those provided by natural quartz and will transmit light of even shorter wavelength—down to 185nm (nanometres).

Once the synthetic crystal has been grown to its desired size, it is examined for defects and must then be cut into thin wedges according to the application intended. There is a critical optimum angle of cut which must be made with respect to the crystallographic axes of the quartz, in order to ensure that the right properties are obtained for the cut crystal, according to the purpose for which the quartz is intended.

The important electrical property of quartz crystal that has pushed it to the forefront of modern electronics was first recognized in the 1880s. It was discovered that when quartz was deformed an electrical charge was generated. From the Greek word 'piezo', meaning pressure, the word *piezo-electric* was derived for this curious property. The effect is now exploited in low-cost crystal microphones, vibration sensors and cheap record player pickup cartridges, where vibrations cause the crystal to bend in sympathy and produce an electrical signal.

The beauty of quartz in its natural state *(right)* and in synthetic form *(below)*. This attractive mineral is—in spite of its appearance—formed from an oxide of silicon, like sand. Being automorphic, quartz grows to a shape that reflects the internal lattice structure of its atoms, provided that growth is unimpeded. This produces an extremely elegant six-sided translucent prism.

The effect was found to work in the opposite direction too, so that if an electrical signal was applied across the crystal it would expand and contract in certain directions at a specific and perfectly maintained frequency. It is this latter property—of perfectly maintained frequency when excited—that has made quartz so important.

Protection

The crystal is, potentially, a highly accurate device which can be relied upon to maintain its frequency over a long period of time, even when operating in electronic equipment under conditions that may be arduous due to extreme heat or humidity. In general, the crystal must be protected from the outside environment by a strong case, which will both transmit electrical signals to the device from external circuits and protect it from mechanical shock.

To avoid contamination from dust and dirt, the crystal unit must also be fully encapsulated in an hermetically sealed holder which may be filled with an inert gas, or evacuated of all air.

Quartz crystals are usually incorporated into electronic *oscillators* and *filters* to improve their characteristics. An oscillator provides a continuous electronically generated signal, and oscillator circuits are used widely in all forms of electronic equipment. Where fixed and stable frequency is required in the signal produced, usually for reference purposes, a quartz crystal is incorporated into the oscillator and this prevents it from 'drifting' in frequency.

Filters are designed to accept signals in one frequency band, whilst rejecting those in another frequency band which might cause unwanted interference. They must often discriminate sharply between the two bands—wanted and unwanted—and the quartz crystal is often used to improve this important characteristic of filters.

It is in its role as a highly stable oscillator that the quartz crystal has now become so widely known, especially in 'quartz' watches, 'quartz-lock' hi-fi turntables and tuners and 'crystal-controlled' transmitters and receivers.

There are various requirements for crystal oscillators, but the most important one employs the quartz property of *frequency stability*. Defined as the ratio of the variation of crystal frequency achieved in practice to the specific design or 'centre' frequency, it takes into account all the factors that add up to cause such variation. These are the effect of temperature change, ageing, variations of oscillator supply voltage, changes in the load an oscillator circuit might experience—and environmental factors such as vibration.

Variations of stability with time may be quoted in different ways. Short-term stability is that achieved over a period of a second or less, and 'daily ageing' and 'long-term ageing' of a month or year may also be quoted. Long-term ageing is normally measured at an elevated temperature such as 85°C, where ageing occurs at a faster rate.

Ageing rate is measured in *parts per million,*

Right Optically finished quartz crystal blanks have their electrodes plated on, using automatic evaporators.
Below Heat sealing a finished crystal into its glass bulb container.

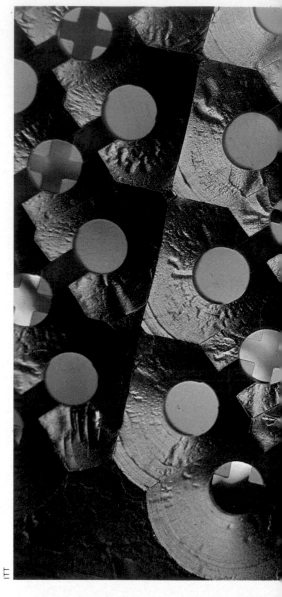

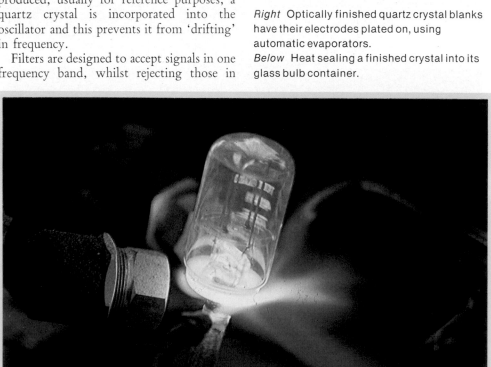

or *ppm,* which is a convenient way of quoting very small ratio changes. Crystals in a solder-sealed case typically age by up to 10 ppm per year, equivalent to a change in operating frequency of just ten-millionths per year. Crystals in glass or resistance-weld packages age less than 1 ppm per year after the first month of life. Good short-term stability is also a frequent requirement.

There are three basic types of crystal oscillator. (Crystal oscillator is often abbreviated to XO, X representing crystal and O oscillator.) The simplest form is known as *SPXO,* short for *Simple Packaged Crystal Oscillator.* More sophisticated is the *Temperature Compensated Crystal Oscillator,* or *TCXO.* For extreme stability there is the *OXCO,* or *Oven Controlled Crystal Oscillator.*

The SPXO combines a basic crystal with discrete electronic components to form a

Temperature Compensated Crystal Oscillator, or TCXO, may be suitable. These use the same crystals as SPXOs, but additional electronic circuits are incorporated in the associated electronics to produce frequency compensation that is equal to, but opposite in direction to, the crystal's own temperature drift. The two drift characteristics—that of the crystal and that of the electronics—cancel each other out to yield a great improvement in stability. Typical stability values for TCXOs are 1 ppm between 0°C and 60°C, 2 ppm between −20°C and 70°C, and 3 ppm between −40°C and 84°C. As TCXOs are used mainly in mobile radio applications, their power consumption is kept low (around 7mA maximum). With this improved stability, ageing becomes more important in the frequency tolerance. As a result, TCXOs generally use glass-encapsulated crystals.

Ovens

Employing a heated temperature-controlled container or oven, the Oven-Controlled Crystal Oscillator, or OCXO, again normally uses a simple crystal. The temperature of the oven is set to the upper turnover point of the crystal, where its change in frequency for a given temperature change is minimized.

To minimize heat losses, and hence the power consumption, the oscillator is usually thermally insulated. Because of this, the OCXO is larger than the TCXO and consumes considerably more power, but the temperature stability of the oscillator can be improved from several parts in a million to a few parts in one hundred million. Further improvements can be achieved by the use of a double oven.

The nature of the quartz crystal industry has largely been determined by the requirements of radio communications. The design of radio systems, and particularly the choice of channel allocations, has been strongly influenced by the stability and selectivity of quartz-controlled oscillators and filters. This trend will continue through the next decade.

Improved stability of 1 ppm, due to all causes over periods of many months, will become standard for temperature-compensated oscillators in the future, with 0.1 ppm possibly available in premium products. Filters will be built with defined low distortion levels, and narrow bandwidths will be available in filters for exacting radio use.

Although radio communication technology is the prime influence on the crystal

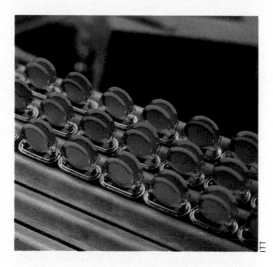
Above Lines of crystal blanks, mounted on their bases, emerging after plating.

Above Crystals being prepared for cleaning, before having their electrodes plated on.

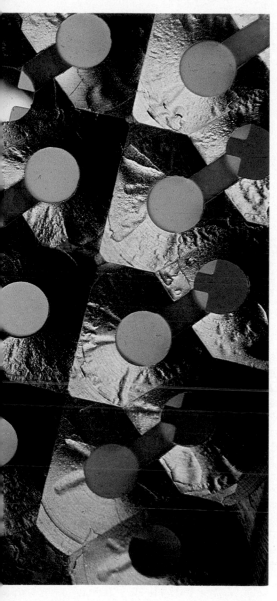

complete working oscillator, the whole being encapsulated in an epoxy-resin package for protection. Alternatively, a thick-film integrated circuit may be married to the crystal and the whole packaged in a can. This type of arrangement provides the smallest type of complete working crystal oscillator, power consumption being dependent upon the type of output available.

Frequency stability of an SPXO is in the range of 1 ppm to 100 ppm, depending upon type and the temperature range over which it is expected to operate. Angle of cut of the crystal is chosen to give optimum performance over the particular specified temperature range of the unit. Simple Packaged Crystal Oscillators are typically used as clock oscillators for electronic synchronization in many different types of circuits.

If better stability is required, then a

industry, due to its unique combination of technical stringency and high-volume usage of crystals, other applications will continue to account for more than half the market value of crystal products sold.

Whilst the word 'quartz' has become commonly associated with certain electronic household products, quartz crystals are still used heavily in other commercial fields, within electronic equipment.

Clocking oscillators, for example, provide a steady synchronizing signal to other electronic circuits and are widely used in data processing equipment. At the beginning of the 1970s, the electronic data-processing manufacturers generally bought crystals in standard holders and provided their own oscillator circuitry. Mounting the crystal holders on computer boards, however, was not particularly convenient, as the leads all

Quartz

Synthesized tuning is already becoming commonplace in hi-fidelity tuners and will be introduced into ordinary transistor radios in the future, eliminating mechanical tuning mechanisms. They will be replaced by automatic electronic tuning, or push-button station selection. Rapidly plummeting cost and steadily increased complexity of microchip circuits, plus the stability of quartz crystals, combine to allow many new electronic techniques to be developed. *Frequency-synthesized tuning* is just one of them. Previously, such circuits have been too complex and costly to exploit, but microchip miniaturization has changed this.

The block diagram shows the role of the quartz crystal in a synthesized tuner, how such a tuner works, and its complexity compared with traditional tuning systems where crystals were not used. Previously, *manual tuning* altered the frequency of a variable oscillator within a radio (its 'local' oscillator), in addition to tuning other circuits.

With synthesized tuning a quartz crystal frequency reference is combined with an electronic circuit arrangement known as a *phase-locked loop*. The radio's local oscillator is now *voltage tuned* (it is a Voltage Controlled Oscillator or VCO) by a Phase Detector, via a filter, and takes its tuning commands from a Programmable Divider, which is controlled by the user. The Crystal Oscillator acts as a highly stable frequency reference source.

If, for example, a user wishes to tune a station which requires a local oscillator frequency of 100MHz, by making appropriate commands

CRYSTAL CONTROLLED OSCILLATOR

QUARTZ CRYSTAL

frequency-stable 100kHz reference signal

PHASE DETECTOR

LOW PASS FILTER

error tuning voltage

VOLTAGE CONTROLLED LOCAL OSCILLATOR (VCO)

local oscillator output to tuner

frequency divided VCO signal

PROGRAMMABLE DIVIDER

pre-set tuning

tuning commands to other circuits

Left A modern quartz-referenced tuning system. *Right* Simple mechanical tuning.

into the Divider, it will divide the local oscillator output down by 1,000. The Phase Detector produces an error voltage that tunes the local oscillator when there is a difference in frequency between the crystal reference signal and the output from the Divider. Since the Divider is receiving a divide by 1,000 command, the Phase Detector must tune the local oscillator to a frequency 1,000 times greater than the crystal reference, or $1,000 \times 0.1$MHz, which equals 100MHz—the required frequency. When the VCO (Voltage Controlled local Oscillator) reaches 100MHz, output from the Divider, 1,000 times lower, reaches 100kHz, which is identical to that of the Crystal

reference Oscillator. With no error between Divider output and the reference, the Phase Detector ceases to produce a tuning voltage and the system then locks into a steady state, reliant for stability upon the stability of the Crystal reference Oscillator. In this state the system is said to be in *phase-lock*.

Obviously, frequency synthesis is a complex method of radio tuning and yet it is rapidly being adopted in all types of receiver, commercial and domestic, for many reasons. In addition to complexity, a frequency synthesized tuner will only tune in steps: 100kHz being a common value on vhf/fm radios. However, most broadcasting stations are accurately spaced at such

Left The web of connections within a Temperature Compensated Crystal Oscillator, or TCXO. The circuits are complicated by the use of electronic temperature compensation.

appeared at a single end of the package. Thus, when integrated crystal oscillators in special compact dual-in-line (DIL) packages appeared, they gained rapid acceptance, and may now be found in equipment ranging from electronic cash registers to computers.

These compact DIL oscillators cover a wide range of frequency, possess a variety of specifications and output circuits. They can provide signals from 250 kHz to 65 MHz, with accuracy ranging from 1,000 ppm for general use to 40 ppm for specialized use. Within a few years they will become capable of operation over the entire military temperature range. By the middle of the decade,

intervals, since their transmitters are also crystal controlled. Because of this, such tuners will tune into broadcast transmitters with a degree of accuracy impossible to achieve by other systems.

This is of great importance in improving programme quality, hence the adoption of radio synthesis techniques for hi-fi tuners.

However, there are many other advantages. Because synthesized tuners hold frequency with the stability of the crystal reference, they cannot drift off-tune. But of more impact in the future however is their suitability to computer control. Already, hi-fi tuners have a memory installed, and pressing an appropriate button causes the memory to tell the synthesis system to tune to the required station. Models now appearing will also switch on and off at pre-programmed times. They can control an accompanying amplifier and cassette deck as well. In the future, mechanical tuning may well disappear altogether.

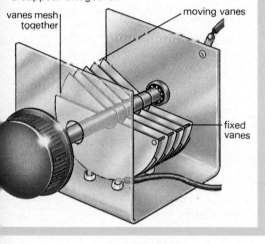

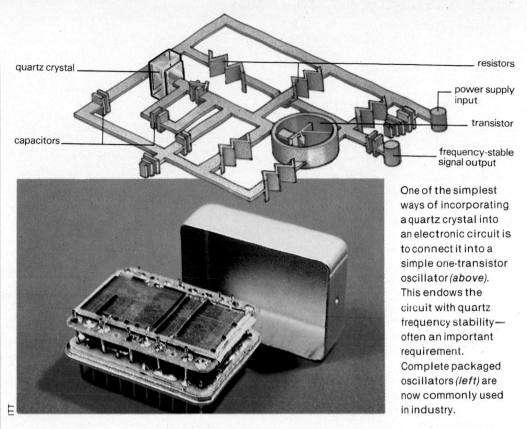

quartz crystal

capacitors

resistors

power supply input

transistor

frequency-stable signal output

vanes mesh together

moving vanes

fixed vanes

One of the simplest ways of incorporating a quartz crystal into an electronic circuit is to connect it into a simple one-transistor oscillator (above). This endows the circuit with quartz frequency stability—often an important requirement. Complete packaged oscillators (left) are now commonly used in industry.

the package will be standardized, probably as a 'resistance weld' design. Frequency range will be extended to cover all logic speeds, and precisions down to 10 ppm will become available for commercial temperature ranges.

Soon, the DIL packaged oscillator will probably appear in temperature-compensated form, with a stability of 1 ppm, and will find application as a miniature oscillator for single side-band (SSB) portable radios.

Telephone trunk systems currently use significant volumes of crystal filters for selecting channels in their channel-division systems. The filters are dedicated to their relevant channels so they are arranged as a comb of filters. The variety of centre frequencies produced may be regarded as the telecommunications equivalent of the mobile radio's 'channel' crystal. Just as such channel crystals will eventually be displaced by fre-

quency synthesizers, so will the telecommunications filter be made obsolete by the change to fully digital transmission systems. The next few years will see fairly high volume of production of telecommunications filters, followed by a relatively sharp cut-off in need as the final systems are installed. Crystal filters in the telephone network will then be limited to terminal applications such as in radiotelephones.

The telephone network will, however, need to use highly precise oscillators for signalling and system control. The crystals within accurate master oscillators for such equipment will have to display an ageing rate of less than 0.03 ppm over a year, and temperature stability of 0.01 ppm over the range of 0°C to 70°C. These are very stringent specifications. Other specialized types of crystal oscillator will also be required in digital telephone systems of the future and so, although various current uses for crystal oscillators will not be needed in the future, new ones will be found.

Area navigation systems require reasonably stable ovened oscillators, with short-term stability similar to telecommunications master oscillators, but less stringent temperature and ageing requirements. New systems, based on satellites, will require better long-term stability, with an ageing specification approaching 0.1 ppm per year.

The clock, watch and television crystal markets are characterized by high volumes of the product at low prices, and are dominated

by Japanese manufacturers. Clock and TV crystals typically operate at around 4 MHz and offer stability comparable with that of other types. Watch crystals are generally of a tuning-fork design operating at 32,768 Hz and are significantly less temperature-stable than clock and television crystals.

The trends for watch crystals have, in the past, been largely towards smaller devices, designed to overcome market resistance encountered by the first, bulky, crystal-controlled watches. Recently, miniaturization has appeared to be almost an end in itself, with quartz watches less than 2 mm thick now on offer. There have also been some thicker experimental watches using specially cut crystals, and also pairs of crystals, to improve timekeeping to chronometer standards.

The effective stability of watch crystals will probably be improved by incorporating temperature compensation on the electronic oscillator circuit chip, as marketing emphasis changes from attracting first-time buyers to providing improved replacements.

As the demand for all types of electronic equipment increases in the commercial and domestic markets, and greater complexity and precision are called for to achieve new functions, quartz crystal will be used in ever greater quantities. There seems little doubt that this crystalline mineral, which appears at first sight so out of place in an electronic circuit, will broaden its role as the perfectly accurate electronic heart.

The electronic companion

During the last ten years, modern technology has entirely transformed not only our means of manipulating numbers, but also our attitude towards mathematical calculations. Today's electronic calculators are a remarkable feat of miniaturization and performance. And they are available at outstandingly low cost.

Whereas a popular general-purpose computer of the 1950s—the IBM 650—weighed almost three tons and required five to ten tons of air conditioning, today's programmable calculator equivalent works many times faster, requires one-hundred-thousandth of the power—and can fit on the palm of one hand.

Calculators have been designed to play popular card games, predict horoscopes, aid slimming by counting calories or simply play melodies for the musically minded. More seriously, users can buy calculators that can be trained, through programming, to do calculations to the purchaser's specifications.

Yet, in operation, calculators rely on the use of a very simple mathematical process. They can only add and subtract, using a simple code of one's and zero's in a system known as *binary logic*. There are ways of carrying out most mathematical computations through long routines of adding and subtracting, all of which would be impracticably tedious for human use. In the ultra high-speed electronic signal processing circuits of a calculator, however, such lengthy calculation techniques are handled in a very

Theo Bergstrom/IMC

short space of time (although programmable calculators do require a short degree of 'thinking' time when faced with complex operations).

The binary system of counting that all calculators use has long been familiar to Man, but there are significant differences between the way a calculator and a person tackles a simple mathematical sum. Counting in groups of ten is convenient to a ten-digit species—one that has ten fingers. But the system has limitations, not the smallest of which is the problem of recording any quantity greater than ten on the fingers of both hands. This problem is easily solved by

noting down each ten-group on paper or some convenient material.

People discovered long ago, however, that it is possible to count in groups of five on the fingers of one hand and record each group on the fingers of the other hand. Using this technique they could record five groups of five, or 25, on one hand and one five on the other hand with their separate fingers. This brought the total count that could be recorded up to 30. Obviously, other counting systems have their advantages.

Counting in groups of ten is mathematically termed working to the Base of Ten. Counting in groups of five is called working to the Base of Five. Electronic circuits have no fingers to count on and so Base Ten or Base Five are not innately suitable Bases for them. However, both electrical and electronic circuits do have two easily defined states—'off' and 'on'. In the 'off' state, no current flows in the circuit; in the 'on' state, current does flow. This makes calculators best suited to counting to the Base of Two (the binary system) rather than to the Base of Ten (the decimal system). *Zero* is an 'off' state in a circuit, *one* is an 'on' state, and *two* (or a group of two in fact) is recorded in another counting register as an 'on' state on the circuit.

Silicon blocks and chips

Many separate circuits are required to register enough groups of two to construct a useful system, and one of the first electronic calculating machines, entitled ENIAC, was constructed in 1946 using no fewer than

18,000 valves (tubes). In comparison, domestic radios of the period used six valves.

It was not until the 1960s—when new semiconductor technologies allowed hundreds or thousands of transistors to be squeezed into a small block of silicon, in a process known as Large Scale Integration, that the task of building a small electronic calculating machine became feasible.

The Large Scale Integrated circuit, abbreviated to LSI, is a form of silicon chip. Current types contain tens of thousands of transistors which may be designed to perform various functions, from mathematical calculations to memory storage, using binary logic. One of the measurements of a 'chip's' performance is the number of active element groups (AEGs) it has, each of which may be composed of one or many transistors arranged to fulfil a specific function. Current chips may possess well over 50,000 AEGs on less than a $\frac{1}{2}$ sq cm of silicon, and it is this density of packing, which has improved dramatically over a relatively short period of

Below This Sharp calculator uses a liquid crystal dot matrix alphanumeric display. It can depict the entire alphabet, in addition to numbers and symbols. A maximum of 24 characters may be stored in its memories and are held when switched off. This allows the calculator to store short notes or messages, and they are recalled as a running display that moves from right to left across the screen in a continuous running loop of information. Future calculators will hold more information and speak their messages.

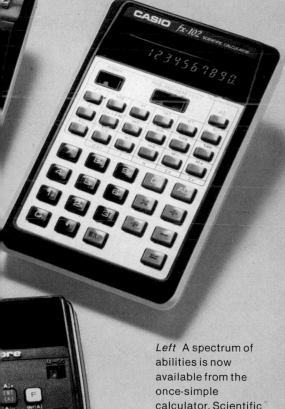

Left A spectrum of abilities is now available from the once-simple calculator. Scientific models possess multiple memories and use powerful algorithms. Small pocket types play tunes and tell the time. Some can even teach simple sums. Red *led* and green VF displays are being replaced by black liquid crystal displays that use less power.

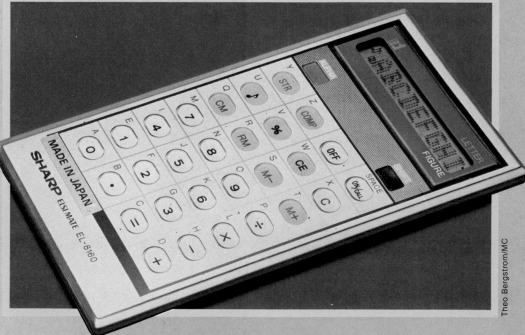

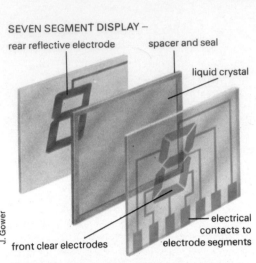

SEVEN SEGMENT DISPLAY –
rear reflective electrode
spacer and seal
liquid crystal
electrical contacts to electrode segments
front clear electrodes

J. Gower

Above The seven-segment display is capable of depicting all numbers from one to nine by activating combinations of the segments. This type of numeric display will be superseded by dot-matrix presentation, which can additionally display letters and symbols.

development, that is the prime factor in the rapid advance of calculator ability.

Theoretically, as more AEGs are incorporated onto a chip, the cost of each AEG declines. It has already fallen from around £5 UK in 1960 to less than 0.005 pence per AEG in 1979, or to one hundred thousand times less in a period of 19 years.

One of the most striking features of these chips is their speed of operation, coupled with low power consumption. Without the latter, it would never have been possible to build calculators that run from batteries alone. The transistors on LSI circuits consume around one-millionth of a watt in operation. New types coming into use will consume even less.

Mathematical operations are carried out in the logic circuits of a calculator by routing digital signals through a series of electronic gates, according to the number of pulses sensed. Input signals come from the keyboard, which is linked to a section of the calculator that converts the decimal keyboard numbers into binary coding. They then become known as *binary coded decimal signals*, or *bcd* for short.

By binary addition, or a process of subtraction known as complementary addition, the calculator carries out even advanced mathematical conversions, through routines known as *algorithms*. These are the particular mathematical processes it has to use in order to solve complex problems by addition alone. For example, a calculator cannot remember sets of answers, as a human does with, say, multiplication tables, even though it pos-

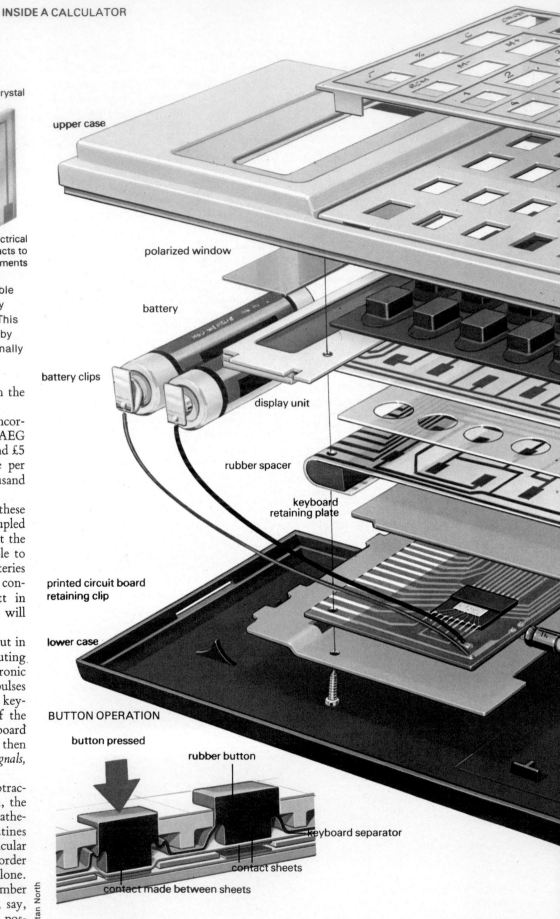

upper case

polarized window

battery

battery clips

display unit

rubber spacer

keyboard retaining plate

printed circuit board retaining clip

lower case

BUTTON OPERATION

button pressed

rubber button

keyboard separator

contact sheets

contact made between sheets

Stan North

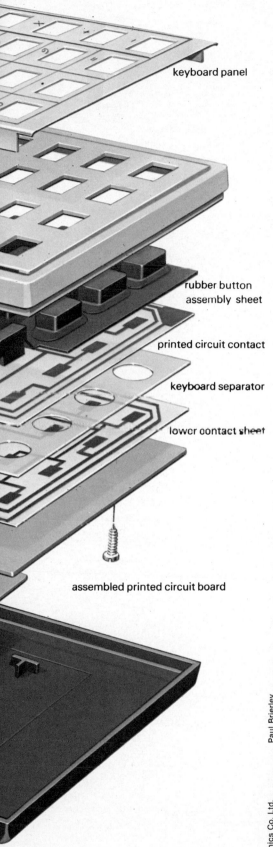

keyboard panel

rubber button
assembly sheet

printed circuit contact

keyboard separator

lower contact sheet

assembled printed circuit board

Above An exploded view of a typical calculator. Little space is needed by the single integrated circuit.

sesses memories. If asked to multiply 8 by 9 (8 × 9) for example, a human will remember that the answer is 72, but a calculator adds eight nines together to produce the same answer. For a human, this process would be intolerably slow, but a calculator can achieve the task in milliseconds (thousandths of a second), or less.

More difficult mathematical processes build upon the four basic ones of addition, subtraction, multiplication and division, squared functions and reciprocals being in the next stage up. In this manner complex mathematical functions can be handled with accuracy through the use of algorithms.

Memories

In addition to logic circuits, all calculators possess a memory of one type or another. At the simplest level, memories store information from the keyboard before mathematical processing takes place. They are also used to store intermediate results during calculations, and instructions to the calculator on how to carry out algorithms, for example. Again, the advent of memories was due solely to the ability to accommodate tens of thousands of transistors on a silicon chip. In order to store, or remember information, a calculator only has to record logic states of zero (off) and one (on) and this is achieved with circuits commonly called *flip-flops*. These simple but ingenious circuits will

take up and hold one state or another, so remembering any state fed to them. Thousands of such circuits can be fabricated onto a silicon chip memory in order to store large quantities of information.

There are a variety of memories available, all of which achieve different functions. *Read-Only Memories,* or *ROMs,* are incorporated into calculators with instructions already recorded on them. They are pre-programmed with various instruction sets for the calculator, in order to tell it how to achieve various tasks, such as performing algorithms for example. A ROM may typically contain 16,000 bits of information, or more.

Data from the keyboard or from calculations may be stored temporarily in *shift registers* or in a *Random Access Memory (RAM)*. The latter type of memory is more flexible in use and is becoming progressively more popular, allowing information to be entered and retrieved without difficulty.

Below The anatomy of a programmable calculator. In the background is the 'thin film' programmable calculator circuit carrying six 'chips', packaged together to save space. In front is a programmable Casio calculator with cassette interface unit to allow information to be stored on tape.

Paul Brierley

Casio Electronics Co. Ltd.

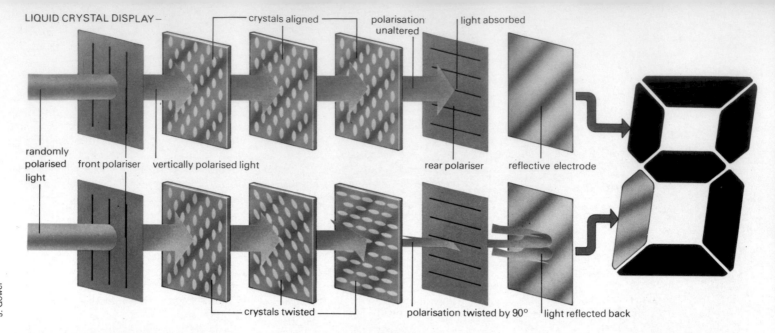

LIQUID CRYSTAL DISPLAY—
crystals aligned
polarisation unaltered
light absorbed

randomly polarised light | front polariser | vertically polarised light | rear polariser | reflective electrode

crystals twisted | polarisation twisted by 90° | light reflected back

J. Gower

Programmable calculators have quite powerful RAMs, but the simplest calculators on the market that perform only basic mathematical functions rely on the use of shift registers for holding data during calculations, whilst a ROM directs their mathematical processes.

Electronic calculators are organised into various functional blocks. The input unit converts decimal keyboard entries into binary code and may hold it until the information can be fed into the *Central Processing Unit (CPU)*, which carries out all calculations by addition. Instructional entries, such as a 'divide' command, are routed through to the ROM, which then instructs the CPU how to process the data fed in from the keyboard. Finally, an *Output Unit* holds the answer from the CPU, decodes it from bcd to a format that will drive a display and also provides suitable signals with which to drive the display devices, since they will not work directly from the logic circuits.

Not all developments in calculators are directly connected with their mathematical computing power. Portability is a very important feature, but requires the use of small batteries of limited power. Red *light emitting diode (led)* displays were popular for their low cost and brightness, but they used a lot of power and ran down batteries.

Attractive green *vacuum fluorescent (vf)* displays followed, but offered little in the way of power saving. *Liquid crystal displays (lcd)* have become popular for small portable calculators since they consume one hundred times less power than glowing types, even though legibility is poorer and back-lighting is needed in the dark. Slow decay time causes lcd displays to fade out instead of disappearing immediately when activating signals are removed, which can also be annoying. Such displays rely on the properties of nematic

crystals in a grease-like fluid that is trapped between a clear front plate and a shiny reflective rear plate. Applying a voltage between the front plate segments of the display and the rear reflective plate causes the fluid to become cloudy, altering the reflection characteristics to produce a visible bar. As with *led* displays, seven segments are used to form a figure 8 and by activating appropriate segments numerals 0–9 can be formed.

The most recent development is the *liquid crystal dot matrix* display that can depict alphabet characters in addition to numerals, allowing calculators to deliver messages, display formulae, ask questions and use Greek characters that occur in mathematics.

Development

Calculators have already reached the point where their complexity and abilities, in mathematical calculations at least, fully meet the requirements of the market. In practice, most users require little more than basic mathematical functions and these are now available at very little cost. Making a calculator more complex in mathematical ability, usually by addition of a powerful ROM and RAMs, is of little use to most people since a fairly substantial knowledge of mathematics is necessary to appreciate it.

Programmable calculators allow highly complex mathematical computations to be carried out repeatedly with great speed, but they must first be programmed by the user and this task is in itself problematical and lengthy. The answer is to provide users with programmes that have already been formulated by the manufacturers to suit the needs of certain categories of users such as engineers, accountants, mathematicians, etc. Programme libraries are available from some manufacturers, containing either lists of programme instructions for the user to enter

Above The operation of a liquid crystal display. With no energizing signal, light from the front polariser is twisted through 90° by the crystals, passes through the rear polariser and is then reflected back out. When energized, the crystals straighten, light is absorbed by the rear polariser and the segment turns black. The front crystal electrode is clear, but the rear may be reflective, or clear if back-lighting is used.

into a programmable calculator or a preprogrammed module that simply plugs into the calculator. All that is required of the user is to enter the data that the calculations require and the answer is presented by the calculator. If many solutions to a problem exist, many answers may often be recalled.

The popularity of this form of calculator is obviously limited, and in consequence manufacturers have tended to find other attractive features that might broaden appeal to beyond those who simply want to add, multiply and divide.

Reduced power consumption and smaller batteries have enabled the size of even complex mathematical calculators with memories to be reduced drastically. As memory capacity is increased in the future from thousands of bits to millions of bits, and displays present information in the form of words and symbols, in addition to numerals, the role of calculators will be greatly expanded. They will develop the ability to store messages, dates, addresses, tell the time, give warning of events like an alarm clock and generally store many sorts of useful information.

In fact, the calculator might become less of a mathematician in the future, but more of a general purpose everyday aid. The ability to understand speech, and to speak, is within the scope of computers. When this technology reaches calculators they could become intelligent electronic companions.

Fiat/CDP

What future for the robot?

Robots will revolutionize the production lines of the world during the next few years. Strong, accurate and reliable, they can handle dangerous or monotonous work at a constant high standard, releasing human workers from hot, noisy and dirty environments. And they can do sophisticated tasks too—already they are used for assembling products as varied and complex as motor cars and wrist watches.

Robots, of course, are still machines, still dependent on their human instructors. But the amount they can be taught is increasing dramatically. For example, systems are now available which automatically recognize different components on a conveyor belt and allow the robot to select the correct program for assembling them.

Similarly, the conveyor drive mechanism can provide signals to synchronize the robot's movements with the speed of the conveyor—faster or slower as required. In loading and unloading a metal forging press, for example, the robot has to pick up a white-hot billet from the chute and place it in the press. Then, after the forging, it has to remove and place the billet in a trimming press for the next operation. A signal tells

the robot that a hot billet is available at the pick-up point. The robot takes the billet and moves to the front of the press, then stops. Then it signals the press 'are you open?'

If the answer is 'no', the robot waits at that point in the program. If the answer is 'yes', the robot places the part in the press. Now it says to the press 'don't close, I've got my hand in'—and the press is prevented from operating. Once the part has been placed in the die and the robot arm has withdrawn, the robot tells the press that the forging cycle can be started. A similar routine is followed to remove the part from the forging press and load it into the trimming press.

Teaching the robot

In another widely-used application, spray painting, the robot will probably be sited in a booth, in front of a conveyor on which will be hung components. A skilled operator teaches the robot the programs necessary to paint the various manufactured parts and it is then up to the robot to carry out the spraying. To do this the robot must have a number of channels for incoming and outgoing commands. For example, the robot must be told when to start painting. This is usual-

ly done by a switch mounted on the overhead conveyor which is operated by the hanger which supports the component.

Once the 'start' signal has been sent, the robot itself commands the painting system to turn the spray gun on and off at the appropriate points in the program, and sends instructions to change the colour of the paint.

Compared with the robots of science fiction, perhaps, such machines seem elementary. But as micro-technology develops, especially in the electronics and computing fields, rapid advances must be anticipated. Communications between robots and other machines in the work place (known as 'interlocking') is done by means of microswitches and other sensors. This kind of apparatus is becoming more ingenious so rapidly that it seems to change every month.

The first industrial robot made its debut in industry in the United States in the early

Above On a production line the robot 'arm' must be able to operate along three axes: extending and retracting horizontally; revolving about its own base; and tilting, to raise and lower the arm.
Inset Robots can perform repetitive jobs such as paint spraying with pinpoint accuracy.

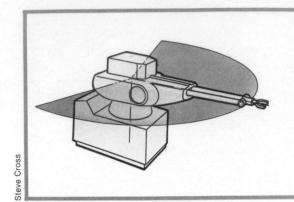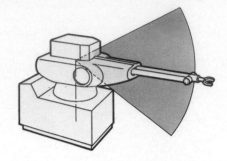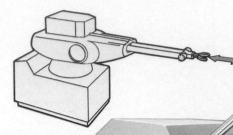

Steve Cross

The three basic axes of a typical robot are comparable to those of the human body. The 'waist' action *(above left)* permits horizontal swinging from side to side. The 'shoulder' *(above centre)* allows the arm to be raised and lowered. The 'arm' *(above right)* can both extend and retract.

1960s. Two companies, Unimation Inc. and the American Machine and Foundry Co., designed and built a number of machines for manufacturing companies, with special interest being shown by the automobile industry. These two pioneer companies are still in business today, with Unimation Inc. now the world's largest supplier.

In general, the design of robots themselves has not changed greatly over the two decades. The industrial robot basically comprises three parts: a hydraulic power unit, a mechanical unit and a control system. The hydraulic power supply may be integrated into the body of the mechanical unit or be a separate, free-standing unit. It provides the energy to operate the various axes or 'arms' of the robot. The hydraulic piston may drive directly, through gears, levers or other linkages, or a hydraulic motor may be used if direct rotary motion is required.

A reservoir containing some 90-110 litres (20-25 gallons) of hydraulic fluid usually forms the base of the hydraulic power unit. A motor-driven pump produces the high pressure and delivers hydraulic fluid at 70 kg/cm^2 (1,000 psi) to the mechanical unit. Delivery is always through a replaceable filter of very fine mesh, usually about 10 micron (one micron is one millionth of a metre), to ensure that no particles of dirt reach the mechanical unit. The fluid returns from the mechanical unit at low pressure and is returned to the reservoir via a radiator. There the heat generated by moving the various axes of the machine is removed from the fluid, by either a fan or by a more elaborate water-cooling system.

Advertising Arts/Hall Automation

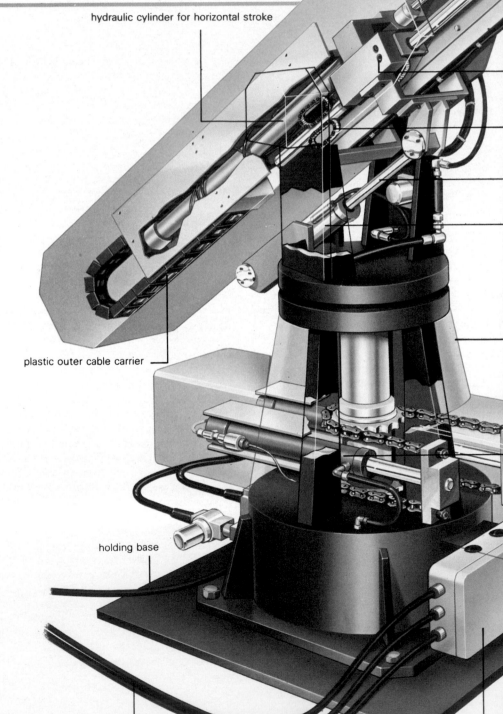

hydraulic cylinder for horizontal stroke

plastic outer cable carrier

holding base

hydraulic lines

hydraulic receiv

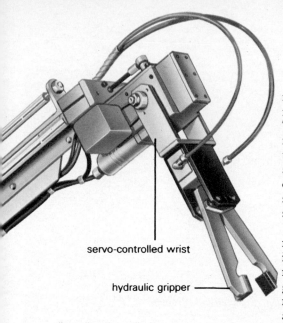

servo-controlled wrist

hydraulic gripper

linear bearing adjusters

linear bearings

hydraulic cylinder

tilt cylinder

outer casing

swing cylinder

Incorporated in the hydraulic power unit is a regulating valve, whose purpose is to maintain the pressure at the required level. The hydraulic power unit is generally also fitted with such safeguards as a float switch to indicate a low level of fluid in the reservoir and a temperature switch to indicate excessive fluid temperature. Both these instruments automatically shut down the system when such levels are reached.

The 'machine' part of the robot, the mechanical unit, provides the means of picking up and manipulating components or manoeuvring various tools such as spray guns or arc welding torches. A typical mechanical unit can operate along three main axes. On the horizontal plane the 'arm' of the machine both extends and retracts; a swing motion, like the waist of a man, permits the arm to revolve about the base; and the tilt axis, the 'shoulder' of the robot, allows the arm to be raised and lowered.

At the end of the horizontal arm are three further axes, which provide the same motions as the human wrist. A gripper attached to the wrist works like the human hand and can take many forms. A simple two finger unit can open and shut to pick up simple parts, while vacuum cups can pick up sheet components such as glass. Special tools such as spot-welding guns or other devices can also be attached to the wrist.

The servo valve

Each of the axes of the mechanical unit is operated independently under its own particular control channel. The main component in each circuit is the servo valve, which is electrically controlled and serves to direct oil at high pressures to either side of the hydraulic cylinder, thus producing motion in the desired direction. Every time the robot's position needs to be adjusted, oil is pumped at high pressure through a regulating valve into the servo valve. If the valve spool is in the forward position, oil will flow into the hydraulic cylinder and apply high pressure to

the rear of the piston, moving it forward. Fluid displaced in front of the piston flows at low pressure back to the hydraulic reservoir. If the valve spool is in the rear position, oil flows to the front of the piston and forces it backwards.

The servo valve is a critical component in the robot system, for the position of the spool not only controls the direction in which the fluid passes but also the volume of flow, and therefore the speed with which the piston moves in the desired direction. And, by applying suitable signals to the servo valve, both the exit ports can be closed, stopping all oil flow and all motion.

The control unit

The 'brain' of the industrial robot, the control unit, is usually housed in a separate cabinet and connected to the mechanical unit by electrical cables. Earlier machines made use of magnetic tapes to store information but today's industrial robots employ solid state silicon electronics and increasingly make use of micro-computer technology. This has eliminated the mechanics of tape drives, and consequently improved the reliability of robots.

The control system is made up of three main parts: an amplifier to drive the servo valve; a logic or instruction element to control the robot's functions; a memory element to store program information; and feedback elements which provide information about the position of the various robot axes.

The operation of any robot can be reduced to two fundamental modes, known as 'point-to-point' and 'continuous path'. In the first, the path by which the robot travels from one position to another is unimportant;

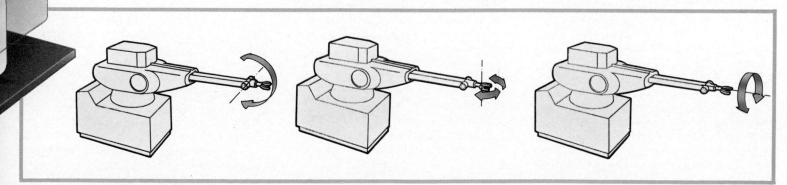

At the end of the arm, three further axes are provided by the 'wrist', which can move vertically (left) and horizontally (centre) as well as rotating (right). A variety of grippers can be fitted to the wrist—for example, vacuum cups or 'finger' units.

the quickest route is the best. In the latter, the route is all-important—for example, when the robot must follow the contours of an object, as in paint-spraying.

For point-to-point operations, the control unit is required only to record the points to which the robot must move; the exact path and speed are determined by the robot's own operating limits. For continuous path operations, however, the control unit is required to record in detail the path of the robot hand and also its speed.

So, for an application such as paint spraying, it is necessary to record the operator's movements continuously as he moves the spray gun. The operator 'teaches' the robot the task by actually moving the mechanical unit arm, with the spray gun attached to the wrist, to perform the task. All his movements, including the trigger actions as he turns the spray gun on and off, are faithfully recorded in the control system memory.

If the robot is going to load a machine, the operator drives the arm by means of a 'teach' control to the point where the part is to be picked up. When he has carefully positioned the hand at this point, he operates a record button and the 'address' of that point is recorded in the control system memory. He moves through the entire program in this manner, leading the robot hand to each desired point and pressing the record button each time to record the address. In playback the robot arm will move to each of these points in sequence.

The feedback element

All this 'teaching' information is transmitted to the control system by a 'feedback element'. This is linked to the mechanical motion of the particular axis, by gears, belts or some other means, and sends the control system an electrical signal proportional to the position of each axis. Thus, when the robot is stationary at one address, that information is passed to the control system and the address of the next point in the program is sent back down the line. The servo amplifier responds to this signal by driving the spool of each servo valve in the appropriate direction to cause the correct motion of each robot axis.

Even as the axes move, the feedback elements are also moving and sending back to the control system positional information. This is processed in the logic element and as the address of the next point is approached, the command signal diminishes and the servo valves progressively close, slowing down the movement of the axis. When the address

ASEA

Above This Swedish robot is programmed to move through 200 set positions in the course of polishing stainless steel sinks.
Left At Volvo's assembly shop, a robot applies glue to car body components.

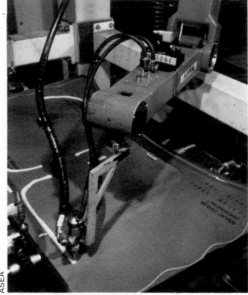

ASEA

received from the feedback element in each axis exactly matches the address for that point as stored in memory, the command signal is reduced to zero, the servo valve is centralized, hydraulic fluid ceases to flow and the axis motion is stopped.

To give some idea of just how fast this happens, a typical robot will swing about its rotational axis at a maximum speed of some 90 degrees per second. When the address is nearly reached, it will slow down in just 10 milliseconds (10 thousandths of a second)! If you consider that a typical horizontal arm length of 1.25 m (4 ft) might hold a load at the end of some 20 kg (44 lb), the enormous power and forces involved in the robot's mechanical and hydraulic systems can easily be imagined.

This then is the industrial robot of today. Many hundreds of these machines are in operation around the world—painting, arc welding, spot welding, unloading die-casting machines, loading lathes, handling sheet glass, hot metal, blocks of dry ice, and many other tasks.

What of the future? The mechanics will probably not change greatly, since they are mainly designed to fit a broad range of tasks, but the robot will become much cleverer; the micro-processor will see to that. The ability to process more and more information at faster and faster speeds will give the robot an increasing level of intelligence. Vision will enable the robot to 'look' at a part and to recognize, for example, which way up it is or to discriminate between good and bad components. Touch sensors fitted to the gripper fingers will allow the robot to vary the force with which it holds each object so that delicate objects can be handled safely. Improved software and programming techniques will give robots more precisely defined 'personalities'—and voice recognition techniques will make them react to such rudimentary but vital commands as 'STOP!'

Right Human labour has been virtually eliminated from Fiat's production lines.
Inset A powerful robot swings a 20 kg (44 lb) blank cylinder against a furnace hammer, in order to forge a high-pressure seal.

Memories that never forget

The fickle nature of Man's own memory, with its annoying capacity to forget or even alter facts that have apparently been remembered, is a weakness he has long struggled to overcome. Through much of 'recorded history', pen and paper have been the most useful and durable alternative to 'remembering'. Even now, the sum total of Man's knowledge through the centuries is stored on paper, even though it ages and is highly inflammable. But the advent of electronic memories has opened an astonishing new chapter in the complex task of keeping track of knowledge.

Before the pocket calculator became so highly developed and popular, engineers used slide rules to speed their calculations and they had to write down intermediate calculations because only one calculation could be performed at a time and the slide rule possessed no method of remembering the results.

The calculator is able to remember intermediate answers in calculations because of the development of the electronic 'microchip' memory, which can store thousands of pieces of information at very high speed, using very little power from the calculator's battery. The memories used in computers also fulfil both long- and short-term needs. Information that has to be kept for long periods, or moved from place to place, tends to be stored on magnetic 'disks' (spelt with a K in the computer world) or reels of magnetic tape. An example of this is a company's salary records or its list of customers.

Disks and tapes also store permanent information that tells a computer how it is to treat the information fed into it. Information in this format is called a program and it is referred to as *software,* to distinguish it from the *hardware* of resistors, transistors and integrated circuits in the computer itself. There is also an intermediate category called *firmware*—a program within the computer itself—built in by its designers and essential to its operation. Of these categories, memories are the hardware of a computer.

Even the most powerful computer Central Processing Unit (CPU) can only handle one calculation at a time. Once it has carried out a calculation it has to store it, then perform another calculation, store that somewhere else, and so on. It may have to carry out a thousand or more calculations, all in a fraction of a second, before it needs the result of the first, but when that first result is needed it must be available instantly. This means the memory must be constructed so that information can be put into and taken out of any part of it at will, without having to sort through the circuits (as you would the pages of a notebook). This sort of memory is called a Random Access Memory (RAM) and it relies on clearly addressing and labelling each piece of information it receives or sends.

Computer language

A digital computer deals with numbers that are encoded from our everyday decimal system into the binary system, which uses only *ones* and *zeros* corresponding to 'on' and 'off' states within the circuits. Long trains of *ones* and *zeros,* representing binary coded decimal numbers, are presented to the memory one by one through a single wire. Alternatively, the entire number, as a 'quantity of information', may be presented in parallel, with a separate wire for each digit. For example, to store a letter of the alphabet in a memory it must first be made to correspond to a digital number. There are 26 letters in the alphabet and, in consequence, 26 numbers are needed to represent them. Identification of space between letters, or words, is also needed. Allowance for punctuation is necessary. All such data must be represented by numbers in the computer. In all, a total of over 60 numbers may be required for this task.

The digits of a binary number increase in powers of two and it needs no less than six digits or bits (2 to the power of 6 or 2^6) before 63 values can be defined in the binary system. This compares with just two bits— 6 and 3—of the decimal system. If, for example, the thirteenth letter of the alphabet, m, is to be identified by binary code, the resultant number is 001101. This can be changed back to the number 13 by adding the three relevant binary values of 1×2^3, 1×2^2 and 1×2^0 (the power raising 2 in each instance being taken from the position of the 1 digit).

This can be presented to the memory one digit at a time, followed, after a time interval, by another train of six digits corresponding to another letter, in what is known as *serial addressing.* Alternatively, it can be presented all at once along six parallel wires to the memory, each of which corresponds to one of the six parts of the binary number. This is known as *parallel addressing.* Parallel inputs generally accept 4, 8 or 16 bits of information each in the computer.

Whichever way the number is presented, each 0 or 1 will be stored, using at least one transistor. The memory simply consists of row after row and column after column of separate transistor circuits, all on a tiny silicon chip. The chip must also carry circuitry needed for checking, sorting, sending, seeking and finding information entering and leaving the memory. Such memories have now reached the point where over 65,000 bits can be stored on a chip and the manufacturers of these memories could make them even smaller if the size of the physical connecting wires and tabs could be reduced.

There are two types of semiconductor RAM: *static* and *dynamic.* The dynamic RAM loses its signal after a time and so has to be 'refreshed' by a regular clocking signal. This means that extra circuits are needed to refresh the memory, but each memory cell or storage unit on the chip is smaller than that of a static cell, which requires at least two transistors to store one bit of information.

The reason that the dynamic memory loses its signal is that the *zero* or *one* is stored within stray capacitance of each transistor memory cell and all capacitors lose their charge eventually, through leakage.

In the static memory, the signal is stored by a device called a transistor flip-flop, a

IBM

Below left An IBM Display Typewriter which employs powerful memory systems. An electronic dictionary can verify the spelling of about 50,000 words and a memory can store 224,000 typewritten characters.
Below An Intel bubble memory with 1 megabit storage. *Bottom* A Rockwell microcomputer with two 256 kilobit bubble memories (blue).

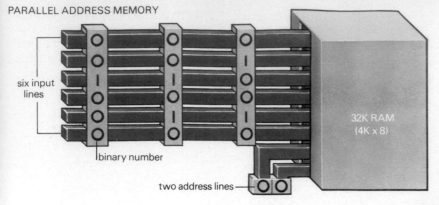

PARALLEL ADDRESS MEMORY

six input lines

binary number

two address lines

32K RAM
(4K x 8)

Below In a serial addressing memory, all information, including that used to determine its address, enters on a single line and is sorted by logic circuits.

Above A parallel address memory accepts the binary information digits simultaneously on parallel input lines. An address code for the information, indicating where it is to be stored, enters the computer on another two lines.

SERIAL ADDRESS MEMORY

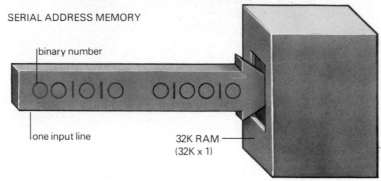

binary number

one input line

32K RAM
(32K x 1)

Ricky Blakely

In a memory, the gate is connected to a 'row select' line which turns it on if commanded to do so, and data then charge up the capacitance of the cell. In a serial-addressing type memory, data enter along a single line, and each binary number is accompanied by a two-bit column-select address that will determine which of the four data columns it is to enter. A logic circuit sends it to the right number.

In a parallel-addressing type memory there is no need for such circuitry. But whether serial or parallel addressing is used, the storage principle is the same: the row address must be given in order to direct the information bits to the correct storage cells. In a serial-addressing memory, for example, as soon as a number appears on a data input line, every transistor along the line will be presented with it. But only the transistor in the correct row will be turned on (by the row-address information) to direct the signal

switch that can only be on or off, so there is no need to keep a cell charged up.

The type of transistor most commonly used for memory storage is the Metal Oxide Silicon or Metal Oxide Semiconductor (MOS) transistor. This is a *field-effect transistor*, or *fet*, of which thousands can be fabricated onto a square centimetre of silicon. These devices are a bar of semiconducting material along which an electrical current is passed, via electrodes at each end of the bar (known graphically as the source and drain). Fused into the bar is a third electrode which

can control the current passing along it, and in accord with its function this electrode is called the gate.

In a memory, the *fet* is used as a switch which can be turned on by sending a signal to its gate. Once in the 'on' condition, current may pass through the device to charge up the memory storage capacitor or 'cell' associated with the particular *fet*. This capacitor can take up either a state of *charge*, seen as a binary 'one', or a state of *discharge*, seen as binary 'zero'. So, by this technique, binary information may be stored.

Read Only Memory, or ROM *(right)*
KEY
1 ROM chip
2 Current-limiting resistors
3 Internal diodes
4 Power line for display
5 Ten-line decimal input
6 Seven-segment display
7 4-to-10 line decoder
8 Decimal output lines
9 Decoder chip
10 Binary number
11 Four-line binary input
12 ROM-controlled power lines

THE OPERATION OF A READ-ONLY MEMORY, OR ROM

Read Only Memories, or ROMs, perform a whole host of functions. The following example of how a ROM might be used in practice involves the problem of supplying information to a seven-segment display used to portray a number on a calculator.

A truth table is first drawn up which indicates the output needed to light appropriate segments of the seven-segment display from the four-line (four bit) binary input from the calculator's logic circuits.

In this example, four lines can carry the binary digits that represent decimal 0 to 9, but ten lines are needed to carry the decimal numbers 0 to 9 into which this binary information must be converted.

The problem of conversion is solved by use of a 4-to-10 line decoder, which is an array of logic gates. When the binary

number 0000 (decimal 0) is received by the decoder for example, it puts a 0 out which is in effect no-voltage on the first decimal 0 line—and a number 1 on all other lines (1–9). This 1 may typically be represented by a steady 5 volt signal.

If a binary number 001 (decimal 1) is received by the decoder, a 0 goes out on the second line representing decimal 1, whilst the other lines all carry a 1. This process is repeated for all the other decimal numbers, the number concerned being identified on the decoder's output by a 0, or no voltage.

There is a simple reason for this arrangement of using 0 as an indicator. It means that the decimal signal line concerned is connected to ground and will short any signal line to which it is connected to the ground also. This is used to cut off power to

a segment of the display, a task organized by the ROM, in order to ensure that the segment is not illuminated. Selective blanking of the segments in this fashion allows decimal numbers 0 to 9 to be obtained from a seven-segment display.

The task of the ROM is simply to ensure that the appropriate display segments are blanked by an output from the decoder. With, for example, a 0 on the first decoder output line, representing decimal 0, it is necessary to blank the centre-horizontal g section of the display, in order that it will display a 0. So a diode is left connected between the first decoder output line and the g power-supply line, so connecting them together. With a 0 on this decoder line, which is effectively a short circuit, the power-supply line to display section g is

to the correct point in the matrix of cells. In a parallel-addressing memory, the four bit number will be presented at all four data lines. A 'one' signal along a row-address line will store the number in that row.

When the time comes to read the number out of the memory, the request for the number is passed to a row-address. When the signal addresses a row, the gates along that row are turned on and, if there is a 'one'—meaning a charged capacitor along any of the data lines in the row—the pulse from the capacitor will be passed to the data line. If a cell along that row has a 'zero' there will be no effect on the data line. Thus the number stored along that row appears at the output of the data line or lines.

Using similar techniques, large amounts of storage can be built up on a single chip. Refreshing is carried out by applying a 'topping up' signal in turn to each column of cells. This is done about 500 times a second.

In the static memory this 'topping up' is not necessary, but the static memory may consume more power because one or other of the two transistors in each cell is conducting the whole time. The conduction also produces heat, which needs to be dissipated and tends to add to the space needed for a cell. The cells are relatively large because they may use four or more transistors. Dynamic RAMs have now reached a capacity of 64,000 cells (64K) per chip, compared with 16K for the latest static RAMs.

Polysilicon

In ordinary MOS static RAMs, actual resistors are not widely used. Instead, a transistor is put in place of 'load resistors', but with gate connected to drain, to create a small-area non-linear resistor. However, one company, Mostek, has started to use actual resistors again, employing ion-implanted polysilicon, which, they claim, occupies little more space than the old technique and also achieves lower power consumption.

Not all RAMs use MOS technology, even though the most rapid developments in memories have taken place in the last ten years using MOS. Bipolar memories are also used, but they tend to be designed in the same way as the MOS static memory, with pairs of flip-flops and a consequent increase in the area needed per memory cell. Bipolar memories, unlike their MOS static equivalents, can be much faster than dynamic memories, though MOS dynamic memories remain the cheapest. But MOS memories may need more than one power supply, and they have to be supplied with circuitry that will allow them to work with the very fast logic of the central processor, which is likely to use bipolar circuits.

A dynamic cell consumes power whether it is being used or not, since it has to be refreshed regularly. Although the load of a

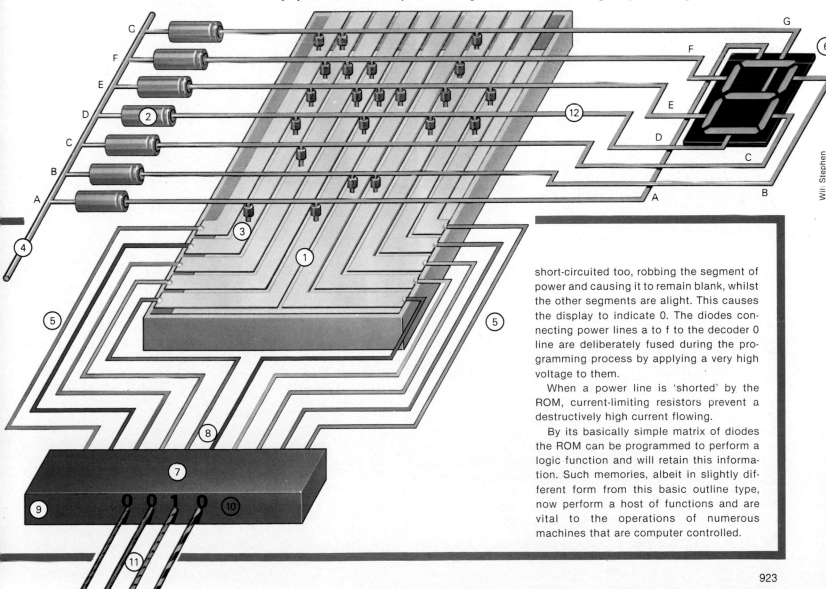

Wil Stephen

short-circuited too, robbing the segment of power and causing it to remain blank, whilst the other segments are alight. This causes the display to indicate 0. The diodes connecting power lines a to f to the decoder 0 line are deliberately fused during the programming process by applying a very high voltage to them.

When a power line is 'shorted' by the ROM, current-limiting resistors prevent a destructively high current flowing.

By its basically simple matrix of diodes the ROM can be programmed to perform a logic function and will retain this information. Such memories, albeit in slightly different form from this basic outline type, now perform a host of functions and are vital to the operations of numerous machines that are computer controlled.

MOS or bipolar static RAM is dissipating power all the time, parts of the memory (say, one or more of the chips on a computer board) can be put on a much lower stand-by voltage. This reduces the power consumed significantly but maintains the signal in each cell. A variant of the MOS static RAM is the Complementary MOS or CMOS RAM. In this the load resistors of the crossed transistor pair in a flip-flop are replaced by transistors which are switched in the opposite way to the actual switching transistors. The resistance presented by these two load transistors is, in theory, infinite. So no current flows and no power is dissipated. CMOS static RAMs consume far less power than other types as a result, which is why they are used extensively in space research. However, they tend to be far slower in operation.

Sapphire substrate

Although theoretically, in CMOS, no current is passed by the transistors, in practice there is a current surge during the short period when the flip-flop is switching over from one state to another. This is caused by stray capacitances but has been countered by RCA, who have developed a method of building CMOS memories on a sapphire substrate. This technique speeds up this operation by about four times, but sapphire is expensive compared with silicon, and though the technique is still being researched, its future is by no means certain.

One disadvantage of all semiconductor RAMs is that they are 'volatile', even though a number of cunning ways of overcoming this drawback have been developed. If power to the memory is removed, all the cells lose their information. This necessitates an elaborate system of back-up power supplies for use in the event of a power failure,

Right An experimental IBM silicon chip that is just 1.5 mm square. In spite of this, it can be made to house 250,000 bits of memory in the small black rectangular area. Making connections to a device of this size presents some problems, since they take up a large proportion of the total space occupied by the chip. Additionally, testing integrated circuits such as these must be carried out after they have been attached to their carrier packages. *Below right* An array of fine probes manages this task by injecting and tapping-off test signals.
Below The characteristic appearance of an Erasable Programmable Read-Only Memory chip, usually abbreviated to EPROM, with its window for ultra-violet light erasure.

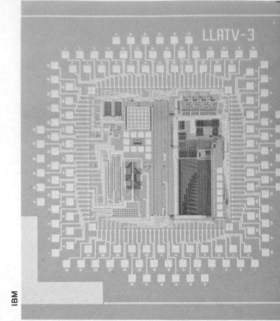

IBM

Paul Brierley

RCA

COMPUTER DISC AND RANDOM ACCESS MEMORIES

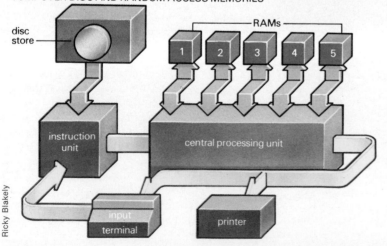

Ricky Blakely

Left The interior organization of a computer. Data are fed in through an input terminal keyboard and acted upon by the central processing unit in accordance with instructions from the disc store memory. Intermediate answers are held temporarily in the RAMs. Solutions are then displayed on a printer or a video terminal.

or the transfer of important information immediately to some form of magnetic storage, such as disk or tape. All magnetic memories are non-volatile. If they were not, music cassettes would be erased simply by turning a tape machine off. But there are other types of semiconductor memory that are non-volatile. These tend to be used to store programmes indefinitely, perhaps for a TV game or to direct operations in a computer, or to instruct a calculator on how to perform a mathematical process. Since such memories only supply information in use, they are termed Read Only Memories, or ROMs—in the jargon of the computer industry. Essentially, such memories consist of arrays of diodes in a matrix and programming is

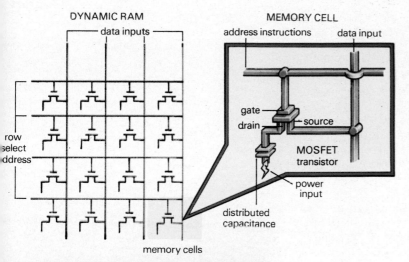

DYNAMIC RAM
data inputs

row select address

memory cells

MEMORY CELL
address instructions   data input

gate
drain — source

MOSFET transistor

power input

distributed capacitance

Left The structure of a dynamic Random Access Memory, or RAM. This type of device houses thousands of MOS field-effect transistors, or MOSFETs. Each one, with distributed capacitance, forms a memory cell. The MOSFET acts as a switch, allowing the capacitance to be charged by incoming data.

achieved by removing unwanted diodes, by fusing them, through the application of a high voltage. The resultant lack of connection between lines may then be used to indicate a logic state of 'one' or 'zero', according to the way the ROM is connected. The simplicity of a ROM is the reason for its low cost and high information density.

Early ROMs were incapable of alteration once they had been programmed. But soon there appeared devices that could be 'wiped clean' by exposing their internal diode arrays to ultra-violet light, through a small window in the top of the device. These are called Eraseable Programmable Read Only Memories (EPROMs), as opposed to the programmable ROMs (PROMs).

The EPROM uses the characteristics of certain semiconductors whereby, without destroying any part of the whole array, each part of the matrix could be programmed by applying something like 30 volts to it. If, however, ultra-violet light were shone over the whole array, this program could be completely wiped out. Such EPROMs have now reached densities of 64,000 bits, and work is well advanced on 128,000 and 256,000 bit versions of these devices.

Another variation of the ROM is the electrically alterable ROM or EAROM. With these it is not necessary to wipe the whole memory, since application of larger-than-normal voltages will individually re-program cells. Thus one memory cell can be changed at a time without the need to re-program the whole matrix.

There are many other types of storage for digital data, many of which, although promising, have yet to make their impact on the memory market. Among these the most important examples are magnetic bubble memories and charge-coupled devices (CCD) memories.

In both these memories the problem is access time, since they are not truly random-access devices and, like the tape loop, depend on continual repetition of stored data until that which is required becomes available. Consequently, such memories are likely to become substitutes for disk storage.

In the case of the magnetic bubble, information is stored by using the properties of certain crystals which, under the influence of a magnetic field, store what look like bubbles of magnetic polarity. Alignment of the bubbles determines whether a 'one' or 'zero' is stored. An alternative is to store either a bubble or no-bubble, signifying a 'one' or a 'zero'. Such devices are capable of storing

vast amounts of information, but the information has to be continually circulated around the equivalent of a magnetic tape loop for the device to function properly.

As a result, a special device is needed to 'write' information into the store, rather like the recording head of an audio tape machine, and another is needed to 'read' it out. The latter can only read the information out when it is available during re-circulation. This makes access time rather long, but, like other magnetic devices, bubbles do have the advantage that their memory is non-volatile and packing density is something like ten times that of a semiconductor memory. Such memories are now beginning to be used by companies like IBM, and some of the Japanese computer makers.

Charge-coupled memory devices, use a semiconductor technique which behaves like a very long shift register. That is to say, the information is stored in a long train and, like the bubble memory, accessed when it reaches special read-out heads.

The physical principle involved is that the CCD memory is a very long field-effect transistor with a number of taps or electrodes between one end and the other, rather like a garden hose with a large number of holes in it. The amount of charge flowing along this transistor can be varied in a more regular way than that of water flowing through a garden hose, and each electrode is presented with a complete picture of the series of signals that were presented to the input of the device.

Clock signal

A clock signal has to be applied to keep the information ticking around the device, and the access time to any bit of information is higher than with other comparable techniques, but cost is said to be low. As a readily-useable technique of memory storage this once-promising method has yet to prove itself suitable for widespread use.

Before the advent of semiconductors, the sheer size of valve memories made the future of such things look limited. Now electronic memories inhabit a world much like that of the human brain: small compared with equipment housing them but crucial to its behaviour. The most significant difference between the two, however, is that unlimited development can be foreseen for electronic memories, aided by computers. They will be as crucial to the behaviour of powerful machines in the future as Man's own brain has been in his ability to develop such electronic wonders.

Teaching machines to talk

In the realms of science fiction, man has for years been talking to machines. No modern science fiction film is complete without its wise and articulate — if sometimes malfunctioning or downright disobedient — computer. Hal, the smooth-voiced controller of the Jupiter probe in the film *2001: A Space Odyssey,* **is a perfect example of the breed.**

But what is the reality? Are machines like these ever likely to be built?

There are, in the real world, two separate problems: first, getting the machine to recognize and respond to speech; and second, getting it to convert its response into speech. The second objective — making a machine that talks — is simpler than the first.

Modern electronic speaking machines use two basic principles. In the first, *concatenated speech synthesis,* a computer is used to store a

Kobal Collection Metro-Goldwyn-Mayer-Inc.

Above Hal, the computer from the film *2001: A Space Odyssey,* is a classic example of a talking machine from sciene fiction.

'dictionary' of spoken syllables — all possible vowel-consonant and consonant-vowel combinations — in the language concerned. To form a word, the computer draws the required syllables from its store and strings them together in the right order — 'com-pu-ter', for example. Then it adds pauses and intonation to make the output intelligible.

The concatenated speech method is very rigid, and produces low quality speech. But it works quickly and is cheap to build

The other method, *synthesis by rule,* does not store sounds or combinations of sounds, but rather a guidebook of the electronic 'rules' which are necessary to generate those sounds, rules determined from detailed analysis of natural speech. Synthesis by rule produces a rather better quality speech, but because many elaborate rules are needed to produce even a short string of speech sounds, the machine takes a long time to synthesize the speech.

Using concatenated speech synthesis, several kinds of speaking machines are already commonplace, especially in the US. Talking calculators can speak each number as

HUMAN AND MECHANICAL SPEECH

nasal cavity
tongue
palate
control keys
air inlet
lips
lips
reed chamber
palate
air inlet
teeth
tongue
soft rubber lining
pharynx
larynx
vocal fold

Early mechanical speech machines copied the human vocal tract. Air flowing past a reed generates sound, and control keys alter the shape of the artificial vocal tract to modify the character of the sound produced.

Nigel Osborne

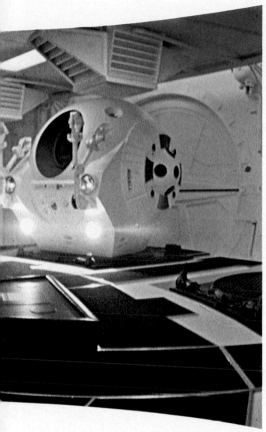

unpredictable pronunciations. How is the machine to match your spoken request with its written — or rather, computer-coded — data? Indeed, how is this machine, or any other, going to understand your speech.

A vital tool for scientists trying to discover how machines might recognize speech was developed early in World War 2. This is the *sound spectrograph,* an electronic machine which analyzes sound waves and produces a permanent, written 'picture' of them. Just say a word, and the machine will produce an accurate, if complicated, *spectrogram* of the acoustic waves your voice produces. The importance of spectrogram patterns is that they can be stored (in digital form) in a computer memory. Theoretically it should be a simple matter for the computer to recognize that word again by comparing the spectrogram of any received word with the pattern in its memory.

The catch is that no two people, even with the same regional accent, say the same word in exactly the same way. A man's voice, for example, has a much lower pitch than a woman's or a child's. His vocal tract is bigger and of a different shape. So the sounds he produces cannot possibly be the same as those produced by a woman or child. And there are quite marked differences between the sounds of individual men, individual women, individual children — that's how we 'recognize someone's voice'.

This problem bothers humans only occasionally. From babyhood, when we learn to talk, we start developing a mechanism for recognizing quite different sounds as meaning the same thing.

Exactly how this happens, scientists do not know. But the human system is highly flexible; even when we hear a particular word in an accent quite different from our own we can usually identify it accurately.

A machine, however, is not nearly as versatile. If you say the word 'pair' to it, for instance, it must first make a spectrogram pattern of the sound you produce. Then it must search through its computerized memory bank and try to find an identical, or nearly identical, pattern. It is highly likely to make mistakes. A man's voice saying 'pair' may produce a spectrogram quite similar to the same voice saying 'pier' — and quite different from a woman's voice saying 'pair'. There are hundreds of problems like this: identical words, spoken by different people, which produce markedly different and confusing spectro-

you punch it on the keyboard, and then read out the result. Small speaking machines, slightly bigger than a calculator, help teach children spelling. Another small machine not only plays a very good game of chess but speaks the moves as they are made.

A 'speaking' machine is the basis of a system used in the US by the Bell Telephone Company to improve the efficiency, and reduce the cost, of some services normally provided by human operators. A keyboard similar to a typewriter, but smaller, is plugged into a telephone set. To find a subscriber's number, you dial directory enquiries and tap out his name and address on the keyboard. The computer at the other end of the line searches its memory bank, finds the number and, using machine speech, transmits the answer to the telephone.

The next stage of development, eliminating the keyboard and allowing you simply to ask the machine for information, is a great deal harder. Any telephone directory is full of people's and companies' names, of various national origins, and many with

Right The silicon chip is an essential element of any speech recognition machine. Only computers have the power to analyze and compare human speech patterns.

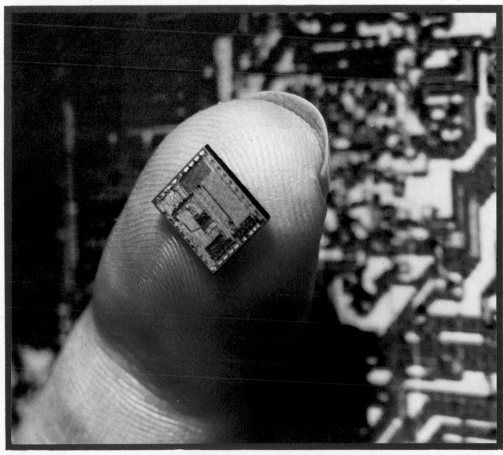

Intel

927

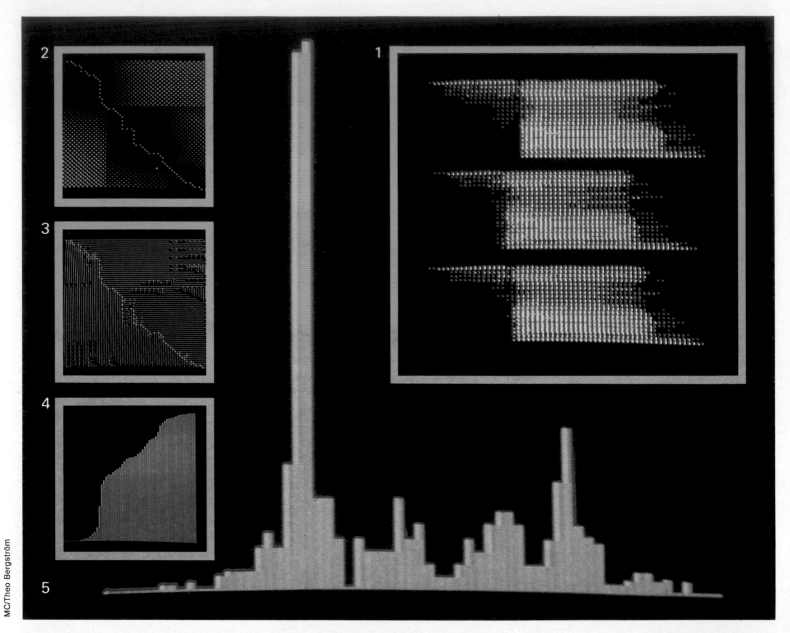

MC/Theo Bergström

gram patterns.

The most efficient way so far found of overcoming this problem is to 'train' the machine, in much the same way that infant humans train themselves. You start with a group of about 100 speakers, all with the same regional accent. Each speaker speaks all the words that the machine is intended to recognize, and the various spectrogram patterns are fed into the computer. The machine then analyzes all the spectrogram patterns of the same word and produces an 'average' pattern, or template, for future reference. It also calculates by how much a speaker's sound pattern is likely to differ from the template.

From then on, the sound pattern of any word spoken to the machine is compared with the various templates stored in its memory, and the expected variation is taken into account. The templates can be 'stretched' in time to match words spoken at different speeds. When a template is found to match a received word, then the machine has recognized the word.

Systems working like this, and which claim to recognize words coming from any speaker of the same language, can achieve success rates of 80% to 90%. For the speakers who 'trained' the machine, the scores can go as high as 98%.

This makes it feasible to construct machines which will understand simple commands and act accordingly. An example might be a drinks-dispensing machine, to which you could say 'tea', 'milk', 'sugar'

Above A computer analyzes the sound of the word 'sum'. (1) The top two patterns represent two versions of the words by the same speaker, the bottom one is an average 'template' generated from the upper two. (2) to (4) These patterns tell scientists how the two versions differ—in fact the sounds 's', 'u' and 'm' are more or less the same, but at the transitions *between* the sounds (between 's' and 'u' and between 'u' and 'm') there is considerable difference. These two points of maximum difference are indicated by the peaks on display (5). What these displays show is that the *same* word spoken by the *same* speaker can sound different to a machine, so it will have considerable difficulty recognizing words from different speakers.

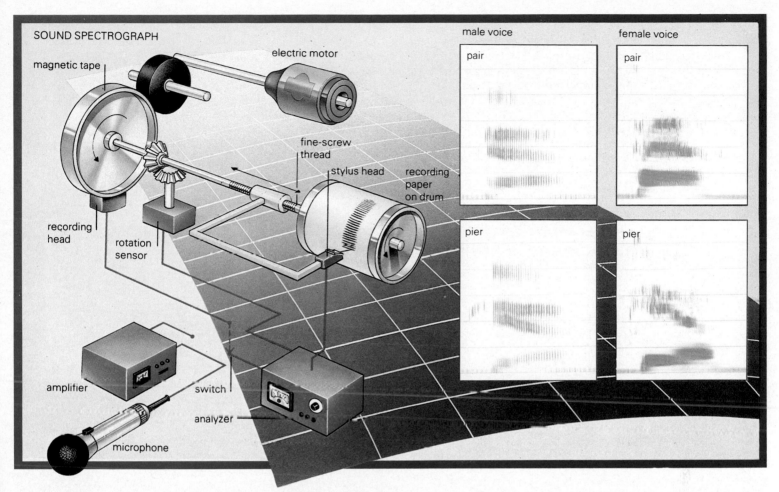

SOUND SPECTROGRAPH

magnetic tape

electric motor

fine-screw thread

stylus head

recording paper on drum

recording head

rotation sensor

amplifier

switch

analyzer

microphone

male voice

pair

pier

female voice

pair

pier

Frank Kennard

and receive — after the machine had pondered your words — a cup of tea with milk and sugar.

But the step upwards from understanding isolated words to understanding connected speech — whole phrases and sentences — is so big that few researchers have tackled it. It poses tremendous problems.

In 1970, for example, the US Defense Department asked a group of scientists to specify what could reasonably be expected of a speech-recognition system. They concluded that the system should work on only a small vocabulary of 1,000 words (someone with a high school education has a vocabulary of between 5,000 and 10,000 words). It should have a very simple syntax (phrase and sentence structure). The task of the system would have to be very specific, like data management or computer programming. The speech input should be continuous (no spelling out words — that would be 'cheating'). The machine should accept speech from a large number of operators, but all in the same regional accents. Finally, each speaker would have to speak clearly, into a high-quality micro-

phone, in a quiet room.

By the end of the 1970s speech-recognition machines (or rather, word-recognition machines) were nowhere near these apparently modest targets. The machines' vocabularies were extremely limited. Instead of the recommended 1,000

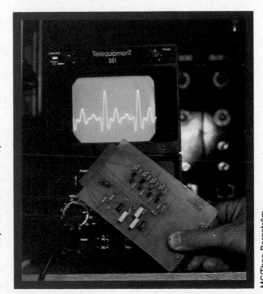

MC/Theo Bergström

Above How a sound spectrograph produces a spectrogram. The sound is first recorded on the magnetic tape. Then it is played back again and again through the analyzer. The stylus marks the paper electrically according to the strength of the signal it receives. To make a complete spectrogram, the drum rotates many times and the stylus creeps across the drum. At each revolution the rotation sensor causes the analyzer to respond to a slightly higher frequency. The four insets are spectrograms of the words 'pair' and 'pier' spoken by male and female voices. Computers use patterns of this sort (in digital form) when comparing sounds. The similarities between the patterns are more marked when the voice is the same than when the word is the same. Consequently, machines find it hard to recognize words spoken by different voices. *Left* Another method of graphically displaying a sound. Here a vowel sound is traced on the cathode ray tube of an oscilloscope. In this instance, the sound has been generated electronically by a series of plug-in circuit modules like the one shown.

words, two of the best systems could use either a 39-'word' vocabulary (26 letters of the alphabet, ten digits from 0 to 9 and three command words) or a 54-word vocabulary of computer programming words.

In 1980 a team of IBM scientists announced that they had achieved the 1,000 word goal. They use a System 370 computer to transcribe spoken sentences into print. But there are still drawbacks. First of all, each speaker to use the machine has to spend two hours speaking 900 training sentences into a microphone. Secondly, the process is slow — a 30-second spoken sentence can take 100 minutes to transcribe. And finally, the machine still makes mistakes.

Plainly there are problems in speech recognition that the 1970 committee failed to foresee.

First, even a short piece of speech — a conversation, perhaps, or a news broadcast — contains hundreds of small speech particles, or *phonemes*. The machine must identify each of these accurately. Then it must sort them out, deciding where one word ends and the next begins — a problem compounded by the fact that most of us run our words together.

Next, even a machine which could reliably break up sentences into their individual words would still have daunting problems. It would have to cope with words that sound the same, but mean different things. We know whether a word is 'pair' or 'pear' or 'pare' because the context (the words around it) tells us. The same applies to 'peer' and 'pier', 'mean' the verb and 'mean' the adjective, and hundreds of combinations like them. But how is the machine supposed to know which is which?

One way would be for the machine to have some sort of dictionary in its memory. This is not practicable because it would need an enormous computer to store all that information. And it would waste a vast amount of time searching through its huge dictionary for every combination of sounds, every word, every combination of words and so on, until it found the right one. It might be accurate — but it would react awfully slowly.

Another way, which would be far more economical, would be to teach the machine not the *whole* language, but the *rules* of the language — how phrases and sentences are constructed from individual words. It would need to know what kind of word you can find next to a noun, what kind next to a verb, what kind at the beginning of a sentence, what kind at the end, and so on.

The trouble with that idea is that there are no reliable rules in English, or in any other language. One simple rule you could teach the machine, for instance, is that no sentence in English ends with the word 'the'. That rule would be correct 999 times out of 1,000, but not invariably —′ as the sentence you have just read will prove.

Vast improvements on present-day speech recognition machines will certainly be made, perhaps by using methods quite different from those on which scientists are now working. But it seems that the sort of machine that could reliably separate connected speech, recognize words, distinguish words that sound alike and sort

out errors and inconsistencies — a really first-class speech machine — would actually have to *understand* speech.

The only reason that we understand speech is that we are familiar with our own environment, the world about us. Without that knowledge, speech would be just a jumble of noise. To understand speech, a machine too would have to understand its environment. It could do this by collecting data through electronic sensory devices, just as babies do through their eyes, ears, noses and hands. It would, in short, have to be a model human being — devoid of feelings and emotions perhaps, but in other ways just as intelligent

Right and below A machine that not only plays a good game of chess but also speaks out the moves as they are made. It uses concatenated speech synthesis — 'particles' of sound are stored in silicon memory chips and electronically put together in the right sequence when the machine 'speaks'. The artificial speech signal is simply fed to a small loudspeaker via an amplifier.

Computer Games Ltd

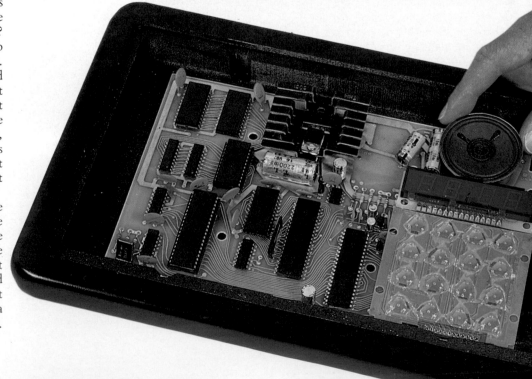

Electronics on display

For the pilot of a supersonic aircraft to take his eye from the view through the windscreen—even for a fraction of a second—could be fatal. As machines of all kinds become faster, the time available to take decisions becomes shorter, and the need for safe manoeuvring greater. If both the view ahead and the machine controls can be seen in the same line of sight, the operator is said to be driving or flying with 'head-up'.

Piloting an aircraft, be it a supersonic fighter or a transport aircraft, requires co-ordination of hand, mind and eye. The eyes are very important. These are human sensors most used to detect what has happened, what is happening and what is about to happen. The pilot's eyes have to look at two very different sets of information. Each requires a different approach. The outside view, through the cockpit glass, for example, varies from day to night conditions. Sometimes visibility is unlimited. Sometimes it is restricted, as for example, when making an approach to land in fog.

Scanning

The inside view of the many instruments and controls requires a disciplined scanning procedure so that none of the essential indicators—attention lights and warning lights—is missed. At times during a flight, the pilot has to divide his attention between the outside world scene and the inside 'instrument' scene. He cannot effectively do both tasks at once. Ideally, during an approach to a landing or when attacking a ground target or in combat with other aircraft, the pilot needs to concentrate as much as possible on the outside scene. There is not enough time for even a brief glance at the instruments and the controls.

From the days of the Wright brothers, until instruments were developed, an aircraft was kept the right way up and heading in the required direction by reference to features on the ground and to the horizon. In poor visibility, and particularly in cloud, the majority of pilots became disorientated and usually finished the wrong way up and close to disaster. Over the years instrumentation was added until it became safe to control an aircraft without being able to see land or sky.

But the special demands of combat flying meant that a pilot still had to look ahead. To do this meant putting more information on-

to the sloping glass of the combiner of the conventional optical gunsight. By the end of World War 2, this achievement gave range of target and automatically allowed for deflection shooting when a target was crossing ahead. However, it took another twenty years of research and development before the cathode ray tube (CRT) and the electronic computer were available. They could then handle enough data for projection into the pilot's line of sight so that he did not have to look down at his instrument panels. The electronic head-up display (HUD) system eventually arrived: a combination of computer, CRT and optics. Today, the modern fighter attack aircraft would be incomplete

Smiths Industries

A British Airways Photograph

Top Pilot's view of a head-up display, with its clear green symbols and numerical information, providing the pilot with essential information about the aircraft's behaviour. By projecting these images onto glass, the pilot's vision through the windscreen is unobstructed, and the need to look down at complex instrument panels *(below)* is reduced to a minimum.

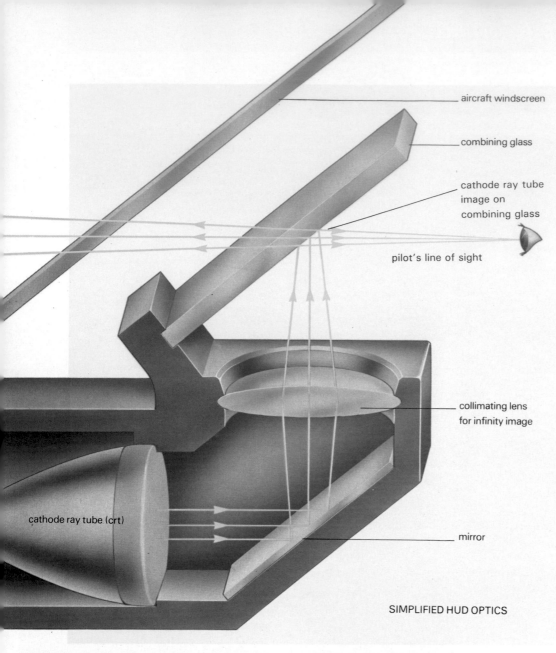

aircraft windscreen

combining glass

cathode ray tube image on combining glass

pilot's line of sight

collimating lens for infinity image

cathode ray tube (crt)

mirror

SIMPLIFIED HUD OPTICS

without its electronic head-up display system. Combining a number of elements, modern systems can tell the pilot not only what he must do to find and attack a target, but also what will happen if he continues with a particular course of action. The amount and type of information which can be projected onto the line of sight is virtually unlimited in theory. But, obviously, the information displayed at any one time has to be limited to that needed to complete a specific task. For example, during the flight out towards a target, the head-up display will show just the symbols and numbers needed by the pilot to steer and climb the aircraft to the cruising height and then to keep on track towards the target. It also keeps him aware of the estimated time of arrival at navigational way points. When he spots them on the ground, he can punch-in a 'fix' on a keyboard. Updating the computer program, the fix will verify that the actual track over the ground is the same as the programmed track. Closer to the target, the HUD will direct the pilot to steer a programmed course which will give the best approach to the target and avoid known hazards like SAMS (Surface-to-Air-Missiles). But only those figures and numbers relevant to the immediate task will be displayed.

Left A simplified view of a head-up display (HUD) optical system. The image from a cathode ray tube (TV screen) is reflected up through a lens onto a special glass combiner. This may present navigational data *(bottom left)* to the pilot through both a computer and an optical system *(bottom right).*

Kuo Kang Chen

A NAVIGATION DISPLAY

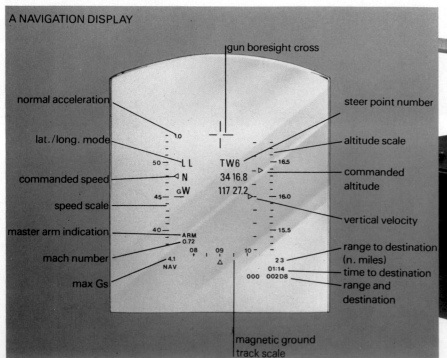

gun boresight cross

normal acceleration

lat./long. mode

commanded speed

speed scale

master arm indication

mach number

max Gs

steer point number

altitude scale

commanded altitude

vertical velocity

range to destination (n. miles)

time to destination

range and destination

magnetic ground track scale

LL TW6
N   34 16.8
GW 117 27.2

ARM
0.72

4.1
NAV

50
45
40

16.5
16.0
15.5

08   09   10
△   23
      01:14
000 002 D8

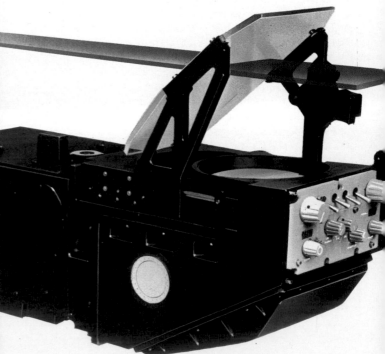

When close to the target, the information will switch to the attack mode. Now, only data needed to aim the aircraft and its missiles in the right direction will be shown. As long as the pilot controls the aircraft to keep an aiming symbol on the glass superimposed on the target, he need not worry about when to release the missiles. The computer does this for him, provided he has previously selected 'FIRE' or 'ARM' on the weapons control panel. Once the missiles fire, another set of symbols takes over and directs the pilot to steer the aircraft clear of the target and back to base.

On the way back, the computer watches the fuel remaining and will instantly flash a warning if there is a problem. A possibility for which the computer has been programmed is a change in tactics. Over the radio the pilot may be told to return to another base or to attack an additional target. All he has to do is punch-in the appropriate code on the keyboard and the symbols on the HUD will change, to give all the information needed to get to the new target.

For the pilot, procedures are fairly simple. The information he needs at any one time is always available. If he needs other data, he has only to interrogate the computer's

memory and the required information will appear on the head-up display's combining glass directly in front of him.

As in all computer systems, the working is somewhat more complex than the result. Essentially the HUD consists of two parts. There is an electronic unit made up of a computer and a CRT, and an optical system which transmits information through a series of lenses and a combining glass into the user's line of sight. The electronic system consists of a computer and a symbol generator. Like all computers, it can be programmed to produce information based on its input data. These signals are then converted in the symbol generator so as to drive the deflecting coils of a CRT. This input is fed from the normal airspeed indicator system and appears, after processing by the computer and the symbol generator, as numbers or symbols on the CRT. The optical system then transmits the airspeed information into the pilot's line of sight. Dynamic information, such as gyrogenerated weapon-aiming information is handled by the HUD computer system in a similar way, so that the pilot sees symbols on the combining glass which indicate how to engage a target successfully.

Digital techniques

Complex circuits have been developed over many years to achieve HUD systems which are reliable and rugged enough for use in vehicles such as combat aircraft. Without the advent of the digital computer, HUD systems would not have been developed to such a high standard of effectiveness.

The optical system of a HUD consists of a series of lenses which convey symbols and characters on the face of the CRT to the combining glass in front of the user's eyes. One of the lenses adjusts the line of the light rays, so that the displayed symbols and characters appear to be focused at infinity. This feature avoids the need to re-focus the eyes when reading the display information. In the basic optical system, the majority of the optical components are used for folding the rays so as to reduce the overall size of the display unit. The combiner is a flat or curved glass screen mounted on a strong rigid frame and set at an angle which 'combines' the information projected from the optical system with the pilot's view ahead. The combiner glass has particular optical properties which give the maximum reflectance of the symbols and characters without reducing the transmittance to a level which obscures the forward view.

The same head-up display system is used in civil aircraft. Instead of weapon-aiming and tactical data, the HUD in a civil aircraft presents the type of information the pilot needs during the approach to a landing. The 'civil' HUD gives a representation of both the approach light pattern and the runway, which will match the real lights and runway as they come into the pilot's range of vision. Other vital data, such as warning and attention-getting symbols, as well as aircraft energy in relation to changes in wind speed, direction and gusts can be displayed 'head-up'. This information appears on a combining glass which can be extended from its stowage position so as to be in the correct viewing position for the pilot's eyes.

Future displays

The pilots of future aircraft will get the majority of their visual information from electronic displays. Some of this information will be presented 'head-up', and therefore in their normal line of sight when looking forward. The remarkable flexibility of the digital computer enables identical information to be shown both 'head-up' and 'head-down', on separate displays.

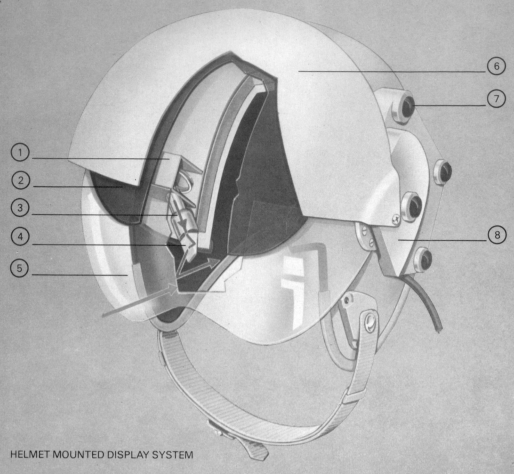

HELMET MOUNTED DISPLAY SYSTEM

Above A helmet-mounted display system.

1 helmet electronics
2 sun visor
3 *led* array
4 prism
5 clear visor combiner
6 visor cover
7 helmet position-sensor *leds*
8 visor latch mechanism

The helmet-mounted display projects images onto the pilot's visor. They are formed by a matrix of light-emitting diodes *(leds)* connected to electronics in the helmet. The images are then projected onto the dichroic-coated polycarbonate visor by a prism.

THE HELMET OPTICAL POSITION SENSOR SYSTEM (HOPS)

target kept within cursors

Right The Helmet Optical Position Sensor System—HOPS for short—can direct a missile along the pilot's line of sight, acting in effect as a weapons-aiming system. The pilot must keep the target image within cursors on the visor *(shown above)* for it to be effective. The helmet position is sensed by cockpit cameras and helmet-mounted *leds*.

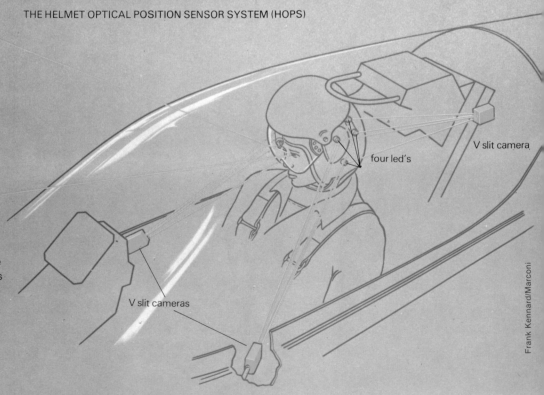

V slit camera

four led's

V slit cameras

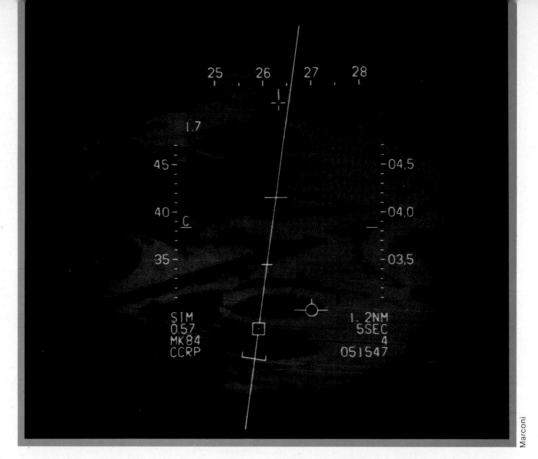

Right Night-time aerial attack on storage tanks, using infra-red imaging of the target on the HUD, overlaid by a green information display. Running vertically is a 'bomb fall line' with its square aiming symbol.

Marconi

The ability to interrogate the outside view of the real world through an aircraft's sensors and computers so as to 'fill-in' missing information will be a standard feature of the flight deck of the 1990s. Filling in the missing parts of the pilot's forward view can be achieved by either using an infra-red sensor or a low-light television. Both methods have the important advantage that they are able to 'see' through fog, rain, snow and darkness. The modern electronic head-up display can combine computer-generated symbology and alphanumerics with infra-red or low-light television pictures. This feature is of particular value for civil aircraft making an approach to land in poor visibility and when taxiing on the ground. The pilot of an attack aircraft with one of these systems is able to find and engage a ground target despite poor visibility around it.

Road and Rail

Head-up display principles can be adapted for both road and rail vehicles. In the 1960s, the Road Research Laboratory in the UK considered the important statistic that seven per cent of all passenger car accidents were caused primarily by excessive speed. Therefore, the RRL asked industry to develop methods of displaying vital information—such as speed—head-up.

Tests were made in the early 1970s to develop a HUD for cars which would be simple in design and low in cost. To achieve these two criteria the designers had to avoid the expensive electronics and optical systems of the aircraft HUD. One solution was to project illuminated numbers and symbols on the semi-reflective panel which formed part of the normal windscreen. The need for a HUD for cars was emphasized when Donald Campbell was attempting the world land speed record in *Bluebird* on a flat desert in Australia. Too much, too soon and the wheels would spin. Therefore an indicator was devised which projected information on to a semi-reflective panel in the windscreen. This showed two vertical pointers moving in the horizontal plane; one indicated the upper limit of power which the wheels could transmit to the surface without slipping; the other showed the actual power being transmitted to the wheels. As long as Campbell kept the power pointer just below the slip-

danger pointer, then he was accelerating his vehicle at the maximum safe rate.

Although technical and cost factors prevented the car HUD experiments of the 1960s from being successful, it is likely that the advent of the microprocessor and cheaper optical systems will make it possible for cars of the future to have HUD.

A major feature of the experimental car HUD was the provision of information about how close the driver could safely come to another vehicle in front. Two vertical lines on the HUD moved together or apart in relation to speed. When moving slowly, the lines were wide apart to indicate that it was safe to come close to the vehicle in front. As speed increased, the vertical lines came together and, provided the driver kept the apparent size of the vehicle in front within the two lines, then there was a safe space between the two vehicles.

Emergency vehicles

The electronic head-up display when applied to road vehicles will be extremely useful at airports. Service and emergency vehicles can be equipped with HUD, which will enable their drivers to 'see' through fog and snow because the electronically generated head-up display will present a view of all obstacles as though on a radar screen.

The combination of electronics and optics which enables information to be presented in an operator's line of sight can be applied to all forms of vehicle. The technique is particularly suitable when it is essential, for both safety and efficiency, that the operator, driver or pilot keep looking ahead, when there is no time to look down at the instrument panel. Head-up display systems have even been considered for Britain's Advanced Passenger Train, or APT.

However, unless the combining glass in front of the driver's eyes and the optical system allows him to move his head from a fixed position, such displays are not really practicable for railway use. In an aircraft the pilot can be expected to keep his head in one position relative to the combining glass.

A particularly useful feature of the electronic head-up display for all types of vehicle is its ability to present different types of information. Whereas the format of conventional instrumentation with pointers and numeral counters cannot be varied, the electronic HUD system can be switched by the user to show only the information that is needed to complete a particular task.

For example, in an airport service vehicle, the head-up information on the windscreen in front of the driver might only show symbols or alphanumerics indicating where the controller wants the vehicle to go. In low visibility the driver could select a synthetic display of the view ahead. This principle has been long established in aircraft head-up display systems, where the pilot is able to select a number of different operating modes such as navigation, target acquisition and attack.

The quietest crime: computer fraud

BBC

Left This scene from a BBC-TV re-enactment of the biggest fraud of all time, in which the Equity Funding Corporation of America produced two billion dollars' worth of phoney insurance between 1969 and 1973, shows the Chief Actuary of the company (left) plotting with a computer programmer who became involved in the crime.
Below right This plan of a computer installation shows some of the security measures that would-be defrauders have to beat. Many others are hidden from view.

1 Finger-length check
2 Main entrance with security guard
3 Input area
4 Pass-through area
5 Surveillance cameras
6 Magnetic volumes library
7 Reader input
8 Control desk
9 Central processor
10 Emergency exit
11 Alarm
12 Flashing light
13 Printers
14 Mirrors
15 Buzzer
16 Magnetic volume /card output
17 Locked output boxes

'There's not a computer installation in this country where if I were allowed in for a day I couldn't set up something that would defraud them substantially'. This quote from a computer crime expert highlights the vulnerability of most computer systems used by business today. As money, bonds and securities are increasingly transferred electronically —a deduction from one computer record and an addition to another—so computer fraud is on the increase.
Computer fraud is a particularly attractive form of crime for several reasons. Thefts are rarely discovered until a good while after they have been committed. Furthermore, the defrauder knows that although his or her crime may be detected, he may never be prosecuted; if the details of a serious fraud involving, for instance, a bank, were made public during a court case, the loss of confidence among investors would cost more than the original theft.

Also, many people are encouraged to commit a computer crime either because they simply regard it as a challenge or, more commonly, they do not feel that they are actually robbing a person, merely a machine.

Committing a computer crime is, in theory at any rate, a relatively simple operation. As an example, take the fraudulent misuse of a computerized payroll system such as that operated by many large employers. In such a system, the details of each employee—name, payroll number, tax details and gross salary—will be held in a computer file on a paper tape or card with punched holes or on a magnetic disc tape. As with any computer system, these devices, together with the actual instructions given to the computer by the operators are known collectively as 'software'. In a large company, the software, together with the operators, will probably be located in a 'machine room' (which may or may not have some form of security around it), while throughout the rest of the building there will be 'terminal rooms'.

Breaking the barriers
Computer terminals consist of a keyboard and, usually, a visual display unit (VDU), which resembles a television screen. By operating the keyboard, the information held in the computer can be made to appear on the screen and processed as required.

Jeremy Gower

To gain access to a particular file, the would-be defrauder is likely to have to break through two security barriers. Firstly, some terminals require a special 'key' to be inserted before they can be used. Secondly, before access to any information is granted, the operator will have to identify him or herself to the computer. This is done by means of a numerical code of about eight digits which, in theory at any rate, is known by authorized personnel only.

Once the relevant file is 'on line' a listing of the whole payroll could be obtained. With the information displayed on the VDU the computer criminal could then overwrite details of his own or another person's pay record simply by typing in new values to the

file. He could, for example, add a zero to his annual gross salary and give himself an effective 1,000 per cent increase. A little less obvious a fraud would be to insert additional names on to the payroll with completely bogus, but quite reasonable, pay details. A further alternative would be to deduct a small sum from every other employee's salary and accumulate all these deductions into one additional pay packet.

While such frauds might at first sight appear ridiculously obvious, the volume of such crime reported in the press is growing constantly. The very nature of computer systems makes such frauds possible. With written, ledger-based accounts, few people have access to the books and any alterations

are immediately visible. Furthermore, auditors are well trained to inspect the books in full. By contrast, many people have access or can gain access to computer files, records can be overwritten at will, leaving no trace of the amendment having been made, and auditors are often poorly trained when it comes to computerized book-keeping. In addition to all this, there is a widespread tendency to regard any information issuing from a computer as gospel truth.

Crime by telephone

In many cases, the defrauder does not even have to be anywhere near the computer building; a computer can often be used by someone hundreds of miles away simply by

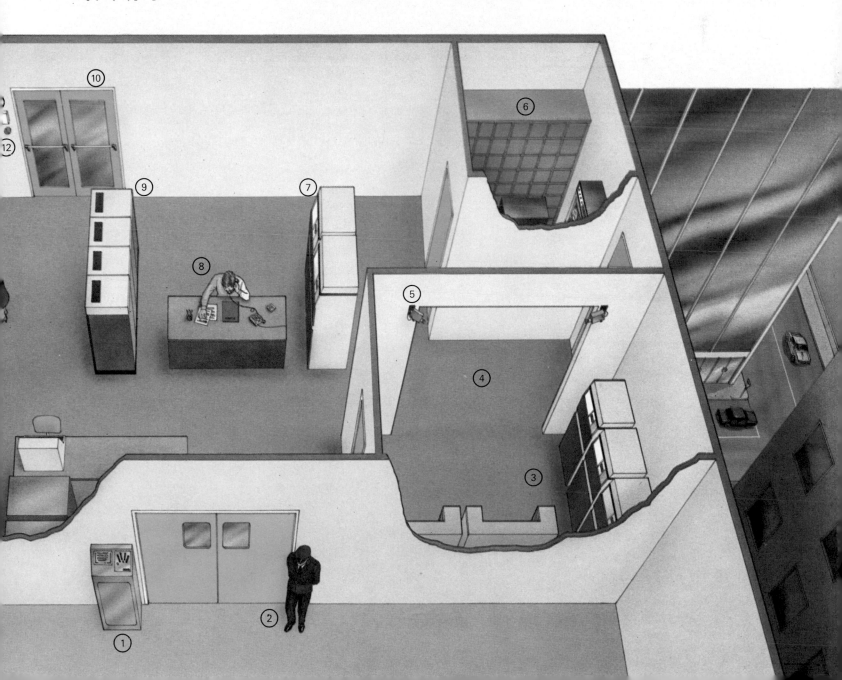

dialling in on the telephone. The ease with which security can be broken in this way was shown recently when three New York schoolboys broke the access codes of a computer in Canada and examined the electronically stored records of 21 companies.

A world-wide problem

Another form of computer crime which is equally hard to combat is the illicit transfer of large sums of money between computer systems, perhaps in different countries, for many computers now transfer information all over the world. One machine calls up another and a rapid exchange of information takes place.

Just as the burglary of a building involves breaking locks or getting somebody on the inside to let you in, so defrauding a computer company involves finding out the numerical code which will allow you entry into the computer. This code is usually very complex. It may, for example, be the sum of various other codes each of which describes, in turn, the transmitting bank, the receiving bank, the sum of money involved, the type of currency and the date. It is on this code that the security of computer-held information largely rests. Unfortunately, while the codes are often unbreakable, the people who are in possession of them are not.

There is a well documented case of an employee of the Bank of America who divulged computer security codes for financial gain to a gang of criminals. The thieves were then able to send a cable to a Hong Kong bank instructing it to pay out 60,000 dollars to a Seoul trading company, where they collected the money. Slightly more devious was the case in Britain where five computer operators at Heathrow airport manipulated the cargo control computer to allow them to smuggle cannabis worth two million pounds into the country. At the other end of the scale, a 19-year-old department store computer clerk was caught after she had amended her husband's account to imply that £270 worth of furniture, which had actually been delivered to her home, had been returned to the warehouse.

The extreme contrast provided by the last two examples serves to highlight some of the problems associated with preventing or tackling computer crime; the increasing use of computers opens more doors to both major and petty criminals alike, as well as attracting the type of person who would not be likely to commit a crime under different circumstances. In addition, the fact that a crime has taken place at all is often not ad-

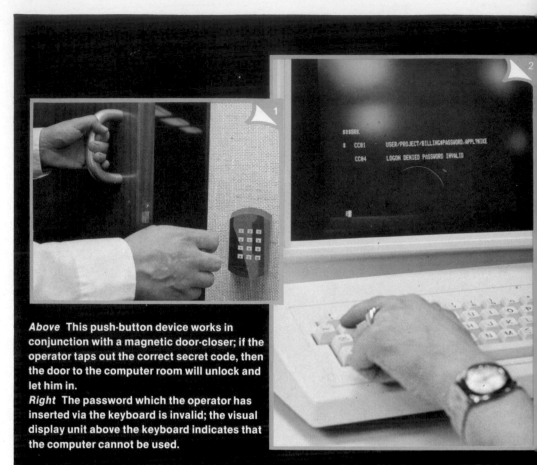

Above This push-button device works in conjunction with a magnetic door-closer; if the operator taps out the correct secret code, then the door to the computer room will unlock and let him in.
Right The password which the operator has inserted via the keyboard is invalid; the visual display unit above the keyboard indicates that the computer cannot be used.

Above Even though the operator has reached the stage of securing access to the magnetic tape on which this computer's information is stored, he cannot obtain the information he wants because a protective block on the tape unit stands in his way.

Right This computer has detected another error by the operator and refuses to let him go any further with his enquiry. Such checks are made at every stage of the data processing operation to help deter the would-be computer criminal in his fraudulent activity.

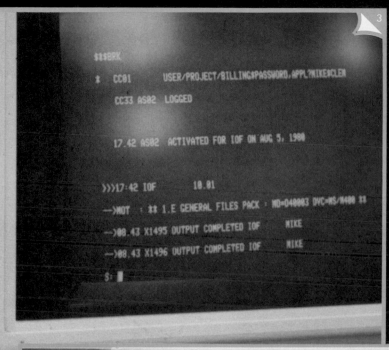

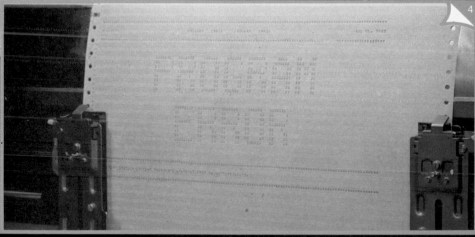

Left Here the operator has used the correct password and is able to receive information from the computer's store. Known (in theory at least) only to authorized personnel, the password is changed frequently to help baffle defrauders. Each operator has a different password. **Below** This computer print-out indicates that the operator fed in a faulty program, which has been refused by the computer concerned.

mitted, to avoid a loss of public confidence in a firm. This often leads to some rather questionable 'cover-ups'; in several cases, employees (often senior and very well-paid ones) known to have committed serious computer frauds have been merely asked to resign rather than being prosecuted.

Stealing programs

Computer crime does not involve just the theft of money or material goods. In many cases, the software associated with a computer system is of far greater value than the hardware (the computer machinery). Some computer programs are extremely long and complex and may require a great deal of testing before they can be put to practical use. It is not uncommon for programs to take a programmer a year or more to develop, and taking into account the high salaries paid to these skilled employees, the investment in software can easily run into millions of dollars. Furthermore, software thefts can be very profitable, as the magnetic tape on which such information is often stored is easily copied. If, for example, an unscrupulous firm could steal a program that had taken five experienced programmers two years to write, then it could have saved about half a million pounds in development cost. Proving a program's ownership is, at present, an almost impossible task. The best means of software protection (from damage as well as theft) is simply the copying of every piece of information and the safe storage of the copies at some remote location from the possible criminal.

Invasion of privacy

Another, more sinister area of computer crime depends on the fact that computer systems contain and process a wide range of personal information which could harm a large number of people if it fell into the wrong hands. For example, an American computer programmer fed information into a computer so that all new job applicants who were coloured would be immediately discarded from the selection process. Details such as credit worthiness, political or religious persuasion, criminal records and so on are all forms of information which are increasingly being stored in computers and which could be made use of by the unscrupulous.

As computer systems become smaller and cheaper, with the help of silicon chip technology, they will be used by more and more companies—and will attract more and more defrauders. It is even possible that computers will be used to defraud computers. For example, a computer might be programmed to explore possibilities so rapidly that it could crack another computer's code.

As computer fraud has become big business, so has computer security, and many firms have sprung up to help the computer user and help make their systems less vulnerable to the would-be thief.

Most of the computer security companies operating at present offer the same approach to reducing the amount of computer crime. Usually their first suggestion is that the number of people who have access to the computer be reduced to as few as possible, ideally one. Obviously the larger the company, the more difficult this will be. A more effective method, and one in wide use already, is the process of logging each computer operation. The effect of this is to note each and every operation, what actually took place, the time and date and the operator who carried it out. An efficient logging system with the records kept secure is one of the best deterrents to crime.

Counteracting computer crime

A similar approach to logging is the duplication of all computer records. Duplicate tapes are then stored at a secure, remote location. Working such a system is not as easy as it sounds, however, as great discipline is required to ensure that the copies are rigorously updated. This is usually done by the *generation number system*. In this system, each time a tape is changed in some way, a copy of the tape is made which carries all the amendments. The new file will then carry the same name as the original but will have a different generation number. In most systems three generations of tape are usually held and these are usually referred to as 'grandfather', 'father' and 'son'. By this system of tape copying and remote storage criminals can be discouraged as they will know that while they may still be able to commit a fraud, sooner or later it will be exposed.

By ensuring that the 'son' generation is as error-free as possible and randomly using it to 'audit' the program in use, fraud can be discouraged even more effectively. Alongside the above measures it is good practice to make all the systems as *hierarchical* as possible so that each operator's code limits him or her to as few data as is practically possible. Such a system then makes it very difficult for any complex fraud to be committed, where the ability to draw on a wide range of functions would be required.

Although many computer users are fully aware of the need for such protection, there is a case for suggesting that efficient security systems should be legally required. For with the proliferation of computers into all walks of life it is not only money that is at risk but information of a highly personal nature, the safeguarding of which should be given the highest security.

Above Computer crime is on the increase with the proliferation of computerized accounting. This mini-computer, as used by banks, mail-order firms and many other businesses, cannot be used until the key (bottom right-hand corner of console) is inserted; this key would be issued to authorized personnel only. Other, electronic, checks help to prevent any fraudulent misuse of the installation.

The two-way radio revolution

Following the earthquakes that devastated parts of Italy in 1980, officials had to rely on two-way mobile radio to co-ordinate the movements of rescue teams, police, doctors, medical and food supplies, and temporary shelter equipment. Even amateurs, using citizen's band (CB) radio, performed an invaluable service by using their sets to organize loans and exchanges of commodities not given priority by the official emergency services, or boosting morale simply by talking to the victims. It is evident that the part played by good communications in responding to natural disasters is of absolute importance. Unfortunately, this is the very time when the most accessible form of communication—the telephone system—is likely to be out of action: lines are cut, underground cables flooded, telephone exchange buildings swept away or burnt down. It is then that two-way mobile radio comes into its own.

Properly speaking, mobile radio covers a whole series of different activities ranging from communicating with ships at sea by satellite to alerting personnel carrying 'bleepers'. However, more than 80 per cent of the market is accounted for by equipment installed in motor vehicles (or carried by individuals) for terrestrial two-way voice communication over fairly short distances.

Remote control

Mobile radio can also be applied to telemetry, the use of radio signals to activate and control distant machines, though in this case not carrying voice traffic; it requires only a sender at one end of the link and a receiver at the other. Telemetry is used for firing propulsion jets on satellites and spacecraft and to alter their position; more prosaically it is used by the modeller to direct 'radio-controlled' cars, boats and planes.

While the advantages that two-way mobile radio offers to emergency services in circumstances less dramatic than those of natural disasters are fairly obvious, new applications are continually being discovered. For instance, a British ambulance system now uses portable equipment which monitors the heart of heart-attack victims, transmitting signals over mobile radio direct to the base hospital. Here a print-out of this information is produced, providing medical staff with an accurate reading of the patient's condition long before they see him or her.

By far the greatest majority of two-way mobile radios are installed in motor vehicles—many of them providing an invaluable aid to emergency services. This US police patrol *(above)* is reporting an incident to the communications centre *(right)* where it can be relayed to other units nearby.

Private operators of mobile radio include radio cab firms, transportation and freight companies, and many types of industrial and manufacturing organizations in which rapid communication between individuals far apart is essential, and the provision or use of a telephone line is impracticable. Rerouting a truck driver or summoning a key worker for new instructions can save huge sums. A 1977 study calculated that private mobile radio was then saving the UK at least £200 million annually. Today, savings are likely to be three or four times as much.

Military and civilian mobile radio users alike require that messages should be unambiguous. Here the similarity between the two ends. The military communicator may lose radio contact through deliberate jamming, or lose secrecy through eavesdropping. In a civilian context, both of these are merely irritating, but for the military, they may make the difference between a mission's success and failure. In addition, the soldier must ensure that his own radio transmission does not betray his whereabouts. Jamming, eavesdropping and locating enemy communications, while protecting one's own, is a major part of military technology.

One widely used protective measure is 'frequency hopping'. The starting point for all radio communications is the generation of a high frequency alternating electrical current in a transmitter. This current continually varies in amount and periodically reverses direction. Frequency is the number of cycles taking place in a given limit of time, and the unit of frequency equal to one cycle per second is known as one Hertz (Hz). Most mobile radio communication takes place at frequencies from about 26 to 900 million cycles per second (26 to 900 MegaHertz or 26 to 900 MHz).

The basic idea of frequency hopping is that all radios communicating in a particular area are synchronized to change to different frequencies while transmitting. This is done rapidly—with perhaps 10,000 changes of frequency per second—making it very difficult for the enemy to lock onto a signal, a technique essential in jamming, locating and eavesdropping. Frequency hopping requires some form of computer control in the radio set, and this is also useful for encoding messages into streams of meaningless babble. The correct decoding formula is then required to 'unscramble' the message.

Mobile radio is also used for the convenience and amusement of private individuals. Telephones in cars, known as Mobile Automatic Telephone Systems (MATS), transmit signals from the moving vehicle to a base station which interprets them as a number to be dialled on the normal telephone network. When a connection is established, conversation may proceed.

A MATS brings people into contact who would otherwise be incommunicado, for example, on a motorway. However, connecting a call to, as opposed to from, a moving vehicle is difficult when its location is not known, as there are limits to a base station's transmission range.

From base to base

A number of solutions have been proposed, most of them involving some form of 'registering' the vehicle in a 'home' area under the control of a particular radio base station. As long as the vehicle remains within the home area, all telephone calls are routed through the home base station; when it moves out of it, the mobile radio set automatically, and without the intervention of the driver, sends out signals to register its presence with the local base station. This in turn informs the home base station of the change. Thus, if a telephone call is made to a vehicle out of the home area, the home base station knows where to re-route it.

If a call is in progress when the vehicle travels into a new area, the base station initially handling the call automatically detects a weakening of the vehicle's signal, and broadcasts a message to all surrounding base stations to listen for it. When one of the stations picks it up it informs the first station which then knows where to route the call when its own reception becomes too weak.

Left Like all other modern fighting forces, the Israeli army relies on two-way mobile radio for communications between field units. *Below* Equipment in an ambulance transmits ECG information to the base hospital by two-way radio.

ZEFA

Pye Telecommunications Ltd

Above The Mobile Automatic Telephone System (MATS) allows a private motorist to keep in touch via the public telephone network. It is powered by the car battery. The aerial is usually mounted on the roof and the transceiver is stowed in the boot.

Below A versatile, compact and lightweight two-way mobile radio. A similar type *(inset)* is being used by a driver on the road.

Cable & 'Wireless Ltd

A mixture of convenience and pleasure is involved in citizen's band radio. The idea that CB is also useful—as in the aftermath of the Italian earthquake—is obviously supported by manufacturers and users alike, but is not accepted by some of the authorities controlling the sharing out of air space. The main charge is that it interferes with other more important radio-type services. This is because most CB sets currently in use operate on a frequency of 27 MHz AM. Illegal in countries such as the UK, AM, or amplitude modulated, frequency causes severe interference on television, radio and hi-fi—as well as jamming frequencies used by aeromodellers with often expensive, and possibly dangerous results.

In theory, the usable radio frequency spectrum extends from 10 kHz to around 300 GHz. In practice, however, as the top of the spectrum is approached, the necessary transmitting and receiving equipment becomes more complex and expensive. Satellite communication at 14 GHz is at present the highest practical frequency in use, although experimental work for satellite broadcasting at up to 86 GHz has been carried out in the United States.

Mobile radio has to compete for its air space with broadcast radio and television, fixed radio links, radio navigation and location systems, radio astronomy, maritime and aeronautical authorities, satellite communications and many other users. In the 80 years that the radio frequency spectrum has been exploited, the growth in the volume and variety of traffic has mushroomed and shows no sign of abating. It is now recognized that, just like oil, the radio spectrum is a scarce and finite resource. Today, it is nearing the point at which it will not support any new traffic, and a 'frequency famine' is predicted. The effects are felt more acutely in developed countries, but no part of the world can remain immune. If a solution exists, it must lie in new technology or else in finding new ways of managing the radio spectrum.

A number of technological approaches to the problem of a potential frequency famine have been made by mobile radio manufacturers and users. A common goal is to pack more users into the same air space. Considerable progress has already been made in this direction by reducing the width of radio channels, and by minimizing the length of time actually spent 'on air'.

Dynar Electronics

MOBILE TWO-WAY RADIO TELEPHONE

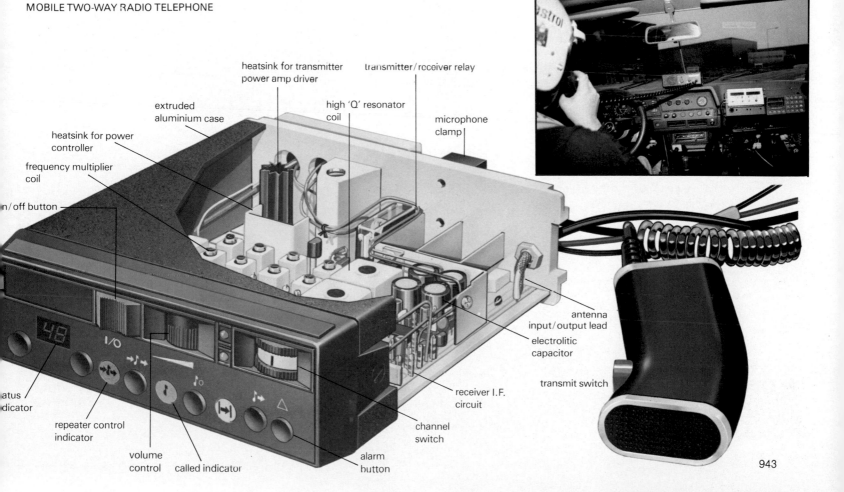

heatsink for transmitter power amp driver

transmitter/receiver relay

extruded aluminium case

high 'Q' resonator coil

microphone clamp

heatsink for power controller

frequency multiplier coil

on/off button

status indicator

repeater control indicator

volume control

called indicator

alarm button

channel switch

receiver I.F. circuit

antenna input/output lead

electrolitic capacitor

transmit switch

Radio channels were 100 kHz wide in the 1950s; today channels 12.5 kHz wide are quite normal, and systems with 6.25 kHz channels have been shown to work. But while more users can be accommodated in the same air space, reducing channel width does impair the quality of transmission. Communication on a 6.25 kHz channel has been likened to listening to a conversation over a dozen pans frying potato chips. Some experts maintain that this is acceptable because the sense of the message still gets through; others disagree.

It is clear that if each user of the air waves reduces transmission time, more people can use that particular frequency. One of the earliest methods of accomplishing this was to use codes for routine messages. Thus, as every fan of American television police programmes knows, '10-4' means 'previous message received and understood' but takes only a fraction of the time to say. More sophisticated methods dispense with the human voice altogether: a button is pressed on a radio set and, almost instantly, the resulting signals are received by another set which converts them into numbers or letters on a small viewing screen. Again, the code represents some standard message.

Such procedures can be used to hasten the identification process between users. In a taxi company, for instance, each cab could have a code number which appears on the controller's viewing screen on contact.

Trunk calls

Another technique is known as 'trunking'. A limited number of radio channels are shared between a large number of potential users on the assumption that they will not all wish to use the system at the same time. A user requesting a call is automatically allocated a channel, but only for the duration of his conversation. On completion the channel is returned to a 'pool', ready for reallocation. Trunking systems may accommodate up to six times as many users as conventional systems.

Low-frequency signals of, say, 30 kHz have a massive wavelength of some 10 km (6 miles). Since these waves travel close to the Earth's surface and are big enough to 'jump' over tall buildings and even mountains, their range is potentially very great. Mobile radio, however, operating as it does at very-high, or ultra-high, frequencies has quite a small range. One of the frequencies proposed for CB in the UK is 930 MHz, giving a wavelength of about 32 cm which is insufficient to 'jump' over even modest buildings.

In theory, at these frequencies, the effective range should be limited to within 'line of sight' of the transmitting station. In practice, since the wavelengths are so short, they tend to 'bounce' off many objects, bend slightly, and to a considerable extent pass through buildings. Provided the general area

Right A new language has evolved between mobile radio users such as truckers so that transmission times can be greatly reduced. *Below* Broad-waveband links require 'line of sight' between aerials, and masts are used to clear obstructions. The short wavelengths used in two-way radio communications will pass through buildings to some extent, and will bend slightly over minor obstacles. Due to multiple reflections, the waves are scattered and will 'bounce' from buildings. Provided the general area is in line of sight, the signal will be adequate in both directions.

John Hillelson Agency

SIGNAL PATHS IN BUILT-UP AREAS

Jerry Banks

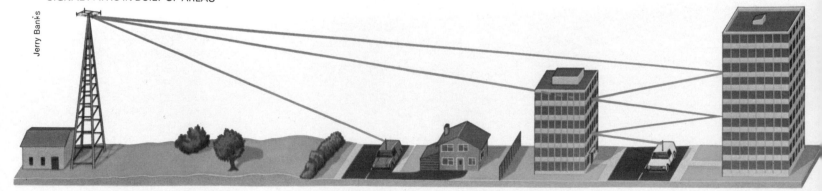

is in line of sight of the transmitting station, signals will be adequate in both directions. Indeed, this limitation on effective range has its advantages: mobile systems in two cities, say, 60 km apart can use the same frequency without causing interference to each other.

Whatever the distance, the very weak signals arriving at the receiving station must be amplified and played out through some form of loudspeaker. For two-way voice communication, both transmitting and receiving facilities must be incorporated in each station (hence the term 'transceiver').

More intelligent and cheaper

If the threatened frequency famine can be averted, the future prospects for two-way mobile radio are exciting. The silicon chip has not only reduced the size of the equipment, it has made it more powerful and more 'intelligent' and yet still cheaper. Already a portable satellite Earth station is in use with some army units, capable of being carried by only two men. And radio facsimile systems, in which documents are electronically scanned and the resulting signals transmitted to moving vehicles for re conversion and printing out, have been tested by police forces.

Vehicle location systems have also appeared in a number of forms; one of the most sophisticated continually scans the rotation and steering of a vehicle's wheels from a known starting point and relays this information back to base over a mobile link. The signals are interpreted to pinpoint the vehicle's position on an illuminated map to within a metre inside the coverage area of the mobile link. The latest versions of car computer terminals, fitted with typewriter keyboards and viewing screens, can transmit or receive a full screen of information in 0.6 seconds; 300 suitably equipped vehicles can operate on a single channel and a print-out of each message can be kept as a permanent record if required.

Manufacturers are now seriously investigating personal wristwatch-type radios communicating via satellites launched by vehicles similar to the American Space Shuttle, and mobile radio computer terminals no larger than a paperback. Work is also speeding ahead on silicon chips which can simulate, recognize and respond to the human voice. Soon, controlling and questioning all types of previously 'unintelligent' machines over mobile radio may be possible. The implications of such a development could be among the most far-reaching of any technological advance for some time.

Above The 'briefcase communicator', developed by NASA, can send or receive signals over vast distances by satellite relay, using a small 'hook-on' aerial. Air-sea rescue services can act swiftly when the base radio room picks up distress signals *(right)* Mobile terminals *(below),* fitted with a keyboard and a viewing screen, are useful to vehicle fleets. Up to 300 suitably equipped cars can be operated on a single channel.

Bugs for professional snoopers

Electronic surveillance equipment is now the most significant tool in the world of espionage. With its aid, agents of Intelligence services can eavesdrop conversations across oceans. Government security services in most developed countries use its techniques to monitor the activities of dissidents, terrorists and suspects. Industrial and business spies employ it to pry out company secrets. Though no estimate can be made of the extent of its use, it is on the increase in both the West and East. The telephone is the electronic snooper's most valued ally. It is universal in homes and offices, has a power source and is connected by wires to an exchange, often many miles away, with which the eavesdropper can tamper. He may use the voltage present in the system to power his electronic devices. He may use the wires themselves to carry audio signals that have been converted into electrical signals. He may even use the telephone instrument itself as an eavesdropping microphone.

Telephone techniques

The *direct wire tap* remains a favourite system. It involves a direct connection to the two wires carrying the telephone conversation. The connection can be made anywhere between the instrument and the exchange, but because wires cluster more thickly close to the exchange the difficulty in locating the target pair increases. Most direct wire taps are, as a result, connected to a point before the pair of wires from the telephone reach a junction box. Advantages of the direct wire taps are many. The agent has no need to enter the target premises. The direct wire tap, easy to attach, provides a reliable output signal which can be readily recorded or transmitted by radio to a listening post.

Once any telephone tap is connected, the eavesdropper has two methods of relaying the audio signals to a convenient listening post. Preferred is a *hard-wire* connection which, properly installed, is reliable, gives good audio levels and is fairly secure from detection. The alternative is a radio transmitter. It can be concealed or disguised and thus need not be extremely small, permitting the use of larger batteries to power transmissions over a range. Such a transmitter, being outside the target premises, can be easily resupplied with batteries, and so operated almost indefinitely

Above A wide selection of bugging devices is currently on the market. Their size is dependent on the duration and range of the transmission required. The greater these factors, the larger the power supply needed.

Only the methods outlined above are professionally known as 'telephone tapping.' Alternative means of using the telephone for electronic surveillance are known as 'bugging'.

Bugging requires access to the target premises—gained illegally or surreptitiously. Bugs are miniaturized radio transmitters, purpose-built in many shapes and sizes, with and without built-in microphones. Bugs for use in telephones require no microphone: the audio signals they intercept are already in electrical form. Many have no batteries since they use power from the telephone system.

As a result such bugs can be highly miniaturized. The smallest type currently in use is about the size of a grain of corn and has two small wires, each about 9.4 mm (0.37 in) long, protruding from it. These wires are connected to the terminals inside the telephone instrument which connect with the *talk pair* of wires. The bug transmits on VHF (Very High Frequency), operating between 148 and 170 MHz. It can transmit over a distance of 91.5 m (300 ft) or more, depending on the installation. The device is barely visible when properly installed and can even be hidden inside the casing of one of the telephone wires.

The electronic eavesdropper can pursue his craft even while the victim's telephone remains on its rest. Conversations in any room containing telephone wires can be captured and fed back down the wires, once the phone

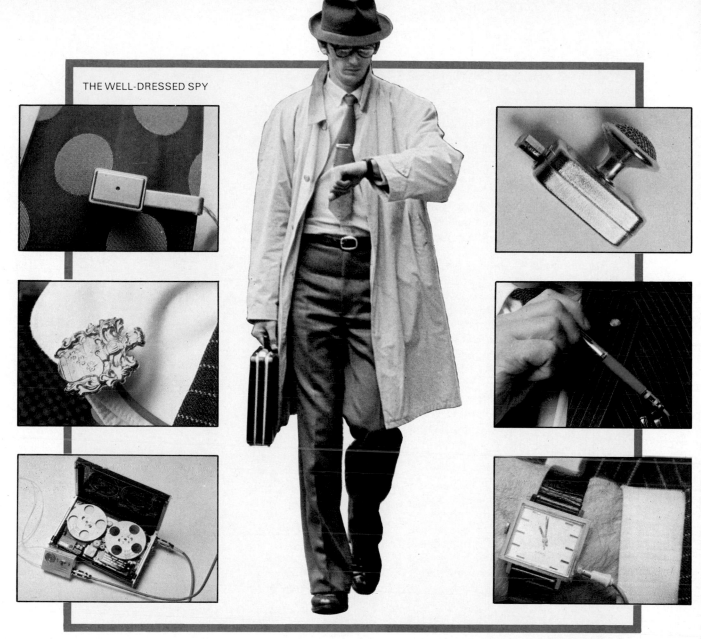

Right Miniaturization of electronic equipment has enabled bugs to be concealed in almost any everyday item. The industrial or business spy may have at his disposal receivers located in tie-pins, cuff links, fountain pens, lapel badges or wrist watches. Audio signals can be transmitted to a third party, or the bug can be connected by a wire to a concealed tape recorder. If signals are transmitted by radio they can be detected by a countermeasure system. This normally operates by scanning the range of wavelengths likely to be used by a bug's transmitter. 'Feedback' betrays the bugs presence. *Below* Pipe-shaped microphones can be passed through walls or keyholes. Even *hard-wire* devices can be concealed in very small spaces.

THE WELL-DRESSED SPY

is equipped with a microphone to convert sounds into audio electrical signals.

A frequently used telephone bug for this purpose is the so-called drop-in mouthpiece. It is simply installed in the telephone handset after removing the original mouthpiece unit and is connected up in the same way as the original. Since many drop-in mouthpieces differ in design from the standard microphone mouthpiece, they may easily be identified as a bug by visual inspection.

More sophisticated is the *infinity transmitter* or 'harmonica bug.' This device is not strictly a transmitter but a tone controlled switching device, usually coupled with a microphone and audio amplifier. The microphone is activated by the switching circuit when it receives an electrical signal of a specific frequency over the telephone line. The frequency initially used was about 440

Right Bugs designed to be carried on the body usually have a built-in capacitor microphone which will not transmit side-tones such as those caused by walking. An off/receive switch conserves power.

Hz, equivalent to a 'C' note on a harmonica, which is how it acquired its popular name.

The infinity transmitter is so called because the original manufacturer claimed that it could eavesdrop on a room from an unlimited distance over the telephone system wires. This claim has proved inaccurate because of differences in the various telephone systems used around the world. In the British system, for example, the *called* telephone controls the circuit which is only disconnected when the called telephone is replaced on the 'hook switch' in the cradle of the instrument; in the United States, it is the

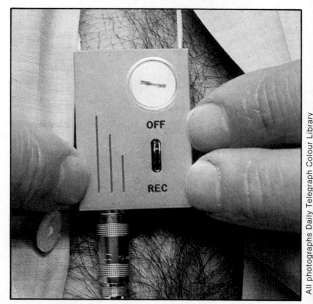

All photographs Daily Telegraph Colour Library

calling telephone which controls the circuit. This makes a considerable difference to the operation of the infinity bug.

An eavesdropper using an infinity bug on the British telephone system would call the number of the telephone in which the bug had been planted. When the telephone was answered, he would pretend to have called a wrong number. Then, before the telephone he was calling was replaced on the handset —cutting the connection—he would blow a tone whistle or place a small electronic tone producer beside the mouthpiece of his telephone. At the other end of the line, the switching circuit 'recognizes' the tone, switches on the microphone and keeps the line open even when the handset has been replaced. From then on, the eavesdropper can continue to monitor conversation and sounds in the room where the target telephone is until he decides to hang up his own telephone.

Compromising

Techniques, known professionally as 'telephone compromising', exist to transform the telephone itself into a bug. Either the mouthpiece, with its carbon microphone, or the earpiece, with its diaphragm and magnet, may be used as a room microphone if the hook-switch is shorted or bypassed. Use of the earpiece requires higher amplification of the signals than does employment of the mouthpiece microphone. In either case, the telephone wires carry audio signals from the target room to a convenient point between the target and the first switching exchange. At this point a direct wire tap is made and the audio signals from the target area fed to a listening post either by a hard-wire connection or by a radio transmitter.

Despite the new technology, one of the oldest methods of eavesdropping remains the most reliable: the use of a concealed microphone connected by 'hard-wire' to a listening point. This is the method used in Eastern bloc hotels where certain rooms have their hidden microphones well positioned to monitor selected areas through tiny pin-holes in the plaster. The microphones, remotely controlled and powered, can be switched off to reduce the possibility of detection during an anti-bug sweep of the room. They have an additional advantage—instead of going to the trouble of installing bugs in the room of a target visitor, the target visitor is simply placed in the already bugged room.

Where a hard-wire installation is not possible because of difficulty in concealing connecting wires, it is sometimes possible to

use silver paint connections—at least in the immediate target area. The silver paint acts as a conductor for electrical signals and can be concealed under another layer of paint. Another, more commonly used, alternative is a radio transmission system. In such a system, the bug comprises a microphone, a power source, a transistorized transmitter circuit, and an aerial. Advances in electronic circuitry and miniaturization enable the necessary components to be packed into a space smaller than a sugar cube.

A typical example currently available measures 15 × 11 × 5 mm (0.6 × 0.4 × 0.2 in), gains sufficient power from a 1.4 volt mercury battery to transmit for 25 to 30 hours up to 27.5 m (90 ft) without an aerial and up to 228.6 m (750 ft) with an aerial. Despite its small size, the microphone is sensitive enough to pick up very low intensity sound within 8.3 to 36.6 m (60 to 120 ft) and will cover the sound spectrum from 100 to 7500 Hz. Human speech has a spectrum of the order of 90 to above 7000 Hz.

Within limits, bugs can be as small—or as large—as the job requires. The greater the

transmitting range required and the longer the bug is required to continue operating, the greater the power supply must be. The greater the power supply required, the larger the bug must be. If the bug has to be concealed in the target premises, there is a limit to how big it can be while still remaining easy to conceal.

The problem can be overcome, however, if the bug can be built into the room or its furnishings. In 1975 when alterations were being carried out to the conference room of the British Communist Party's headquarters in London, a bug was discovered built into a wooden beam supporting panelling surrounding the platform. It measured about 14 × 3.8 cm (5.5 × 1.5 in) and was powered by long-life mercury batteries. A switching circuit enabled the bug to conserve energy by being switched on and off by radio signal.

One of the most ingenious such devices is a bugged electric light bulb. It can be inserted in a desk lamp or other bulb holder to function normally while at the same time bugging the room. Fitted with a small transformer, it draws its power from the

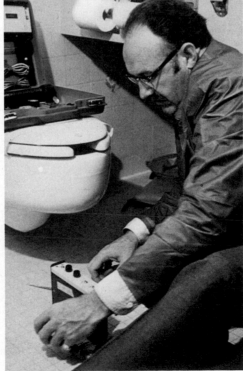

Left A really unfortunate businessman may discover his company's secrets being threatened, literally, on all sides. The tape recorder (1) has a *hard-wire* connection to the microphone under the carpet (2), of the type shown below. A radio receiver's aerial (3) picks up signals from a transmitter implanted in the wall—conceivably even by a dart (4). Another receiver (5) can send a strong signal via the roof aerial to activate a bug concealed in the wallchart (6). This may have lain dormant for any length of time, making it particularly difficult to detect. Such a device has been discovered in the U.S. embassy in Moscow, hidden in a wall plaque—a 'present' from the host nation. Speech patterns making minute vibrations on the window pane can be conveyed through a laser beam (7) connected to a receiver. But such devices are still experimental. Small transmitters on the same, or varying, wavelengths are concealed in the light fitting (8), the ashtray (9), the television (10) and a pen (11). Both telephones are bugged—one by a microphone in the handset (12), the other by a mini-transmitter (13). The handset bug can be activated by a frequency generator through an outside line, even when the telephone is not in use. Any one of these devices, especially if used in conjunction with a bug concealed in the visitor's clothing, would enable all conversation in the room to be overheard.

Above A telephone 'tap' may employ a transmitter (inset) no larger than a grain of rice. It can be hidden in a telephone wire and transmit to a receiver up to 100 m away. *Right* Gene Hackman in *The Conversation.* Even in the world of films, electronic surveillance is not necessarily glamorous.

Left Pipe-shaped microphones can be passed through walls or floors. Unlike transmitter bugs like the one in the cigarette packet, they are *hard-wire* listening devices.

mains. The microphone and transmitter are concealed in the base of the bulb and continue to operate as long as the light bulb continues to function in the socket.

Another form of 'wire-less' installation can be inserted into an electric wall plug. It consists of a microphone and transmitter which sends its signals over the electric power lines. The listener plugs his receiver into an outlet socket which must be on the same electric circuit but may be in another room in the same building, or even beyond. Provided the receiver is plugged in at a point before the power lines reach a transformer relay station, it may be at any distance from the target area.

Considerable publicity has centred on the use of laser beams for audio surveillance. Claims have been made for a device said to direct an invisible laser beam at a pane of glass in the window of a target room. The glass, according to the story, vibrates in relation to the audio vibrations in the room. These vibrations were said to have been carried back along the reflected laser beam to a receiver which turned the vibration signals back into audio. First reports of this device appeared some ten years ago but experts in the field of audio surveillance deny seeing a laser device which will operate satisfactorily except under extremely controlled conditions and then only over very short ranges.

The practice of detecting audio surveillance devices and providing protection from them is known as *audio countermeasures* or, more popularly, 'debugging'. It is a practice which requires skill and highly sophisticated electronic equipment.

There is no way to determine conclusively the existence of a properly installed wire tap on a telephone line, short of visual inspection of the whole line between the suspect instrument and the telephone exchange. Most devices which purport to check telephone lines for taps work on a system of voltage measurement. When the telephone handset is in the cradle and not in use, the voltage of the system is around 48 volts; when the handset is lifted up and the telephone is in use, or in the 'off-hook' mode, the voltage drops to about a quarter of this level.

A simple voltmeter can be used for this check. If a variation is revealed in the normal voltage level of the system, something may be amiss. But such a voltmeter can only detect crudely installed line taps which affect the system voltage, or the presence of crude, voltage-actuated tape recorder taps or improperly installed devices drawing excessive current from the system. In any case, the electrical characteristics of any pair of telephone wires can be changed simply by changes in the atmosphere, such as weather, temperature or humidity variation, or by the distance from the exchange and equipment additions within the systems. These changes can be greater in extent than those produced by the introduction of an efficiently installed quality wiretap.

Most countermeasure systems for detecting radio systems are more productive, since the transmitter, in the operational mode, gives off a radio signal which can be detected by some form of radio receiver tuned to the correct wavelength. Since the wavelength of the transmitter (if there is one) is not known, this is accomplished by a specially

Right An inconspicuous attache case reveals a device to monitor eavesdropping activity on up to seven lines at once.
Left The telephone user can also be alerted by a bug detector fitted into the handset.

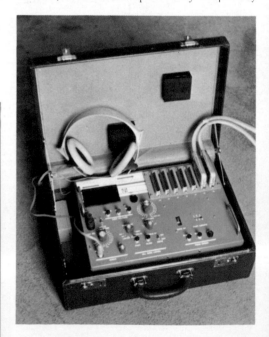

Left This telephone will operate normally when all the switches are in the off position. It can be activated to detect a bug in the vicinity of the telephone using a radio frequency (RF) detector, or to reveal electronic eavesdropping on the line. An audio monitor can record both sides of a conversation during normal or 'secure' use.

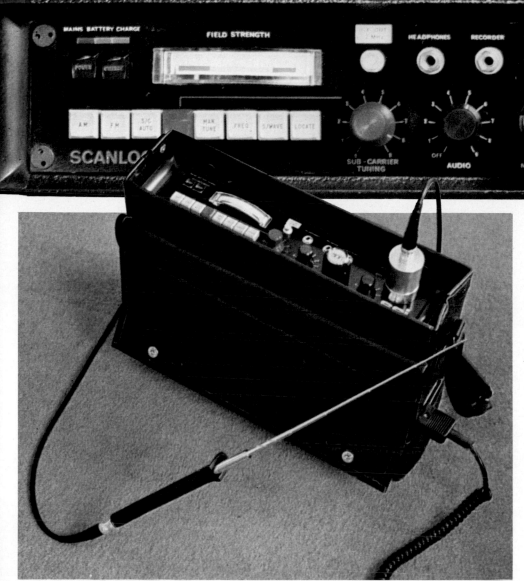

Peter Burt/M.C.

Above and left Ever more sophisticated countermeasure systems are being developed. This detector can sweep a high range of frequencies and lock on to a bug, or be left in an 'on-guard' state, or tuned manually.

constructed receiver which is automatically tuned to sweep the full range of frequencies on which a bug might transmit.

When such a receiver picks up a signal from a nearby transmitter, the audio signal received is picked up by the transmitter's microphone, resulting in howling produced by *feed-back* – the familiar noise from a poorly installed public address system.

Other debugging techniques involve the use of *field strength* measurement equipment. This device contains an aerial, a diode detector, and a sensitive amplifier connected to a meter or other form of indicator. The meter measures the relative radio frequency energy intensity detected by the instrument, which may vary over a very small area. In some cases the more sophisticated devices can be adjusted to give a 'nil' reading for the particular location, ignoring, as it were, the high radio frequency background normally found in the centre of a built-up area. Any variation from the nil reading indicates a nearby low-level transmitter, such as a bug.

This device can also be provided with a speaker which, with the aid of high level amplification of the output of the field strength meter, can produce the feedback howl when a transmitter is nearby. Another sophisticated variation uses two aerials, each of which gives a field strength reading. If the two aerials give a different reading from each other, the one giving the higher signal strength is closer to the transmitter. By skilled use of the aerials a knowledgeable expert can quickly locate an electronic bug.

Complete countermeasures

An indication of the seriousness with which many individuals and companies regard the bugging threat is the presence on the market of a complete electronic countermeasure system built into four lightweight suitcases—at a cost comparable to the price of a luxury saloon car. This equipment can detect transmitters using power lines or telecommunication lines as signal paths; adaptations or by-pass systems which com-

promise telephones; the presence of microphones or unidentified wires; and radio transmission bugs. Microphones discovered by the system can be activated so that their location can be pinpointed. Similarly, wire pairs buried in walls or other structural features can be traced with a metal detector to their termination or point of origin.

In the world of electronic surveillance there exists a constant battle of surveillance measures and countermeasures. The infinity transmitter, for instance, was made more secure by using two or more tones—rather than one—to activate it. Now the countermeasure experts have developed multi-tone sweep devices. These use simultaneous frequency sweeps arranged and timed to activate any infinity transmitter on the line. Each new development on the side of the eavesdropper is, before long, overcome by the counter-measure experts.

And although eavesdropping equipment may become more and more sophisticated, the countermeasure experts look forward to the day when telephone systems will use optical fibres to carry vast numbers of calls and will make it extremely difficult, if not impossible, for the wire tapper to operate. Similarly, the use of high frequency radio signals conducted inside buried, metal pipes to provide larger capacity communications systems, will also present the eavesdroppers of the future with complex problems.

For his part, the eavesdropper will welcome the development of new microcomputer processor techniques of the type provided by the micro-chip in pocket calculators. These are being used to develop new radio signal processing capability which will almost certainly result in reducing power needs, while at the same time boosting transmission range and making the signal, bearing stolen secrets, more difficult to detect.

A computer behind the wheel

Every minute of the day someone, somewhere, dies in a road accident because a driver has made a human but fatal mistake. And every day of the week major arterial routes become clogged and congested as thousands of motorists join the queues leading to the big cities. With more traffic on the roads today than there has ever been, many traffic engineers long for the day when each car will be controlled by a computer and the driver can sit back in his seat knowing that the computer will steer him safely clear of all potential accidents and traffic jams. But just what are the possibilities for in-car computers?

Ideally, of course, cars would be totally automatic. Information supplied by cables buried in the road would be received by the car's central computer, or computer terminal, and the car would then be automatically guided to its destination swiftly and safely. All the driver would have to do is switch on the car and tell it where he wanted to go.

If all this seems rather remote, remember that even now ground staff at Houston are able to guide a spacecraft with pinpoint accuracy millions of miles through space without the aid of signposts. In the light of this kind of achievement, fully-automated vehicle guidance takes on a more realistic appearance. Indeed, much of the basic technology needed already exists and, while they are still only used to monitor and control a few specific functions, silicon chip microprocessors are becoming more and more common in production cars.

With the advent of the chip and its continuous refinement, there is no major technical barrier to equipping a car with a central computer small enough to be practical but sufficiently powerful to allow complete automation of all the car's functions. The real problem lies in the 'interface' between the car and its computer.

For the computer to do its job, it must always have accurate and up-to-date information on all the factors that will influence the final decision. The large number of functions in the average car and the larger number of different factors which affect their operation make this a less than simple task. The microprocessor for a fuel injection system, for instance, needs to have information on engine speed and load, throttle setting, air temperature and air density, if it is to provide the correct mixture under all conditions—and even this may not be enough. Each of these quantities requires a separate sensor, each of which must be tied in to the microprocessor. These sensors need not be particularly complex—a simple cam is all that is needed for an engine speed sensor—but it is clear that the more functions that are computer-controlled, the more sensors there must be.

Additionally, all these sensors must be properly 'interfaced' with the computer. Whether the sensor registers as a varying

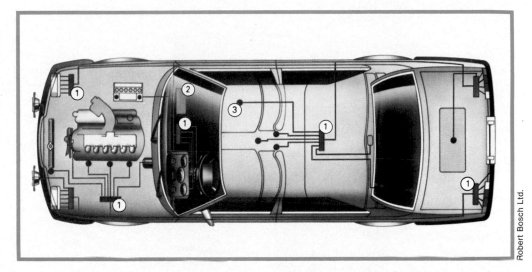

Robert Bosch Ltd.

Above left A Series 7 BMW car fitted with Bosch's 'Motronic' system which monitors and controls many of the car's functions. Because the system operates digitally, a single wire can carry commands to several different units (for example headlights, sidelights, indicators and horn). This does away with the need for the complex and expensive wiring harness found in most modern cars.
Left The 'Motronic' main unit and the BMW engine it controls. The main advantage of using microchips in this application is that the engine can be made to operate at maximum fuel efficiency.

BMW (G.B.) Ltd.

electrical resistance, as a signal from a photocell or whatever, it must be translated into a form that the computer can understand. The problem is that computers operate 'digitally'—that is, by shunting small discrete electrical charges around their circuitry. Ultimately, a digital computer is just an elaborate series of on/off switches and the information must be provided in such a way that it activates the right switches in the right sequences. Unfortunately, the information from the sensors is rarely digital. It is usually in 'analogue' form where the output changes as the quantity it is measuring changes, giving a continuously varying signal rather than a sequence of separate impulses. Converting from analogue to digital form is not an insurmountable problem but it adds further to the complexity and expense.

Interfacing

The problems of interfacing information into the computer are in fact considerably less taxing than the problems of the outward interface, the one that changes the computer's decisions into actions on the car. This is where one advantage of the chip—its very low power consumption—turns into a big disadvantage. The problem is that a signal of a tiny fraction of an ampere must operate a big fuel valve or apply the brakes with enough force to stop the car. Of course the signal has to be amplified and in some cases to a very large degree. Like hi-fi amplifiers, the signal is increased by using the small signal to switch on a larger one. Unfortunately, quite apart from introducing inaccuracies, this may create unwanted oscillations in a closed-loop system. These oscillations may arise whenever a correction starts chasing an error without catching it up. In

Above How a computer controls a car. Sensors (3) are located throughout the car to monitor such things as oil pressure, engine temperature and amount of fuel. The information they gather is fed to a nearby 'subscriber station' (1) which puts it into digital form and transmits it along a single wire to a central controller (2). The data are processed, and digitally coded control 'messages' are sent to the various subscriber stations where they are decoded and acted on—for instance the ignition timing may be advanced or retarded.

the worst case the oscillation can diverge and over-correct. In a hi-fi system this simply produces a shrill scream: in a car's braking system it could be fatal.

Even with a large clear signal and oscillations well-damped, the signal may not yet be in the right form. The brakes on the cars of today, for example, are often hydraulically actuated. So the interface for the area of the computer controlling the braking system must not only include an amplifier but also a means of converting an electrical signal into a hydraulic movement. Again this is not too difficult in most cases and can be achieved by the use of such things as solenoids (acting as valves in conjunction with a central hydraulic pump). Solenoids are coils which when electrified become magnetized and induce a mechanical movement. They are already commonplace on cars—in the starter motor, for instance, or in the flow regulators of computerized fuel injection systems—and indeed are older than the car itself. But once again they add complexity and expense.

It is clear that the problems of interfacing are far more of an obstacle to the development of the computerized car than the computer itself. Computer systems are already

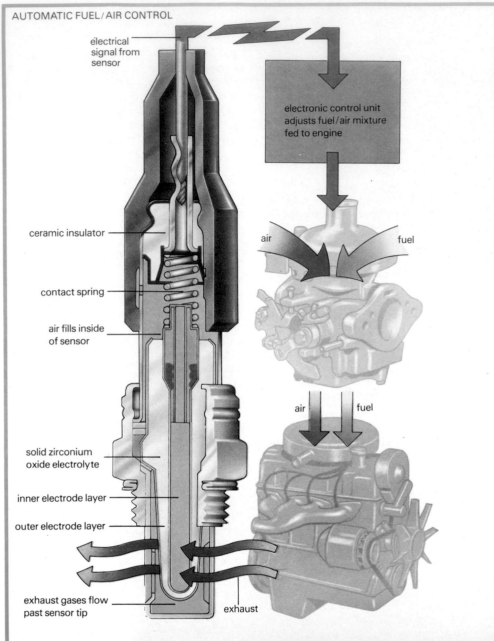

AUTOMATIC FUEL/AIR CONTROL

electrical signal from sensor

electronic control unit adjusts fuel/air mixture fed to engine

ceramic insulator

contact spring

air fills inside of sensor

air

fuel

solid zirconium oxide electrolyte

inner electrode layer

outer electrode layer

air

fuel

exhaust gases flow past sensor tip

exhaust

well developed and their capabilities known and it is the perplexities of the interface that are absorbing the majority of the research into electronic systems for cars.

Until recently there has been very little serious research into the possibilities of a fully-computerized car. Computerization in cars has concentrated rather on providing a few separate functions with their own microprocessors rather than any overall concept and central 'brain'. This is understandable because nearly all research and development of cars is carried out by manufacturers who are normally only interested in improvements when these are going to maintain or increase sales. Total computerization can only realize its potential when every car on the road is so equipped.

Three main functions

Naturally, then, the use of computers in cars has been on a relatively limited scale until now. Manufacturers have incorporated microprocessors into the more expensive models of their range in order to give them a performance lead over their rivals, or merely as a selling point. Hence we have seen the appearance of processors in connection with three main functions: fuel injection, ignition and instrumentation.

Development of fuel supply and ignition controls in particular has been given an extra boost by the introduction of strict laws governing exhaust emissions, notably in the United States and Japan, and the probability of these becoming more and more severe. These laws specified the amount of polluting gas—unburned hydrocarbons, carbon monoxide and oxides of nitrogen—that may be present in the exhaust, and the limits could only be met by controlling fuel delivery and ignition timing far more ac-

Above and right An electronic fuel control system. At the heart of the system is a sensor *(right)* which produces an electrical signal in response to the amount of oxygen in the exhaust gas. The electronic control unit then alters the fuel/air mixture to give better economy and cleaner exhaust.

Robert Bosch Ltd.

curately than had previously been the case.

Most American cars and many from other countries are now equipped with electronic ignition. Although the majority are still timed mechanically, a number of more recent systems have begun to incorporate microprocessors to improve the accuracy with which the ignition is timed to fire the spark plug. Data is supplied to the microprocessor concerning engine speed, the position of the piston (taken from the crankshaft position), and the load on the engine. The microprocessor then compares the information it receives with instructions imprinted in its memory during manufacture. The comparison enables it to decide when ignition should occur, and it issues instructions accordingly. The ignition is duly fired at the correct moment. This sort of facility had already been shown to improve performance and reduce toxic emissions. In particular, Volkswagen have shown how it is possible to hold the engine's idling speed steady by varying the timing.

These systems, however, are only a starting point and the potential for processor control is much greater. For instance, by the end of the 1970s a number of manufacturers had introduced a 'knock sensor'. This is valuable because engines could be more efficient if they ran on higher compression ratios. At present, petrol engines rarely have compression ratios of more than 10 to 1, and restrictions on the lead content in petrol often make them much lower. If they were any higher, the engine would be subject to detonation or 'knocking'. When this occurs some of the mixture explodes violently before it is ignited rather than simply burning. Knocking not only wastes power, but can wreck the engine. The knock sensor detects knocking as it is about to start, since incipient knocking shows up as an unusual variation in crankshaft's speed before audible knocking begins. If knocking is detected, the processor retards the ignition to prevent it.

Fuel injection

Fuel injection systems have been using microprocessors for some years. Originally only varying the mixture according to engine speed and throttle setting, many systems now include sensors for monitoring such things as engine temperature and air pressure. More recently, the German manufacturer Bosch introduced a 'lambda sensor' to detect the amount of oxygen in the exhaust which is an indication of the efficiency of combustion. It works simply by checking the electrical conductivity of the

exhaust gases. Any deviation from the set limits can then be corrected by altering the fuel injection settings via the microprocessor. In a simple form, this system is already incorporated in many cars for the American market because of the reduction in exhaust hydrocarbons it can give.

It may seem that once the engine runs lean (with the least possible fuel in the mixture), at very high compression and with a very clean exhaust due to the use of microprocessors, electronics can do little more to improve engine performance. But some people see much greater scope for the microprocessor. The valves are one area that have great potential for electronic technology. Lucas–CAV in Britain have already developed solenoid-operated valves while in the USA a prototype engine with hydraulic valves had been built by 1980.

Electronic operation of the valves opens up an enormous range of possibilities. It would,

for instance, allow the driver to change the characteristics of his engine at will. At the flick of a switch, he could change the car from a highly 'tuned' sportster to an economical slogger with plenty of pulling power simply by altering the timing and duration of the valve opening. Or he could shut off the valves altogether, turning the engine into an air compressor for efficient engine braking. Such a system would certainly 'stretch' a microprocessor system more effectively than many of today's applications, which often leave a good deal of the processor capacity unused.

Below Robert Bosch's computerized ignition system. Sensors measure such things as engine load, speed and temperature, crankshaft position, throttle position and air temperature and pressure. The information is fed to a central processing unit which then fires the spark plugs at the right moment.

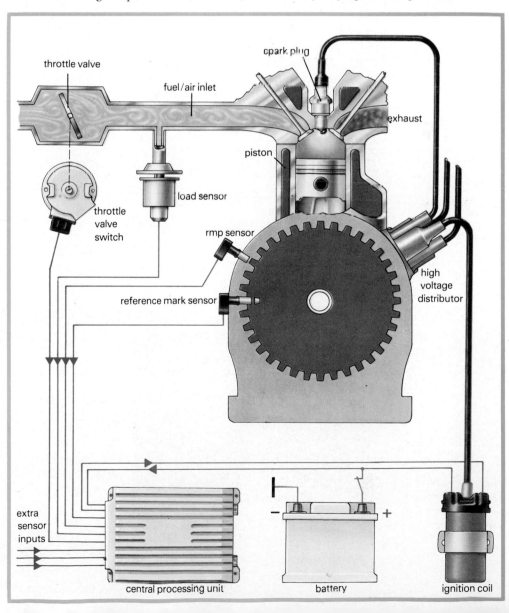

The engine is by no means the only part of the car which can benefit from electronics. Another promising area is the transmission and some engineers believe that, in the near future at least, there is more to be gained in economy from controlling the transmission electronically than by applying microprocessors to the engine. In such a system, the microprocessor would have an imprint of the engine characteristics and would change the gear ratio automatically to match these with the prevailing speed and load conditions. The final proof will probably have to await developments in the field of continuously variable transmission (CVT). In either case, microprocessor control will make for more accurate matching of drive ratio with power and speed.

Nevertheless, in many ways the most obvious sign of the presence of electronics in the expensive cars of today is in the instrument panel. Since the Lagonda emerged from the works of the refurbished Aston Martin company in 1976, complete with solid state instrumentation and a full digital display panel, a number of other similarly equipped cars have appeared, each with a dashboard looking like something from a starship. There is no advantage to these displays in themselves—the conventional

analogue dials show speed, engine rpm, fuel level and so on just as well, if not better—but they do open the way to the introduction of a vast array of other functions. Some cars already have a display which gives a continuous display of fuel consumption and average speed.

Researchers suggest that a digital display can be projected on to the windscreen to give the driver all the information he needs without having to move his eyes from the road. These head-up displays (HUD) have been familiar to the pilots of jet fighters for some years, but as yet no one has been able to develop a format small enough or cheap enough for use in cars, or worked out how to provide the illumination needed to provide displays bright enough to be easily visible on sunny days.

One area that has received only scant attention is the car's braking and suspension system. Now, however, BMW and Mercedes offer an anti-lock braking system on their top-of-the-range models. Considering that aircraft have had such systems for more than 20 years, it is, perhaps, surprising that this development has taken such a long time to arrive. The trouble was that while a hydro-mechanical sensor and control system worked well enough for a big, heavy aircraft

wheel, it did not respond quickly enough with the much lighter braking system of the average car. Electronics provided the answer. In an electronic anti-lock system, the sensor detects that the wheel is slowing down too quickly—that is, it is about to lock—and the circuitry and output releases the brakes quickly so that the wheel does not lock and then equally quickly re-applies them.

Suspension system

Some engineers feel that if the brakes can be controlled in this way through their hydraulic pipes, then so could a car's suspension system. Systems have already been demonstrated which prevent a car rolling on bends, but a fully electronic system could do much more. Not only roll, but also 'pitch' —end-to-end rocking of the car due to braking and acceleration—could be controlled, and the quality of the ride could be varied to suit the driver. Most conventional car suspension systems are restricted by the design of their springs. An ideal spring should act progressively, being soft at the beginning of its travel to absorb small bumps, and becoming steadily stiffer towards the end to react to large bumps. This is difficult to achieve with mechanical springs, but less so with high pressure air/liquid springs and these, in conjunction with a microprocessor controlling a valve to either let air out of the spring or pump it in, could give excellent suspension.

With microprocessors playing an increasing role in the control of individual functions —fuel intake, exhaust, ignition, transmission, brakes, suspension, steering and instrumentation—now is the time to assess the chances of all these separate systems becoming linked together and governed by one central computer in a fully-automated car.

Although a full scale electronic car has not yet been built, there are signs that the concept is not as far off as might be thought. A number of years ago, the German firm of Bosch introduced their Motronic fuel injection and ignition control system which was installed in the BMW 732i. The significance of Motronic is that it has a single processor to control both ignition and fuel intake, and has spare capacity to serve several other functions, given the interface and extra programming. The processor is of the very large scale integrated (VLSI) type. This is important because, in the past, car electronics have tended to employ simple, purpose-built processors for each function. The VLSI, however, although expensive and time-consuming to produce, is far more versatile

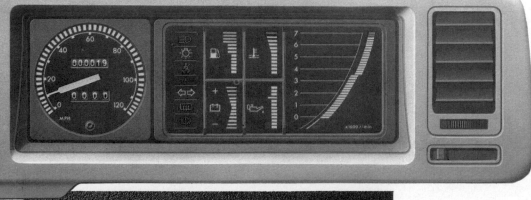

In car instrument panels solid state displays replace traditional dials, and 'trip computers' work out such things as fuel consumption and estimated arrival time. Having no moving parts such systems are likely to prove more reliable than their mechanical predecessors.

Smiths Industries Ltd.

Robert Bosch Ltd.

and shows the way forward to centralization. With a VLSI, the memory can be programmed to act as an ignition timer, gear changer, automatic headlamp dipper and any other task that is selected. In other words, a single VLSI can do the same job as a whole collection of purpose-built processors. So although the Bosch Motronic and similar systems are only programmed for fuel injection and ignition timing, there is no reason why numerous other functions cannot be controlled by its electronic brain once the various interfaces have been developed.

Once all the car's functions come under the control of a central computer, it becomes more feasible to allow the computer to react to external features such as traffic and route deviations. One possibility is the incorporation of a microwave radar linked to the computer. This would detect the positions and speeds of other cars and pedestrians, even in dense fog. It could then link automatically with the car's braking system to keep a safe distance or stop. It could also tell the driver whether it was safe to overtake on winding roads by 'seeing' round the bend.

Already in the late 1970s, a number of people, particularly in West Germany, were trying out systems whereby information about traffic conditions and alternative

Above This idea for a solid state instrument panel includes a 'head up' display — data are projected on to the windscreen.
Right This central unit gathers and processes data from all parts of the car and can control ignition, fuel injection, cruise speed and anti-locking brakes. It also operates warning lights and acts as a car computer.

routes from a central computer could be picked up from the road and displayed on a console within the car. Again there is no reason why this cannot be developed to provide much more detailed information and perhaps, eventually, take over control of the car's central computer entirely within crowded city areas.

It is clear that, although there are many problems to overcome, the basic technology for a fully-computerized automatic car is already available. For it to become a reality, however, massive investment in roadside monitors and in research and development to overcome the detail problems is essential, and perhaps the concept will need the stimulus of government legislation. Nevertheless, in the near future we can expect to see microprocessors taking over the control of more and more individual functions in the car and improving safety and performance.

Norden Systems

Radar and Stealth

The ability to navigate gargantuan oil tankers through crowded, fog-bound channels, land aircraft during thunderstorms at night, and maintain effective anti-missile defences all depends on the technology of radio detection and ranging, more commonly known as radar.

Like most good ideas, the basic principle of radar is very simple. Radio energy transmitted from an antenna is reflected by the target and the echoes are then picked up by a receiver and analyzed. Simple in concept, this principle has been developed and expanded in an almost unbelievable number of ways so that radar now stands as an important branch of electronics technology in its own right.

A radar system in its simplest form, as found on many merchant ships, comprises three components—the antenna (scanner), the transceiver and the display unit.

The antenna, typically 200–300 cm (6–10 ft) wide, focuses pulses of very high frequency radio energy into a narrow 1–1.5 degree, vertical beam. The frequency of the radio waves is usually about 10,000 MHz (10,000 million cycles per second), which corresponds to a wavelength of three centimetres. The antenna is rotated at a speed of 10–25 revolutions per minute and scans up to a range of 80 km (50 miles). To achieve maximum range and avoid obstruction by the ship's superstructure, the antenna is mounted as high as possible.

A vital feature of all radar is the harmony achieved between the transmitting and receiving elements of the transceiver. Everything depends on the 'trigger' which determines an accurate measurement of the time elapsed between pulse transmission and the return of its echo. Typical radar frequency is about 10,000 pulses per second, but this can be varied to suit requirements. Short pulses are more suitable for short-range work, while longer pulses are better for long range where as much energy as possible must strike the target.

Controlled oscillation

An important part of the transceiver is the modulator circuit which 'keys' the transmitter so that it oscillates for exactly the right length of time, usually a few microseconds. The transmission power is generated in a valve called a *magnetron.* This is capable of

Above Familiar radar display with radial trace. The antenna *(inset)* is synchronized with the display, so that when it points dead ahead the trace is at 0° on the screen.
Below A civil aviation radome, located well away from undesirable sources of clutter.

Jerry Mason

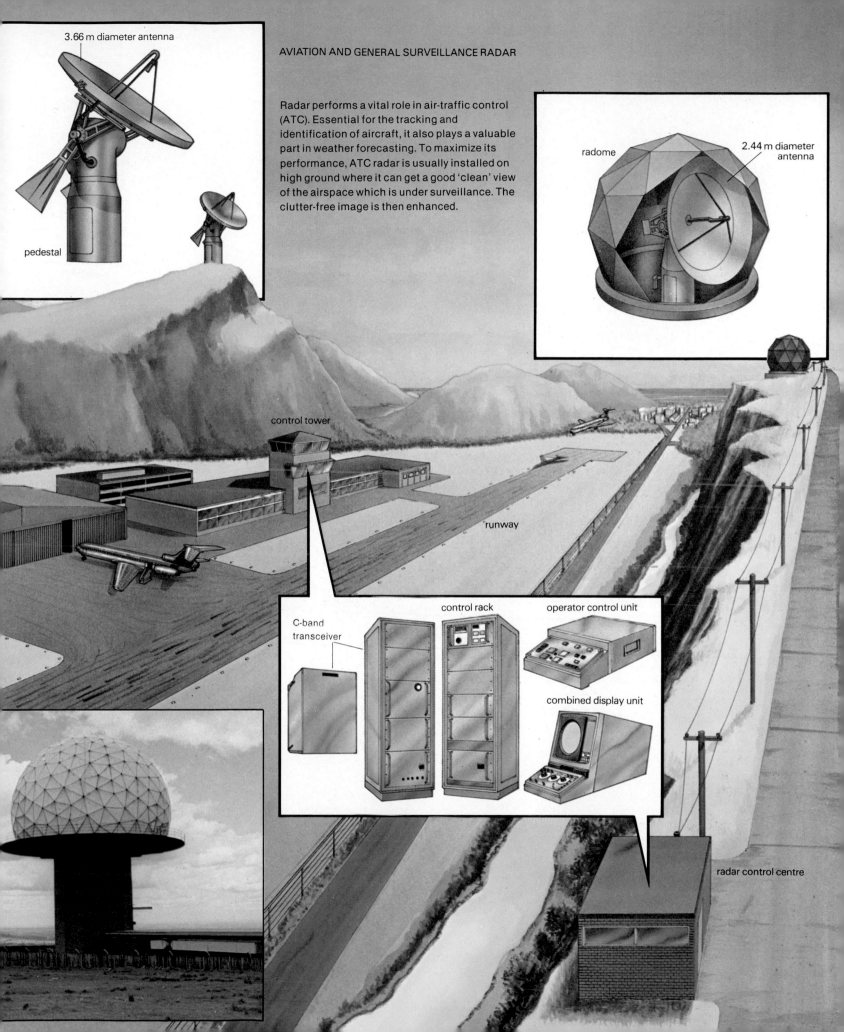

AVIATION AND GENERAL SURVEILLANCE RADAR

3.66 m diameter antenna

pedestal

radome

2.44 m diameter antenna

Radar performs a vital role in air-traffic control (ATC). Essential for the tracking and identification of aircraft, it also plays a valuable part in weather forecasting. To maximize its performance, ATC radar is usually installed on high ground where it can get a good 'clean' view of the airspace which is under surveillance. The clutter-free image is then enhanced.

control tower

runway

C-band transceiver

control rack

operator control unit

combined display unit

radar control centre

handling both very short pulses and very high frequency oscillations.

The magnetron exploits the fact that electrical currents at very high frequencies tend to travel on the surface of a conductor, and when they are made to travel along the inside surface of a tube the electromagnetic/electrostatic radiation they cause can be trapped inside the tube and made to travel along it. In this way, transmission pulses are fed to the antenna to be 'squirted' out in a narrow beam. These tubes, or *waveguides,* are meticulously engineered: the inner dimensions of the waveguide directly correspond to the wavelength, and therefore the frequency, of the transmission. Any obstruction or distortion in the waveguide would reduce the strength of the beam.

Between each pulse the transmitter is switched off and isolated. The incredibly weak echoes from the target are picked up by the receiver and fed down the same waveguide used for transmission, but deflected into the receiver, amplified and then passed to the display unit.

The display unit, which usually carries all the controls, houses the cathode-ray tube. Similar in principle to a monochrome television tube, radar cathode-ray tubes have a circular screen, 30–40 cm (12–16 in.) in diameter, calibrated in degrees around the edge. Unlike television tubes, in radar displays the electron beam travels radially from the centre of the screen to the edge. The radial motion of the electron beam, known as the *trace,* is synchronized to that of the antenna so that when the trace is at 0° the antenna is pointing dead ahead.

The beginning of each trace corresponds exactly with the moment at which a pulse of radar energy is transmitted. When an echo is received, it momentarily brightens the trace at a point from the centre of the tube corresponding to the time taken for the radar pulse to travel to the target and return. Thus, a 'blip' appears on the screen at the range and bearing of the target. As the trace rotates, a complete picture is built up from the afterglow of the fluorescent coating of the tube. This type of display is called a Plan Position Indicator (PPI), and it is the usual means of presenting radar information.

Signal enhancement

While many refinements have taken place in the analogue components of radar, these have been rather limited. It is in the handling of signal reception that the greatest development has taken place. Signal enhancement is essential for a number of reasons, not least being the need to eliminate unwanted information.

All reflective surfaces struck by a radar beam will return an echo. In the case of marine radar, rough seas will reflect energy, particularly waves close to the ship. These unwanted echoes are called 'clutter'. One way of getting rid of clutter is to reduce the amplification of signals at the beginning of a trace and increase it again once the unwanted signals have been passed.

Another source of clutter is caused by heavy rain or snow, which produce signals in the form of 'smears' on the display screen.

For long-range, land-based, air-traffic-control radar the problem of clutter is much more critical. Any obscuration of the targets—principally aircraft—could be extremely dangerous. To maximize performance, air-traffic-control radar is usually installed on high ground to give a good 'clean' view of the airspace under surveillance and thereby reduce clutter to a minimum.

Aircraft are, however, more easily identified. Due to their high speed they are easily distinguished from background echoes. If a target appears in the same place twice, it can be assumed that it is not an aircraft but an echo which can be suppressed. The device that carries out this function is called a moving target indicator (MTI). With an MTI in operation the air-traffic controller sees a virtually clutter-free picture containing only those targets that are moving.

In the most modern radar, two pieces of equipment—a signal processor (which houses the MTI) and a plot extractor (incorporating the latest computer techniques) combine to enable the data to be transmitted over telephone lines. The signal processor takes the received signals and digitizes them, turning them into a form which can be handled by a computer. The plot extractor then analyzes the digitized signals and creates a computer message which can be understood in the air traffic control centre. Each 'word' of the message gives the range and bearing of a target.

Being many kilometers from the radar, the air-traffic-control displays are not synchronized to the rotation of the antenna. Instead, a computer constructs the picture on the screen in the order the plot extractor transmits its message. This system allows symbols—such as crosses, squares or triangles—to be used to indicate targets in-

Jerry Mason

Left Height-finding radar determines aircraft altitude using trigonometry. By nodding up and down as it transmits its radar beam, the relationship between antenna angle and the distance of the plane can be used to compute altitude.

stead of the traditional undifferentiated blip.

Whether blips or computer-generated symbols are employed, it is important to know the identity of the targets visible. The height of aircraft must also be known so that other aircraft on the same course can be maintained at different levels. To do this, a second radar is used. Not surprisingly, this is called a secondary surveillance radar (SSR).

The SSR differs from the primary surveillance radar in that it requires the 'co-operation' of the aircraft. The secondary radar antenna is usually mounted on top of the primary radar. It transmits a series of pulses which are received by the aircraft's 'black box' or *transponder*, which then

GEC-Marconi Electronics Ltd.

Above Performing a crucial role in extending the horizon of NATOs defence radar is the remarkably shaped Nimrod Mk 3 AEW. Carrying highly sensitive radar in two curious bulges front and rear, these early warning aircraft are operational 24 hours a day.
Below Mobile 3D search radar has a wide range of both civil and military applications.
Left High-power, long-wavelength radar systems require massive antennae. To scan distances of up to 400 km (250 miles) a typical air-traffic-control radar antenna would measure as much as 16 m (48 ft) in width and it would stand at least 5 m (14 ft) high.

AEG-Telefunken

Jerry Mason

transmits a coded message back to the secondary surveillance radar.

The transponder, which all aircraft flying in controlled airspace are compelled by internation law to carry, is linked to the aircraft's altimeter. When an aeroplane flies into a controller's zone, the pilot is radioed and advised of an identification code allocated to him and this he dials into the transponder equipment. Thus, when the secondary radar 'interrogates' the aircraft, the transponder responds with a coded message comprising the aircraft's identification and height.

The data transmitted by the aircraft are then displayed on the controller's screen as a 'tag' on the symbol generated by the primary radar. The controller now has range, bearing, height and identification of the targets flying in the airspace under his jurisdiction.

If the object of civil air-traffic control is to keep the targets on the radar screen apart, then air defence requires that the targets are brought together! That is to say, the air defence controller's role is to guide interceptor aircraft or missiles to the enemy.

This task requires the same basic information—range, bearing, height and identity. Range and bearing are acquired in the usual way by primary surveillance radar, and a similar technique to the civil secondary radar is used— IFF (Identification, Friend or Foe).

To ascertain the height of unidentified—hence unco-operative—aircraft, special height-finding radars are used. Two methods of height-finding are employed: 'nodding' radar and '3D' radar.

A nodding height-finder can be likened to a conventional primary radar standing on its side. When the air-defence controller wants to find the height of a target, the height-finder is rotated to the bearing of the aircraft, whereupon the radar 'nods' up and down. The height of the aircraft can be determined from the angle of the antenna when its beam strikes the target.

3D radar

A number of nations are currently introducing three-dimensional radar into their radar defences. Instead of using two kinds of radar—primary plus height-finder—the two functions are combined in a single antenna system. There are several different ways this can be done. One system uses a special type of antenna called a *planar array*. Incorporating 60 receiving elements, it radiates in the usual way for horizontal coverage, but for vertical coverage it transmits a continuous wide-angle beam and uses its receiving elements to create a beam-forming network.

A simple analogy to its operation is the human eyeball. Each beam of light falling on the retina excites one or more rods and cones

which then transmits a minute electrical current along the optic nerve. The size of the current depends on the intensity of light. The brain then analyzes this, together with the response of adjacent rods and cones, to ascertain both the nature and the position of the object being observed.

In a similar way, 3D radar checks the echo received by one or more of its receiving elements against the reception of adjacent elements in order to ascertain the precise angle from which the echo was received. As soon as the angle is known, a simple calculation of angle against range is all that is required for the precise height of the aircraft to be established.

Hostile military aircraft, unlike their civil counterparts, are unlikely to be co-operative. Indeed, they will most probably try to either 'jam' or outwit defence radar using electronic countermeasures (ECM). One of the earliest countermeasures developed, and used to good effect in World War 2, is what was then called 'window' and is now called 'chaff'. This consists of thousands of strips of aluminium tapes ejected from the aircraft to act as a 'curtain' and conceal the aircraft from the radar's beam.

An intruder is also likely to try and 'flood' defence radar with energy, blotting out accurate reception. Another form of jamming, called 'gate-stealing' or 'spoof' jamming, at-

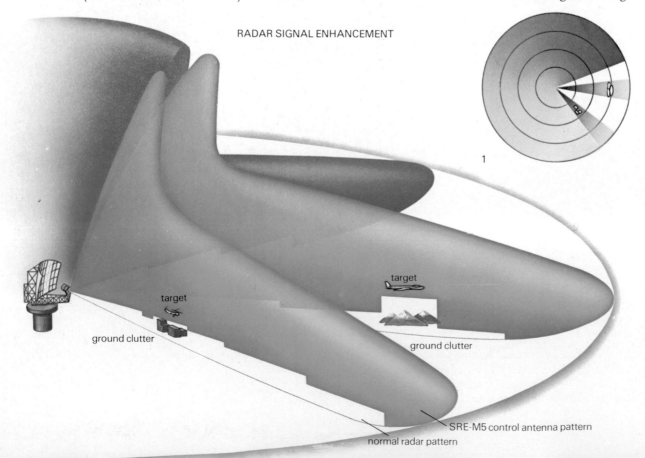

RADAR SIGNAL ENHANCEMENT

1

2

target

target

ground clutter

ground clutter

SRE-M5 control antenna pattern

normal radar pattern

The SRE-M5 radar employs a 'clutter-adaptive' antenna which allows for the local clutter environment. Stationary clutter, which might cloak a taget (1), is overcome by the radar beam hopping over it. Targets can thus be detected even when flying over ground clutter (2).

AEG-Telefunken/Martin Woodford

tempts to introduce fake targets on the radar screen to mask the genuine hostile target.

To remain effective under this barrage of electronic warfare, modern defence radar responds with electronic counter countermeasures, (ECCM). To defeat chaff, the radar can switch-in its MTI. Because chaff drifts at wind speed and aircraft travel considerably faster, the two may be differentiated. To counter spoof jamming, the radar can 'hop' from frequency to frequency with each pulse it transmits. Such agility requires the jamming signal to be spread over a very wide frequency band, reducing its power and effectiveness. Frequency hopping, which is random, prevents spoof jamming as it is impossible to predict the frequency the radar is operating on.

Stealth Project

The twighlight world of countermeasures and counter countermeasures is becoming ever more sophisticated, yet the defence world was stunned by ex-President of the United States, Jimmy Carter's revelations of the Stealth Project—involving the development of aircraft 'invisible' to radar. Very little precise information about the Stealth aircraft has been revealed, except that it would have a very low radar profile, being coated with radar anechoic paints developed to absorb rather than reflect radar energy.

Once the furore had died down, radar experts began to consider the scant information released. It soon became clear that while anechoic paints and a low radar profile would enable an aircraft to escape detection at the extreme ranges of defence radar, the plane would nevertheless be detected eventually. But, of course, the extra miles gained before detection could make all the difference to the outcome of the aircraft's mission, and radar engineers around the world are already at work developing countermeasures to Stealth-type aircraft.

Military radar, of necessity is often established in very remote areas. To protect it from the elements, it is often housed in geodesic domes—called radomes—which look rather like Brobdignagian golf balls. Designed to be lightweight and very strong, these domes are virtually 'transparent' to both transmission beam and radar echoes. Radomes can also be found on board ships, where they protect their specialist radars from the corrosive effects of both salt-laden atmosphere and the sea itself.

The nose cones of many military aircraft are, in fact, radomes. Two rather startling-looking aircraft developed recently are the

Nick Farmer

Creating radar-invisible aircraft remains a dream. But combining the use of anechoic paint with a delta-wing design could result in aircraft with very low radar profiles.

STEALTH – INVISIBLE DEATH

Boeing E3A AWACS (Airborne Warning and Control System) and the BAC Nimrod Mk 3 AEW (Airborne Early Warning) aircraft. These surveillance aircraft extend the radar horizon of a country's defence system and are better at detecting low-level aircraft. The antenna on board the E3A is housed in a mushroom-shaped radome mounted on top of the fuselage. A different approach has been taken in the Nimrod AEW. In this aircraft, 'half' of the antenna is mounted in the bulge at the front and the other 'half' in the bulge at the rear.

Radar is not only used for surveillance. It can also be adapted for the guidance of missiles. Guidance systems can be fitted on board the missile, on the ground, or on a ship close to the missile launcher. On-board missile guidance falls into two categories—active and semi-active.

Active seekers

Active radar seekers give a 'fire-and-forget' capability to the missile. A complete radar transceiver and antenna enable the missile to search for, and to be guided to, its target. Semi-active seekers have only a receiver fitted and the target has to be 'illuminated' by an associated radar transmitter mounted on its launch platform.

Another form of missile guidance is 'command-to-line-of-sight', also known as 'differential tracking'. Here the radar, either ground- or ship-based, tracks both the target

and the missile simultaneously, and commands the missile to bring it's direction onto a collision course with the target.

As well as helping defend national sovereignty, radar has a role to play safeguarding highways and byways. However, police radar traps, unlike most radar, do not determine the precise location of targets. Rather, they show the rate at which the target's position is changing.

To do this, the radar employs the Doppler principle. Radar energy reflected from a moving vehicle changes in frequency as the target approaches. The rate of change in frequency is directly proportional to the speed of the vehicle.

Early speed traps required the police to stand the antenna on a tripod at the side of the road where sharp-eyed motorists could often see it. With the advent of microelectronics, the size of the radar has been reduced to a small, hand-held radar 'gun'. But just as countermeasures have been developed for defence radar, quick-thinking electronics entrepreneurs are now marketing radar-warning receivers for motorists!

Of the myriad uses to which radar has been put, safety is the keyword—safety from would-be aggressors, safety at sea, safety in the air and safety on the highway. Continuing developments, keeping in step with the rapid advance of electronics technology, will bring radar and its benefits into areas at present unconsidered.

New lights in the dark

Few inventions have had a more radical effect on the way people live, and then been so taken for granted, than the electric light. The domestic electric light bulb has changed little in appearance, or principle, since it was first demonstrated just over one hundred years ago. But this does not mean that the initial brilliance of lighting technology has dimmed. There now exists a greater variety of electric lighting than ever before, and advances are continuing to be made.

Although there are a bewildering multiplicity of electric light sources, there are only two basic types. The most widely used is the familiar *filament light*. This has remained largely unchanged for several decades, but demand for specialized applications has renewed activity in the filament light field. The other principal source of illumination is the *discharge light*. Still evolving, the discharge light offers exciting new possibilities as scientists work to improve its efficiency, increase it acceptability and reduce its energy consumption.

Filament lamps

The filament light works on a simple—and not highly efficient—principle. An electrical current passes through a thin wire, causing it to heat up until it becomes incandescent and gives off light. If the filament were exposed to the oxygen in the air, it would rapidly burn up. So it is enclosed in a bulb—traditionally of glass, although other materials which exclude air but allow light to pass through will suffice.

The light works better if the bulb contains an inert gas—rather than a vacuum. During manufacture, the air is is pumped out and the bulb refilled either with a mixture of argon and nitrogen or with krypton at low pressure. The gas reduces the rate of evaporation of the filament and allows the bulb to function at higher temperatures.

Today, the filament is generally made of tungsten. The earliest versions used carbonized cellulose thread. Tungsten is an extremely brittle metal, but it has the high resistivity needed for incandescence—and with a melting point of 3,410°C, it can withstand the 2,500°C at which the modern light bulb operates.

The filament is free-standing, supported only at its two ends by rigid lead-in wires which carry the electrical current. These are connected to the electricity supply through the lamp socket.

To achieve useable light values, the filament needs to be far longer than the internal dimensions of the bulb, so a coiled filament is used. The filament of a 100w bulb, if uncoiled, would extend for more than a metre.

Technically, the primary effect of a light bulb is to produce heat. Only ten per cent of the energy produced by the filament is in the form of light. The rest is dissipated as heat.

But the filament lamp, inefficient though it is, remains popular for domestic use. It is cheap to make, easy for the ordinary person to carry and install, needs no elaborate wiring or switching gear, and has an acceptable life span. Furthermore, it gives a pleasant pale yellow light which the human eye finds aceptable, with adequate—if not accurate—colour rendering.

For bigger installations needing a greater output of light for a given input of power—and where technicians are available to install the lighting—the filament lamp has given way to the other main product of lighting technology, the discharge lamp.

Fluorescent tube lighting

It has long been known that the discharge of an electric current through a gas at low pressure produces light. A flash of lightning is an example. St Elmo's fire, the glow that fishermen see at the masthead of their fishing boats, is another. As early as 1810, Humphrey Davy demonstrated that a discharge could take place between two carbon electrodes. But development along that path lapsed in favour of filament lamps, and it was not until World War 2 that discharge lamps again came to the fore. The first type was, in fact, a separate branch of lighting technology and led to the familiar fluorescent tube.

Certain chemicals occurring in nature—the *phosphors*—give off bright light in response to ultra-violet rays. To make use of this feature, the electric light manufacturers coated the inside of a glass tube with 'fluorescent' chemicals and passed an arc through the tube between two electrodes. The ultra-violet radiation generated by the discharge energizes the phosphors, and they

Left Only ten per cent of energy produced by filament bulbs is in the form of light. The rest is heat.
Right Sodium lamps along low-lying, fog-prone stretches of motorway can cut the accident rate, while neon lights at night *(far right)* form the shape of buildings. At night, too, building continues *(inset)* under powerful floodlights. When the lights went out in New York *(top right),* life came almost to a standstill.

Denis O'Regan

Photo Library International

Thorn Lighting

MBZ Joint Venture/Thorn Lighting

give off light. The result was the fluorescent-tube lighting with its eerie pale violet glow that illuminated offices and factories in the immediate post-war period.

Fluorescent-tube lighting has advanced since then. Modern tubes give a far more accurate colour rendering, as well as a more comfortable light. This is mainly the result of the different combinations of chemicals used to coat the inside of the tube.

By mixing different phosphors, producing the artificial phosphor halophosphate, and adding small quantities of other rare elements, it is possible to produce a range of mercury vapour tubes giving 30 or more different lighting qualities, including several varieties of 'white'. From this range, the type, or combination of types, which is most acceptable can be selected.

Eventually, the phosphors lose their capacity to absorb ultra-violet rays and give off useable light, and the tube must be replaced. But a significant advantage of this type of lamp is a working life which, on average, will last 7,500 hours.

The other main advantage of fluorescent lighting is the amount of light produced. The fluorescent cylinder converts almost all of the electrical energy it receives into light energy. The result is a light output of the order of 80 to 90 lumens per watt, compared with an average of about 12 for the filament bulb.

1,000 hours of useful life

Modern fluorescent lamps also have greatly improved acceptability. The cylinder can be made to any length up to about two and half metres (8 ft). And, unlike the filament lamp, it does not throw out its light from a single source. The result is fewer and softer shadows, ideal for working environments such as kitchen or office.

The other line of development of the fluorescent lamp is in the selection of appropriate gases to fill it. If metallic iodine or bromine is added to the gas inside the arc

Thorn Lighting/Paul Brierley

tube, the electrical discharge produces a different quality of light.

Metal halide lamps are extremely efficient at up to 85 lumens per watt, and with up to 1,000 hours of useful life they are extremely durable. They are economical, compact—as small as 40 mm (1.5 in) in diameter for the 'spotlight' type of lamp, and they produce an intense white light. This makes them ideal for industrial lighting, both indoors and outdoors; for high-ceilinged areas such as shops, offices, and public assembly rooms; and for floodlighting, both buildings and sports fields. The lamps can be mounted on masts without the elaborate structures needed for heavier types of floodlights.

The fluorescent tube is not a real discharge light, however, but depends on the fluorescence generated by radiation from the ultra-violet end of the spectrum. The true discharge lamp gives off light in the visible part of the spectrum, through the discharge itself. The character of the lighting depends largely on the gas used.

Economic and long lasting

The two most widely used gases are vapourized sodium and mercury, contained within an inner 'bulb'. The arc passes between two electrodes, first vapourizing the metals in the bulb and then exciting the resultant gas so that it emits light.

The standard sodium lamp, working at low pressure, emits a light concentrated at the yellow end of the visible spectrum. Its main deficiency is that under its light most colours tend to appear as yellows and browns. But that is not a significant problem to the car driver, and low-pressure sodium lighting is widely used for illuminating main roads and motorways.

One advantage of discharge lighting is its economy. It has the greatest luminous efficacy of all lamps, producing up to 160 lumens of light per watt. Being relatively inexpensive to make as well, and long lasting, is extremely popular with public authorities.

The most significant recent development in sodium vapour lighting has been the increase of pressure inside the arc tube to approximately one quarter atmospheric pressure. This radically alters the character of the sodium discharge lamp. At the higher

Right The Philips SL lamp uses 33 per cent less electricity than a standard filament bulb but lasts over five times longer. It combines a miniaturized fluorescent lamp coated with phosphors and a thermal cut-out system in case of over-heating.

Grose Thurston/Philips Lighting

PHILIPS SL LAMP

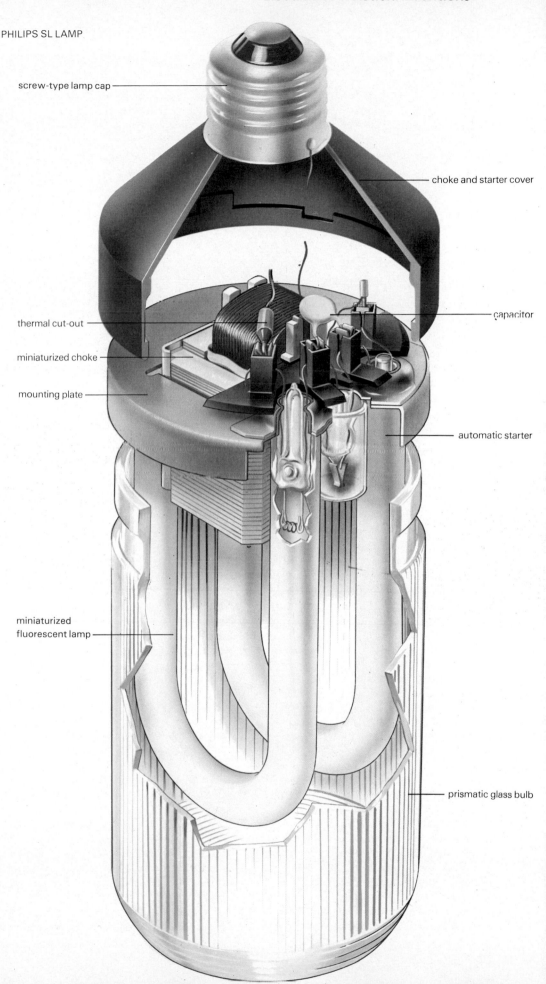

screw-type lamp cap

choke and starter cover

thermal cut-out

miniaturized choke

mounting plate

capacitor

automatic starter

miniaturized fluorescent lamp

prismatic glass bulb

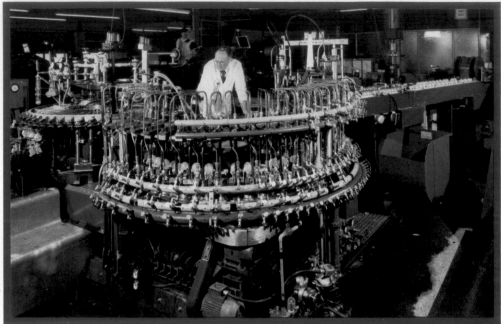

pressure, the excited sodium radiates light over a greater part of the visible spectrum. The result is a warm golden glow which the human eye finds highly acceptable. It is, therefore, widely used in schools, swimming baths, and workshops.

Problems were encountered, however, in producing materials which could contain the sodium vapour at temperatures of 700 to 800°C. The answer was to make the small, inner tube of transparent alumina-oxide.

Although fluorescent and discharge lamps have taken ·over from standard filament lamps for a range of industrial and commercial uses, and for road lighting, there is one type of filament lamp which is bringing exciting new developments to that branch of the technology.

Tungsten halogen lamp

One of the deficiencies of early tungsten filament lamps was that particles of the filament evaporate at high temperatures and blacken the inside of the bulb. The evaporation also reduced the life of the filament.

Scientists discovered the answer to the twin problems of blackening and evaporation in the phenomenon known as the halogen cycle. They added a small quantity of one of the less-corrosive halogens—fluorine, iodine or bromine—to the gas inside the arc tube. When the lamp is operating, the halogen combines with any tungsten vapour at or near the bulb, preventing blackening.

Furthermore, the tungsten halide that results diffuses throughout the bulb and eventually reaches the region near to the fila-

Above At the Thorn Lighting factory at Merthyr Tydfil, Wales, 5,000 light bulbs are produced every hour—the fastest rate worldwide.
Right Like many modern industrial sites, this chemical plant uses high-pressure sodium floodlights for outdoor illumination.

ment. There, the high temperature causes it to split up into atoms of tungsten and halogen. The tungsten settles back on the filament and the halogen recirculates in the filling gas. As a result, the tungsten is both prevented from reaching the glass and is redeposited onto the filament.

Exploiting the halogen cycle offers several advantages. The bulb can be made far smaller for the same lighting values, allowing it to be filled with gas at up to four atmospheres—eight times the pressure in a standard filament bulb. This in turn reduces the rate of evaporation of the filament, and makes it possible either to prolong the life of the bulb or to increase the brightness—or to achieve any combination of the two.

The extra heat generated by the filament—about 250°C at the bulb wall—would melt ordinary glass. Quartz promised to be a durable replacement, but it was difficult to cut. That problem was solved by the laser beam, in one of its earliest industrial applications. It also proved difficult to seal the lead-in wires to the quartz at the point where they entered the bulb, so extremely thin and flexible foils of molybdenum were devised.

The lamp, introduced onto the market as a car headlamp in 1961, was known at first as

a quartz iodine light—from the material used for the bulb and one of the components of the gas. The name rapidly gave way to the more accurate name of 'tungsten halogen'.

Since their introduction, tungsten halogen lights have proved invaluable. They are low in weight, compact, and cheap to buy and run. They have largely replaced large, standard floodlights at sports arenas. They are widespread as vehicle headlights and almost universal in new cine and slide projectors, giving sharp, white light from a compact source—without the high temperatures that call for elaborate ventilation. In photographic applications, bromine is the halogen generally added to the filling gas, since it gives more accurate colour rendering than the alternative iodine.

Self-powered lamps of the future

Scientists are still searching for alternative light sources that will give bright and pleasant light, be compact and convenient to install and cheap to buy and to run. Perhaps the most likely development in discharge lighting is the establishment of a new combination of gases. It has been known for 30 years that the gas xenon produces a remarkably strong light when an electric current passes through it. But, so far, few

ZEFA

Above The intense heat generated by high-powered lamps needed for underwater photography would shatter glass. A quartz covering is normally used instead. *Below* The total time taken by a high-pressure sodium lamp to ignite can be up to half an hour.

applications have been found for xenon lighting, although a lighthouse in New Zealand operates with a small, xenon lamp in a quartz globe. The lamp is as bright as the Sun, and in broad daylight it is visible from a distance of 32 km (20 miles).

Argon and neon mixed together also seem promising as a light source, particularly at higher voltages. This means that a greater light can be gained from a tube of a given length. The gas krypton also offers increases in the light produced for a given wattage, and allows a ten per cent reduction in the electrical load.

An alternative path of development involves the generation of higher frequency electrical discharges, controlled through electronic circuitry. Such discharges excite the gas more efficiently so that it gives off more light. And yet another line is in phosphor research, aimed at achieving the best blend of phosphors for the type of light required.

Today, much of the emphasis is on efficiency, and attempts are being made to produce economical lights. One promising area of study involves a group of phosphors which are excited into giving off light by infra-red radiation. The lamp could thereby be used partly to power itself, greatly reducing the amount of energy needed to drive it.

Thorn Lighting

Electron microscopes

With their ability to magnify by up to one million times, electron microscopes are now showing us an astonishing micro-world previously only dreamt about. By overcoming the severe limitations of traditional microscopes, they can reveal some of the finest details of molecular structure. New developments have now widened their role to embrace specialized forms of chemical analysis. Consequently, the electron microscope has become an invaluable analytical tool widely found in medical and industrial research establishments. In spite of failing to capture the public imagination, it is probably one of the most important single scientific instruments currently available to us.

Healthy human eyes cannot normally see, or distinguish as separate, objects that are smaller, or closer together, than 0.1 mm. This is the *resolution* of the eye. Over three hundred years ago, men started to see smaller and smaller objects by putting together glass lenses, so producing optical microscopes through which light passes. By the end of the last century it was realized that these microscopes could be improved no further in terms of resolution. Structures as small as 0.2 μm (1.0 μm—one micrometre—is one millionth of a metre, or one thousandth of a millimetre) could be resolved, needing a magnification of at least 500 times (0.2 μm × 500 = 0.1 mm) to bring them to a size that could be seen by eye. The reason that optical microscopes cannot be improved further is that their resolution is limited by the wavelength of visible light. When, during the early 1900s, it was realized that electrons are also part of the electromagnetic spectrum and have an associated wavelength, the way for the construction of electron microscopes was opened up.

Max Knoll and Ernst Ruska made the first instruments in Germany during the early 1930s. In most electron microscopes, electrons are produced by heating a thin piece of tungsten wire. Unfortunately, electrons do not move very far in air and therefore the inside of an electron microscope has to have most of the gas molecules removed with vacuum pumps. Electrons produced by the electron 'gun' are focused, using electromagnetic lenses in a similar manner to the way that glass lenses focus light. This allows a fine beam of electrons to be directed against the specimen.

Laboratoire d'Optique Electronique du CNRS

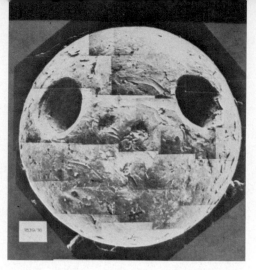

Right Bead of Death. This scanning electron microscope (SEM) picture clearly reveals the tunnel in a minute platinum-iridium ball that contained a lethal dose of ricin poison. It was fired into the leg of Georgi Markov, a Bulgarian broadcaster working for the BBC in London, by an unknown agent. Within a short time, Markov was dead.

Below A transmission electron microscope (TEM) picture, magnification 50,000 times, showing influenza viruses attached to a cell membrane by their external spikes.

Dr S. Patterson/Science Photo Library

Harwell

Far left This high-voltage electron microscope (HVEM) at Toulouse, France, operates at 3,000,000v. The focusing column seen here houses a series of magnetic lenses. However, much of the structure remains out of view. Above the column is an equally large accelerating tube and high-voltage supply. Magnifications can exceed 300,000 times. The extent of the upper structure in an HVEM can be seen in the 1,000,000v instrument at Harwell, UK *(left)*.

There are two main types of electron microscope: the *transmission electron microscope* (TEM), where electrons pass right through very thin specimens, in the same way that light passes through the specimen on an optical microscope, and the *scanning electron microscope* (SEM), where the electron beam is scanned to-and-fro across a specimen. These two types of electron microscope can give very different views of a specimen. They serve important but different functions.

Transmission electron microscopes have extremely high resolving power: that is, they allow us to see very fine detail of specimens. In fact, the best resolution is about 0.1 nm (nanometre)—one ten thousandth of a micrometre, or one million times smaller than can be seen with our eyes alone. Some impression of the formidable magnification provided by the TEM can be obtained when it is realized that the dot on this 'i', magnified one million times, would be 400 metres in diameter!

Of related interest is another form of microscope—the *field ion microscope*—which produces extremely high magnification (about a million times) of metallic specimens which are used as the anode in a high-voltage vacuum apparatus. This is not really an electron microscope at all, but because it is possible to determine the positions of atoms in the rather beautiful images that can be produced, this instrument gives information complementary to that from TEMs.

Using very high magnifications, TEMs can provide detailed information about structures such as viruses—most of which are far too small to be seen at all with a normal optical microscope. Frequently, the TEM is actually used to identify viruses in specimens from diseased patients, allowing appropriate treatment to be used. For example, smallpox virus can be distinguished by electron microscopy from chickenpox virus, which is very similar. It is also possible to examine some of the different macro molecules that make up living cells and, for example, to study different forms of DNA (deoxyribonucleic acid)—the material that contains information for determining exactly what is made within a cell and how it operates. Increasingly, in the medical field, electron microscopes are involved not only in basic research on cells and tissues but also in routine pathological examinations of specimens so that improved diagnosis of ailments can be achieved.

In the case of very small objects such as viruses and isolated molecules, it is possible to prepare them for examination by simply mixing them with a stain that provides a contrasting background. However, thicker specimens cannot be penetrated by the electron beam, and it is then necessary to make very thin preparations either by cutting sections of embedded material using an *ultramicrotome,* or by grinding hard materials down to very thin wafers. An idea of the thin 'slices' required is obtained by realizing that between 1,000 and 2,000 sections could be cut from the thickness of this page.

Studying atoms

Using such techniques, the finest details of biological cells can be observed, while the arrangement of atoms and molecules in metal and other materials can also be studied—a factor of the greatest importance to materials scientists and engineers, who are concerned with the structure and strength of different materials. Electron microscopes are therefore used both to give information about the microstructure of new materials as they are being designed (for example, new alloys or polymers), and also to help in the analysis of failures of materials. This is a major branch of materials science and involves the examination of components that have broken, or in some way 'failed'. Whether in the crash of an airliner from metal fatigue, the shutting-down of a nuclear reactor due to the development of cracks in its pressure vessel, or the closing of a pipeline on the North Sea through leaks from welds, the electron microscope has an important role to play in helping to determine causes of failure.

Computers now often aid analysis of high-resolution pictures obtained from electron microscopes so that, particularly if the specimen has a regular structure such as that found in natural or synthetic crystals, it is possible to reconstruct images with far greater detail than could be seen in the original micrograph. This can lead to the construction of models which show three-dimensional relationships at very high resolution.

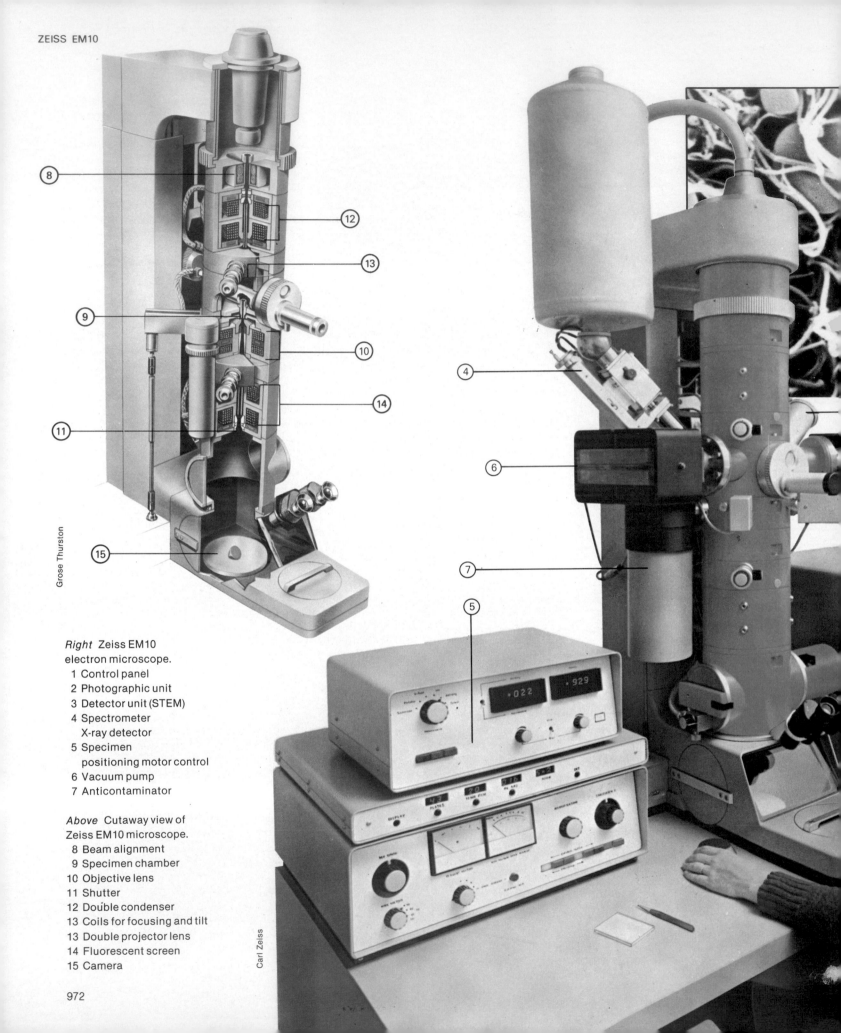

ZEISS EM10

Grose Thurston

Carl Zeiss

Right Zeiss EM10
electron microscope.
1 Control panel
2 Photographic unit
3 Detector unit (STEM)
4 Spectrometer
  X-ray detector
5 Specimen
  positioning motor control
6 Vacuum pump
7 Anticontaminator

Above Cutaway view of
Zeiss EM10 microscope.
8 Beam alignment
9 Specimen chamber
10 Objective lens
11 Shutter
12 Double condenser
13 Coils for focusing and tilt
13 Double projector lens
14 Fluorescent screen
15 Camera

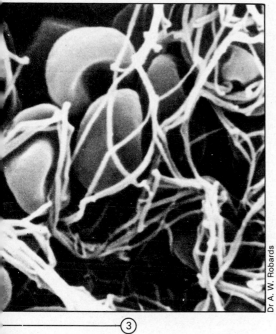

Dr A. W. Robards

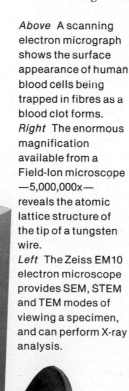

Most TEMs operate at accelerating voltages in the range 50–100,000v. This is the force imparted to electrons as they are fired towards the specimen. Even at such relatively high forces, electrons only pass through thin pieces of material—usually less than one ten-thousandth of a millimetre in thickness.

Some electron microscopes have been made which have much higher accelerating voltages—1,000,000v or even 3,000,000v! Such instruments are monsters because not only is it necessary to have a relatively enormous electron gun, but the electromagnetic lenses must also be much more powerful—and hence bigger—to focus the more energetic electrons. These instruments can shoot electrons through greater thicknesses of material, and it is therefore possible to study the three-dimensional arrangements of structures rather than the limited two-dimensional views that are obtained from much thinner specimens.

Powerful microscopes, such as those using 1,000,000v or more to accelerate the electrons, provide highly energetic radiation sources. They are therefore used to study the effects of radiation on different materials. The advantage of carrying out such experiments in a high-voltage electron microscope

(HVEM) is the ability to watch the specimen continuously as it is irradiated, and even record events on a videotape recorder. Such studies have been extremely useful in the development of components for use in high-radiation environments, such as nuclear reactors.

Scanning electron microscopes (SEMs) produce an extremely fine beam of electrons which is swept to-and-fro across the specimen in what is called a 'zig-zag raster'. This is exactly the same motion as the electron beam that is deflected to-and-fro across the screen of an ordinary television set. When an electron hits the specimen, a number of things can happen. The electron might bounce back, becoming a 'back-scattered electron', or it might knock another electron out of its orbit. This is a secondary electron.

Electron detectors

Such events depend upon the nature of the specimen, and the electrons produced are 'seen' by electron detectors. The signal produced is then used to generate a picture on a cathode-ray tube, again in the same way as that on a domestic television, but by using the electron signals rather than broadcast signals from an aerial. Scanning electron

Above A scanning electron micrograph shows the surface appearance of human blood cells being trapped in fibres as a blood clot forms.
Right The enormous magnification available from a Field-Ion microscope —5,000,000x— reveals the atomic lattice structure of the tip of a tungsten wire.
Left The Zeiss EM10 electron microscope provides SEM, STEM and TEM modes of viewing a specimen, and can perform X-ray analysis.

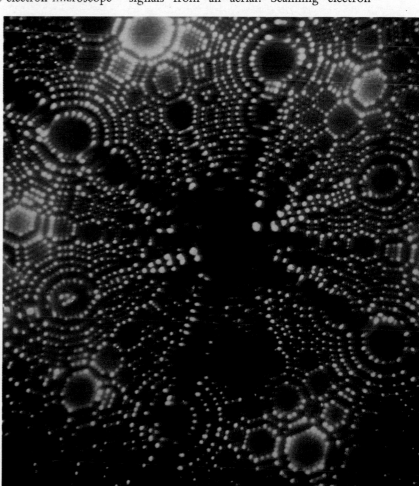

Paul Brierley

microscopes are thus very good for looking at the surfaces of objects and although they are not used at the very high magnifications achieved by transmission electron microscopes, they provide a completely different range of information.

In the biological sciences, special techniques for preparing specimens mean that it is possible to look at the internal surfaces of cells that have been fractured open, as well as looking at the normal outer surface. Consequently the scanning electron microscope is extremely useful in studying details and contours of different surfaces.

For example, one of the most important features of many cancer cells is the way that they interact with their neighbours. The SEM allows detailed pictures to be obtained of these cell surfaces, and aids research into the understanding of such diseases, suggesting how they might be treated. Blood cells can also be looked at in the SEM and differences seen as they grow old—something that is very important in relation to storing blood for transfusion. The formation of the fibrils that lead to a blood clot may be viewed in detail in studies of clotting. The SEM provides many other striking views of plant and animal cells that cannot be obtained by different means.

In the microelectronics industry, scanning electron microscopes have proved to be an equally great asset. It is possible to use them to look in detail at the microcircuits that are

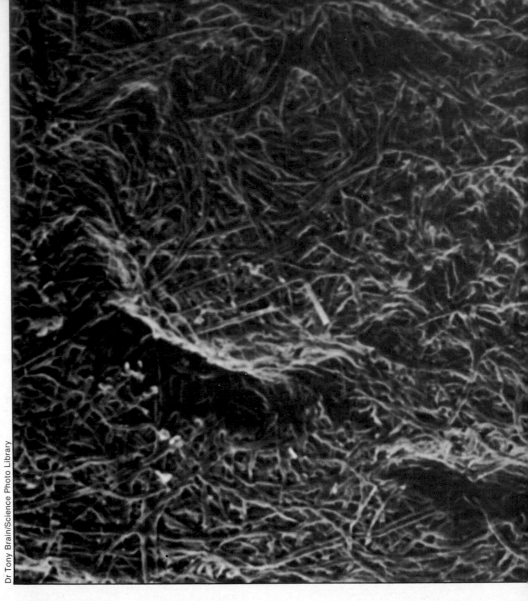

Dr Tony Brain/Science Photo Library

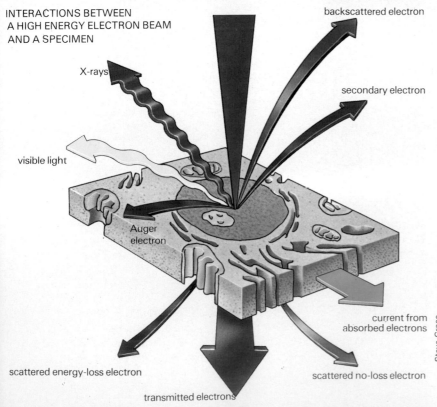

INTERACTIONS BETWEEN
A HIGH ENERGY ELECTRON BEAM
AND A SPECIMEN

high energy electron beam

backscattered electron

X-rays

secondary electron

visible light

Auger electron

current from absorbed electrons

scattered energy-loss electron

scattered no-loss electron

transmitted electrons

Steve Cross

Above The imprint made on paper of a typewritten letter S, seen by scanning electron microscope. *Left* When an electron beam hits a specimen, a number of interactions occur. An electron might bounce back, becoming a backscattered electron, or it might knock another electron out of its orbit, producing a secondary electron. For material analysis energy loss from electrons can be measured, using both Auger and energy-loss electrons. X-rays and visible light may also be measured.

now constructed on a tiny silicon chip, and also to operate the microscope as an instrument to fabricate circuits by using the electron beam as a 'writing' tool, controlling it by computer so that the required circuit is produced on a special surface.

The image in an SEM is produced by signals from the surface of a specimen. There is not the same limitation of space that exists in a TEM, where electrons have to penetrate very thin specimens which must be within a millimetre or so of the lens, due to the extremely short focal lengths of electromagnetic lenses. It is thus possible to have specimens as large as half a brick in some SEMs. This can be very useful to materials scientists, and others, who want to look at large objects. For example, by studying fractured pieces of metal, such as a steering rod from a car or a broken bolt from a crashed aeroplane, the SEM can help determine whether component failure was the cause of an accident or the result of it.

In forensic science the SEM also makes a considerable contribution, for example, by

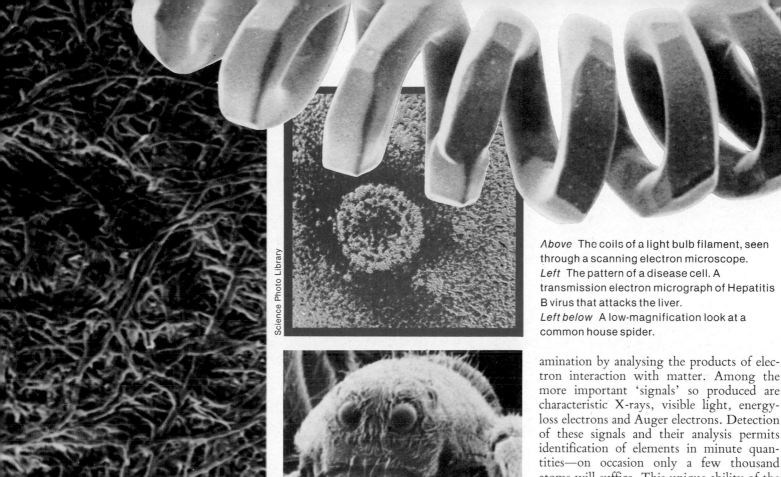

Above The coils of a light bulb filament, seen through a scanning electron microscope.
Left The pattern of a disease cell. A transmission electron micrograph of Hepatitis B virus that attacks the liver.
Left below A low-magnification look at a common house spider.

allowing the detailed study of bullets or by providing high-magnification pictures of the tiny poison-impregnated pellets that have recently been used in political murders.

Because the specimen chamber is so big it is possible to carry out dynamic experiments and to observe specimens while they are subjected to certain treatments. In this way it is possible to look at the wear of cloths and other materials as they are abraded by weaving machinery, or to study the wear of teeth as they are attacked by dental implements. Such examples serve to illustrate the vast range of experiments that can now be carried out, recorded on video tape, and subsequently analysed in detail.

In the scanning electron microscope, an image is produced from the specimen surface, but there is no reason—provided again that the specimen is thin enough—why the electron beam should not be scanned to-and-fro across the specimen, but with the image formed from electrons going through the object. This is the *scanning transmission electron microscope* (STEM) and it has some ad-

vantages over the conventional TEM. It can give very high resolution even from relatively thick specimens, and has even allowed the observation of single atoms!

In principle, there is no reason why a microscope should not be operated in either scanning reflection or scanning transmission mode at the will of the operator and, indeed, it is this type of combined instrument that is becoming more widely used. In such a way it is possible to look simultaneously, through different viewing systems, at the surface structure of an object (SEM) while also looking through it (STEM).

Chemical analysis

Although electron microscopes now give many different types of structural information from specimens of every conceivable type, this is nowadays by no means their only use. Because high-energy electrons can interact with specimens in many different ways, each producing characteristic 'signals', it is often possible to determine something of the chemical nature of the material under ex-

amination by analysing the products of electron interaction with matter. Among the more important 'signals' so produced are characteristic X-rays, visible light, energy-loss electrons and Auger electrons. Detection of these signals and their analysis permits identification of elements in minute quantities—on occasion only a few thousand atoms will suffice. This unique ability of the electron microscope has won it a place in investigative laboratories around the world.

Instruments, in general, are becoming more versatile, although some are confined to very specialized use. For example, many TEMs now offer the opportunity for TEM, STEM or SEM observational modes as well as analytical facilities—all on one instrument.

Computers are being used to drive electron microscopes, and they are of equal importance in helping to analyse the data accumulated from them. In image reconstruction, data handling from analytical techniques and analysis of image features, the computer is becoming ubiquitous and indispensable in its supporting role.

Electron beams can now be made extremely powerful, both in terms of accelerating voltage as well as beam current. A great amount of energy can be dissipated in the specimen and this can—indeed, often does—lead to radiation damage.

In the TEM, special procedures for 'low-dose' exposures, usually incorporating image intensifiers in the recording system, are used to obtain pictures at the highest resolution. Radiation damage may also be reduced by lowering the temperature of the specimen.

These developments of the electron microscope will in the future extend its analytical power, bring it into new fields of use and provide a better understanding of the smallest structures in the world about us.

From stylus to speaker

The world's greatest classical composers were fated only to hear their symphonies rarely, at concerts. To turn on music at will, day or night, reproduced to a standard almost indistinguishable from a live performance, is the gift of hi-fi technology. It has revolutionised popular and classical music making and led to an explosion in the number of people who can now listen to music every day of their lives. New developments, including indestructible discs played not by stylus but by laser beam, are in active progress. The story of hi-fi is one of continuing improvement to a few basic techniques.

A hi-fi system's task is to convert the coded input it receives—from disc or tape or via a radio tuner—into audible vibrations in the air—sound waves.

Originally hi-fi was a term coined to describe the best available record players, designed to reproduce music with a high degree of fidelity to the original sound at a live performance. However, in the context of today's advanced electronic studios the term has changed its meaning and now

B. H. Morris & Co. (Radio) Ltd.

defines the ability of the system to faithfully reproduce the coded message it receives.

The change of meaning is subtle but significant. Technology can now *record* a live performance accurately. Whether this recording is reproduced properly in the living room, without distortion, depends on the quality of the hi-fi system.

In addition, to speak about high-fidelity to an original performance means little when that original may never have existed. In the case of electronic music or music recorded on a multi-track tape recorder, each instrument is recorded separately and the complete work does not exist as a whole until all its elements are assembled in a mixing studio. Indeed, most of todays pop music is so recorded.

Music can be encoded and preserved in a

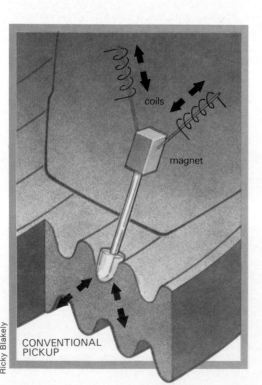

Ricky Blakely

CONVENTIONAL PICKUP

coils

magnet

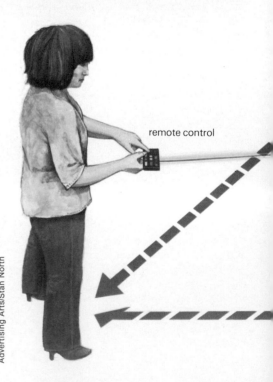

Advertising Arts/Stan North

remote control

Left Variations in the record groove make the stylus oscillate. In turn this causes the magnets in the cartridge to vibrate and induce a current in their coils, which is then passed to the amplifier. The tone arm *(above)* should be very light and with low pivot friction.

variety of ways. But the two most popular storage mediums are disc and tape.

On disc, music is encoded in a vinyl surface as a wavy groove whose shape corresponds to the waves the microphone receives as sound waves from the original source. The sound of many instruments playing together produces a complex and tortuous groove path that can contain information about the acoustic space in which the recording was made, as well as its tones and harmonies.

With the right equipment the spatial aspects of the music can be reproduced at home. The simplest way of doing this is stereo, where two speakers are fed with separate signals amplified separately, in order to mesh acoustically in the listening room and create a sound 'image'.

How the human ear perceives stereo is a complicated subject, but what is important is that the input to the system must consist of two separate signals recorded simultaneously by different sets of microphones.

On discs these two signals are encoded in one groove. The movement of one wall of the groove up and down corresponds to one signal and the movement of the other wall corresponds to the second signal.

Before these groove wall patterns can be transformed into sound they must first be converted into corresponding electrical signals, as varying patterns of current; then amplified to a strength at which they can drive a pair of loudspeakers. The speakers convert the electrical signals back into high energy sound waves.

At each stage of conversion and amplification, distortions and overtones can be in-troduced into the signal by the mechanical and electronic process involved. A good hi-fi system will have elaborate circuits to correct these unwanted additions to the signal. The provision of such policing circuits distinguish a good sound system from a bad one.

The greatest distortions are introduced where one form of signal is converted to another, whether from a mechanical form to electrical (as in the disc pick-up) or vice versa (as in the loudspeaker). As a result, correct design of the pick-up and loudspeaker is crucial to correct translation of the original signal on the disc.

In most modern pickups the groove vibrations are transferred to a tiny magnet linked to the stylus armature. This magnet is surrounded by very fine coils of electrical wire. As the magnet vibrates and moves in and out of the coil, the magnetic field cutting across the coils changes and cause a current to be induced in the wires. This is fed to the input of the amplifier and then the loudspeakers.

In order to get two signals out of the one groove the magnet is linked to two coils. If

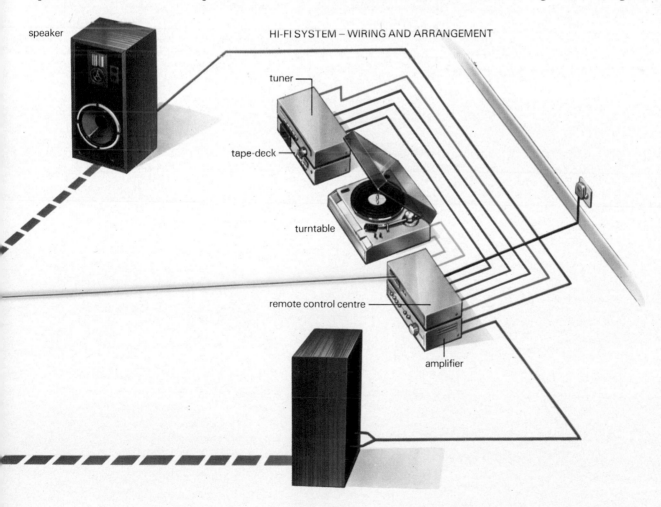

speaker

HI-FI SYSTEM – WIRING AND ARRANGEMENT

tuner

tape-deck

turntable

remote control centre

amplifier

Above left A typical magnetic cartridge showing the stylus, coil and magnet arrangement. The accuracy with which the cartridge translates the oscillations of the stylus defines the quality of the signal passed to the amplifier.
Left A hi-fi system wired for use. With the most sophisticated systems even the shape of the room has a significant effect on the sound created.

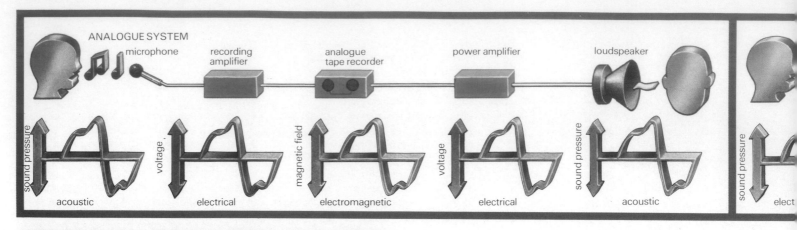

ANALOGUE SYSTEM

microphone — recording amplifier — analogue tape recorder — power amplifier — loudspeaker

sound pressure | voltage | magnetic field | voltage | sound pressure | sound pressure

acoustic — electrical — electromagnetic — electrical — acoustic — elect

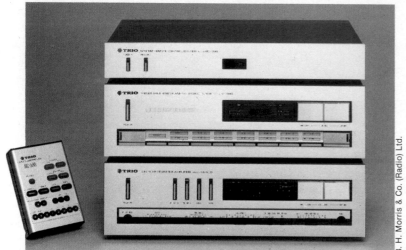

Left Top flight hi-fi equipment replete with remote control. *Below* One of the most advanced turntables, the Kenwood L-07D, features a highly rigid construction aimed at eliminating signal loss due to vibrational movement. This should allow a wider frequency range to be appreciated free from distortion.

B. H. Morris & Co. (Radio) Ltd.

the diamond stylus receives no signal on wall A of the groove, the motions of wall B will vibrate the magnet in coil B, but there will be no motion along the axis of coil A and so no signal will come from that coil, from motions due to wall B. Similarly wall A will only produce a signal in coil A, and will have no effect on coil B. This ingenious device is at the core of the hi-fi system, and if it does not function efficiently the best equipment in the world will not rescue the resulting signal.

The principle of separating the two signals is simple, but there are many subtle distortions that can be introduced if the stylus does not exactly follow the shape of the groove as it tracks the waves. Elliptically shaped stylus

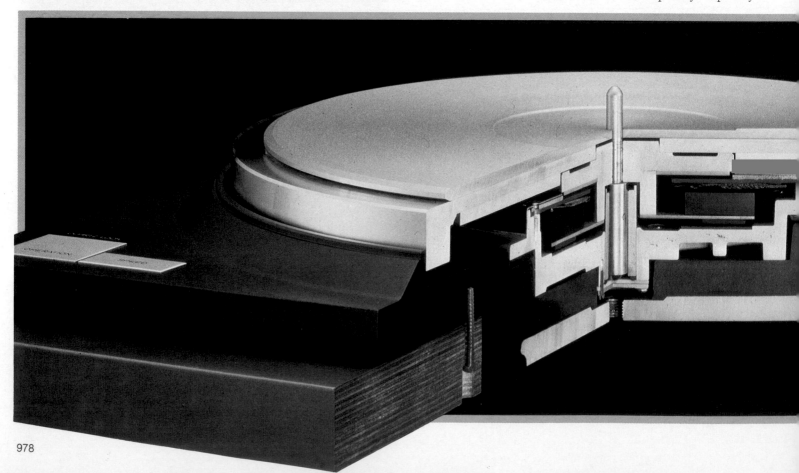

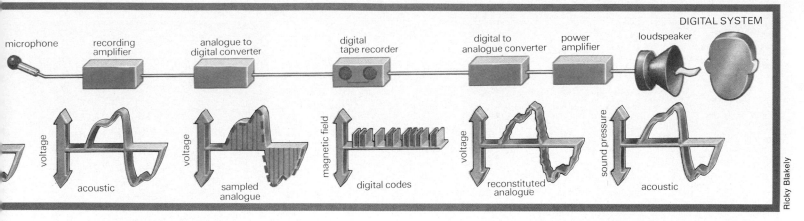

DIGITAL SYSTEM

microphone | recording amplifier | analogue to digital converter | digital tape recorder | digital to analogue converter | power amplifier | loudspeaker

voltage — acoustic | voltage — sampled analogue | magnetic field — digital codes | voltage — reconstituted analogue | sound pressure — acoustic

Ricky Blakely

points create a greater area of contact with the groove than can spherical points. An elliptical point sits firmly in the groove even when required to vibrate at frequencies up to 20,000 cycles per second.

Although good design of magnets and coils minimizes intrusion of one signal into the other coil, there is always some leakage.

The mechanics of groove/stylus interaction dictate that the mass of the pick-up arm, the mass of the stylus tip, and the springiness of the arm which connects it to the rest of the cartridge are all linked by a set of mathematical equations.

Since low-mass arms and low stylus tip-mass are expensive to engineer there are a variety of pick-ups available to suit the many different qualities of arms found in domestic record players. Top flight models have low tip mass, high springiness and require fitting in low loss, low pivot friction pick up arms.

Pick-ups

The way in which the stylus support is connected to the magnet (and the shape of the support itself) is critical if all the information in the record groove is to be extracted. Many types of pick-up are available. Some link the movements of the stylus to coils rather than the magnet; others rely on changing the electrical capacitance of internal elements to produce the signal. But all are trying to achieve the exact conversion of the groove pattern into an electrical signal.

Above A comparison of traditional recording technique with the newer digital mode. Digital recording has the principal advantage of producing a signal which is much less susceptible to both noise and distortion.

Music is encoded on tape as stripes of varying magnetization in an iron oxide film which is coated on an acetate base strip. By running the tape past the pole of a U-shaped playback head the currents generated by the varying magnetic fields on the tape traversing the pole piece are picked up by a coil behind the U piece.

Large open reel tape recorders, popular before the advent of the music cassette, used wide tape and recorded up to four tracks on the tape. Two tracks in one direction provided the two signals necessary for stereo, two in the other direction enabled the playing time of the tape to be extended. Most open reel tape decks offered a choice of three running speeds, permitting economical recording at low speeds, or better reproduction at higher speeds.

Cassette revolution

The advent of cassettes, originally intended as quick load, convenient capsules for dictating machines, effectively ousted the open reel recorder, except in the field of live music recording. The width of a cassette tape is half that of a standard open reel deck and is only one eighth of an inch wide. Crammed on this are four narrow tracks—which presents a challenge to efficient signal transcription.

The narrower a recording track, the bigger, relatively, is the size of the iron oxide particles that carry the magnetization. As a result, there is less available signal relative to the inherent noise on the tape (which comes from the random magnetizations in the oxide particles). In brief, the track width narrows so the noise on the tape increases.

Furthermore, cassettes run at a slow speed, 4.76 cm/sec (1.87 in/sec), which

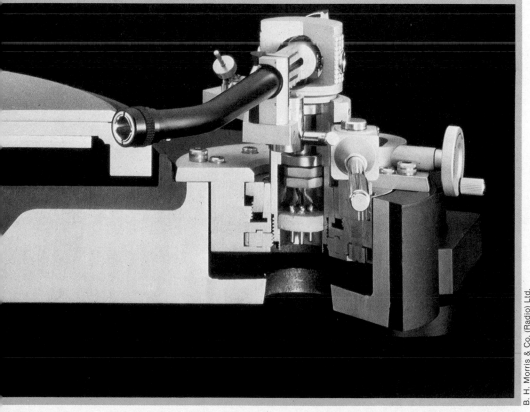

B. H. Morris & Co. (Radio) Ltd.

Philips laser-read Compact Disc. The disc is 115 mm wide and is played on one side only—yet it can carry up to one hour of music. This is stored *within* the disc in the form of flats and pits which represent a binary encoding.

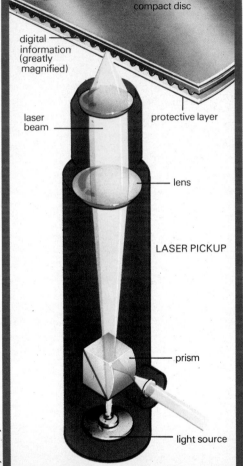

compact disc

digital information (greatly magnified)

laser beam

protective layer

lens

LASER PICKUP

prism

light source

The compact disc laser *(left)* scans the disc from the centre outwards, detecting the sequence of pits and flats at a rate of 1.6 million per second! This information is then passed to a converter *(below)* before being amplified and heard through speakers.

means that the high frequency response of the tape is limited. Very high frequency signals stripe the tape at intervals which are nearly as small as the size of the oxide particles at this speed. This means that the sound of instruments with very high frequencies in them, such as cymbals and violins, will tend to get lost in the background of tape hiss.

Dolby system

Several systems have been developed to reduce background noise: the most popular is the *Dolby system*. In essence, a Dolby circuit artificially boosts the high-frequencies in the signal as the music is recorded; then diminishes them on replay. As a result, tape hiss is reduced on replay, along with the artificially boosted music frequencies, but sounds at its proper level. The Dolby circuit has other refinements to ensure that the tape hiss is eliminated without affecting the music signals. All other noise reduction circuits operate in a broadly similar fashion.

In addition, to aid reduction of background noise, some manufacturers developed tapes with special oxide layers that become more strongly magnetized than regular tape. Others developed tapes employing oxides of chromium, some of them mixed with iron oxides.

The magnetic characteristics of these new tapes required certain circuits inside the tape deck, called the *bias* and *equalization* circuits, to be modified for each type of tape. The bias circuit provides a high-frequency transfer signal that is mixed with the music signal as it is fed to the tape to ensure that distortion introduced by the transfer is kept to a

Measuring only 13.2 cm × 8.5 cm × 3 cm, the Sony Stowaway is remarkable for its high quality of sound reproduction.

minimum. The equalization circuits control frequency response characteristics. On modern cassette decks two multi-position switches vary the circuits' characteristics for the different types of tape available.

Tape technology is still in a state of development. But new tapes and advances made in noise reduction circuits have enabled even middle quality cassette decks to equal most of the earlier open reel decks.

With the growth of the VHF radio network the demand for high quality FM tuners has increased greatly. The tuner is thus a key part of the hi-fi system.

An FM (frequency modulated) signal is quite different from the signal which drives an ordinary transistor radio. AM (or amplitude modulated) signals, receive on the medium wave tuning scale, vary the strength

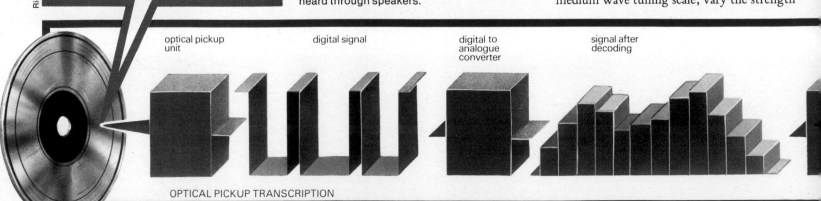

optical pickup unit

digital signal

digital to analogue converter

signal after decoding

OPTICAL PICKUP TRANSCRIPTION

of the carrier wave to carry the music signal. FM, or frequency-modulated signals vary the actual frequency of the carrier wave in time with the music signal and thus dispense with the interference from atmospheric sources to which AM signals are prone. As a result, a low noise, high quality signal can be obtained from a tuner.

The amplifier

At the heart of any system is the amplifier. Its function is to detect the appropriate input, whether disc, tape or FM, and amplify the signal received until it is strong enough to drive a pair of loudspeakers. A pre-amplification stage incorporates all the control functions—among them, treble and bass, loudness controls, volume and balance.

Again, the quality of modern amplifiers depends more on refinements to the amplifying process than any other factor. The way in which the amplifier's circuits treat the incoming signals can introduce unwanted or spurious signals which result in a distortion of the sound. Feedback circuits and good design can eliminate signal by comparing the profile of the amplified signal with that of the incoming signal and eliminating any differences.

The loudspeaker, like the pick-up cartridge, converts the signal from one form (electrical) to another (mechanical movements of the speaker cone). The simplest loudspeakers consist of a coil of wire between the poles of a magnet. The coil is attached to a cone of paper or plastic which is anchored at the edges to the speaker frame. As the signal current passes through the coil it interacts with the magnetic field between the poles of the magnet and moves the coil in sympathy with the flow of current in the wire. Thus the paper cone vibrates and sound waves are produced.

The situation is complicated by the inability of one size of speaker cone to handle the efficient dispersion of all frequencies of

Daily Telegraph Colour Library

Left A speaker undergoing testing in a specially designed acoustic chamber. As it is impossible to design a single speaker cone which can handle the whole range of audible frequencies, most speakers *(below)* have two, three, or even four cones to allow them to reproduce music tones more faithfully.

sound. In order to reproduce all the frequencies in the music spectrum loud speakers use up to three or four separate drive units, one to handle the bass notes, another the mid range frequencies (such as voices and most instruments) and one or two units to take the very high frequencies.

Inside the speaker is a *crossover* circuit which divides the incoming signal into the the frequency bands most suited to each drive unit.

Like the pick-up cartridge, a loudspeaker can introduce a great deal of distortion—mainly through resonances in the material of the speaker cones, which artificially 'colour' the sound output. These distortions are kept to a minimum by making the cone as stiff as possible (by plastic coatings or reinforcing laminates) and by attention to the way in which the edges of the vibrating cone are anchored.

In addition, speaker cabinet design is important to the final result. For example, the bass response of a speaker set can be extended to very low frequencies by 'loading' the back of the drive cone with columns of air trapped in specially designed boxes and labyrinths. This loading drives down the resonant frequencies of the box and bass drive cone and

JVC (U.K.) Ltd.

leads to a very deep, smooth bass sound.

There is no doubt that the future of hi-fi will be revolutionized by micro-chips and the digital technology associated with them. Digital processing of a music signal means that many of the distortions introduced by the components described above can be eliminated.

Discs encoded into digital signals on light reflective surfaces can be read by laser beam and so become virtually wear free since there is no physical contact between disc and reading device. Such discs can also be used to carry video signals and it is likely that in the future hi-fi and video will develop into one 'home entertainment' field. On the near horizon is the advent of discs that will provide sound through a hi-fi system and matching pictures on the TV screen.

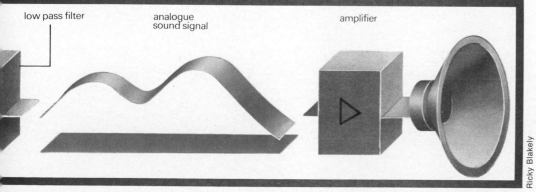

low pass filter   analogue sound signal   amplifier

Ricky Blakely

Record makers' magic

Vast arrays of knobs, sliders and dials make recording and mixing appear very complicated. The process is, however, relatively straightforward; a fascinating interface between art and science where the creativity of musicians and the technical skill of recording engineers blend together.

Perhaps the best way to appreciate the process of recording and mixing is to follow it through the stages which begin with a musician and a song and ends up with a disc or tape. Once all the basic arrangements have been made such as obtaining a recording contract, booking a studio and hiring a producer, engineers and any additional musicians, the business of recording begins.

Reaching for perfection

It is very rare for all the component parts of a song to be recorded simultaneously. Instead, the music is divided up into parts, each of which are recorded separately. The tape recording machines generally used have as many as 64 tracks running parallel with each other and it is possible to record or play back any of these independently or together through the studio monitor speakers or headphones. For example, it may be advisable to record the bass guitar and drums first and listen to them a few times before adding the other parts of the music.

It is therefore possible to listen to tracks which have already been recorded while recording on a blank track. In this way the engineer can ensure that each part of the music is absolutely perfect before moving on to the next instrument or group of instruments.

Backing tracks

A typical song might require a drum backing, a bass guitar part, two rhythm guitars, a piano, a brass section and three vocal parts, plus a short orchestrated section in the middle. The producer draws up a chart showing which tracks on the tape will be used for each part of the music. The arranger of the orchestrated section will be briefed and he will produce a score, and perhaps the musicians as well, on the appointed day.

Recording begins with the backing tracks. This is because a good solid bass track and drum track provides a firm foundation for the rest of the song.

In the studio the musicians, producer and the studio engineer decide exactly how they want these instruments to sound. The bass and the drums are usually recorded together and it is essential that the two musicians concerned hear each other when playing. However, they will not be recorded on the same track and it is, therefore, necessary to isolate the musicians from each other so the sound of the drums does not spill over on to the guitar track and vice versa.

This can be done by placing heavy acoustic screens between the two musicians, but because drums are so loud they are more often recorded in a separate booth at the side of the studio. This has a glass window, so that the drummer and the bassist can see each other as they play. The engineer arranges for each of the musicians to hear what the other is playing through headphones, ensuring no leakage of sound between the tracks.

To record drums properly requires more than just one microphone. The engineer may decide to use as many as 14 or 15 to get all the nuances of the drum sound and he may spread the different parts of the drum kit across more than one track on the tape allowing him greater control of the sound at the mixing stage. The producer has probably allowed three or four tracks for the drums

Theo Bergström

Neve

Above This *Neve* console at Air Studios in London is typical of the highly sophisticated machinery used in sound mixing. On many consoles silicon technology has enabled microprocessors to supersede frame wiring. Electronics can be mounted on plug-in cards which are easily accessible for replacement. *Left* For the sound engineer to achieve the result he is seeking, the drummer may find himself surrounded by many more microphones than drums. Here, directional microphones picking up sound from bass, tom-tom drums and cymbals combine to provide the optimum choice of sound blend.

and the engineer decides, in consultation with the producer, how to make the best use of them.

Typically two microphones are hung above the drums to pick up the overall sound of the kit. A second, highly directioned microphone with a specially designed, built-in frequency response which attenuates the high frequencies, is placed right in front, or even inside the bass or kick drum. The purpose of this microphone is to pick up only the bass drum sound and feed it to its own track so it can be altered in sound character to mix in with the rest of the music.

Other directional microphones are used to pick up the sound of the cymbals and the snare drums. The reason for using directional microphones is that each one only picks up the sound of the drum at which it is aimed. Of course, there will be some feedback between each microphone but directional microphones keep this to a minimum.

The output from each of these microphones is fed to the input of the most important piece of control equipment in the studio—the mixer.

The mixer enables the sound from several microphones to be mixed together and

Neil Raphael

Denis O'Regan

Far left The bass guitar sound—here from the group *Jody Street*—can be fed directly into the mixer rather than via microphones. Extra reverberation can be added in the console if needed.
Left A special effects unit can modulate the basic sound of the rhythm guitar. This one is part of the *Thin Lizzy* group's equipment.
Below The sound, controlled by effects units and the mixer console, will be passed into a 24 track master tape. The producer will later mix it down into eight tracks for the final version.

routes the resulting signal to the track assigned to it. The engineer may decide to route each microphone to a separate track or he may mix two or more of the signals together to get the kind of sound the producer is looking for. This part of an engineer's job is, in its own way, just as creative as that of the musicians. He must be able to easily understand what the producer wants and how to achieve the desired sound with his microphones. Such expertise only comes with experience and this is why the best engineers are just as much in demand as the best musicians.

Recording guitars is equally complex. The bass guitar can be fed straight into the mixer from the electric pick-up on the guitar itself, or it can be recorded by microphones placed around it in the same way as the drums. Usually no more than one track is used for such an instrument and not more than two microphones. The reason for using

microphones rather than a straight feed into the mixer is to get some of the reverberant sound from the guitar which is caused by the shape of the studio in which it is played.

Words and music

Feeding an electric instrument straight into the mixer results in a very dry sound with no richness—although artificial reverberation can be added later. This may be what the producer wants, in which case the engineer records the guitar accordingly. If the producer is not sure what he wants, other microphone arrangements may be tried until he likes the sound.

The rhythm guitars usually go on the tape next, the musicians again hearing what has already been recorded via headphones as they play their part. Often a rough vocal track is recorded at the same time as a guide to the guitarists, to help them get the feel of the song. This vocal track will be over-recorded

later when all the other instruments have been recorded.

Very complex musical arrangements can be built up in this way, the microphone techniques and recording levels being altered for each different instrument in order to get the required sound.

The mixer is a very flexible recording tool enabling signals from the tape or microphone to be mixed together in the musicians' headphones, or played through the large monitor speakers in the control room, which is separated from the actual studio by a sound-insulating glass window. Further controls on

undersound. The producer has various frequency controls, rather like tone controls, at his disposal. These *equalizing circuits,* as they are called, can alter the sound of any instrument by boosting or attenuating selected frequency bands in the signal coming from the tape.

For instance, the producer may wish to give the bass drum a more forward sound by boosting the sound of the drum in the 'presence' band at around 500 Hz upwards.

This makes the sound more defined and sharp. The bass guitar sound could also be tailored in this way, but it is more likely it will be done at different frequencies to complement the drums. Just which frequencies to boost and by how much is up to the producer and it requires considerable skill and experience to get it right.

In a similar fashion the producer alters the tonal qualities of all the other instruments, perhaps adding some reverberation to the vocal as well as controlling the relative levels of each instrument throughout the song while bringing up the sound of the guitar for a solo. The whole process can be very intricate involving the alteration of level and tone as the music is actually playing and the

Above left A vocalist may record a rough version of his final performance as a guide to the other musicians. Later he will use the multi-track playback through the headphones to create the final vocal track. He may even overdub this with additional material.
Below Pierre Boulez conducts the London Philharmonic Orchestra in a studio session.

the mixer enable the character of the signal on the tape to be changed or routed to effects processors, such as reverberation chambers or plates, or phase distortion circuits.

Once all the tracks on the tape have been filled up, the complicated process of mixing down to a pair of stereo tracks can begin. All the complex controls of the mixer come into their own here, and again it is the intuitive skill and the ear of the engineers and producer that determine just how good the final mix will sound.

Typically the 'mix down' process starts by tailoring the rhythm track to provide a solid

Neve

producer and engineer will rehearse the operation of the level controls, tone controls and other switches until they can repeat the exercise perfectly for the final recording on two track tape.

In order to get a good stereo image each track is fed to the two track stereo tape recorder, which is recording the final mix, via a *pan pot.* This is a control which splits the signal from the tape between the two stereo tracks. By rotating the pan pot the producer can position the instrument anywhere on the stereo stage. He may decide, for example, to spread the sound of the drums right across the image by putting the bass in the centre, the kick drums at either side and the cymbals left centre. He balances the rhythm guitars symmetrically around the centre of the image and probably places the lead vocals in the centre with the harmonies possibly at either side.

Indeed he may decide to move voices and instruments around in the course of the song. The whole process of mixing down can get very complicated and the latest mixer desks have computer memories which remember how the control functions change in the course of the mix.

Above The *Neve 8108.* Features include: 1 Control of incoming signals from studio or pre-recorded multi-tracks. 2 Filters to eliminate unwanted sound at either end of the spectrum. 3 Parametric equalisers, forming a sophisticated hi-fi tone control. 4 Auxiliary assignment and multi-track controls: mixes of sound are made and signals can be sent to reverberation plates for echo effects. Single tracks or pre-recorded multi-tracks can be relayed to individual musicians through earphones. 5 Secondary fader to adjust approximate sound and multi-track mixes for monitoring purposes. 6 Mix down controls with pan-potentiometers. The operator can 'place' stereo sound—e.g. making the singer appear to be on the right or left of the listener. 7 Primary fader to control the main sound balance and volume. 8 Interrogation switches, using microprocessors to determine which microphones are feeding which tracks. 9 Remote control panel to monitor total routing of signals through the console. 10 Touch pad facilities panel for quadraphonic listening through loudspeakers.

GTO Films Limited

The complexities of modern recording techniques make the producer and engineer as vital to success as the artist. The final mix down of a release by Elton John (below) may take far longer than studio time.
Left Musicians recording over tracks fed to them through earphones or 'cans'. The atmosphere of work in a recording studio is carefully reconstructed in a film about the music industry *Breaking Glass* (below left).

The producer will try out a routine with the engineer and if he likes it, all he has to do is press an 'update' button and the mixer desk remembers all the movements of the controls. He can then run the tape again, sit back and listen to the music mixing itself from the stored instructions in the memory.

This technique can lead to the most involved mixes which would not have been possible without a memory to remember all the intricate control changes.

The process of mixing down the music to a stereo pair will take the producer and engineer a long time, possibly far longer than the original recording in the studio. If, at the end of the mix down, the producer feels he needs extra instruments or effects on the tape, changes can still be made. First he plays the section he wishes to augment and feeds the two tracks to a second tape machine via the mixer. At the same time the extra instruments or sound effects are fed into the mixer and recorded along with the original music.

The new section is then cut out and spliced into the original tape in place of the section used to feed the second tape recorder. This technique can also be used to add extra instruments if all the tracks on the multi-track tape are full and there is no room for more instruments.

It is quite possible that the producer may wish to use one studio and engineer for the recording and another studio altogether for the mix down, possibly on the other side of the world. The standardization of tape calibration levels enables this to be easily done, and it is not uncommon for today's music to be recorded in half a dozen different locations, with half the musicians never seeing the rest of the performers.

Direct cut discs

The standard multi-track approach to recorded music means that the signals are passing back and forth through the mixer several times in the course of the recording process. However good the processing electronics, this inevitably means some loss of clarity and definition in the music.

Also, transferring the final stereo mix to a cutting lathe which cuts the master disc means more loss of definition, and so music recorded in this way will be slightly 'dim' when compared to the original sound. There are ways to overcome this, for instance, by feeding the output of the mixer at a live recording session straight to the disc cutter and missing out the mix down stage. By doing this, at no time is the signal recorded on tape and the clarity and definition which can be achieved in this way are quite stunning.

The difference in sound between a 'direct cut' and the regular type of disc is subtle but nevertheless noticeable, especially when reproduced on high quality stereo systems.

However, very few direct cut discs are made as the technical problems involved in the recording are quite formidable. For instance, all the microphone input levels and the stereo balance must be set up beforehand for the whole orchestra or group. Furthermore they must play for the whole side of the disc straight through, pausing in between each track, and there must be no mistakes, or the whole side will have to be recorded again. Direct cut discs are available, but for the reasons above they are understandably more expensive than normal. But apart from their expense direct cut discs are limited in their application.

In the future the increasing use of microelectronics to record the signal in digital form on the tape will mean multi-tracked music will be recorded with the clarity of a direct cut disc. Once again the micro-chip invades and improves the existing technology.

MEDICAL SCIENCE
PART SEVEN

INTRODUCTION TO PART SEVEN

This volume is nicely balanced. It begins with Military Technology and ends with Medical Science. Science taketh away, but it also giveth. And its achievements in the field of medicine and health are truly marvelous. It is a good place to end our journey in pursuit of knowledge, a journey that began in the first volume with the exploration of space and that ends here with exploration of interior space—our own bodies.

Medical marvels are all around us; we hear of a new one every week, or at least whenever someone close to us becomes ill or must enter a hospital. The surgeon has wonderful new tools, as does the diagnostician. All over the Earth doctors and researchers are seeking cures for human ills, and the rewards are very great if they find them. That is as it should be. Nor is it difficult to see why the search is pressed with such singleminded passion. Doctors are human beings after all, and they are as likely as you or I to fall prey to those ills that flesh is heir to; hence their desire to find cures before that happens.

There is one shadow of concern that falls upon technological medicine. When a physician has the most modern machines with which to work, is he or she more likely to forget the patient? After all, we should not ignore the fact that although there have been many bad doctors before the modern age, there have been good ones, too, who helped people to get well without powerful drugs and sophisticated machines and all the paraphernalia of the specialist of today. The secret those old doctors knew was that people have to *get* well and cannot be *made to be* well. A desire for health is still the most potent medicine.

But it is often not enough, and there is not one of us, when illness or accident happens, who is not glad of the great progress that has been made in recent years. And if we are fortunate enough to be well today, we hope they discover a cure for whatever we get tomorrow!

For with all the progress, we are not there yet, we do not have all the answers yet, illness and pain are still a part of everyone's life. Will it ever be different? Perhaps not. But at least the doctors, bless them, will go on trying to make it so.

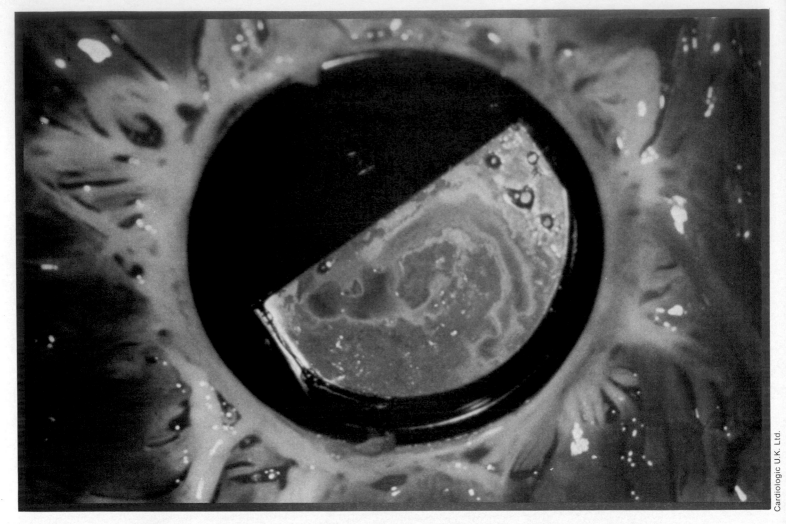

Cardiologic U.K. Ltd.

New parts for the heart

The idea of replacing some function of the ailing or ageing human body with an artificial substitute, or prosthesis, is not a new one; some mechanical devices such as spectacles, false teeth and artificial limbs have been used in this way for many hundreds of years.

Today's devices, however are infinitely more ingenious—and are becoming increasingly common. They range from artificial heart valves and pacemakers to metal and plastic limbs, man-made larynges (voice-boxes), permanent contact lenses and artificial wind-pipes. Even more ambitious substitutes for vital human organs or systems are being developed, tested and perfected.

Many people owe their health or their very lives to the ingenuity of modern prosthetic medicine. Nowhere is this more true than in the case of heart disease.

The heart is an astonishing piece of natural design; weighing only one pound and about

the size of a clenched fist, it pumps the entire blood content of the body through its four chambers every minute, beating over 36 million times a year. It deals with a total of 2,000 gallons a day or 50 million gallons in the average life-time. Its performance is impaired by stress, alcohol and tobacco. It is not surprising that heart failure of one sort or another is a major cause of death in modern societies.

Artificial Heart Valves

One of the commonest heart diseases affects the heart valves which control the flow of blood into and out of the various chambers of the heart. When they begin to malfunction, leaks develop between the heart chambers or narrowing of the valves occurs, causing the heart to act less efficiently.

One of the most commonly fitted artificial heart valves is the Starr–Edwards ball valve, in use since 1960. This was named after the

US heart specialist Albert Starr and the aircraft engineer M. L. Edwards who designed it. The valve consists of a hollow ball made of a very hard, tough chromium-cobalt alloy (in early models it was made of silicone rubber) held inside a metal cage. The cage and orifice (the opening through which the blood flows) are covered with a cuff of knitted polypropylene to encourage the ingrowth of tissue and discourage blood clots from forming. The cuff also allows the valve to be sewn into position more easily.

Various other designs of artificial heart valve have been developed over the years. With many of them, however, problems have arisen because of defects in design or materials. One of the main troubles was that blood clots tended to form on the valves. Another was that the tissue lining the heart often grew into the orifice, reducing the area through which the blood could flow and preventing the valve from opening and closing properly—a situation that could prove catastrophic for the patient.

The other main problem was that the flow of blood caused a great deal of turbulence which was not present in a natural valve, and this led to destruction of red blood cells and consequent anaemia in the patient.

Recently a valve has been developed that seems to avoid these problems. First tested in July 1976, and known as the St Jude Medical bi-leaflet valve, it consists of two 'leaflets' that rotate within a ring, upon which a seam-free sewing cuff is mounted. When the valve is closed, the inner edges of the two leaflets meet at the centre of the orifice—they do not lie flat but at an angle of 30–35°. As the blood forces the leaflets open, they swing out to a maximum angle of 85° from the horizontal, allowing the blood to pass through three openings, a narrower one at the centre and two larger ones at either side.

The design and angling of the leaflets means that minimal blood pressure is needed to open them, so that they open and close more quickly than other designs. The flow of blood through the valve is smooth, so there is little danger of clots forming or blood cells being destroyed. Because the blood flows through on both sides of each leaflet, the leaflet surfaces are completely washed on both sides during each pumping cycle; this also helps to prevent clots from forming.

The pivoting mechanism is enclosed by guards; these prevent the ingrowth of tissue

Right These commonly fitted artificial heart valves consist of a hollow ball made of a very tough alloy housed within an alloy cage. The cuff of knitted polypropylene enables the surgeon to stitch the valve securely in place. Different sizes are for different positions.

which would interfere with the opening mechanism. The valve is made of machined graphite coated with pyrolytic carbon, a very strong material used in prosthetic devices since 1969. Until the development of this design, it had proved technically difficult to make a valve from this material. Pyrolytic carbon can more than withstand the great forces constantly present in the heart; it is three times more resistant to fracture than bone. Although so strong, it is lightweight, and extremely durable, with an almost infinite capacity for wear.

The sewing cuff is made of the man-made fibre Dacron. This was chosen because it allows the rapid but controlled ingrowth of tissue over the whole area of the cuff, which anchors the base of the valve firmly in place. The sutures used to stitch the cuff to the heart are hidden under the material, so the surface of the ring remains smooth and discourages clots from forming.

The low profile of the valve, even with the leaflets open, makes the surgeon's task of fitting it much easier, as it gives him greater visibility when suturing it in place, especially where more than one valve is being replaced.

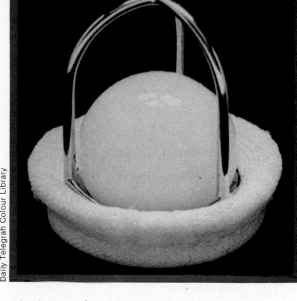

The low profile also does not obstruct the blood vessels leading into and out of the heart, and makes the valve ideal for patients with small ventricular chambers. The valves are made in a wide range of sizes to suit different patients.

Setting the pace
Another lifesaving device is the artificial pacemaker, cunningly designed to deliver regular electric stimuli to a heart unable to beat at the correct rate on its own. To explain how it works it is necessary to outline the way in which the four-chambered heart normally operates. Blood from all over the body,

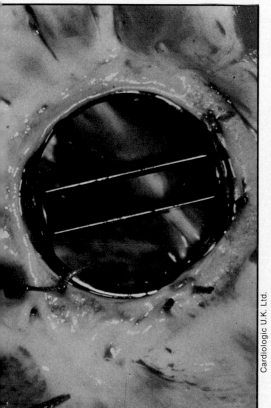

Opposite page After it has been stitched in place, the sewing cuff of the St Jude heart valve is invaded by heart tissue which helps to anchor it more firmly. *Left* A view of the valve with its two leaflets fully open. *Above* Fully open valve before implantation into the heart. *Right* The valve fully closed.

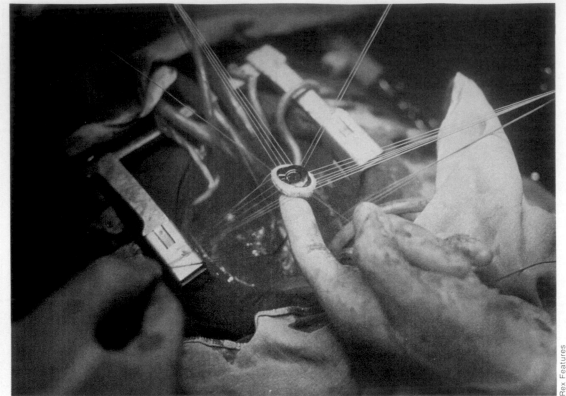
Rex Features

deprived of its life-giving oxygen, and laden with waste carbon dioxide, is transported via the veins to end in the right-hand collecting chamber of the heart—the right atrium. (The plural of atrium is atria).

From the right atrium the blood passes through a valve into the right pumping chamber—the right venticle. The right venticle contracts regularly to force the blood along the pulmonary arteries to the lungs. In the lungs, carbon dioxide is exchanged for oxygen, and the blood then drains back through the pulmonary veins to the left atrium of the heart. It then passes into the left ventricle, the heart's main pumping chamber. This contracts at a variable rate to drive oxygen-rich blood through the arterial system to all parts of the body. The pumping work itself is done by the heart muscle, or myocardium, which makes up most of the heart's bulk. Unlike other muscles, the myocardium has its own built-in mechanism to maintain the rhythmic heartbeat independently of its nerve connections.

This *pacemaker* is located in a knot of nerve fibres, the sino-atrial node (S-A node) in the rear wall of the right atrium. It generates a brief low-intensity electrical impulse approximately 72 times per minute in a resting adult. This causes the right and left atria to contract, thrusting the blood into the empty ventricles. The impulse now passes quickly to another specialized knot of tissue, the atrioventrical node (A-V node), near the junction of the atria and ventricles. Here the impulse is delayed for about seven hundredths of a second—precisely the right time to allow the atria to complete their contractions.

Next, conducting fibres fan out through the ventricles to carry the impulse to every muscle fibre so that the ventricles contract

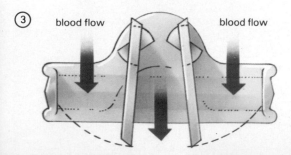
① orifice ring — ears — leaflet — leaflet

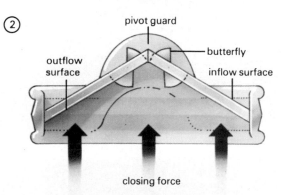
② pivot guard — butterfly — outflow surface — inflow surface — closing force

③ blood flow — blood flow

Above In this operation to insert an artificial heart valve, the sutures which will be stitched into the heart tissue to hold the valve in place can be seen passing through the sewing cuff. This is a pivoting disc valve.

Left Anatomy of the St Jude valve. Seen from above (1) and from the side (2), the two leaflets pivot within the orifice, fitting snugly at the centre when closed. The pivot consists of 'ears' at the ends of each leaflet which fit into grooves called butterflies. The pivot guard protects the mechanism from becoming clogged by ingrowth of heart tissue. Blood flow through the open valve (3) is via three roughly equal areas, so that turbulence is minimal and blood cells are not damaged.

Below Testing the valve in this device proved that it produces very little turbulence.

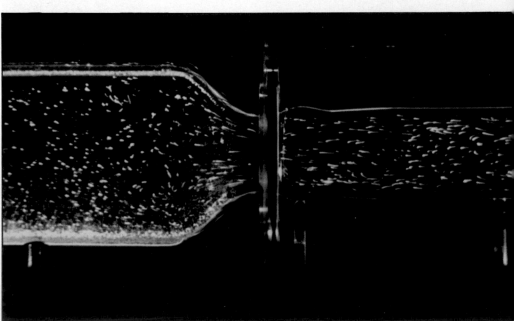

within about six hundredths of a second to pump the blood into the arteries.

If disease or degeneration destroys the S-A impulse, the A-V node can take over its function; although co-ordination between atria and ventricles is lost, the heart continues to do its job quite effectively. Even if the A-V node is damaged, the body can generally cope with the reduced heart rate of about 40–50 beats per minute. However, damage to the conducting fibres (the Purkinje fibres) within the ventricles is much more serious. Although the heart continues to beat, it does so at a much slower rate—about 30 beats per minute or even less—and there is little or no feedback via the nervous system so that it cannot step up its pumping action to meet the demands of exercise. Also, this condition, known as *Heart Block,* may seriously impair the workings of other organs, especially the brain and kidneys, and the heart may cease to beat entirely for several seconds, resulting in

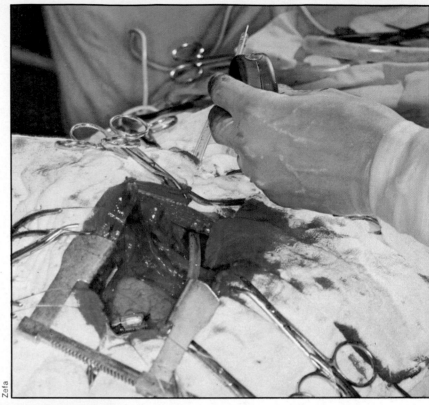

Right Implanting an artificial pacemaker. The surgeon is holding the pulse generator, which he will insert into a cavity cut in the chest wall. To the left of the generator is the lead which conveys the electrical impulses to the heart. The lead has been threaded through a large vein and the electrode at one end sewn into the heart wall. The other end of the lead will be attached to the pulse generator. *Below* This X-ray shows in situ generator and lead.

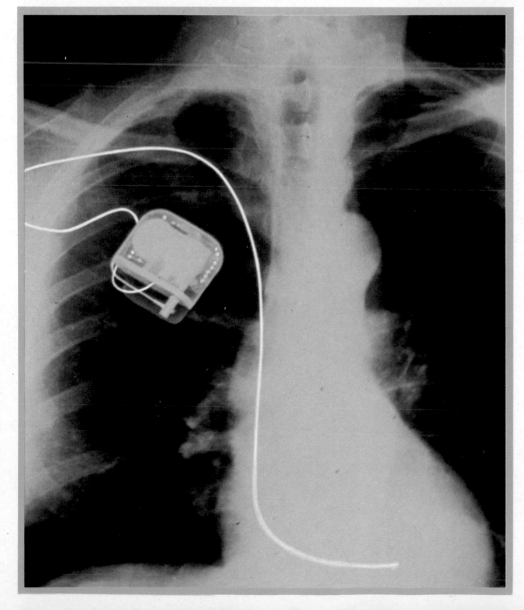

fits and sometimes complete loss of consciousness.

In the late 1950s medical science came to the rescue of such unfortunate patients by developing the artificial pacemaker. Although pacemakers cannot correct the underlying disease, they can enable patients who formerly would have been disabled or at risk of dying to live a normal life.

External pacemakers

External pacemakers were the first type to be developed. Although they have the advantage that battteries can be changed or recharged without recourse to surgery, they suffered from the great drawback that the wires connecting the pacemaker to the heart eventually gave rise to infection where they passed through the skin, and so they were abandoned for use as permanent pacemakers. However, external pacemakers are invaluable in hospitals as a stop-gap measure to tide the patient over a temporary block, as may occur during a heart attack or, much more commonly, during the implantation of a permanent pacemaker.

Implanted pacemakers

Modern pacemakers are entirely implanted in the body, complete with batteries. Although they have the disadvantage that they must be implanted surgically and periodically be replaced or recharged, they have replaced the external sorts for the reasons described above.

Each pacemaker consists of two main components: the pacemaker itself, often known as

995

the pulse generator, which contains the power source and circuitry to produce the right electrical impulse; and the lead, an insulated wire which carries the impulse from the pulse generator to the heart. Some types of pacemaker need two leads. Both pulse generator and leads are generally implanted under local anaesthetic only.

The surgeon makes a single incision across the upper chest, and gently threads the lead through a large vein, usually the cephalic vein, until it passes through the right atrium, through the tricuspid heart valve and into the bottom of the right ventricle. Here the terminal of the lead is firmly wedged into the thick bands of muscle lining the ventricle.

Now the surgeon checks that the leads are correctly positioned, using an instrument called a fluoroscope, which shows an X-ray picture of the chest on a fluorescent screen. When he is satisfied, he stitches the lead securely into the vein.

The next task is to cut a pocket between the subcutaneous (under-skin) tissues and the underlying bands of pectoral muscle above the left breast. The surgeon connects the lead to the pulse generator, and sutures the latter in place so that it cannot move about and cause discomfort or damage. The body reacts by gradually forming a fibrous sheath around the pulse generator and lead, which further helps to anchor the device.

At present, the most commonly fitted pacemaker is the stand-by or demand type, which is particularly useful for the patient who suffers from intermittent heart block. This ingenious little machine incorporates a device that monitors the patients' own ventricular impulse and cuts in only when this natural message fails.

Demand pacemakers

Most modern demand pacemakers are programmable—that is, the heart specialist can alter both the pulse rate and the duration of each pulse (the pulse width measured in microseconds). He does this by means of a programmer containing a digital silicon chip circuit controlled by an extremely accurate quartz crystal timing device: This can change the pulse rate or width via an electromagnet which is held about 5 cm (2 in) above the patient's skin over the buried pacemaker. It sends out digitally coded pulses which are picked up by the pacemaker. The programmer is small enough to be hand-held, yet it can change the pulse rate over a large range—between 30 pulses per minute and 119 pulses per minute in one model—and also the pulse width—between 0.1 and 1.9 microseconds. Moreover, change of rate or width takes only 2.5 seconds.

A further, even more remarkable develop-

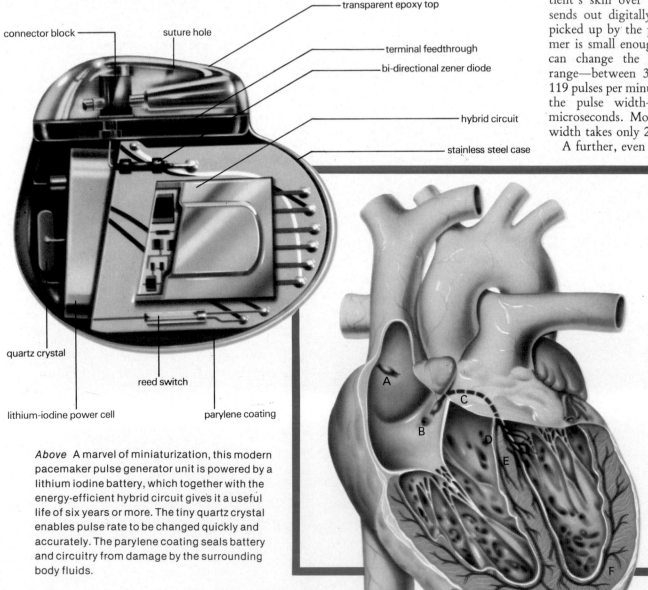

HEART PACEMAKER PULSE GENERATOR

connector block — suture hole — transparent epoxy top — terminal feedthrough — bi-directional zener diode — hybrid circuit — stainless steel case — quartz crystal — reed switch — lithium-iodine power cell — parylene coating

John Bovosi

Above A marvel of miniaturization, this modern pacemaker pulse generator unit is powered by a lithium iodine battery, which together with the energy-efficient hybrid circuit gives it a useful life of six years or more. The tiny quartz crystal enables pulse rate to be changed quickly and accurately. The parylene coating seals battery and circuitry from damage by the surrounding body fluids.

Left In the heart's natural pacemaker system, impulses arise in the sino-atrial node (A) at a rate influenced by nerves and hormones. The impulses radiate across the walls of the atria, making them contract. Next the atrio-ventricular node (B) relays the impulse via another bundle of fibres (C) into the wall between the two ventricles. Here it passes into right (D) and left (E) branches, one for each ventricle. Each branch conveys the impulse to the ventricular muscle via a network of fibres (F).

Above The slim rounded contours of this pulse generator make for good anchorage and minimal discomfort when it is implanted.
Below This system saves the pacemaker wearer frequent trips for check-ups. He hangs a transmitter on his chest (1), calls the doctor (2), then places his telephone mouthpiece over the transmitter speaker (3). Another method uses a fingertip electrode (4) so that the transmitter need not be held on the chest. On taking the call, the doctor places his phone earpiece in a coupler (5). This feeds the signal into a receiver which produces a digital readout and an electrocardiogram (6).

ment has been that of a patient follow-up system, by means of which the doctor can check that the pacemaker is working correctly and diagnose potential problems without the patient having to move from his home—a boon for pacemaker wearers who are elderly, infirm or live far away from the surgery. The patient hangs a small transmitting device on his or her chest, dials the number of the surgery or hospital department, gives his name and pacemaker number, then turns the transmitter on and places the telephone mouthpiece over the transmitter's speaker.

When the doctor or his assistant receives the call, he records the patient's name and number, switches on a receiver and places his telephone earpiece on a device called a phone coupler. The unit converts the patient's telephone signal to a digital readout of pacemaker rate and pulse width, and also gives a permanent record in the form of an electrocardiogram—a curve traced on graph paper showing the action of the heart.

Small enough to be stored in a file drawer, the receiving unit can be easily carried home if the doctor needs to remain on call. He does not need to attend to the receiver constantly; he can attach it to a telephone answering device. He can also couple the receiver to other equipment, such as an oscilloscope, for more detailed analysis of the patient's signals. The doctor can communicate with the patient via the telephone while the receiver is producing its readout.

Solving the Power Problem

The development of transistors and integrated circuits meant that pacemakers could be made both rugged and accurate, but the problem of developing a reliable, long-life power source remained. Early pacemakers used mercury batteries, with an average life of only two to three years; these emitted hydrogen gas, so the power unit could not be completely sealed, which meant that body fluids could enter slowly and reduce the life of the battery. A few years ago, it seemed as if nuclear-powered batteries would provide the answer to the problem, but their safety has not been conclusively established, and they have generally been abandoned in favour of the lithium-iodide fuel cell.

First developed in Canada in 1973, this is a solid state battery which does not produce any gas, so that the power unit can be hermetically sealed against intrusion by body fluids. The battery maintains a steady output which does not decrease markedly until it needs replacing, and will last in some cases for eight years or more.

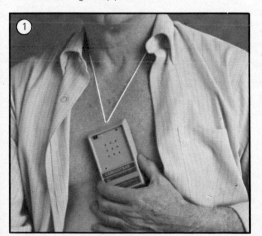

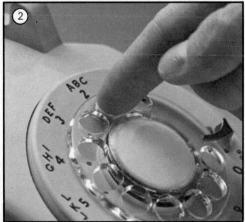

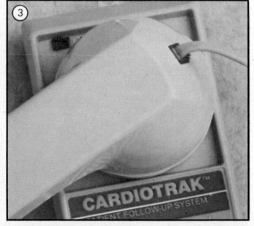

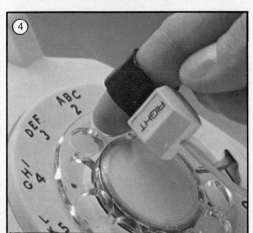

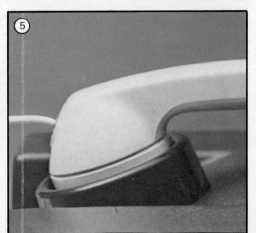

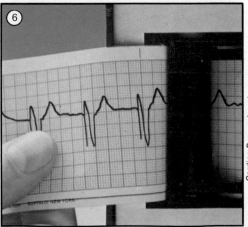

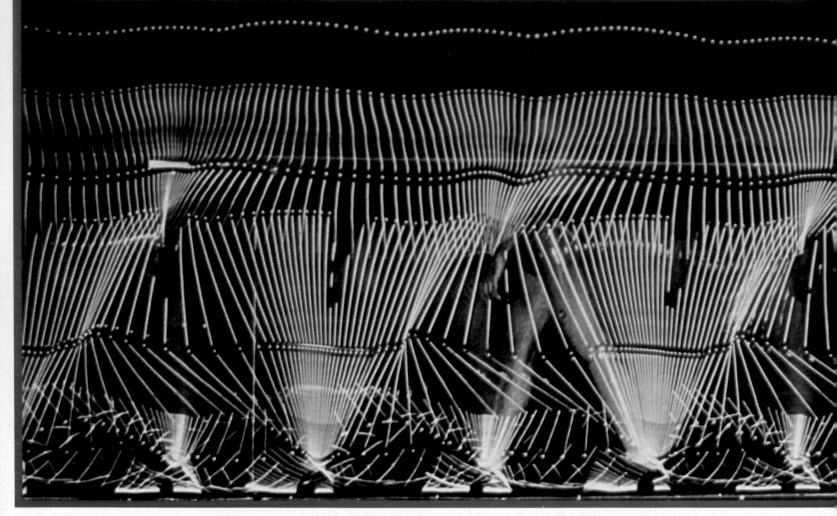

Replacing lost limbs

To lose a leg or arm, in times past, meant condemnation to a limping, limited future. Today's amputees can expect to don a device that looks and works like the lost limb. Skilfully engineered, the new leg will stride at normal walking speed on level ground—without undue mental or physical effort on its owner's part, and with little or no limp. An artificial arm will permit its owner to compete in most occupations—including driving. And no able-bodied person has the amputee's luxury of a spare limb, provided to permit adjustments and servicing without inconvenience to the client.

Design of *prostheses,* as the replacement limbs are termed, rests on sound appreciation of the natural structure's remarkable talents. Bone, for example, has an ultimate tensile strength of nine tonnes per square inch. It can withstand a compression of up to 11 tonnes per square inch and stretching up to a 1,300-tonne limit. Compared with synthetic

materials, the performance of bone can only be matched by tough metal alloys, structural woods and plastic laminates.

Less of a challenge is the matching of human joints with a mechanical equivalent. In machine terms, a body joint is no more than a bearing. At most, the likely loads on it will not exceed a few hundred pounds—and the maximum speeds of a body joint in action present no problem to the engineer (though the best artificial joint cannot match the natural version for low-friction performance).

Replacing the role of muscles is mechanically straightforward too. Among their functions, skeletal muscles act as *damping devices* when lengthened under load and as *actuators* when shortening under load. Operating at forces of some 50 lbs and speeds of some five feet per second, they are regularly replaced in an artificial limb by hydraulic and pneumatic devices.

More complex is the replacement of the body's internal signalling system: the motor

signal that the nerves transmit to the muscles from the central nervous system, feeding back information on limb position, velocity and pressure sensations. But these signals, transmitted at the rate of a few hundred feet per second, can easily be replaced by modern electronic switching control systems.

In theory, then, each part of the damaged body system can be matched by an equivalent artificial device. Yet obstacles still lie in the path of those seeking to provide the perfect prosthesis. The hindrance comes at the point where man and machine meet. The complex of nerves that were sliced through at amputation could theoretically be replaced by electronic controls capable of manipulating a limb in the huge variety of ways enjoyed by the natural body. But no system has been

Top Gait analysis aids design of comfortable artificial limbs. Non-handicapped volunteer walks between accurately spaced marker dots, while sequence of photographs is taken. Then dots on pictures are joined to make diagram.

devised to enable the patient to operate such an array of controls. And no suitable power supply is yet available to energize it.

As a result, prosthetic limbs remain limited in their range: the knee control must rely on gravity or another applied force to operate. But, within the limitations of their craft, designers can go a long way to provide a limb that will function reliably, be comfortable to wear, and look as much like the real thing as possible.

The task is likely to remain a medical challenge for the foreseeable future. At any one time, in the Western world, one in 10,000 citizens is an amputee. Each year, some 65,000 new patients enter the grim statistics. In the West as a whole, some 770,000 men, women and children must live with the disability. A conservative estimate by the World Health Organization indicates that each year over 40 million new prostheses are needed worldwide.

The vast majority of new limbs are needed in Third World countries, where malnutrition compounds the problems caused by the ravages of war and disease, and rehabilitation

Right Gait analysis has given limb designers a precise picture of how people walk.
Below This machine tests the endurance of artificial feet and ankle joints. It can simulate up to five years of walking.

Mike Courtney

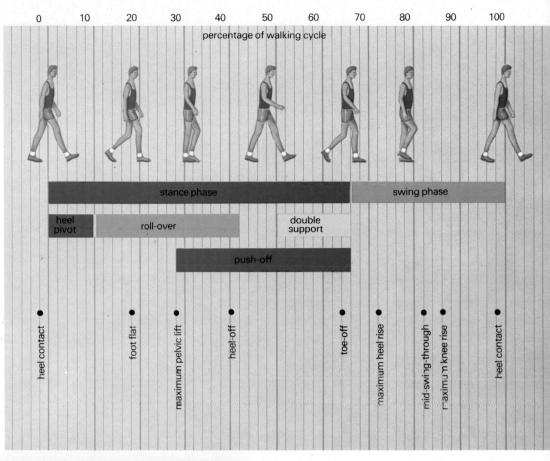

THE WALKING CYCLE

percentage of walking cycle

stance phase / swing phase

heel pivot / roll-over / double support

push-off

heel contact · foot flat · maximum pelvic lift · heel-off · toe-off · maximum heel rise · mid-swing-through · maximum knee rise · heel contact

J. E. Hanger & Co. Ltd.

of amputees becomes a major task there.

In every cause of damage resulting in amputation—except for accidents in the home—males outnumber females by five to three. As a group, men are more likely to suffer the wounds of war, industrial accidents and road injuries. In many countries, too, breakdowns in the body's blood vessel system affect men at twice the rate for women—and such breakdowns may result in unavoidable amputation. Relatively few arms are amputated today—less than one-twentieth the number of legs removed.

Artificial legs

Design of artificial legs depends on a careful examination of the action of the healthy limb, using a variety of *gait analysis* techniques, which give the prosthetic engineer valuable information on the different phases of the walking cycle and the various stresses and strains with which the artificial limb must cope. This involves such factors as arm swing, trunk movements and changes in the position of the head, as well as the movements of the legs themselves.

Whenever an amputation is carried out,

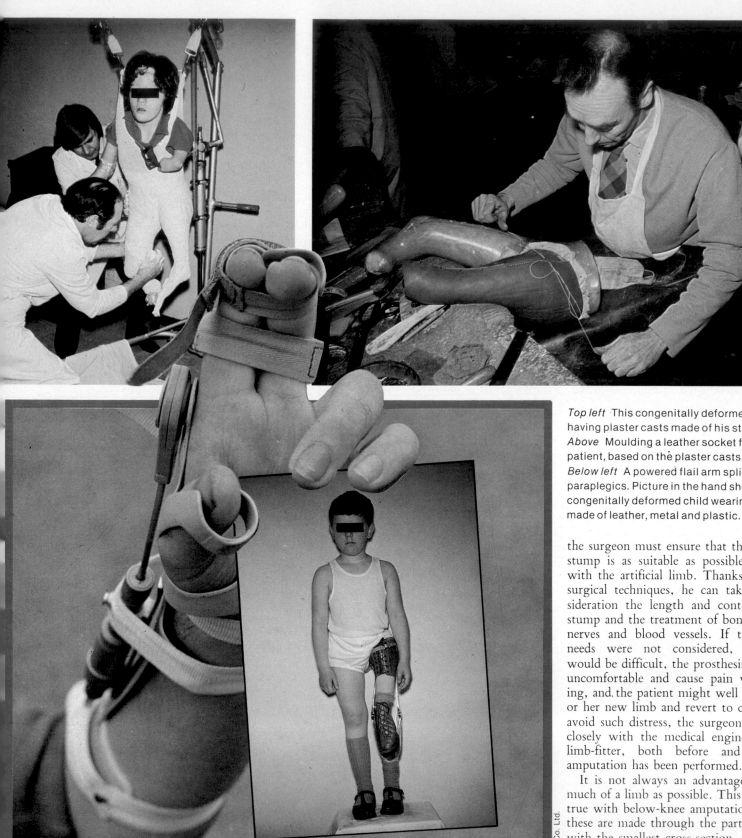

Top left This congenitally deformed child is having plaster casts made of his stumps.
Above Moulding a leather socket for the same patient, based on the plaster casts.
Below left A powered flail arm splint for paraplegics. Picture in the hand shows a congenitally deformed child wearing a limb made of leather, metal and plastic.

the surgeon must ensure that the remaining stump is as suitable as possible for fitting with the artificial limb. Thanks to modern surgical techniques, he can take into consideration the length and contours of the stump and the treatment of bones, muscles, nerves and blood vessels. If these crucial needs were not considered, limb-fitting would be difficult, the prosthesis would feel uncomfortable and cause pain when walking, and the patient might well abandon his or her new limb and revert to crutches. To avoid such distress, the surgeon co-operates closely with the medical engineer and the limb-fitter, both before and after the amputation has been performed.

It is not always an advantage to save as much of a limb as possible. This is especially true with below-knee amputations. Because these are made through the part of the shin with the smallest cross-section, problems of blood supply may arise, causing the stump to waste away and eventually necessitating a second amputation at a higher level. On the other hand, the stump should not be too

J. E. Hanger & Co. Ltd.

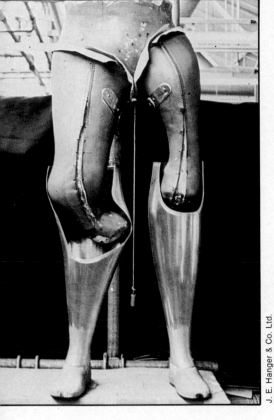
J. E. Hanger & Co. Ltd.

short, or the imbalance of the remaining muscles may pull it to one side.

If possible, surgeons try to *disarticulate* a joint by separating the bones in the amputated part of the limb from those on the healthy side of the joint, rather than cutting through the bone. Disarticulation is less of a shock to the patient, and the wound heals more quickly. Disarticulation usually leaves a healthy stump which can bear the necessary load when the artificial limb is fitted. It also involves little impairment of hip function.

During the period immediately after surgery, while the wound is healing and the stitches are being removed, bandaging of the stump is of crucial importance. At this stage, the stump is still swollen and filled with fluid. So the patient must use a temporary prosthesis, or *pylon,* for which careful measurements must be taken, until the stump has shrunk to a stable shape and size. Then the prosthetist measures the patient for the permanent prosthesis, and takes casts of the stump. He also arranges an appointment for the patient to try out the new limb.

Stump and socket

At the trial, the prosthetist tests the limb for correct *fit* and *alignment.* 'Fit' refers to the accuracy with which the stump fits into the socket of the limb. 'Alignment' indicates the geometrical relationship between the various segments of the limb. If the trial is satisfactory, a date is set for delivery; if not, a

Above left Third stage in making limbs for the child at the top of the opposite page. Now the basic socket shape is established, the prosthetist can add the lower limbs.
Above The finished limbs, ready for fitting.
Below Born with no left hand and only part of her forearm, this six-year-old child has had her life transformed by a myoelectric hand.

re-trial is necessary. When fitting is completed, the patient must return the limb for the finishing touches to be made.

To make a comfortable socket, the prosthetist must identify the positions of all the major arteries and veins involved, and have a good understanding of the general blood and lymphatic systems, too. He must also be familiar with the muscles and nerves of the stump, and find the exact location of the bony areas, which may be used for load-bearing and for stabilizing the socket. Finally, he must understand the structure of the skin and the workings of the sweat gland system to make a comfortable socket.

The socket has several important functions. For the lower-limb amputee, it must be able to bear the total weight. Even more crucial is the need for the socket to cope with the very great forces developed during walking—these may reach two or three times the body weight. The forces involved at the knee joint are much greater than those in the drive of a powerful car.

The socket must maintain a close fit

around the stump at all times. This feature is called *suspension.* The prosthetist may ensure suspension by building into the socket either a suction device or an adhesive section. Alternatively, the suspension may be external to the socket—pelvic suspension for an artificial leg or shoulder suspension for an artificial arm. It is often necessary to use two or more methods of suspension.

The action of the stump during walking is to extend and flex the knee via movements of the hips. This produces bending movements at the point where stump and socket meet. The stump then acts as a lever to position the shin and foot. Any wasted energy in this front-rear plane results in an awkward gait and also produce painful pressure points. Bending stresses and torque (rotational forces) are also involved in the stump-limb mechanism.

The socket must give the patient feedback as to the exact condition, position and functioning of the prosthesis. This is achieved by trial-and-error learning during training. The patient learns to become aware of the very small changes in pressure, as well as the slight displacements and vibrations that are transmitted to the socket by the prosthesis.

Artificial arms and hands

There are two main types of artificial arms. There are strong, purely functional arms into the end of which the amputee can fit tools such as screwdrivers or hammers;

Syndication International

1001

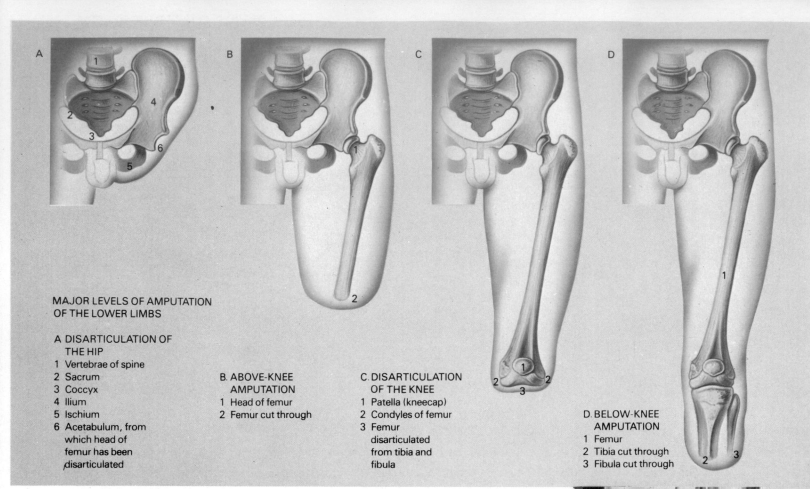

MAJOR LEVELS OF AMPUTATION
OF THE LOWER LIMBS

A DISARTICULATION OF
THE HIP
1 Vertebrae of spine
2 Sacrum
3 Coccyx
4 Ilium
5 Ischium
6 Acetabulum, from
which head of
femur has been
disarticulated

B. ABOVE-KNEE
AMPUTATION
1 Head of femur
2 Femur cut through

C. DISARTICULATION
OF THE KNEE
1 Patella (kneecap)
2 Condyles of femur
3 Femur
disarticulated
from tibia and
fibula

D. BELOW-KNEE
AMPUTATION
1 Femur
2 Tibia cut through
3 Fibula cut through

these are also supplied with a wide variety of hooks for general manipulation. The second type of arm, used on social occasions, is the cosmetic arm, fitted with a surprisingly realistic as well as functional hand.

Many artificial arms are powered by compressed carbon dioxide gas stored in cylinders. Small valves control the flow of gas which opens and closes the hand or moves the elbow joint. Other arms are *myoelectric.* They work by means of sensors placed over the remaining muscles of the stump. These pick up the minute electrical signals which are transmitted from the central nervous system and cause the muscles to contract. The signals are integrated over preset time scales (usually a few milliseconds only) and are used to open and close the hand. By moving certain muscles in the stump, the amputee can send signals to the hand's controls and produce a graded grip from the lightest pinch to a power clasp of several pounds of force. Another great advantage of myoelectric hands, particularly for below-elbow amputees, is their lack of any suspensory harness.

Regarding artificial legs, all the major movements—at the hip, the knee and the ankle, together with those governed by the toe joints—can be adequately replaced, but the search for improved mechanisms and lighter and stronger structures continues.

Above Most common types of leg amputation. Disarticulation separates bones at joints, rather than cutting through them.
Below right A machine compresses artificial feet in order to test their strength.

Replacing joints

A typical hip joint replacement for a high-level amputee incorporates a single-hinge joint, which gives freedom of flexion and extension of the hip. Over-extension is prevented by an adjustable stop. Flexion is controlled by a *limiter mechanism* which the wearer can over-ride by using a *delimiter mechanism,* which allows him to sit naturally and comfortably. The limiter mechanism is adjustable, allowing the wearer to vary the length of his stride with ease.

A typical knee joint replacement may be a simple-hinge joint or a comparatively sophisticated *polycentric* device which imitates more closely the complex movements of the natural knee joint. This device may incorporate a hand-operated knee lock or a semi-automatic one for older wearers who feel they need the added confidence of walking with a locked knee, together with the simple facility of flexing the leg for sitting or kneeling. The semi-automatic lock automatically resets itself when the patient stands up and straightens his or her leg.

More sophisticated devices are prescribed

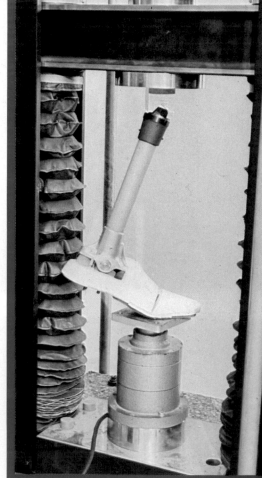

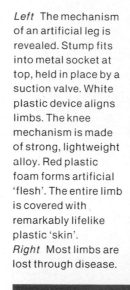

Left The mechanism of an artificial leg is revealed. Stump fits into metal socket at top, held in place by a suction valve. White plastic device aligns limbs. The knee mechanism is made of strong, lightweight alloy. Red plastic foam forms artificial 'flesh'. The entire limb is covered with remarkably lifelike plastic 'skin'.
Right Most limbs are lost through disease.

CAUSE		MALE	FEMALE	TOTAL
UK AMPUTATION STATISTICS 1976				
TRAFFIC ACCIDENTS	Pedestrians	32	20	52
	Vehicle users	156	17	173
	Rail	16	4	20
	Other	19	6	25
		223	47	270
OTHER ACCIDENTS	Industrial	144	11	155
	Home	22	22	44
	Recreation	16	6	22
	Armed Forces	34	2	36
		216	41	257
DISEASE	Blood vessel system	2,268	915	3,183
	Metabolic	553	351	904
	Infection	52	25	77
	Cancer	133	123	256
	Nervous system	8	7	15
		3,014	1,421	4,435
	Congenital deficiencies	112	72	184
	Overall Total	3,565	1,581	5,146

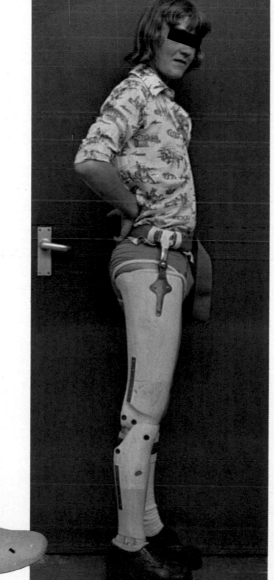

for the younger, more active amputee. These include *stance phase* controls which produce a high-friction movement above the knee when load is applied to the leg, thus preventing the knee from inadvertently collapsing but retaining all the phases of walking. *Swing phase* knee control mechanisms often incorporate hydraulic or pneumatic damping devices as well as simple friction devices. These ensure that the artificial knee moves in a remarkably natural way. They are designed to take energy from the swing of the leg at the appropriate moment, thus enabling the patient to walk at various speeds without the need for re-adjustment. An amputee wearing one of these limbs may be slowly walking along the road at one moment, and the next running to catch a bus.

Ankle joints are adjustable, being controlled by rubber springs around a central pivot. Also, the forces that produce inversion (pointing inwards) and eversion (pointing outwards) of the foot are controlled in a similar fashion. Torque absorbers or *rotators* are also used in ankle joints. These give great benefit to the active as well as the geriatric amputee and are particularly useful for golf-playing patients.

With the advances made by medical science, more and more disabled people survive. A recent US study revealed that loss or disablement of the legs accounted for over half of the totally disabled. Prosthetic engineers are constantly working to improve life for those unfortunate enough to form part of these statistics.

Left An above-knee amputee tests his new leg. The four-bar knee-linkage will replace the whole range of movement which he lost when his leg was amputated after an accident.

J. E. Hanger & Co. Ltd.

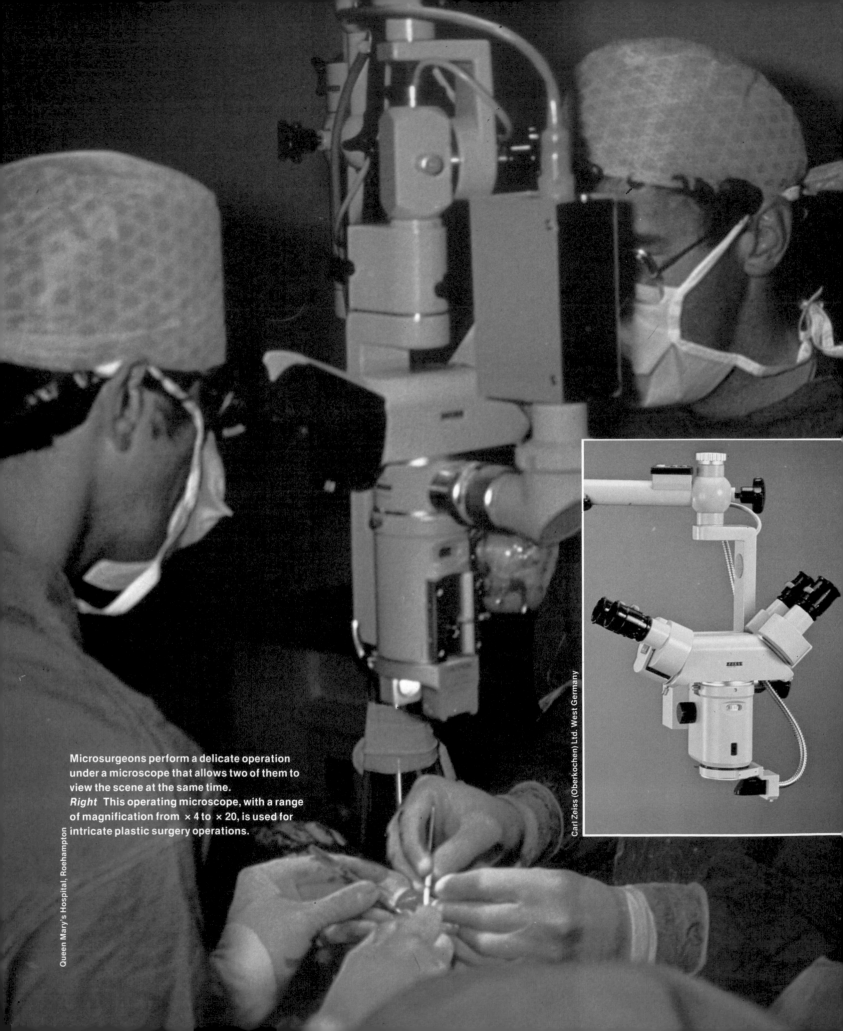

Microsurgeons perform a delicate operation
under a microscope that allows two of them to
view the scene at the same time.
Right This operating microscope, with a range
of magnification from × 4 to × 20, is used for
intricate plastic surgery operations.

Queen Mary's Hospital, Roehampton

Carl Zeiss (Oberkochen) Ltd. West Germany

The small miracle of microsurgery

An emergency leading to microsurgical replacement of a severed limb begins in a high-speed drama; and ends in one of the lengthiest and most exacting operations of medical science. From the scene of the industrial or motorway accident that sliced off a hand or foot, the patient must travel fast—his lost extremity preferably packed in ice. He may then face up to 19 hours on the operating table as, bent over their microscope, surgeons work to repair the damage. Their task will include stitching blood vessels only half a millimetre across.

New microscopes—with up to three binocular eyepieces, enabling more than one surgeon to work at once—have enhanced chances of success. So have finer stitching materials and improved anaesthetic techniques. But microsurgery is not as recent a

miracle as many suppose. Severed hands were replaced in the USA and China in 1964. And microsurgery has been almost routine in Australia since the early 1970s.

Use of a microscope as an aid to surgery dates back further still. In the 1920s, the first operations were performed with its aid in the cramped confines of the middle ear. Eye surgeons soon followed their colleagues' example and begin to repair and reconstruct parts of the front chamber of the eye with the help of microscopes. But to the layman, the precision and skills involved in such operations remain incredible.

The best candidate for *replantation microsurgery*, as doctors call the process, is the victim of a clean cut. Chances of replacing a hand shorn off by an industrial guillotine are as high as 90 per cent. Road accident victims—whose veins and tendons may have

been wrenched rather than cut apart—have diminished chances. Though a clean cut can be established by preliminary surgery in some cases, the repaired limb will inevitably be shorter than it was before.

Younger patients have better chances of recovery than the elderly. Among other factors, they have more stamina to face lengthy anaesthesia. Stamina is the key factor in the surgical team's approach to the decision of whether to operate or accept the amputation. And those long hours at the operating table are expensive.

Is it worth it?

Costs of a six-strong surgical team, backed by nurses, for a day-long session are high. Aftercare in hospital, followed by months of physiotherapy, add to the bill. In the final analysis, the patient may have a limb that functions only imperfectly—and has a high sensitivity to pain. Thus, confronted with a candidate for microsurgery, the initial decision must be whether an artificial limb would serve the patient (and the budget) better than a reconstructed one.

With the decision to rejoin the divided body taken, two teams go into action. The severed hand is a 'patient' in itself—and, at room temperature, has only a short span of independent life. Estimates vary between 20 minutes and five and a half hours before limb death occurs. But cooling to near zero centigrade can prolong life to some 30 hours. Packing in an ice bag—but not refrigeration—is the best method of preservation. Nevertheless, every moment that the hand is detached reduces its chances of survival.

Matching nerves

While the patient is being prepared for operation, his severed hand goes under careful analysis. Twenty-four tendons, for example, join the wrist to the hand. Ten are crucial: the extensors and flexors that stretch and clench the fingers. How many more the surgeons will prepare for rejoining depends on the time available. They must shave back the ends of the two bones involved (the ulnar and radius) to remove contaminated fragments. They must identify the arteries and veins for correct matching. Most complex task of all: they must match up the severed nerve ends correctly. To get 'wires crossed' in this process risks attaching a *motor* nerve (directing movement) to a *sensory* nerve (recording feeling)—an error that would

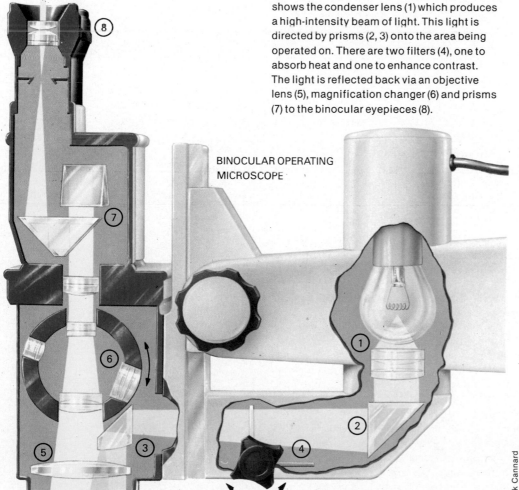

Below Cutaway of an operating microscope shows the condenser lens (1) which produces a high-intensity beam of light. This light is directed by prisms (2, 3) onto the area being operated on. There are two filters (4), one to absorb heat and one to enhance contrast. The light is reflected back via an objective lens (5), magnification changer (6) and prisms (7) to the binocular eyepieces (8).

BINOCULAR OPERATING MICROSCOPE

Frank Cannard

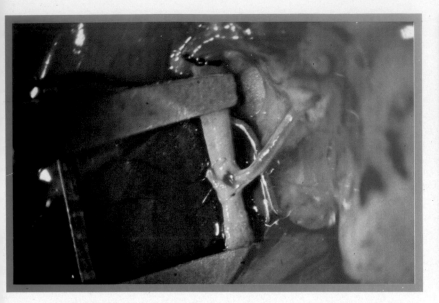

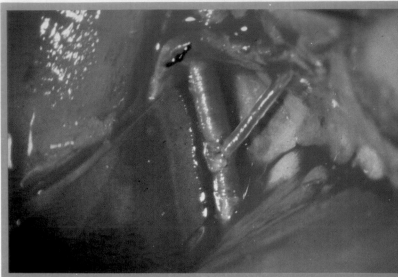

John Watney

Above In an arm operation, two arteries are being joined, using nylon thead far thinner than a human hair. This is visible to the left of the arteries. Microclips on the large artery interrupt blood flow until the vessels are joined *(above right)*. Then the clips are removed and the arteries turn red with blood.

result in a complete loss of function in both cases. Correct matching will depend on a painstaking comparison of cross sections of nerve bundles, in hand and wrist, before stitching can commence.

The first stage in bringing patient and hand together again is the rejoining of the bones. Plates or pins secure the bones, to provide a firm base for the operation—there is little point in trying to carry out intricate stitching of tiny blood vessels and nerves on a limb that is moving about because the bones are not yet fused. Unfortunately, however, this essential task does bring drawbacks to the patient. One of the results of joining the bones is that the hinge movement of the wrist is impaired, limiting the repaired hand's range of possible movements in the future.

Tendons and veins

Having secured the bones, the surgeons turn their attention to the extensor tendons and the veins in the back of the hand. They repair any damage to these vital parts and then re-unite them with the corresponding parts in the wrist and sew up their incisions.

Next, the surgeons turn over the hand so that they can rejoin the major blood vessels and also the flexor tendons. Once the blood vessels have been reconnected, the tourniquet can be removed and the blood flow restored.

Having rejoined both blood vessels and tendons, the surgeons must attempt to restore sensation in all parts of the hand by rejoining the intricate network of nerves—using the tiniest stitches of all.

When all incisions have been closed up, the arm must be surrounded by a plaster cast and immobilized in a sling to give it a chance of healing undisturbed. The fingers are left protruding from the cast so that the doctors can see that the blood is flowing properly to and from the hand. After a few days, the nurse removes the plaster cast to see that no blood is leaking into the surrounding tissue and collecting there.

The patient must then wear the plaster cast again for about three weeks until the doctors are satisfied that the repaired arm is ready for gentle exercise. Several months of intensive physiotherapy help the patient to regain the use of his hand, but it may take as long as 18 months for full movement and sensation to return.

The principles of replacing arms, legs or feet are similar to those involved in rejoining severed hands, although individual surgeons have their own special techniques. It is easier to join blood vessels in the upper parts of the limbs, but the larger the section of limb lost, the less the chance there is that proper function and sensation can be restored. The supply of nerves is more complicated nearer the

Right At a hospital in Shanghai, China, a surgeon examines a patient's new hand that was created by microsurgery. After he lost both hands, the man's right hand was rebuilt by transplanting his two second toes onto the stump. He needed six months of physiotherapy to ensure that his new hand functioned well.

body trunk and the task of trying to repair this intricate network is extremely difficult.

One country whose surgeons boast notable successes with replantation microsurgery is China. Safety regulations in factories are less rigorous there than in many Western countries, and a considerable number of limbs are severed each year in industrial accidents. Chinese surgeons have, of necessity, become highly skilled in restoring lost limbs. At some hospitals, microsurgery teams claim 90 per cent success rates. One or two Chinese surgeons even go so far to remove an arm deliberately to clear surroun-

Xinhua News Agency

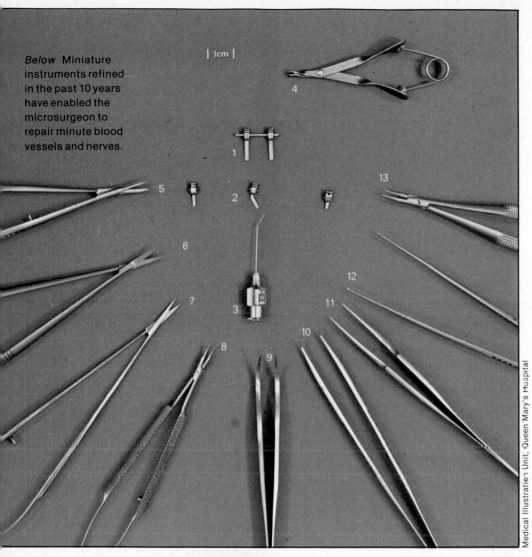

Below Miniature instruments refined in the past 10 years have enabled the microsurgeon to repair minute blood vessels and nerves.

| 1cm |

Medical Illustration Unit, Queen Mary's Hospital

1 Double clamp
2 Three single clamps, one of them angled
3 Micro-irrigator (fits on a syringe, irrigates tissue with saline solution)
4 Clamp applicator
5, 6 Curved scissors
7 Straight scissors
8 Clamp
9, 10 Tweezers
11 Jeweller's tweezers
12 Toothed tweezers
13 Curved tweezers

a hand created from pinned-together bones.

If such non-living material is inserted into the body, however, there is always a chance that the living tissue will 'reject' it completely or that infection will set in.

Replacing severed limbs or digits is not the only use to which orthopaedic and plastic surgeons put microsurgery. Another problem being overcome by this technique is the repair of limb fractures which are so severe that whole pieces of bone—as much as 23 cm (9 in) long—have been lost. Road accidents are the most common causes of this mutilation, in which a leg may be left hanging by the tendons. Previously, the only chance of saving the patient from the massive infection that often resulted was to amputate the remains of the limb.

Transplanting ribs

Now, surgeons can remove one of the patient's uppermost ribs together with its blood supply, muscles and associated skin, then re-insert it in the gap between the two broken pieces of leg-bone.

Finally, microsurgeons attach pieces of *cancellous* bone to either end of the transferred rib-bone. Tough but spongy in texture, this type of bone is found at the ends of the body's 'long' bones, such as the one that has been broken. Because it has the advantage of being capable of continued growth, the cancellous bone transplants encourage the rib bone to fuse smoothly with the remains of the damaged bone. Often, it is necessary to insert needles right through the damaged leg, to ensure that the bones remain immobile and firmly attached to each other while they are knitting together.

Within a matter of weeks, the patient can start to use the repaired leg, as long as he or she does not put too much weight on it. Before this operation became possible the patient would have lain immobile for months, and still suffered the risk of amputation if infection set in.

The 'replugging' of blood vessels has also been carried out to help fight a wide range of diseases, including strokes. Often, after a stroke, the blood supply to part of the brain

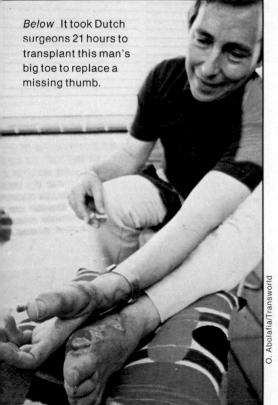

Below It took Dutch surgeons 21 hours to transplant this man's big toe to replace a missing thumb.

O. Abolafia/Transworld

ding tissues of cancerous growths and then replace the healthy limb quite successfully.

Another operation performed by Chinese microsurgeons has brought new hope to unfortunate patients who have had all their fingers severed. Rather than wearing a metal hook or an artificial hand, some of them prefer to lose one toe from each foot and have the two toes attached to the fingerless hand. Although they have the appearance of a lobster's claw, these hands may work surprisingly well. Some patients are able to pick up peanuts using chopsticks—a feat many Westerners are unable to perform with all their fingers and both thumbs intact.

Operations are sometimes performed in which two toes are removed from each foot. Although this makes walking difficult, it is necessary if the patient has lost all the fingers on both hands. Some surgeons choose to remove the bones from the toes before they stitch them onto their new home. This enables them to insert a metal device shaped rather like a fork to give the new hand greater strength than could be obtained from

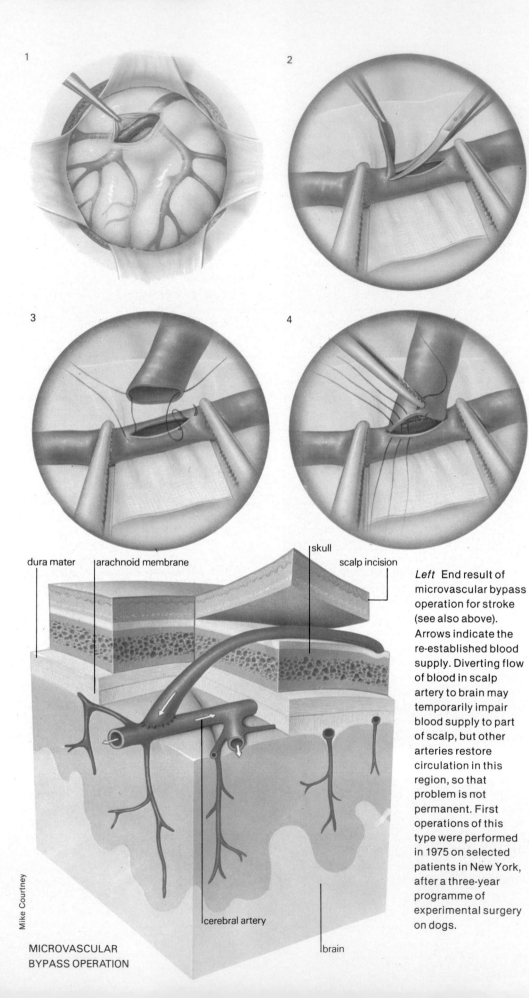

1

2

3

4

5

Left In an operation on the brain of a stroke victim, the blocked cerebral supply artery is by-passed by a scalp artery, which assumes its task. Two incisions are made in the scalp, to reveal the scalp artery and the site at which a small circular portion of skull will be removed, to reveal the supply artery. When this has been achieved (1), the surgeon, working with a microscope, cuts an opening in the supply artery (2) and temporarily anchors the scalp artery to it with two sutures (3). More sutures are inserted to hold the vessels securely (4), and are then tied (5). By this method blood supply is restored by a vessel that passes through the opening made in the skull (6). Because of the technique, this is called a bypass operation.

dura mater | arachnoid membrane | skull | scalp incision

Mike Courtney

cerebral artery | brain

MICROVASCULAR BYPASS OPERATION

Left End result of microvascular bypass operation for stroke (see also above). Arrows indicate the re-established blood supply. Diverting flow of blood in scalp artery to brain may temporarily impair blood supply to part of scalp, but other arteries restore circulation in this region, so that problem is not permanent. First operations of this type were performed in 1975 on selected patients in New York, after a three-year programme of experimental surgery on dogs.

is so severely restricted that the brain cells die. Microsurgeons can repair even the smallest blood vessels, which form the last stage in the transport of the blood through the body to the stricken tissue.

They carry out intricate operations to re-connect these minute vessels to larger arteries which still have a clear passage through to the part of the brain that has become cut off.

Such *bypass* operations can not only prevent a stroke in patients who might otherwise be likely to suffer one; they can also help those who have already had a stroke by preventing a further attack. For some stroke victims, the operation may even lead to a partial recovery of lost brain function.

Microsurgery faces a formidable challenge in attempting to deal with diseases in which nerves have degenerated or become completely functionless. All too often, the damaged nerves resist attempts at repair, but a few surgeons have achieved successes in routing these, the tiniest vessels of all.

Facial palsy, or Bell's palsy, is a distressing condition in which the nerves on one side of

Xinhua News Agency

must rely on microsurgery to repair their inactive or diseased reproductive systems.

In many countries, one in ten women who are infertile suffer from some blockage or disease of the Fallopian tubes. To remove the blockage requires more than pushing a rod up the tube to renew the cavity through which the egg passes. Instead, a dye is injected into the womb and when the surgeons can see that its flow up the tube has been blocked, they begin to cut out the piece of blocked tube and sew back together the remaining ends. (Similarly, if the surgeon finds that there is a blockage or damage to the tube at the ovary end, he removes the damaged tissues.)

Inevitably, this procedure shortens the tube, which in itself reduces the chances of the woman becoming pregnant. After conception, the growing embryo takes four days to travel down the Fallopian tubes—sufficient time for both it and the womb to prepare themselves for *implantation* (the attachment of the embryo to the womb). If the tube has been shortened, the travelling time is reduced, and the womb will not be

Left The Chinese claim great successes with microsurgery. After suffering an industrial injury that severed his hand, this Algerian was treated by a visiting team of Chinese microsurgeons. They re-connected the hand so succesfully that he is now able to manipulate objects quite easily.

ready to accomodate the embryo, which passes straight through the cervix (the neck of the womb) and out through the vagina.

In addition, suspicions that the Fallopian tubes are more than merely a transport system are being confirmed. Gynaecologists now believe that they provide nutrients and other substances that are essential for the embryo's development. Severely damaged tubes may not provide these vital ingredients, even if they have been unblocked.

Despite these problems, the application of the microsurgeon's skills has meant that the pregnancy rate following tubal surgery has been increased from 12 per cent to almost 50 per cent in cases where the tube was unblocked near the womb, and from only five to about 20 per cent where the surgery was carried out at the opposite end of the tube. The same technique—removal of the blockage and repair of the ends—is also used to reverse male and female sterilization.

The next frontier in microsurgery, plainly, is transplantation from a dead donor to a living patient. The surgical technology exists. But our knowledge of immunology is not yet sufficient to prevent the patient's body from rejecting the transplant.

Nevertheless, a challenge such as this is hard to resist, and with continued co-operation between microsurgeons and immunologists, it is possible that limb transplants may become a reality in the not-too-distant future.

the face have stopped working. The sufferer can no longer shut his or her eye on the affected side, and is unable to smile or draw the face into any other expression, because the muscles cannot work without 'instructions' from the nerves.

Some of the nerves involved are capable of regeneration, so the surgeons attempt to re-route a nerve on the unaffected side of the face over to the paralysed side, usually around the mouth or beneath the nose. They then hope that it will grow far enough and provide the missing stimulus to the paralysed muscles for them to function again.

The technique is still far from successful, however. Frequently, the surgeons successfully re-route the nerve only to find that the muscle, deprived too long of stimuli, has withered and died, so that it cannot be triggered into life by the new nerve.

More successful is the microsurgery widely used to help infertile women. The 'test tube' baby technique offers much hope for the future, but until its success rate can be increased, the majority of infertile women

Jeremy Gower

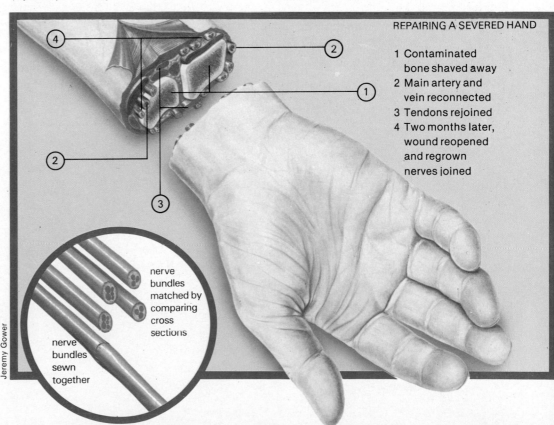

REPAIRING A SEVERED HAND

1 Contaminated bone shaved away
2 Main artery and vein reconnected
3 Tendons rejoined
4 Two months later, wound reopened and regrown nerves joined

nerve bundles matched by comparing cross sections

nerve bundles sewn together

The surgeon's eye

Paul Brierley

The enormous complexity of the human eye has presented an immense challenge to the ingenuity of those working in the field of medical technology. Both the investigation and the treatment of eye diseases demand the use of instruments of astonishing sensitivity and sophistication. And, even after surgery, a patient will probably still need assistance from a Man-made lens. For some people, the conventional spectacle lens or contact lens is impracticable. For them, recent advances in the design of intra-ocular lenses—artificial lenses that are permanently implanted in the eyes—are of immense importance.

Many visual problems, especially in elderly people, are caused by disturbances of the blood vessels at the back of the eye. But gaining access to these blood vessels to investigate the nature of the disorder has proved extremely difficult. The technique of *fluorescein angiography* now enables the specialist to examine the eyes with relative ease. It takes advantage of the physical property of fluorescein in absorbing light of one wavelength and emitting it at another wavelength. Blue light is shone onto blood vessels carrying fluorescein and, by monitoring the resulting emission of green light, the path of fluorescein through the blood vessels can be followed.

The patient's pupils are first dilated with drops and he is seated in front of a special camera while 5 cc of 20 per cent sodium fluorescein are injected as a single mass, or *bolus,* through one of the larger veins of the arm. The circulation time from the arm to the brain is about ten seconds. After this time, the fluorescein dye may be seen coursing through the retinal arteries, capillaries and veins, and a series of photographs is taken at short intervals. From these, the eye specialist can build up a complete picture of the condition of the vascular system of the back of the eye.

The blood vessels of the healthy eye are impermeable to fluorescein, so the dye remains within the vessel walls. However, disease may result in leakage of the vessel's contents and subsequent damage to the retina. This is readily apparent in the photographs of the retina, as the dye leaks out together with the contents of the vessels.

Alternatively, disease may shut off the passageway through the retinal vessels. This, too, can be detected quickly since the dye cannot penetrate the blocked vessels.

Yet another situation often occurs as a complication of diabetes. New vessels form, only to break down and bleed, resulting in loss of vision. The bleeding vessels may also form scar tissue when healing, and the retina may become detached—both conditions seriously affecting vision.

Laser treatment

Having identified the site of the problem affecting the retina, the eye surgeon is faced with the problem of treating it. Obviously, direct access to these very delicate tissues is difficult, but they may be treated with minimal interference by using a laser.

The word *laser* is an acronym derived from Light Amplification by the Stimulated Emission of Radiation, and the light beam which it produces is of far greater intensity than any normal light source. When focused on a very small area, a laser beam produces extremely high radiation densities.

The application of laser technology to the treatment of retinal disease is simple. The patient's pupil is first dilated and then the external surface of the eye is anaesthetized with drops. Next, the surgeon fits the eye with a small contact lens, equipped with three mirrors at varying angles of inclination. These enable the surgeon to focus on any particular part of the retina.

With the patient seated comfortably at a microscope attached to the laser machine, the surgeon can fire the laser at the diseased area of the retina causing a minute, highly localized retinal burn. As many as 500 such tiny burns may be used in a single treatment, and treatments may be repeated.

Naturally, this laser burn causes destruction of both normal and abnormal retinal tissue. This means that pathological lesions

Right An eye surgeon uses a laser to treat a patient with a diseased retina. The laser beam burns tiny holes in the diseased area.
Below Photographs of the retina taken with a special camera. Blood vessels are revealed as fluorescein dye is made to flow through them. Compared with a normal retina (A), that of a diabetic patient (B) has diseased vessels, with fluid leaking from them (whitish blobs) and tiny bubbles, or microaneurysms, in the vessel walls (left side). Another condition (C), is that of branch vein occlusion (BVO), in which new frond-shaped vessels are formed. Causes include high blood pressure and diabetes. Disease may also close vessels; this can be readily spotted as the dye does not penetrate them (D). All these conditions can be treated by laser, as with this retina (E) affected by BVO, in which the laser burns (yellowish spots) are clearly visible.

A, C, D, E Peter Curran B Institute of Ophthalmology

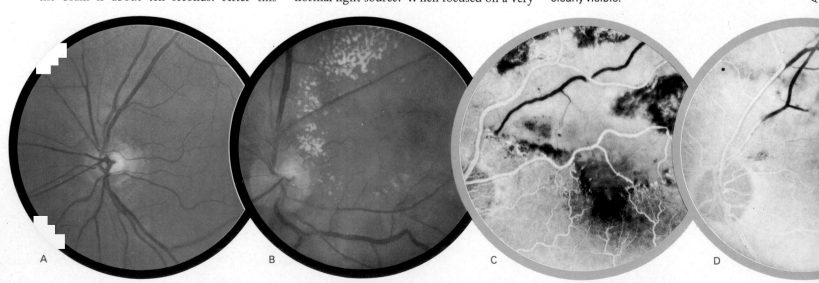

A   B   C   D

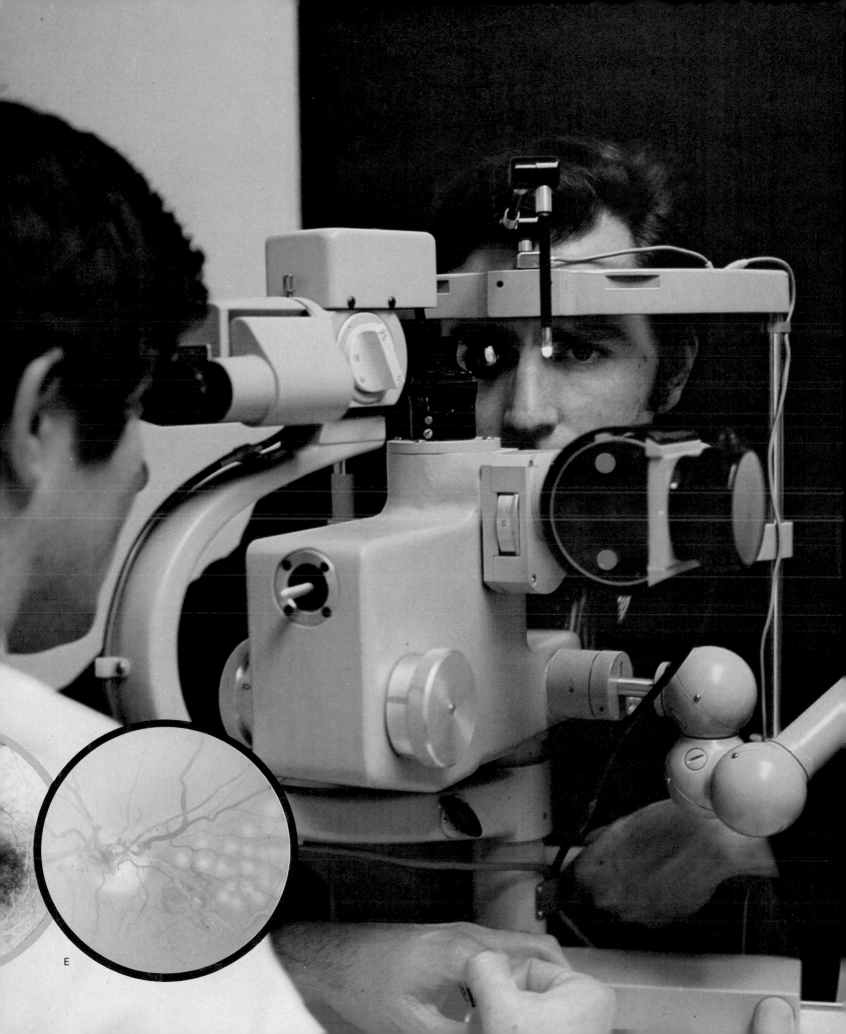

E

REMOVING A CATARACT

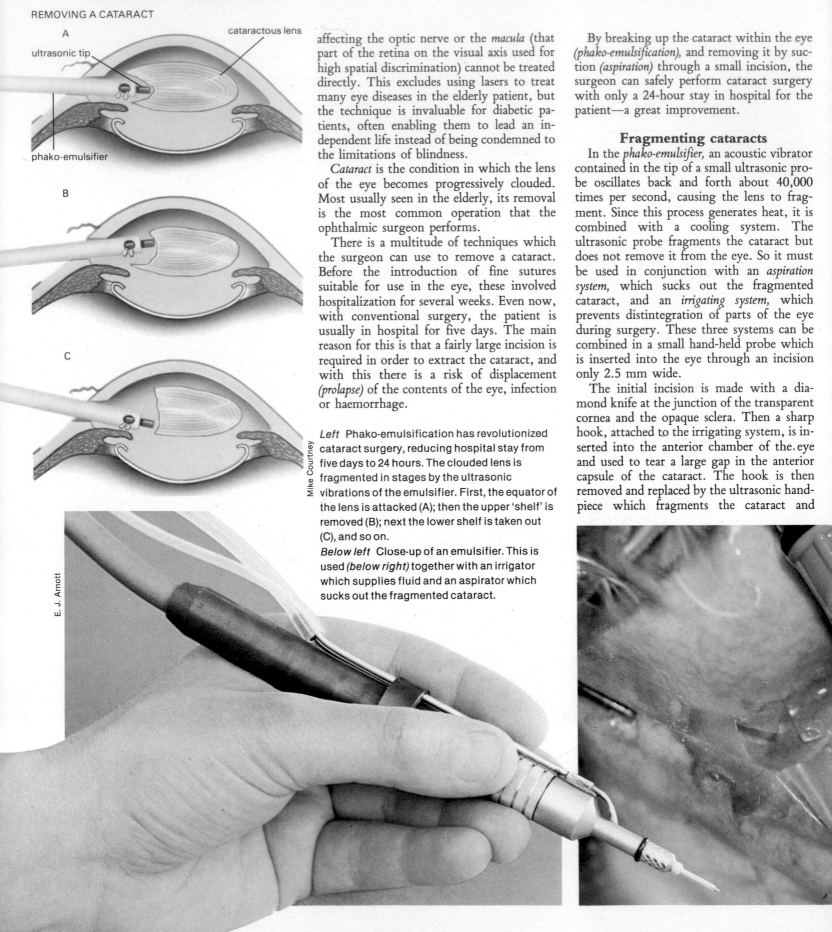

A

ultrasonic tip

cataractous lens

phako-emulsifier

B

C

Mike Courtney

E. J. Arnott

affecting the optic nerve or the *macula* (that part of the retina on the visual axis used for high spatial discrimination) cannot be treated directly. This excludes using lasers to treat many eye diseases in the elderly patient, but the technique is invaluable for diabetic patients, often enabling them to lead an independent life instead of being condemned to the limitations of blindness.

Cataract is the condition in which the lens of the eye becomes progressively clouded. Most usually seen in the elderly, its removal is the most common operation that the ophthalmic surgeon performs.

There is a multitude of techniques which the surgeon can use to remove a cataract. Before the introduction of fine sutures suitable for use in the eye, these involved hospitalization for several weeks. Even now, with conventional surgery, the patient is usually in hospital for five days. The main reason for this is that a fairly large incision is required in order to extract the cataract, and with this there is a risk of displacement *(prolapse)* of the contents of the eye, infection or haemorrhage.

Left Phako-emulsification has revolutionized cataract surgery, reducing hospital stay from five days to 24 hours. The clouded lens is fragmented in stages by the ultrasonic vibrations of the emulsifier. First, the equator of the lens is attacked (A); then the upper 'shelf' is removed (B); next the lower shelf is taken out (C), and so on.
Below left Close-up of an emulsifier. This is used *(below right)* together with an irrigator which supplies fluid and an aspirator which sucks out the fragmented cataract.

By breaking up the cataract within the eye *(phako-emulsification),* and removing it by suction *(aspiration)* through a small incision, the surgeon can safely perform cataract surgery with only a 24-hour stay in hospital for the patient—a great improvement.

Fragmenting cataracts

In the *phako-emulsifier,* an acoustic vibrator contained in the tip of a small ultrasonic probe oscillates back and forth about 40,000 times per second, causing the lens to fragment. Since this process generates heat, it is combined with a cooling system. The ultrasonic probe fragments the cataract but does not remove it from the eye. So it must be used in conjunction with an *aspiration system,* which sucks out the fragmented cataract, and an *irrigating system,* which prevents distintegration of parts of the eye during surgery. These three systems can be combined in a small hand-held probe which is inserted into the eye through an incision only 2.5 mm wide.

The initial incision is made with a diamond knife at the junction of the transparent cornea and the opaque sclera. Then a sharp hook, attached to the irrigating system, is inserted into the anterior chamber of the eye and used to tear a large gap in the anterior capsule of the cataract. The hook is then removed and replaced by the ultrasonic handpiece which fragments the cataract and

CURING CATARACTS

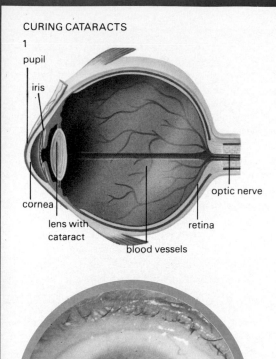

1
- pupil
- iris
- cornea
- lens with cataract
- blood vessels
- optic nerve
- retina

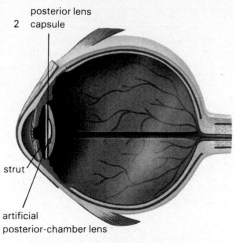

2 posterior lens capsule
- strut
- artificial posterior-chamber lens

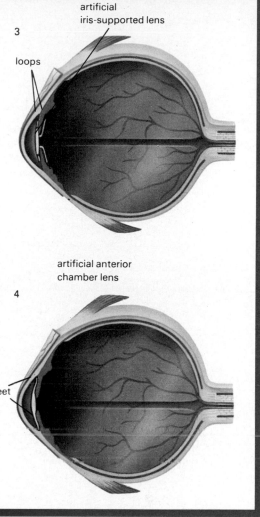

3 artificial iris-supported lens
- loops

4 artificial anterior chamber lens
- feet

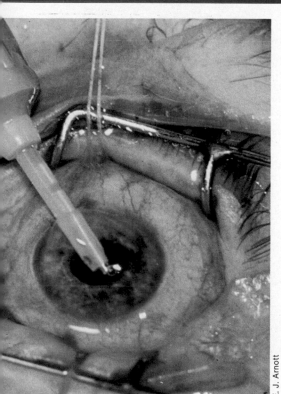

At first, only the centre of the lens is affected by a cataract (1) and drugs that dilate the pupil help the patient to see better. In a 'mature' cataract *(left)* the lens becomes completely opaque and the pupil looks greyish. If left untreated, blindness may result, but modern surgery and the development of artificial intra-ocular lenses are very successful at restoring vision. Extra-capsular extraction removes the clouded lens fibres, but leaves the lens capsule in place, and is commonly followed by fitting a posterior chamber lens (2). In the other type of extraction—intra-capsular—both lens and capsule are removed. This can be followed by the insertion of an iris-supported lens (3) or an anterior chamber implant (4).

Mike Ccurtney

aspirates the debris at the same time.

This technique removes the cataract but leaves its posterior capsule intact. It is known as *extra-capsular extraction,* as distinct from *intra-capsular extraction,* in which the posterior capsule is removed. The advantage of the former technique is that the remaining posterior capsule provides mechanical support for the vitreous gel of the eye, lessening the long-term risk of retinal detachment and cystoid macular oedema—two complications leading to serious visual disturbance.

Older techniques of extra-capsular extraction failed to remove all the cataractous material, which resulted in inflammation of the eye and an increased incidence of complications. By using the irrigating-aspirating system of the probe under microscopic control, all the cataractous material can be removed, leaving the eye uninflamed.

The posterior capsule is usually transparent, but often contains the ends of fine cells. These may develop into fibrous cells and lay down opaque scar tissue, thus undoing the good done by the operation. To remove these cells, a diamond-coated probe is

used to polish the capsule and thus ensure its transparency. In about 15 per cent of cases the posterior capsule becomes opaque after a period of time and has to have a small hole cut in it, but this is a minor surgical procedure. The wound is then stitched with microscopic suture material, finer than a hair, which causes no irritation and can safely be left in the eye.

Restoring vision

Removal of the cataract does not, on its own, restore vision, since the lens was in the patient's younger days the main focusing mechanism of the eye. To bring back sight after the operation an optical device must be used. There are three main types: cataract glasses, contact lenses and intra-ocular lenses (plastic lenses inserted into the eye).

The disadvantage of glasses is that they cause 25 per cent magnification of objects and a constricted field of vision. Although contact lenses give a much better quality of vision, learning to wear them for the first time at the age of 80 is a daunting task. A solution to the problem is to insert a replace-

E. J. Arnott

1013

Right The most popular type of anterior chamber intra-ocular lens is that designed by the British surgeon Peter Choyce. The four feet, two on each side, fit snugly in the angle between cornea and iris, anchoring the implant firmly in position.

Far right The Little Arnott posterior chamber lens. Little is a US surgeon, Arnott a British one. The two loops become fixed by scar tissue into the lens capsule that remains after an extra-capsular extraction.

Peter Curran

Rayner Intraocular Lenses Ltd.

ANTERIOR CHAMBER IMPLANTS

Rayner Choyce
Mk VIII

Rayner Choyce
Mk IX

PUPIL POSITIONED IMPLANTS

Rayner Binkhorst

Rayner Federov

Rayner Binkhorst
Iridocapsular

Rayner Binkhorst
(3 loop)

Rayner Federov
(3 loop)

Rayner Federov
with Cheng Iridectomy Clip

POSTERIOR CHAMBER IMPLANTS

Rayner Pearce
(Bipod available)

Rayner Little Arnott

Rayner Boberg - Ans

CORNEAL IMPLANTS

Rayner Choyce
Keratoprosthesis

Left A selection from the wide range of intra-ocular lenses. 'Pupil-positioned implant' is another name for the iris-supported type. Corneal implants, used in the rare cases when a corneal graft fails, are inserted via a hole cut in the diseased cornea.
Right The Little Arnott lens.
Far right Side view of the same lens.

ment lens into the eye at the time of removal of the cataract—this gives good vision.

Many types of intra-ocular lenses have been designed since the first one, used by Harold Ridley in London in 1949. But it is only recently that the design and manufacture of these lenses has been of such consistently high quality that surgeons can use them with confidence.

All these lenses consist of a central optical device—usually a plano-convex lens—with varying types of support on the sides to fix the lens in position in the eye. The lens is made of polymethylmethacylate (PMMA), which is inert within the eye. Some lenses have fixation loops made out of nylon, and there is evidence that over a long period this

is broken down, or degraded, within the eye. However, no serious complications have been observed.

There are three main types of lenses. Firstly, there is the *anterior chamber implant.* This lies in front of the iris and is fixed in position by feet at the angle between the iris and cornea. The most commonly used design of this type is that developed by the British surgeon Peter Choyce. It has the advantage that it may be safely inserted in the eye of a patient who has had conventional cataract surgery but cannot wear glasses or contact lenses.

The second type is the *iris-supported lens,* which is fixed in position by loops lying in front of and behind the iris. This lens is not quite as stable as the first type, and the loops

may rub on the iris, causing a chronic inflammation within the eye. It is used in conjunction with an intra-capsular extraction of the cataract, which involves a higher incidence of retinal complications. The pupil cannot be readily dilated because the iris surrounding the pupil holds the lens in position, and if it is opened too much the lens will fall back into the vitrous gel of the eye (the transparent jelly-like mass that fills the eyeball behind the lens). So examination and treatment of these complications is difficult.

The third type of lens lies in the *posterior chamber*—the small space lying behind the iris and in front of the vitreous gel. It is fixed in place by struts extending into the capsular bag left behind after the extra-capsular

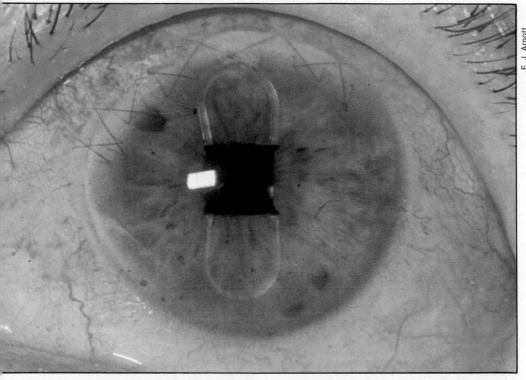

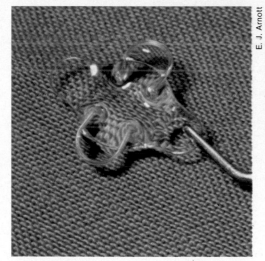

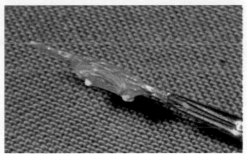

Below The surgeon's technique in inserting an iris-supported lens involves holding the lens by its plastic prong and sliding it into place over a plastic 'glide' which protects delicate tissues. The air bubble is temporarily inserted to protect the cornea, then refilled with body fluids.

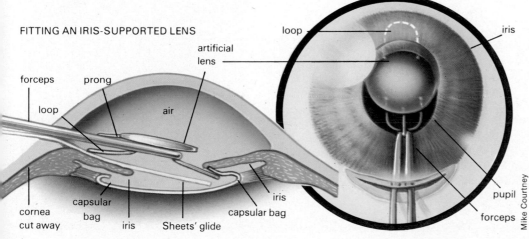

FITTING AN IRIS-SUPPORTED LENS

forceps
prong
loop
air
artificial lens
loop
iris
pupil
forceps
cornea cut away
capsular bag
iris
Sheets' glide
iris
capsular bag

cataract extraction. The advantage of this type of lens is that it lies in the most appropriate physiological position and therefore gives the best optical result. Since an extra-capsular extraction is involved, the posterior capsule is left intact. The lens is fixed by *fibrosis* (formation of scar tissue) within two months, so after this time the pupil may, if necessary, be dilated. The posterior chamber lens is probably the best type at present, but its use does require thorough removal of cataractous material.

Naturally, the presence of an artificial lens within the eye carries the risk of infection or other complications (though this is only slight), whereas wearing cataract spectacles involves no such problems. Set against this slight risk, however, is the fact that an elderly patient coping with the difficulties of wearing cataract glasses is probably likely to have problems balancing and judging distances, and is thus more likely to end up in an orthopaedic ward with a fractured femur.

Coping with problems

Most of the problems with intra-ocular lenses occur at the time of insertion, when there is a danger of damage to the transparent cornea. Corneal transparency depends on the integrity of a single layer of delicate cells on its inner side. Unfortunately, the PMMA of which the artificial lens is made will adhere to these cells if it touches them. If this happens during surgery they are stripped off and the cornea becomes cloudy, resulting in serious loss of vision.

The surgeon avoids this problem, firstly by using special anaesthetic techniques to reduce the tendency of the vitreous gel to push the lens forward, and secondly by inserting an air bubble into the anterior chamber before fitting the lens. The bubble of air acts as a cushion, protecting the cornea as the lens is inserted.

This air bubble is then removed and the anterior chamber refilled with body fluids. The only disadvantage of the air bubble is that it makes it difficult for the surgeon to see behind it, but the recent development of a non-toxic, transparent substance, *Helon*, has made his task considerably easier.

The advantages of intra-ocular lenses outweigh the problems involved in their use, so they remain the treatment of choice for elderly patients with cataracts.

In this and other ways, eye surgeons continue to use the latest advances in technology to improve life for those unfortunate enough to suffer impaired vision.

E. J. Arnott

E. J. Arnott

Mike Courtney

Breaking the sound barrier

One in every twenty adults you meet may suffer from some degree of deafness. One in five over the age of sixty needs amplification to aid their hearing. For many with hearing impairment, however, the future can be louder than the present.

In the normally functioning ear, sound arrives through the air in the form of alternate positive and negative pressure waves called *compressions* and *rarefactions*. (Each complete wave consists of one compression and one rarefaction). The frequency of sound will depend upon the number of complete waves which occur in one second, and its intensity upon the amplitude (size) of the waves. The frequency of a sound determines how its pitch is heard and its intensity determines how *loud* it sounds.

The human ear is divided into three parts: the outer, middle and inner ear. When a sound occurs, air-borne energy is funnelled by the *pinna* (outwardly visible part of the ear) into the ear canal, causing the eardrum and attached ossicles (the three tiny bones of the middle ear—the *malleus, incus* and *stapes*) to vibrate at the frequency and intensity of the incoming sound.

The last bone in the chain, the stapes, moves in and out of the oval window with a rocking motion, setting up fluid-borne vibrations in the delicate fluid-filled structures of the cochlea of the inner ear. Since fluid is practically incompressible, the membrane covering a second opening called the round window (situated beneath the oval window), bulges outwards to compensate for the inward movement of the stapes.

In this way complex wave patterns which code information about the frequency, intensity and length of the sound impulses are set up on the *basilar membrane*. This membrane runs throughout the entire length of the cochlea. The wave patterns excite specialized hair cells in the *organ of Corti* which sits upon the basilar membrane throughout the length of the cochlea. The hair cells in turn transmit electrical impulses to individual nerve fibres which eventually combine to form the auditory (hearing) nerve. The high frequencies are heard at the basal end of the cochlea nearest the stapes; the low frequencies are heard at the opposite end or apex of the cochlea. In the auditory nerve, information is coded by brief electrical impulses which travel via the brainstem to the auditory cortex of the brain. There they are

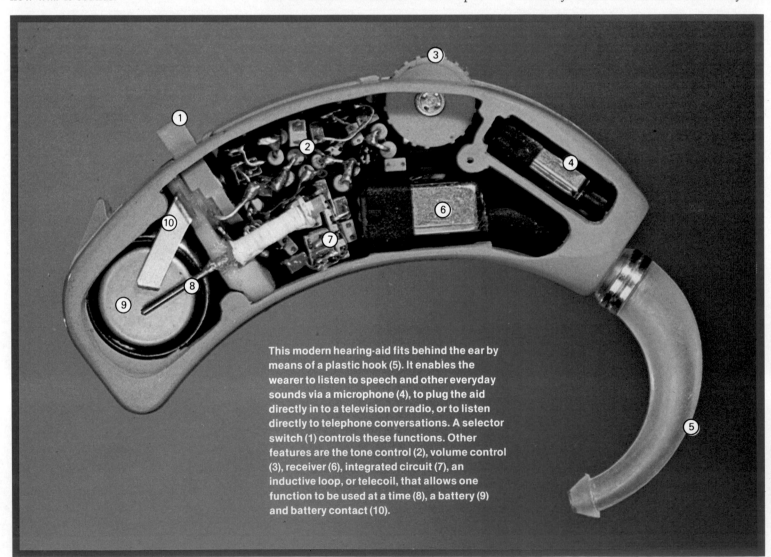

This modern hearing-aid fits behind the ear by means of a plastic hook (5). It enables the wearer to listen to speech and other everyday sounds via a microphone (4), to plug the aid directly in to a television or radio, or to listen directly to telephone conversations. A selector switch (1) controls these functions. Other features are the tone control (2), volume control (3), receiver (6), integrated circuit (7), an inductive loop, or telecoil, that allows one function to be used at a time (8), a battery (9) and battery contact (10).

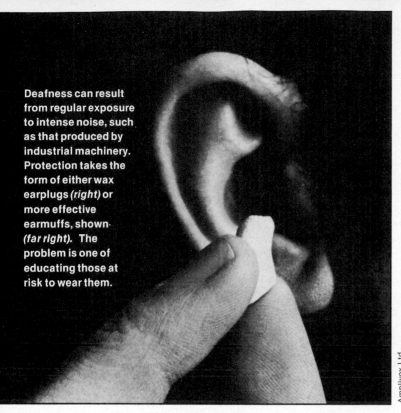

Deafness can result from regular exposure to intense noise, such as that produced by industrial machinery. Protection takes the form of either wax earplugs *(right)* or more effective earmuffs, shown *(far right)*. The problem is one of educating those at risk to wear them.

Amplivox Ltd.

perceived as a sensation of sound. In spite of many years of research, scientists still do not fully understand the exact nature of the information coding which takes place.

Two types of deafness

Two types of deafness afflict Man. A *conductive* deafness is caused by any obstruction to the conductive mechanism of the outer or middle ear which prevents sound waves from reaching the cochlea. A *sensorineural* deafness, as its name implies, results from a loss of function of the sensory apparatus of the inner ear and/or its connecting neural (nerve) pathways to the brain.

Conditions which can cause conductive deafness include obstruction of the ear canal (by wax, foreign bodies, growths or inflamation of the skin which lines the canal). Other causes are infection, injury, malformation of the outer or middle ear and glue ear—a condition common amongst children in which the middle ear, normally air-filled, fills up with thick fluid. Conductive deafness can sometimes be prevented or minimized by correct treatment if expert advice is sought as soon as a problem occurs.

There is as yet no cure for sensorineural deafness—although research continues. This type of deafness may be caused by hereditary factors, maternal infection during pregnancy such as rubella (German measles), accident, injury and infection. Noise damage commonly results in sensorineural deafness—after prolonged work in 'deafening conditions'. There are also certain rare conditions that affect the auditory nerve and there is

Ménière's disease—a condition characterized by episodes of deafness, *tinnitus* (noises in the ears) and dizziness. Because the inner ear contains the organ of balance as well as that of hearing, this type of deafness is sometimes accompanied by a disturbance of balance.

Some forms of sensorineural deafness are preventable. Genetic counselling may help to reduce the risk of hereditary deafness. Immunization of adolescent girls against German measles can help to prevent tragic cases of rubella deafness.

Damaging sounds

Although it has been known since the 1800s that intense noise can damage the hearing, many people exposed to damaging levels of noise still fail to wear simple hearing protection. Excessive noise exposure will result in damage to the delicate hair cell structures of the inner ear. The risk of damage increases with the noise intensity and exposure time.

The most damaging sort of sound is that produced by gunshots and the continuous noise of heavy industry. However, any noise may cause damage if it is loud enough and the exposure time long enough. Roughly speaking, the level of noise which may cause damage is that which is so loud that normal conversation is impossible. Where intense noise reaches 90 decibels—the equivalent to the sound of a road drill at five yards—then ear protection should be worn. A risk to hearing exists at 90 decibels when the total exposure time is eight hours a day. The decibel scale used to measure sound levels is a logarithmic one and when sound intensity

Below Two views of one of the latest models of 'in-the-ear' hearing aids. This type can also be converted for fitting behind the ear. Some modern hearing aids run on a battery that is no bigger than an aspirin.

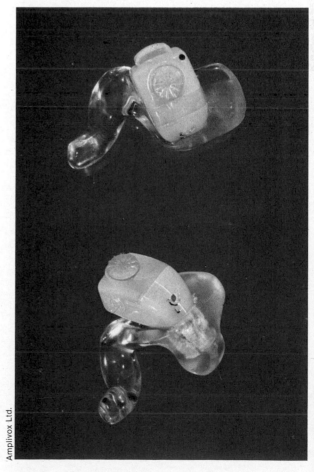

Amplivox Ltd.

increases by only three decibels, the resultant sound energy is doubled, as is the risk of damage. At a level of 93 decibels, a mere four hours exposure per day is hazardous, and so on; for each doubling of sound energy (each three-decibel increment) the permitted exposure time is halved.

The deafness caused by exposure to continuous intense noise may not become evident for some years. The first symptoms are usually ringing in the ears (tinnitus) and a temporary loss of hearing. Both symptoms disappear after a period of rest from the noise initially, but eventually both the tinnitus and the deafness become permanent. The deafness which results predominantly affects the hearing for high frequencies. Sensitivity to low frequencies remains relatively normal—and for this reason the impairment may go undetected for some time. Difficulty

is first noticed with conversation. This is because the most important sounds involved in speech intelligibility (the high frequency consonant sounds) fall with the frequency range at which the hearing loss occurs. Hearing of speech takes on a distorted quality and great difficulty in understanding conversation results. There is no cure for this type of hearing loss although a hearing aid may help to some extent. In later years, noise-induced deafness combined with the progressive loss of high-frequency hearing which occurs with increasing age, creates a handicap that can rarely be helped, even with a hearing aid.

When a risk of noise-induced deafness exists, the best solution is to reduce the noise at source, where possible by replacing noisy equipment or at least reducing noise levels by the use of baffles and sound-absorbing material. Often such sound reduction is pro-

hibitively costly or may not be technically possible. Whenever noise cannot be reduced to a safe level then ear protection should be worn at all times. Removal, even for a few minutes, significantly reduces the protection provided. The measurement and analysis of noise, as well as calculation of the amount and type of protection needed, is an expert task. The earmuff type of ear defence gives the most effective protection from industrial noise, although various types of earplugs are available which may be adequate in some situations. A special type of ear defender is available for protection from gunfire noise.

In certain situations, noise levels are so intense that even the best hearing protection is not enough to reduce incoming sound to a safe level. The noise from aircraft jet engines and diesel and turbine engine noise from industrial sites are health hazards of this kind.

Rod Sutterby

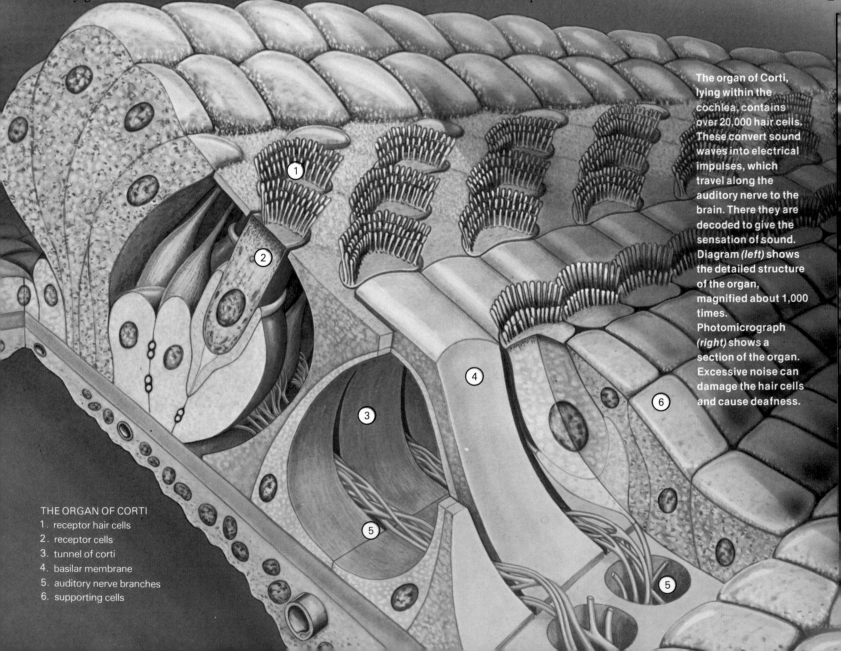

The organ of Corti, lying within the cochlea, contains over 20,000 hair cells. These convert sound waves into electrical impulses, which travel along the auditory nerve to the brain. There they are decoded to give the sensation of sound. Diagram (left) shows the detailed structure of the organ, magnified about 1,000 times. Photomicrograph (right) shows a section of the organ. Excessive noise can damage the hair cells and cause deafness.

THE ORGAN OF CORTI
1. receptor hair cells
2. receptor cells
3. tunnel of corti
4. basilar membrane
5. auditory nerve branches
6. supporting cells

Left The electron microscope reveals the hidden secrets of the organ of Corti. Magnified 2,200 times, this view shows the nerve fibres that contact the delicate hair cells and transmit impulses to the brain. *Sensorineural* deafness may result from degeneration of these nerve pathways. Cochlear implant surgery offers new hope for those suffering from this type of deafness.

Scientists at various research centres are working on a technique of sound cancellation sometimes known as 'anti-sound', designed to eliminate such noise at its source. A computer-controlled system monitors and analyzes the emitted sound and instructs loudspeakers to generate a second sound which is exactly 'out of phase' with the original. Each time a compression (high pressure component of sound wave) is sensed the computer sends instructions for an identical *rarefaction* (low pressure component) to be generated, and vice versa.

Cochlea implants

To aid victims of deafness, surgical research is now in progress at various centres into *cochlea implant surgery*—a technique which may offer considerable hope in the future to those suffering with profound sensorineural deafness.

The cochlea implant is not, as is commonly thought, some kind of artificial ear which is implanted to replace the non-functioning ear. It is an electronic device consisting, usually, of an external unit called a stimulator which converts sound to electrical impulses and relays them to a surgically implanted electrode in or on the surface of the cochlea. Electrical stimulation of the cochlea by this method has induced sensations of sound in the profoundly deaf.

Each centre has its own approach to cochlea implant surgery. In Los Angeles the technique used is to implant a single active electrode into the basal turn of the cochlea with its earth electrode in the Eustachian tube. Also implanted under the skin behind the ear is an internal coil with electrical connections to the implanted electrode. The internal coil, its connections and the implanted electrode form the internal components of the cochlea implant device. The external unit consists of a small microphone held in the ear canal by means of an earmould. A second external coil behind the ear is carefully aligned with the internal coil, and the stimulator device is worn on the clothing like a body-worn hearing aid. The function of the external unit is to convert incoming sound pressure waves to electrical impulses which are then relayed to the external coil. Electrical signals pass from the external to the internal coil by means of electromagnetic induction and thus to the implanted active electrode in the cochlea, where electrical stimulation of the hearing nerve occurs.

In its present state of development, the cochlea implant does not restore normal hearing or even come close to doing so. Im-

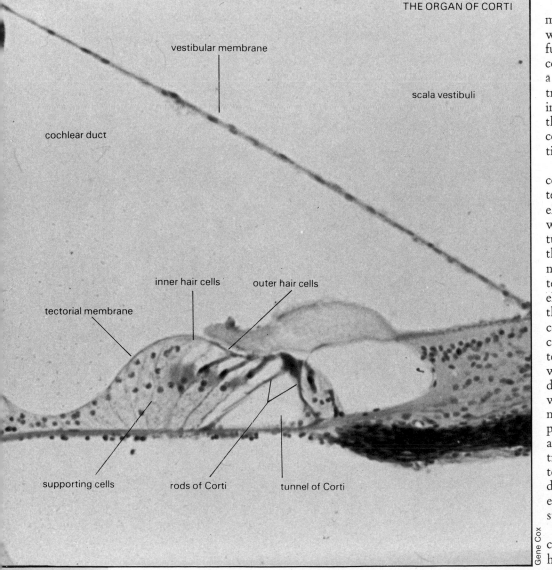

THE ORGAN OF CORTI

vestibular membrane

scala vestibuli

cochlear duct

inner hair cells

outer hair cells

tectorial membrane

supporting cells

rods of Corti

tunnel of Corti

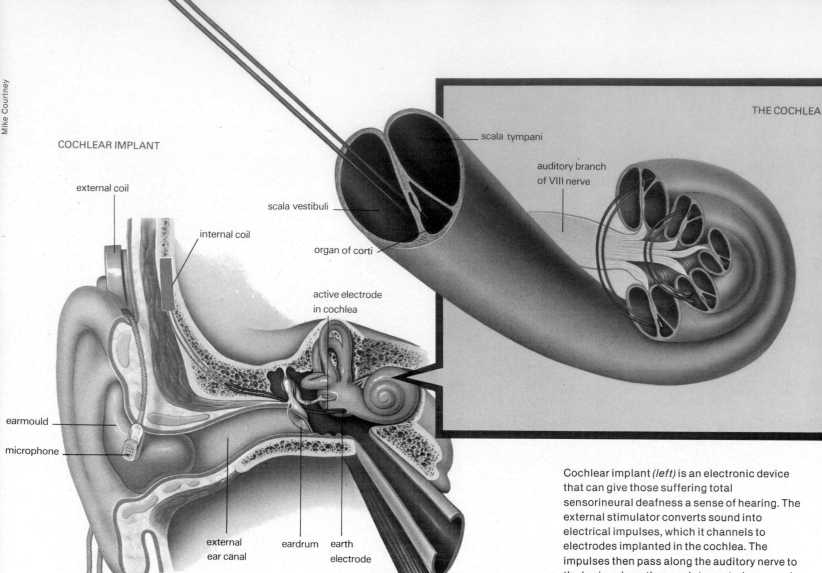

COCHLEAR IMPLANT

external coil

internal coil

scala vestibuli

organ of corti

active electrode
in cochlea

earmould

microphone

external
ear canal

eardrum

earth
electrode

THE COCHLEA

scala tympani

auditory branch
of VIII nerve

Cochlear implant *(left)* is an electronic device that can give those suffering total sensorineural deafness a sense of hearing. The external stimulator converts sound into electrical impulses, which it channels to electrodes implanted in the cochlea. The impulses then pass along the auditory nerve to the brain, where they are interpreted as sound. Close-up of cochlea *(above)* shows details of the electrodes.

plant patients experience only a very limited sort of hearing which is often described as having a 'buzzing' quality rather like the distorted sound of a radio which is not accurately tuned to a station. This effect derives in part from the degeneration of the auditory nerve which occurs in patients with profound sensorineural deafness (such deafness results in a decreased capacity to carry information to higher coding centres of the brainstem and brain). In addition, it is not yet possible to simulate the complex patterns of electrical activity which take place in the normal auditory nerve.

Surgeons in France and America are now working on multiple-electrode implant systems. The electrodes are placed at separate sites along the cochlea. Since the basal end of the cochlea 'hears' the high frequencies, more variety in the pitch of sounds experienced through the implant device should be possible by separate stimulation at different sites along the cochlea. Research into

single and multiple electrode implantation continues. At Guy's Hospital, London, a research team is working on a technique which involves single electrode placement on the surface of the cochlea. This technique offers the advantage of leaving the delicate structures of the cochlea intact for further surgery when implant technology improves.

Awareness of sound

At present the cochlea implant operation is performed only on patients with profound sensorineural deafness and is still very much in its research stages. It may not be successful for all deaf patients, and deaf people who can hear anything at all are still better served by a conventional high-powered hearing aid, as the operation may destroy all residual normal hearing. However, for those with no hearing at all, even the limited information provided by electrical stimulation of the cochlea is a great advance. The cochlea implant user has, for the first time, the advantage of awareness

of environmental sound as well as the ability to hear his own voice. With training he is thus better able to monitor his own speech. The implant also permits an awareness of the conversation of others, and even though the words may not be understood, the pattern and rhythm of speech combined with lip-reading will enhance speech intelligibility.

When deafness cannot be helped by surgery or other treatment, then auditory rehabilitation should commence as soon as possible. The sort of therapy required will vary for each individual and will depend upon many factors. A great number of environmental aids exist which can help the deaf to lead a more 'normal' life. Such devices include television and radio adaptors, and warning systems—usually flashing lights or louder bells for the doorbell, telephone and alarm clocks. Modifications to the telephone can be made for the 'hard of hearing'. The more recent development of the deaf telephone with its visual display, already available in some countries, has been a major breakthrough, enabling even the profoundly deaf to communicate by telephone.

One improvement has been designed to help sufferers from sensorineural deafness whose hearing for the high frequencies is much poorer than in the low frequencies, or whose high frequency hearing has been entirely destroyed. A conventional hearing aid does not greatly assist speech discrimination for such patients. The new aid, called a frequency-transposing hearing aid, supplements conventional amplification with information which is transposed from high frequencies down to low frequencies. This is done in such a way that speech sounds pro-cessed by the device sound relatively normal (except for certain emphasized consonant sounds such as s, sh, ch, t and z). Preliminary trials have been encouraging and have shown beneficial effects on the ability to understand speech and the quality of the user's speech.

Many people with hearing problems, as well as some with normal hearing, suffer persistent noises in the ears or head. For many tinnitus sufferers this is a most distressing condition for which there is as yet no cure. The noises may be very loud and can take the form of a great variety of sounds such as ringing, buzzing, whistling, hissing or other more complicated sounds. In most cases the tinnitus is heard only by the sufferer but it may sometimes be audible to others. The noises are usually less noticeable in a noisy environment when outside noise tends to 'mask' the tinnitus. It is usually much more troublesome when all is quiet, particularly at night when many sufferers experience difficulty in getting to sleep.

Tinnitus is caused by abnormalities of the hearing mechanism which are often microscopic, usually arising from a fault in the delicate cochlea structures or due to the presence of noisy blood vessels. Research into tinnitus has been pioneered in the US and UK. The results of most of the various treatments which have been tried, including

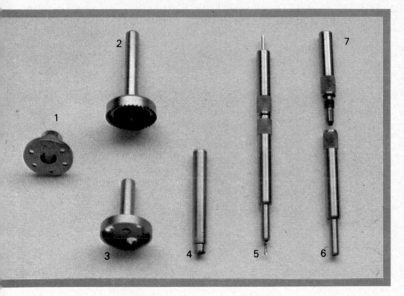

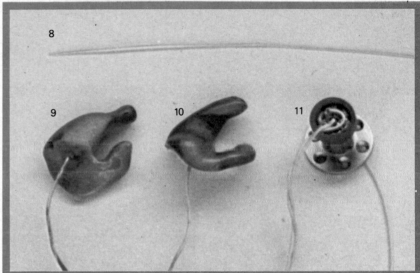

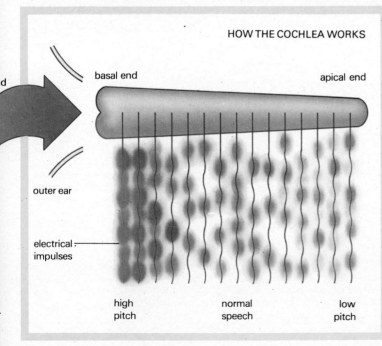

HOW THE COCHLEA WORKS

basal end

apical end

outer ear

electrical impulses

high pitch　　normal speech　　low pitch

Above Equipment used in cochlear implant surgery: stud for electrical connections (1); drills and other surgical tools (2–7); platinum electrode (8); gold-plated earmoulds for patients with no eardrum (9,10); stud with connector (11). *Left* How the cochlea works: nerve fibres running from thick end of spiral cochlea (here shown unrolled) transmit high-pitched sounds, those at other end transmit low-pitched sounds. Cochlear implant tries to mimic this.

surgery, drug therapy and many others, have so far been disappointing. The treatment which offers most hope for sufferers at present is the use of a tinnitus masker. This, though not actually a cure, can help to reduce the problem for the sufferer.

The tinnitus masker is a small instrument which is worn in the same way as a behind-the-ear hearing aid. Unlike a hearing aid, which amplifies incoming sound, the masker generates *white noise*—a sound which is rather like rushing water or the sound of wind blowing in the trees. Many people find this masking noise more pleasant to listen to and preferable to their own tinnitus because the sound now comes from an external source which is controllable—the masking instrument has a volume control—rather than from inside the head. In addition, there is a possibility that the tinnitus may be quieter when the masking noise is switched off—a phenomenon called 'residual inhibition'.

The athlete as machine

Faced with the possibility that human performance may be reaching its ultimate limits, sportsmen and sportswomen are cooperating with scientists and doctors in a study of the human body at its physical peak. Their goal is a new understanding of exercise and training which will lead to radically improved coaching methods.

Athletes make good subjects for medical research because they cooperate well in experimental programmes, and they are likely to have comprehensive medical case-histories and detailed records of their sporting achievements. Also, they provide scientists with a rare opportunity of studying the performance of healthy rather than sick people, to learn how exercise affects growth and ageing. The athletes, in return for submitting to the rigours and indignities of experimentation, hope to learn how to advance their capabilities far beyond present limitations.

Looking for a breakthrough, some researchers focused on the heart, blood vessels and lungs (the cardiovascular system) which supply the oxygen on which the muscles depend. But these organs only really improve significantly in unfit people who start to take regular exercise. For fit athletes, oxygen availability is not a critical factor since they can always breathe efficiently to obtain more air than the muscles actually need. Further improvement would be pointless.

Diet was another potentially exciting area for research which proved disappointing. The diets recommended by some sports coaches have produced no dramatic improvements in performance. As long as athletes eat well, their fitness depends on the amount of exercise they get. Drugs, particularly steroids, do seem to confer short-term advantages, but they are not explored by responsible sports scientists since their use is banned on ethical grounds.

Inside the muscles

Instead, sports scientists have tended to concentrate over the last few years on the workings of muscle itself. Electron microscopy has already revealed the fine structure of muscles as a mass of interlocked protein filaments. Two types—*actin* and *myosin*—form the basic active material of the muscle fibres. Stimulated by nerve impulses, they bunch up into each other, so the fibre—and thus the entire muscle—contracts. To contract, muscles require energy, which enters the body 'locked up' within molecules of food, particularly carbohydrates. They are processed in the digestive system to create long-chain molecules of a carbohydrate called 'animal starch', or glycogen, which is transported to the muscles and stored as tiny granules inside them. From here, the energy is transferred to the muscle cells by a series of chemical reactions in which the various pieces are assembled in the right sequence.

The energy is finally used to create an internal bond in the substance adenosine triphosphate (ATP). Responding to the stimulus of the nerve signal, two phosphorus atoms will drop off it, liberating a pulse of energy inside the proteins making up the muscle fibre and making the muscles contract. If the energy flow stops, they relax.

Repeated exercise makes the body stronger and able to function longer before fatigue sets in, a fact that athletes rely upon. It seems that exercise causes an increase in all the chemical processes involved. There is more actin and myosin to contract, more ATP waiting to disintegrate, more glycogen storing energy and more oxygen and nutrients supplied by the blood vessels. The amounts of enzymes also increase, and it seems that the body can acquire more biochemical pathways to release more energy for a longer period.

However, the process takes surprisingly long to make any physical difference in the size and shape of muscles. It may take six months of constant exercise before muscles

Different types of athletic events *(opposite page)* involve different sorts of muscle fibres. The Olympic decathlon ontrant throwing the javelin *(centre left)* makes full use of his fast-twitch fibres. By contrast, the marathon runner *(near left)* depends on his slow-twitch fibres for success. During the final sprint at the end of such long-distance events the muscles, working without any oxygen, start to fail and the athlete may collapse. This is dramatically shown on the face of the champion UK athlete Brendan Foster, being supported by his team-mate, Mike McLeod, after winning the 10,000 m race in the 1978 Commonwealth Games *(far left)*. *Right* The differing biochemistry of fast- and slow-twitch muscle fibres.

MUSCLE AND LACTIC ACID

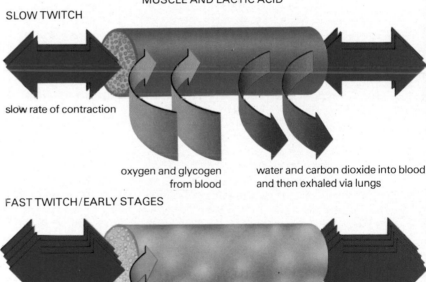

SLOW TWITCH

slow rate of contraction

oxygen and glycogen from blood

water and carbon dioxide into blood and then exhaled via lungs

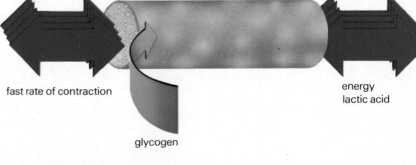

FAST TWITCH/EARLY STAGES

fast rate of contraction

glycogen

energy lactic acid

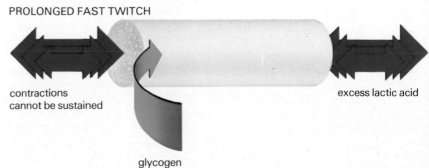

PROLONGED FAST TWITCH

contractions cannot be sustained

excess lactic acid

glycogen

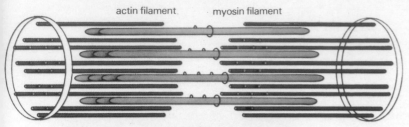

actin filament   myosin filament

relaxed

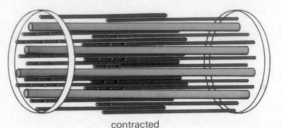

contracted

Left The protein filaments of skeletal muscle (stripes on microphotograph, *far left*) slide between each other, causing muscle to contract.

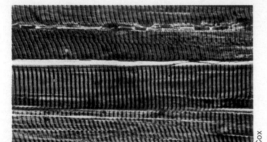

Gene Cox

develop the bulge of traditional fitness.

This whole intricate expansion cannot be manipulated from the outside. Only exercise can improve muscle tone, and the body's response is probably genetically predetermined. So far, this kind of biochemistry has proved more useful to doctors studying degenerative diseases of the muscle. The changes occur under stringent limits. The number of cells does not increase. They will heal themselves, replace themselves and increase in size, but they will not multiply.

Fast and slow twitch

Of vital importance to sports science, the muscle cells are of two distinct kinds—*fast-twitch* and *slow-twitch* muscle fibres. Slow-twitch fibres, as the name implies, work comparatively slowly but are able to contract repeatedly or maintain their contraction until all their stored glycogen is depleted. Fast-twitch fibres operate much more efficiently, but fatigue easily and use more glycogen.

In animals, the distinction is often very obvious, with particular muscles composed wholly of one type and certain animals relying on a predominance of one type. The cheetah, for instance, in its awe-inspiring dash to capture a gazelle, is running a race whose tactics are laid down by the necessities of its muscle structure. Its fast-twitch fibres allow it to sprint at speeds of up to 110 km/h (70 mph) but it cannot sustain a long pursuit. The prey, with a preponderance of slow-twitch muscle fibres, must dodge and manoeuvre until its pursuer tires. The same ritual is fought out between the domestic cat racing for the nearest tree and the lumbering dog hoping to beat it by sheer endurance.

In human beings, the position is more complicated, since we are born with a mixture of fast- and-slow-twitch fibres that is unique for each individual. Sportsmen and sportswomen need a high proportion of high-twitch fibres for power and sprint events and a predominance of slow-twitch fibres for success in endurance sports. In an attempt to discover the details behind these facts, physiologists have examined the way in which muscle fibres change in response to various kinds of exercise.

The basic technique is a *muscle biopsy,* in which the researcher pushes a long hollow needle into the sportsperson's muscle tissue. In the side of the needle is an open window into which a scrap of the muscle bulges. An extremely sharp blade is slid across the window and a few hundred cells are withdrawn for analysis. It is an easy technique to perform, but it is used only sparingly since it does involve some degree of temporary injury to highly trained muscles.

No matter what the individual's training regime, the proportion of each type of muscle fibres remains the same. Certain exercises improve fast-twitch fibres while others stimulate slow-twitch ones, but a fibre cannot be induced to change its characteristics, just as the total number of muscle cells—and thus the absolute limit of muscular power—cannot change.

Some people will never be very strong; others will excel at certain kinds of events only. Many sports do require the use of both types of fibre, however—particularly events demanding quick bursts of effort repeated many times, such as team sports, canoeing and mountaineering.

Muscle research has provided very useful information for designing and evaluating practical training programmes for individual athletes. Training for an endurance event, such as a marathon race, requires exercise regimes which develop steady effort for long periods of time. Training for power events, such as shot-putting or high-jumping, involves fast-twitch muscle regimes which extract maximum effort, followed by a short rest, then a repetition of the cycle to exhaustion or a desired fatigue level. Athletes can avoid overtraining on a kind of regime which will be of no use to them in their particular sport and development programme.

This scientific approach to sports training is a straightforward way of approaching the problem. But attempts to improve performance have been disappointing. So exercise physiologists have turned the problem inside out and ask themselves not why athletes are so good, but why they do not perform better. They are primarily interested in fatigue, and even define fitness itself in terms of the body's ability to recover from standard tasks.

Muscle biochemistry

To find out what happens in muscles as athletes tire, researchers have looked very carefully at the biochemistry of the different kinds of muscle fibre. In one revealing experiment, they took strands of pure fast- and slow-twitch fibre from nonessential muscles in mice and carefully mounted them in baths of nutrient to keep them alive. The researchers then stimulated the fibres with electrical signals which duplicated those from the mouse's nervous system. They could then investigate the chemical changes in the twitching muscles by dissecting the fibres at different stages and examining the nutrient.

These experiments showed that the biochemistry of fast-twitch muscle fibre is

Dudley F. Cooper/Dept. of Movement Studies/St. Mary's College, Twickenham

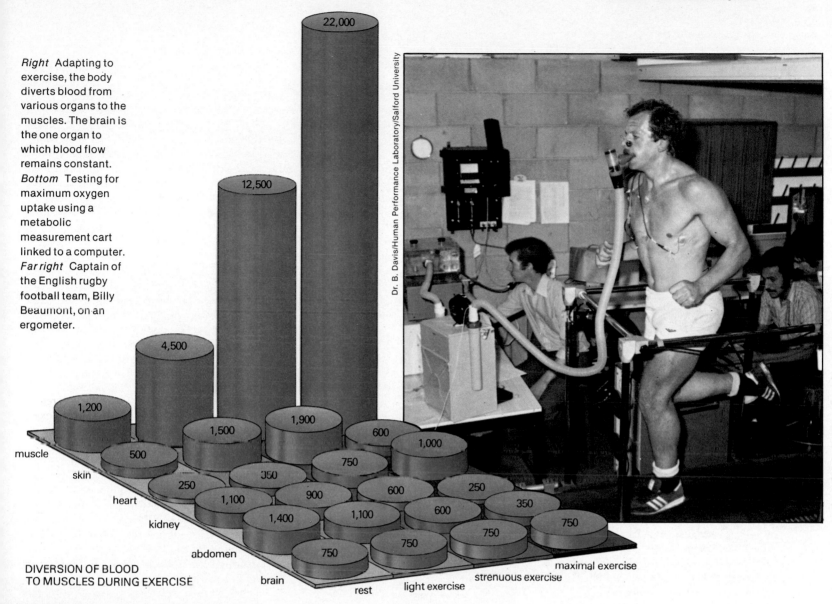

Dr. B. Davis/Human Performance Laboratory/Salford University

Right Adapting to exercise, the body diverts blood from various organs to the muscles. The brain is the one organ to which blood flow remains constant.
Bottom Testing for maximum oxygen uptake using a metabolic measurement cart linked to a computer.
Far right Captain of the English rugby football team, Billy Beaumont, on an ergometer.

DIVERSION OF BLOOD TO MUSCLES DURING EXERCISE

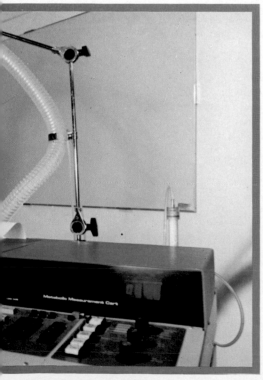

radically different from that of the slow-twitch kind. Most of the energy developed in a fast-twitch fibre comes from the dismantling of the glycogen molecules to form the substance called lactic acid. This *anaerobic* reaction, so called because it does not use oxygen, yields a rapid supply of energy, but as the level of lactic acid increases, the chemical balance in the muscle cells deteriorates, the action of the muscle is inhibited, and the process eventually stops.

Slow-twitch muscle fibre is more versatile and more complicated. Running at slow speeds, it dismantles glycogen molecules in a different way. It uses oxygen in an *aerobic* reaction to carry the process past the formation of acid in a complex series of reactions to yield water and carbon dioxide, which is ultimately breathed out. With no acid formed, the reaction can continue until all the glycogen is exhausted. If the muscle is activated faster, it will switch to behave rather

like fast-twitch fibre, giving more energy but with less efficiency. Again, lactic acid is formed and in time the muscle fails.

Lactic acid, then, emerges as the villain of the piece, the byproduct which pollutes the whole process. Coaches instructing power athletes using fast-twitch fibres pay particular attention to exercises which develop high levels of lactic acid. The body will respond to this punishment by increasing the oxygen supply, developing the scale of the non-lactic reactions which occur just after the initial contraction, increasing the body's ability to remove and process the acid, and providing buffering chemicals to limit its damage on site.

This description of aerobic and anaerobic respiration, the buildup of lactic acid and the role of muscle glycogen is now generally accepted and put to considerable practical use by scientifically minded coaches and athletes. Nevertheless, a number of anomalies is coming to light which suggest that although the theory is close enough to the truth to be of practical importance, it still fails to explain a number of significant factors.

Some individuals can function creditably as athletes while not being able to remove lactic acid from the bloodstream in the normal way. Also, adding lactic acid artificially proved less incapacitating than biochemists had expected. Neurophysiologists have also recently become interested in the performance of athletes, and suggest that the nervous system itself may fatigue and play an important role in limiting athletic performance. Other researchers propose that the role of fat has been underestimated. Both long-distance walkers and professional cyclists seem to metabolize their fat in ways which minimize their dependence on glycogen but the mechanisms involved are poorly understood.

Running on treadmills

Although they have much to investigate in their study of fatigue, exercise physiologists are already deeply involved in the practical training of athletes. The basic tool of their trade is a machine called an *ergometer,* which is a device on which athletes can exert themselves in the laboratory while their metabolic changes are monitored under carefully controlled conditions. Many different variants of the basic machine, such as bicycles, rowing shells and canoes, can be used, but the most familiar kind of ergometer is probably the treadmill. Athletes are instructed to run on a moving floor while various measurements are taken of their performance, and often computerized.

Analysis of changes in the heart rate tells the physiologist when the athlete is working at maximum effort. The amount of oxygen breathed in, the proportion used, and the amount of carbon dioxide and water vapour added to the breath in the lungs, coupled with statistics about how much work is being performed at the time, tell the researcher what kind of aerobic and anaerobic activity is going on in the muscles. If necessary, blood tests and muscle biopsies can provide more direct evidence.

Marathon runners are one group of athletes who find this research particularly useful. They will run on a treadmill for hours at a steady pace, with their pulses slow and unchanging, their breath calm, hardly

Right Measuring an athlete's jump. A good vertical jump is thought to be associated with a high proportion of fast-twitch fibres.

sweating at all. As the treadmill's speed is gradually increased, the amount of excreted carbon dioxide and water also rises until a speed is reached where it begins to drop. At this point, the slow-twitch muscles have been forced into anaerobic respiration and the runners' bodies begin to fatigue. Their heart rates increase, they breathe faster and begin to sweat. At higher speeds still all these functions become erratic. They stagger and eventually collapse—a chilling sight because high-level athletes will drive themselves obediently to the point where their bodies quite literally fail and they fall gasping onto the moving floor.

The skill of the exercise physiologist really lies in the interpretation of the graphs and charts which the computer produces. A potential champion can be detected, although this is less important than it seems because he or she will already have demonstrated practical ability on track or field. Runners' aptitudes for particular events can be tested so they can choose their most appropriate competition distance. By comparing their performances in previous sessions, the coach can decide how they are responding to training, and whether they are likely to improve significantly.

Runners can be advised on how to run a race. The tactics of long-distance running depend chiefly on the manipulation of the aerobic-anaerobic threshold. One runner will attempt to remain just inside the aerobic bar-

rier, while his or her opponent slips over into the anaerobic mode, generating the lactic acid which will corrupt the chemistry of endurance. On the treadmill the runners learn their own rhythms, establishing how fast they can run in the aerobic mode and what risks are incurred by generating brief burst of lactic acid. This is particularly important for marathon runners because they run infrequently and acquire little race experience.

Measuring the amount of fast-twitch fibres available to the slow-twitch athlete turns out to be very simple. All the researcher has to do is measure the time an athlete is in the air performing a standing jump which can be computed against body weight to provide a reliable comparison. Surprisingly, this can be an important factor in deciding tactics for a marathon runner, because the race often finishes with a group of runners tightly bunched together and forcing themselves into a last desperate sprint around the arena. The runners who know they are deficient in fast-twitch capacity must make sure they are far enough in front to compensate in the punishing finale.

A foregone conclusion?

Such research produces a disconcerting conclusion: given the sufficiently detailed statistics about the ergometer performance of the athletes involved, some exercise physiologists claim they can determine before an event what the outcome will be, assuming the competitors commit no tactical errors. Athletes must use their bodies skillfully and learn from their performance on the ergometer, but beyond this, the accurate predictions of the scientist seem to leave very little room for the personality and determination of the athlete.

As Olympic records become increasingly difficult to break and athletes near the absolute limits of human performance, it seems that success in athletics is largely a matter of having enough muscles of the right type, the right-sized lungs, the best body shape to reduce air resistance and so on, factors which are determined mainly by heredity. The sports scientists' research also appears to indicate that the body sets its own limits which cannot be markedly exceeded by training.

If this turns out to be the case, it is likely that sports scientists will concentrate on helping to refine the methods by which athletes reach their individual peaks, and also turn their attention to sports where a high level of skill is involved, or to team sports, in which the players do not compete against absolute limits of time and space.

Dudley F. Cooper/Dept. of Movement Studies/St. Mary's College, Twickenham

When vital organs fail

Specialists in spare-part surgery have never lacked for enemies. In the 1950s, kidney transplants (now running at a routine 800 lives saved a year) were challenged as a costly way to benefit the few while the many lingered on waiting lists. Although the climate of opinion has now altered, the ethics of removing organs from the newly dead to benefit the living have been keenly debated. Many countries will allow the removal of organs only after the heart has stopped beating; but it is still essential, at any rate for heart and liver transplants, to keep the blood moving and respiration going until the organs have been removed.

For the transplant surgeon, one enemy remains paramount: the human body itself. Deep in its programmes for survival lie inflexible, biochemical rules, leading it to reject foreign matter inserted or grafted. Ways to outwit the body's own refusal to cooperate with his efforts are among the surgeon's most challenging problems.

Problems of Matching

Transfer of major organs falls roughly into two categories: those where the donor and the recipient are related and those where there is no link. The latter are much more common—and more problematical.

Clearly, the more closely a new organ is matched to its new owner, the greater is the likelihood of success. Also, if the surgeon removes the organ while the donor's blood is still circulating through it, then less damage is done due to lack of oxygen. Hence the current vogue for removing organs from patients whose brains are dead but whose tissues are still physiologically alive. This means that the blood circulation must be maintained by a machine until the organ has been removed.

The transplant team must take great care to maintain organs for transplant in perfect condition, otherwise the recipient may undergo a

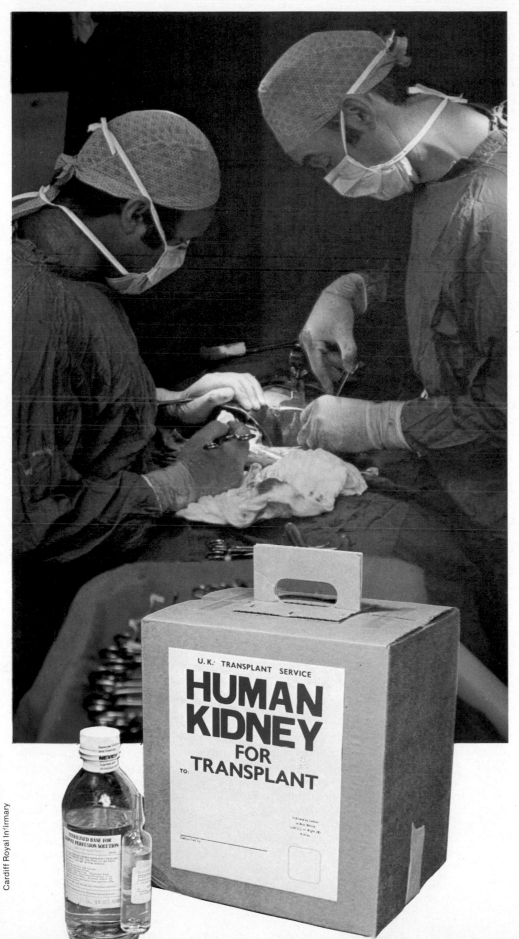

Above left A kidney transplant in progress.
Left This apparatus is used in transporting a kidney from donor to recipient. The contents of the small bottle are added to that of the large one and the solution is passed through the artery of the kidney immediately after its removal. The insulated box preserves the kidney safely for up to 24 hours, wrapped in a sterile plastic bag and surrounded by ice.

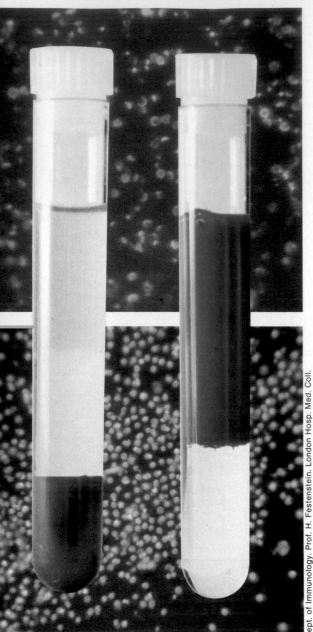

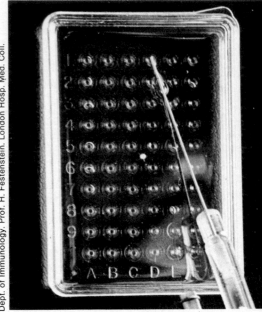

Dept. of Immunology. Prof. H. Festenstein. London Hosp. Med. Coll.

recipient within 24 hours. Livers remain usable for a shorter time still and must be rushed to their new owner within 10 hours. Nevertheless, this is long enough for British surgeons to have successfully transplanted Dutch livers into English patients.

Such successes depend on effective national registers of the tissue characteristics of patients awaiting transplants. Already, in several countries these are held in computers for rapid retrieval of the information. Then, when a suitable organ becomes available, it can be rushed to the patient whose tissue characteristics most closely resemble that of the dead donor.

Fingerprinting blood cells

A whole branch of pathology, known as *tissue typing,* has grown up with the development of transplantation surgery. Just as you cannot pour blood of an incompatible group into a patient during a blood transfusion so you cannot transfer incompatible tissues. For tissue typing, the equivalent of the ABO system of blood grouping is the human leucocyte antigen (HLA) system.

The controllers of this system—HLA genes—are found on the sixth human chromosome. They contain the instructions for manufacturing to the protein molecules known as human leucocyte antigens, which occur on the surface of the leucocytes or white blood cells. These provide the 'fingerprint' of the white blood cells of any one individual. If cells with different fingerprints are put into the body then defence mechanisms start working and the foreign material is destroyed.

This is why it is so important that the HLA molecules of someone waiting for a transplant are 'typed' so that a donor can be chosen whose cell fingerprints most closely match those of the recipient.

The development of this technique is not yet complete since biologists are constantly finding that the HLA picture is more complex than they thought.

Approaching the problem of rejection from the other direction, improved drug treatment has helped to reduce the failure rate of kidney transplants. The idea of drug treatment is to damp down the body's defence mechanisms to just the right degree, so that they do not destroy the new, foreign organ but are not so severely suppressed that the patient falls foul of every trivial infection that comes his way. If the immune system is totally knocked out, then the patient could die from the type of infection we would normally treat with a few aspirins and a day in bed.

Left The technique of tissue typing helps to reduce the chances of a transplant being rejected. The immunologist obtains lymphocyte cells from the blood of donor and recipient to see if their cell 'fingerprints' match. First he pours a blood sample on to Ficoll-Triosil medium in a test tube *(far left).* Then he uses a centrifuge to separate its contents into four layers *(left):* serum at the top, Triosil in the centre and red blood cells at the bottom, with a thin band of lymphocytes between the two upper layers. Background shows lymphocytes being 'fingerprinted'. 'Labelled' fluorescently *(top),* they are identified by seeing whether they die when serum of known type is added; orange die reveals dead cells.

Liver transplants

Liver transplantation has tended to be the poor relation of the transplant family—until recently, that is. The first liver transplant was carried out at the end of the 60s. The biggest series in the UK is being undertaken by the combined Cambridge/London team which pioneered kidney transplants on a large scale. Another major research centre is in the US, at Denver, Colorado.

Because of the hazards of the operation, it is generally the very sick who have not responded to less radical treatment who are chosen for liver transplants. This in turn lessens the chances of success because of the weak condition of the patients—something of a 'Catch 22' situation. Nevertheless, 12 British liver transplant patients are still alive, three more than a year after surgery, and one $5\frac{1}{2}$ years after receiving the transplant.

The problem associated with liver trans-

fruitless and possibly hazardous operation. For this reason the donor, having been certified as dead because of the absence of brain pattern (electroencephalograph), reflexes or respiration, may be given drugs or fluids to keep the kidneys working normally, in addition to blood circulation and oxygenation being maintained artificially.

Estimates made in the US indicate that as many as 30 per cent of organs removed for transplantation prove to be unusable, either because they have been mishandled or inadequately supplied with blood and other vital body fluids, or simply because a suitable recipient cannot be found before the organ's very brief lifespan outside the body runs out.

Once respiration and heartbeat have stopped, the kidneys have to be removed within about an hour. They can then be cooled and kept alive for up to two days, though there is a greater chance of success if they reach the

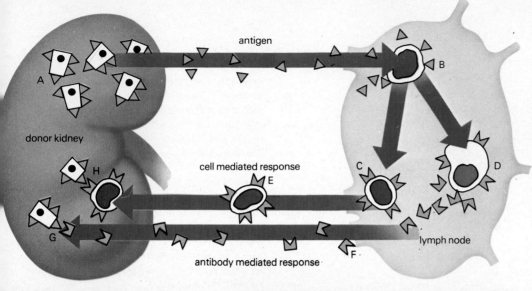

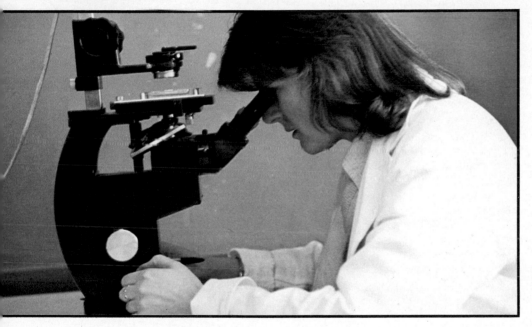

Left Rejection of a transplanted kidney is due to mobilization of the recipient's defence mechanism against foreign tissue. Antigens from the donor kidney (A) pass in the blood to the lymph nodes, where they cause lymphocytes to form T-lymphocytes (C) and plasma cells (D). The latter release antibodies (E) which, with the T-lymphocytes (F) return to destroy the transplant, the antibodies by damaging blood vessels (G) and the lymphocytes by infiltrating tissue (H).
Below An immunologist adds antibodies from a rabbit to samples of a patient's lymphocytes. Analyzing their reaction, using a phase contrast microscope *(below right),* enables the patient's tissue type to be determined.

Many liver specialists are now sufficiently optimistic that this technique will cut down the risks of liver transplantation so that patients can be given grafts at an earlier stage of the disease, before they become very seriously ill. This should increase the chances of full recovery after the operation.

Dealing with diabetes

The other important type of organ transplant, as yet in the early and largely unsuccessful stages of development, is that of the pancreas.

It has been estimated that insulin will run short within the next 20 years as more and more diabetic patients live out their lives to normal lengths instead of dying at an early age from the complications of the disease. The production of insulin by genetically manipulated bacteria provides an exciting possibility of overcoming this shortage. Nevertheless, in spite of this exciting advance a consistently successful technique of transplanting the pancreas would do away with any shortage of insulin altogether.

Earliest attempts at transplanting the pancreas intact met with little success and the problem of rejection seemed overwhelming. Later work focussed on transplanting only those cells of the pancreas which are actually responsible for insulin production—the Islets of Langerhans.

Experiments worked well enough in rats, mice and rabbits, but when surgeons turned their attention to the human pancreas the odds again seemed to be against them. Not only was it difficult, if not impossible, to isolate and grow the Islet cells in sufficient quantities but there was also the problem of digestive juices from the pancreatic duct, destined for the intestines, digesting the Islet cells instead. (In the experimental animals it proved possible to transplant the Islet cells in-

plantation is not so much rejection of the organ as the difficulties associated with the operation itself. The cirrhotic patient generally reaches the operating table in poor condition. His liver has probably failed totally, his fluid balance is likely to be disturbed and his kidneys may be in a bad state too. This means that his blood pressure is probably low—a problem that may be aggravated by anaesthesia at the start of the operation. Haemorrhage during surgery together with severe imbalance of the body's natural salts, or electrolytes, is accompanied by reduced amounts of blood returning to the heart when major veins are clamped during the operation. So an insufficient supply of blood reaches the very areas where it is needed most—the new liver and the already damaged kidneys.

In recent months British surgeons have successfully begun to overcome these serious problems during surgery by connecting the

patient to a machine which bypasses some of the major blood vessels between the heart and the abdomen. In this way blood is taken from the femoral vein which carries blood back from the legs towards the heart, is oxygenated, warmed and then pumped back into the femoral artery by the bypass machine.

Using this technique, the surgeons can reduce the amount of blood which is lost during the operation and hence also reduce the need for blood transfusions, which because of their acidic and slightly abnormal electrolyte levels, are likely to impair the contraction of heart muscle and so lead to disturbances and even stoppage of the heart rhythm.

So the bypass enables the surgeons to clamp the necessary blood vessels which supply the liver and allows them to remove the diseased organ and replace it with a fully functioning liver, without risking the possibility that insufficient blood will reach the new tissue.

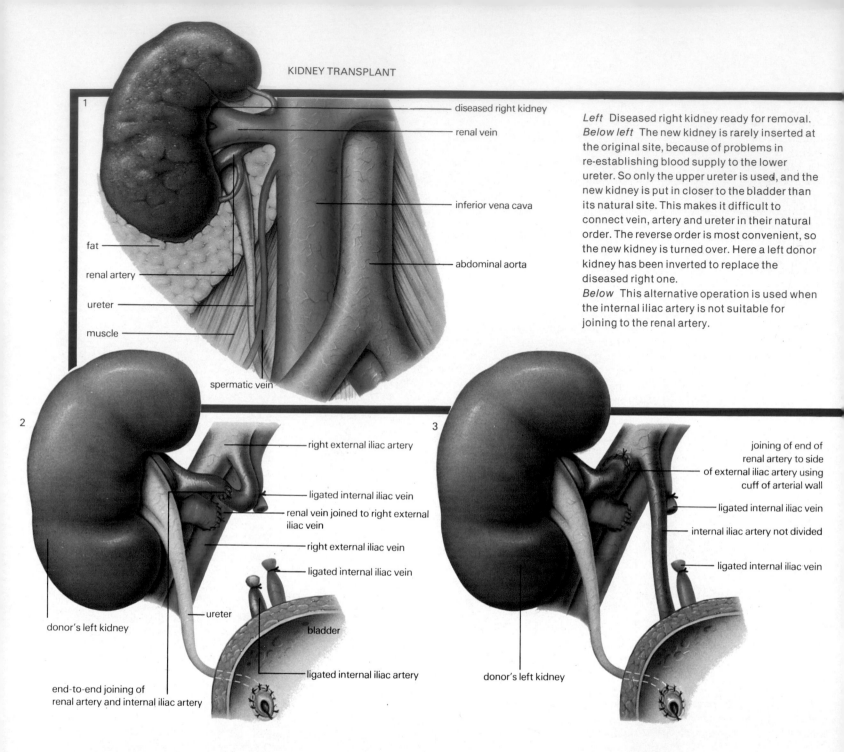

1

- diseased right kidney
- renal vein
- inferior vena cava
- abdominal aorta
- fat
- renal artery
- ureter
- muscle
- spermatic vein

Left Diseased right kidney ready for removal.
Below left The new kidney is rarely inserted at the original site, because of problems in re-establishing blood supply to the lower ureter. So only the upper ureter is used, and the new kidney is put in closer to the bladder than its natural site. This makes it difficult to connect vein, artery and ureter in their natural order. The reverse order is most convenient, so the new kidney is turned over. Here a left donor kidney has been inverted to replace the diseased right one.
Below This alternative operation is used when the internal iliac artery is not suitable for joining to the renal artery.

2

- right external iliac artery
- ligated internal iliac vein
- renal vein joined to right external iliac vein
- right external iliac vein
- ligated internal iliac vein
- ureter
- bladder
- ligated internal iliac artery
- donor's left kidney
- end-to-end joining of renal artery and internal iliac artery

3

- joining of end of renal artery to side of external iliac artery using cuff of arterial wall
- ligated internal iliac vein
- internal iliac artery not divided
- ligated internal iliac vein
- donor's left kidney

---

to the liver or spleen or induce them to grow in the abdominal wall, away from the dangers of the pancreatic duct.)

During the last two years, research into human pancreatic transplants has again changed direction and scientists are concentrating their efforts on transplanting parts of the pancreas together with their blood supply into the waiting recipient. The 'tail' and part of the body pancreas, together with its splenic artery and vein attachment, have been successfully transplanted in more than 20 cases in two centres in Europe. In some operations surgeons have blocked the pancreatic duct with a rubber-like material which sets solidly

in the duct and prevents the escape of digestive secretions.

Fighting leukaemia

While surgeons struggling to replace worn-out organs are battling against the body's natural defence system, then the research workers trying to treat leukaemia patients by transplanting bone marrow seem to be at an even greater disadvantage. For it is in the bone marrow that many of the cells which make up the body's defence mechanisms are produced. So doctors are transplanting the very material which develops into the cells participating in the re-

jection of foreign agents or tissue grafts. Even so, quite promising results have been obtained when bone marrow cells are removed from a close relation of the leukaemic patient and injected into the recipient following orthodox drug treatment to kill the patient's own diseased cells.

But what of those leukaemic patients who do not have a relation with sufficiently similar bone marrow to transplant? Drugs and radiation treatment may help them, but all too often their condition proves fatal.

Recently, however, doctors in Seattle, USA performed a successful transplantation of bone marrow from an unrelated donor to a patient with acute leukaemia. This is believed to be the first operation of its type. Although unrelated, the donor and recipient were sufficiently well matched for their HLA patterns to enable cancer specialists to attempt the transplantation. The patient was a 10-year-old girl. Her own marrow cells were first destroyed by a combination of radiation and drugs and then the transplantation was performed.

Drug treatment continued for more than three months after the operation to suppress the girl's reactions against the new cells. Nearly a year later, she is in good health with normal bone marrow function. The cancer specialists cannot be certain of the complete

Pancreas transplants are still experimental, but surgeons can already transplant the 'tail' of the pancreas *(below),* which contains vital insulin-making cells. Blood vessels at left will be joined to those of the recipient.

success of the transplant for some time, but it does open up the possibility of bone marrow transplants between unrelated people in the treatment of one of the most tragic killer diseases.

Problems of supply

Can man achieve immortality, some speculate, if his body is dealt with like a machine, replacing each organ by a spare part as it wears out? In spite of all the exciting advances described above, the day of the bionic man, with a body full of replacement organs, is far from us. Even if all the problems of surgery and rejection can be overcome, yet another hurdle remains—insufficient numbers of spare parts.

In most places, not enough people carry kidney donor cards to supply the need for renal transplants and even in areas where there are sufficient kidneys, hospitals are often unable to arrange for kidneys to be

removed and matched with waiting recipients until it is too late.

At present would-be donors in most countries must fill in cards to signify that they are willing to donate their organs. Many people who are in theory in favour of leaving their organs for transplant, fail to fill in such cards, either through apathy or forgetfulness. An 'opting out' system would mean that organs would automatically be removed unless people signed a form to the contrary. Such a system already operates in several countries, including Israel.

In the UK, a bill which would change the present system of donating organs from an 'opting in' to an 'opting out' system has been considered during no less than seven parliamentary sessions, but the 'opting in' system still prevails.

A survey of public attitudes was carried out in the UK last year. This showed that while the majority of those asked were in favour of kidney donation, three quarters of the sample were against an 'opting out' system. The general feeling was that the result of changing the law would be an infringement of personal liberty. A person's body would effectively belong to the state unless he or she decided otherwise.

Nevertheless, the present situation is highly unsatisfactory, for unless a better supply of organs for spare part surgery can be ensured then the growing expertise of the surgeons will go to waste, and patients who could have regained a normal life will continue to suffer and die.

Below far left At the start of a kidney transplant operation, the surgeon has cut and sealed off the iliac artery of the recipient.
Below left This pale donor kidney is ready to be connected, with two arteries held by forceps.
Below The kidney's blood vessels are now joined up to the patient's; it is pink with blood.

Renal Transplant Unit. Sheffield A.H.A.

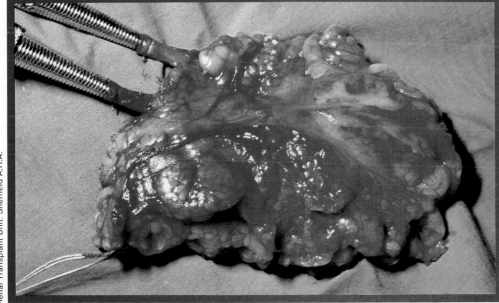

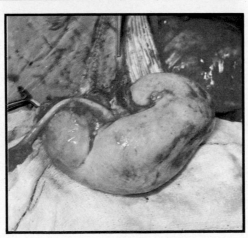

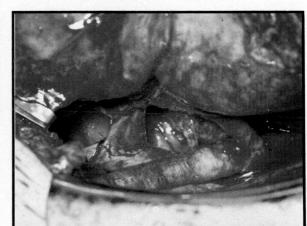

Right and above Although the strain placed on a gymnast's spine is severe, back troubles rarely occur in fit people who warm up before exercise. Problems of the lower back are more often caused by strain on an unsupple spine.

The trouble with backs

The human backbone is a masterpiece of natural engineering that can withstand hundreds of pounds of pressure, yet still remain extraordinarily supple. It is not so robust, however, that it can accept without protest the abuse to which it is constantly subjected. In modern industrial societies, back pain ranks as one of the chief physical ailments. Doctors in many developed countries estimate that as many as eight out of ten people suffer from back pain—usually in the lower back—at some time in their lives. For a distressingly high proportion of them, this affliction proves to be severe.

In the USA, 75 million Americans have back problems. Of this total, five million are categorized as partly disabled and two million as unfit for any sort of work. In Sweden, 'backaches' are the largest single cause of worker absenteeism, and they cost the UK 18 million working days each year, with a daily one million pound loss in production. Many doctors believe that the problem is worsening; surveys indicate that it claims seven million new victims each year.

The body's girder

The 24 bones, or vertebrae, of the human spine join together to form a structure designed to daunting specifications. Alone among the mammals, Man habitually stands on his hind legs. This habit puts great demand on the backbone, which acts as a vertical girder, bearing both bending and compression stresses along its length. It must provide suppleness and a wide range of movement, as well as supporting the whole weight of the upper body and acting as scaffolding for the skull, shoulder bones, ribs and pelvis.

Each one of the irregular vertebrae fits as neatly on top of its neighbour as the pieces of a jigsaw puzzle. Although they fit snugly, they are capable of limited movement one upon another—a feature which gives the spine its considerable suppleness. But its intricacy brings problems, too.

Like the rest of the body, the back suffers its share of diseases and congenital weaknesses. The cause of a pain may be

All Sport/Tony Duffy

anything from a malformed skeleton to cancer. But the vast majority of cases spring from an upset in the delicate balance of the spinal structure. The trigger for an explosion of pain which can leave a patient terrified to move can be as innocuous an action as reaching for a newspaper or making a bed.

Researchers into back problems have conducted a survey of such hazards using *radio pills*. There is a connection between intra-abdominal pressure (IAP)—which results from the contraction of the abdominal muscles—and the pressure between the vertebrae and their separating discs. So a measure of IAP serves as a measure of stress upon the spine.

To find out the IAP associated with any task, the researcher gives the volunteer a pressure-sensitive radio pill to swallow. The volunteer then dons an aerial connected to a receiver and chart-recorder around his waist. This system enables the researcher to monitor IAP continuously as the volunteer makes the required movements.

Radio pill surveys have shown that certain ways of performing everyday tasks score very high IAPs. Lifting heavy objects involves a risk of back injury at least as great as that associated with any other activity. But, even for lifting, there is a survival code which offers the chance of avoiding damage.

Any lifting done from a stooping position with bent back and straight legs is particularly hazardous, and produces a far higher IAP than the same lift made with a straight back and bent legs—where the legs take the strain. A straight back means just that, because twisting (torque) can be as harmful as stooping.

In nursing—a profession which, among its many demands, requires an ability to lift patients—there is an alarming incidence of serious back trouble among nurses, often as a result of inattention to lessons in correct lifting techniques. Of an estimated 430,000 nurses employed by health authorities in the UK, 82,000 suffer back pain each year as a result of some physical activity while on duty; 68,000 of these incidents are attributed to moving or supporting patients.

People suffering the agonies of acute back pain face the added misery of being informed that the prognosis for successful treatment is not high. The initial attack can usually be overcome, but the odds are that it will return—probably many times. Although modern medicine can help fight off the three most common causes of lower back pain—muscle strain, 'slipped disc' and facet-joint syndrome—the most effective cure is unspectacular and generally neglected.

Muscle strain, although painful, is probably the least damaging cause of lower back pain. In the course of overwork, muscles simply rebel against the constant contraction and relaxation, particularly when they are not accustomed to it, and they knot up in sustained, convulsive contraction. As a result, nourishing blood vessels constrict and muscle cells die. As they perish, they send out agonizing distress signals via the nervous system. This is the chain of events likely to befall the normally inactive person who rushes into sudden bouts of demanding sport or unusually heavy work.

Scientific evidence

Like most attacks of lower back pain, the first strike of muscle strain usually wears off after a week or two of complete rest. There are hosts of claims that other forms of treatment—from manipulation by osteopaths to heat and massage by physiotherapists—will work more quickly and effectively. These claims may well be justified, but the fact is that there is no hard, scientific evidence for them. The present state of knowledge on treatment for lower back pain is that it will probably just go away if left to itself.

Muscle strain is no light matter, but it is generally less painful and disabling than the most infamous of all spinal ailments: the 'slipped disc'. Discs lie between the vertebrae—hence their full name of intervertebral discs—and serve as the spine's shock absorbers. Each one consists of a fibrous cartilage ring surrounding a soft pulpy part called the *nucleus pulposus*. The discs are quite elastic, thereby helping to give the spinal column its range of movement.

The discs between the lumbar vertebrae of the lower back and between the cervical vertebrae of the neck are much thicker than those between the thoracic vertebrae of the

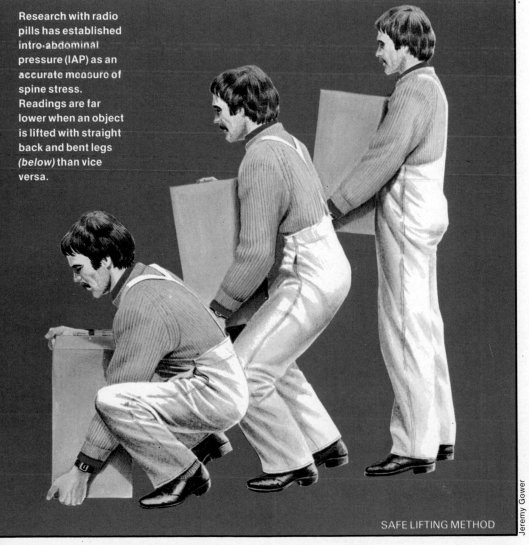

Research with radio pills has established intro-abdominal pressure (IAP) as an accurate measure of spine stress. Readings are far lower when an object is lifted with straight back and bent legs (below) than vice versa.

SAFE LIFTING METHOD

Jeremy Gower

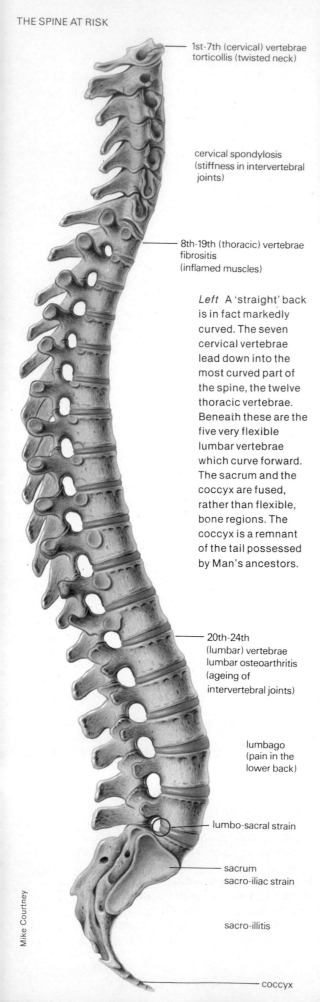

1st-7th (cervical) vertebrae
torticollis (twisted neck)

cervical spondylosis
(stiffness in intervertebral
joints)

8th-19th (thoracic) vertebrae
fibrositis
(inflamed muscles)

20th-24th
(lumbar) vertebrae
lumbar osteoarthritis
(ageing of
intervertebral joints)

lumbago
(pain in the
lower back)

lumbo-sacral strain

sacrum
sacro-iliac strain

sacro-illitis

coccyx

Mike Courtney

Left A 'straight' back is in fact markedly curved. The seven cervical vertebrae lead down into the most curved part of the spine, the twelve thoracic vertebrae. Beneath these are the five very flexible lumbar vertebrae which curve forward. The sacrum and the coccyx are fused, rather than flexible, bone regions. The coccyx is a remnant of the tail possessed by Man's ancestors.

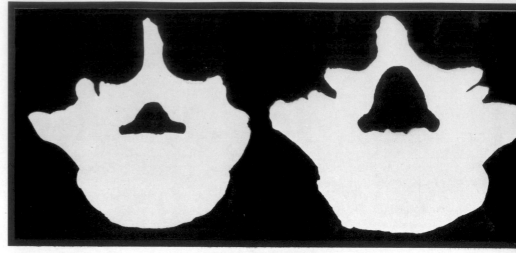

upper back, giving the neck and lower back greater freedom of movement. So it is no coincidence that the lumbar and cervical regions are most likely to suffer from a 'slipped disc', although a disc may 'slip' at any point in the spine.

More serious problems may arise when the ruptured disc presses onto the spinal cord, the main pathway carrying messages directly to and from the brain. In such cases, the patient may suffer paralysis, loss of sensation and difficulty in passing or retaining urine. To many suffering a ruptured disc, medical science can offer little help. Once damaged, a ruptured disc will never be as good as new, although time may bring some natural repair to relieve the painful pressure.

Removing a disc

In some cases, the injury and its attendant pain persist, and surgery may be prescribed. The surgeon will perform a *laminectomy,* or removal of the disc, and perhaps follow that up by fusing together the vertebrae above and below the removed disc with chips of bone taken from elsewhere in the patient's body. A laminectomy operation is often very effective, but it is by no means routinely successful. A disturbingly large number of patients receive no relief from it and some experience more pain. The operation has even been criticized as unnecessary.

Although surgery cannot guarantee a cure for a ruptured disc, research into use of the instrument known as an *ultrasound diasonograph* may help to diagnose potential sufferers of ruptured discs. Study of a wide range of skeletons shows that the spinal canal along which the nerves of the spinal cord run exists in two distinct shapes. In most people, the spinal canal is roughly dome-shaped at the level of the lowest lumbar vertebra. However, for an unlucky 12 per cent or so, the canal at that level has a trefoil (three-leaved) shape with a narrower cross-section.

Above These sections taken from fourth-century skeletons show how the two basic shapes of spinal canal differ from one another. At the level of the lower lumbar vertebra, most canals are dome-shaped *(right).* An estimated 12 per cent of the population, however, have trefoil-shaped spinal canals *(left)* which are more vulnerable to pain from trapped nerves. The innovation of the ultrasound diasonograph has enabled doctors to build up an accurate picture of a patient's spinal canal.

This discovery led researchers to suspect that the 12 per cent of the population with trefoil-shaped spinal canals are more likely to suffer pain from a herniated disc. It appears probable that, when the disc bulges into the spinal canal through its ruptured wall, there is more room to accommodate the intrusion in the large, dome-shaped space than in the trefoil-shaped canal, where lack of space may mean that the ruptured disc pinches the spinal nerves—which send out their distress signals of pain.

Until recently, there was no means of measuring the spinal canal of patients suffering from ruptured discs (short of complicated and dangerous surgery), so the suspicion that these patients had trefoil-shaped canals remained unproven.

The advent of the ultrasound diasonograph has changed this. The machine transmits sound waves and receives their echoes. By minutely timing the interval between transmitting and receiving sound reflected from the bony surfaces of the spine, the machine builds up a picture of the shape of the spinal canal accurate to 0.5 mm.

As the procedure is quick, painless and risk-free, it was used on a number of suitable subjects in a programme sponsored by the British Back Pain Association and the National Coal Board, beginning in 1978. As far as could be determined, the patients suffering from disc lesions did, on average, have

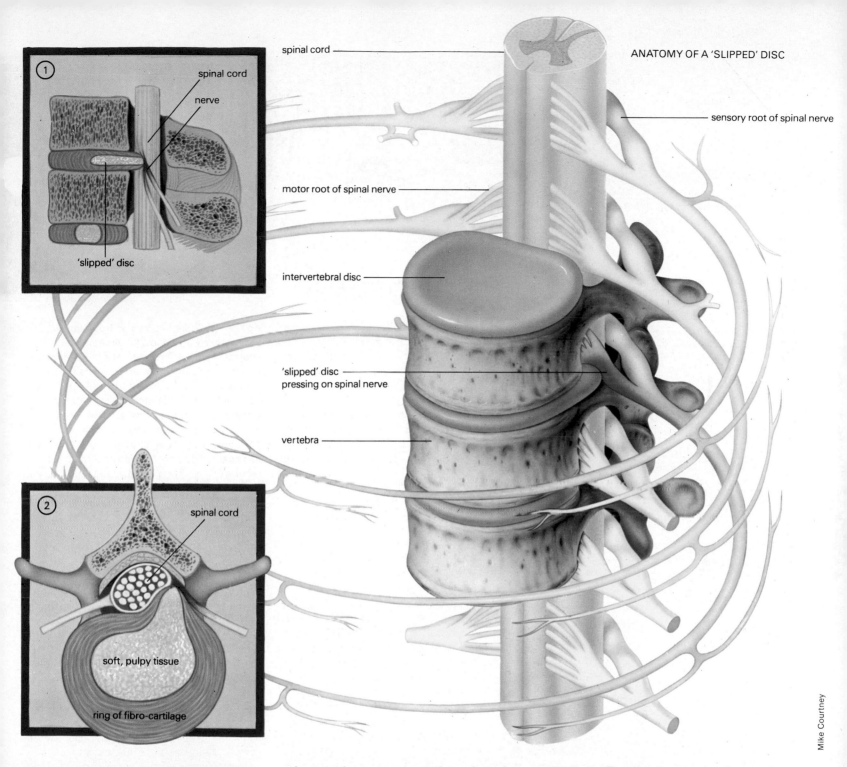

spinal cord

sensory root of spinal nerve

motor root of spinal nerve

intervertebral disc

'slipped' disc
pressing on spinal nerve

vertebra

1

spinal cord

nerve

'slipped' disc

2

spinal cord

soft, pulpy tissue

ring of fibro-cartilage

Mike Courtney

Above A 'slipped' disc is most likely to occur in the lumbar and cervical regions of the spine. The fibrous outer edge of the disc ruptures and presses on one of the network of nerves which swirl about the spine (1). According to the degree and location of the pressure, this can cause severe pain and even disability. A disc in cross-section (2) shows the layers of pulp and cartilage. When the cartilage gives way, the pulp presses on the nerve, causing a 'slipped' disc. As well as being felt in the spine itself, pain is also experienced in the part of the body served by the pressurized nerve—often in the suffering individual's legs or arms.

a very much narrower spinal canal at the level of the fifth lumbar vertebra than was usual among the general population. Other measurements tended to support the view that most of those with narrow canals had the trefoil-shaped canals found in 12 per cent of skeletons.

This does not mean that those with dome-shaped canals are immune from painful discs or that those with narrower canals are bound to suffer from them. What it does mean is that a group of people who are more than usually likely to suffer pain or disablement from degenerative changes in the spine can,

with a little effort, be discovered and warned before they suffer injury. It should not be impossible to use the ultrasound diasono-graph to weed out applicants for those jobs that carry a very high risk of damage to the lower back.

During the 1970s, some doctors began to suspect that a condition known as *facet-joint syndrome* might be as common as cause of back pain as ruptured discs. The main body of each vertebra is a more-or-less cylindrical piece of bone, but it has a bony arch projecting behind it. In a healthy spine, these vertebral arches are lined up neatly one above

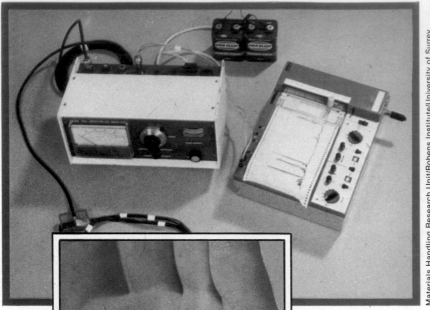

Materials Handling Research Unit/Robens Institute/University of Surrey

the other, and are capped by spinal processes which form the bony protrusions that can be felt running down one's back. Lateral bony outgrowths are attached to either side of the arch. These articulate with their neighbours above and below at the small, flat-surfaced *facet joints*. During a sudden, violent twist, a facet may become dislocated and press on one of the nerves leading from the spinal cord. As with a ruptured disc, this can cause intense pain.

Facet-joint syndrome proved difficult to detect until the mid-1970s, when X-ray techniques were revolutionized by *computerized axial tomography* (CAT). A CAT body scanner can give a detailed picture of both bone and tissue in body sections between 2 mm and 13 mm thick.

This is a great improvement over the traditional X-ray method, by which only bone is made visible with any clarity. Also, since the X-ray method involves the injection of dye into the spinal canal, it can prove uncomfortable or even dangerous to the patient. Unfortunately, however, CAT machines are still scarce, since they cost between a quarter and one million pounds each.

Although diagnostic techniques have become highly sophisticated, treatment of the acute pain that often occurs when a disc

Right The spinal epidural stimulator can bring relief to sufferers from chronic back pain. The control box is small enough to be carried in the patient's pocket or hooked over a belt.
Far right The scarring on this wheelchair-bound patient's skin indicates that the stimulator has been implanted high up in the cervical region of the spine. It is attached to the outside of the spinal cord's sheath and is activated through wires beneath the patient's skin.

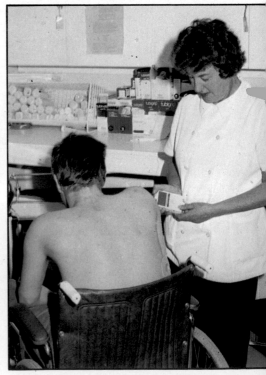

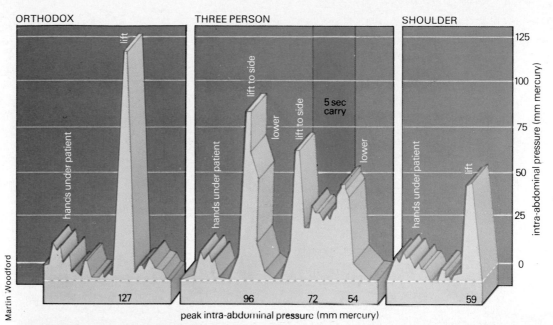

Left Research into back problems has been advanced by techniques using the pressure-sensitive radio pill *(middle left)*. In test conditions, intra-abdominal pressure readings are transmitted from the pill in the volunteer's stomach, through the aerial to the receiver and recorder *(top left)*. This research has been particularly useful in preventing back trouble in people with high-risk jobs, such as nursing. The graph *(right)* shows intra-abdominal pressure records for a male nurse in training. The readings express pressure exerted to lift a 65-kg patient in three different ways. Scientists at the Materials Handling Research Unit in Guildford, UK, have tested several thousand volunteers with the aim of reducing back strain in industry, medicine and sport—fields that involve considerable activity and/or lifting.

ruptures or a facet becomes dislocated remains relatively crude. At its most drastic, it involves blocking off the nerves in the damaged area, leaving it completely numb. Less irrevocable are attempts at stimulating the injured nerves and overwhelming their pain signals with less unpleasant ones.

Battery stimulator

One simple method of treating the acute pain that can occur with back problems is to provide the patient with a battery stimulator which can be applied to the painful area. A more sophisticated alternative is to implant an electrode beneath the skin at the precise point where it will best counteract the pain and connect it to a receiver implanted in some convenient part of the body, such as the chest. The patient has a battery-operated stimulator with an aerial, which he or she can press onto the skin above the receiver, turning it on to alleviate the pain by sending bursts of current surging through the buried electrode. This system has a claimed 65 per cent success rate as a last resort treatment. In the final analysis, prevention is still the best cure. However, if an attack of pain does occur, once it is over the best way of preventing a recurrence is by a constant regime of exercises. A supple spine is far less prone to back injury than a stiff one, even when it undergoes a considerable amount of strain.

Right Complete removal of a disc, in an operation known as a *laminectomy,* can relieve painful and dangerous pressure on the spinal nerves. The 'missing' disc can be replaced artificially, or the upper and lower discs be fused together. In time, damaged nerves grow again and the results are often successful.

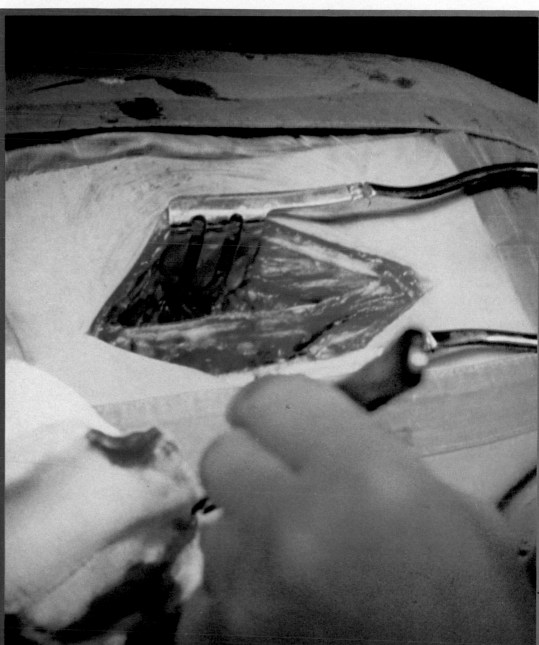

Moulding the human body

Working 2,500 years ago, a Hindu practitioner named Susruta sliced flaps of skin from his patients' cheeks to rebuild a nose or repair damaged ear lobes. The technique remains the basis of all reconstructive surgery. By grafting whole pieces of tissue from one area of the body to another, or by manipulating flaps of skin (still attached at one end to their original site) into new positions, the surgeon can transform the body's appearance.

Many people think that plastic surgery is so called because it involves the use of plastic materials. Although such non-living material is used in reconstruction, the term 'plastic' refers to the reshaping of the body by the surgeon. His task may be to remedy an inherited deformity, repair the ravages of disease or injury—or simply to erase the wrinkles of the aged rich. In each case the surgeon must decide whether the benefits will outweigh the risks of scarring involved.

In some instances, he may have few options. Radical surgery to remove tumours or wounds inflicted in modern warfare may leave disfigurements that only the plastic surgeon can rebuild. Skills developed in tackling such dire cases have helped to make the business of face lifts and 'nose jobs' safely routine.

The decision to move tissue from one part of the body to another either by a free skin graft or by the creation of a pedicle *(skin flap)* depends on the case involved. A free skin graft—normally from an area of the body where some scarring will not be obtrusive, such as the inner arm or thigh—is the simple, less traumatic technique. Again, requirements of the individual case dictate how deeply the knife will slice into the body's complex, two-layered defensive system.

To take a free skin graft, the surgeon first smears the skin with lubricant, then, with the knife flat against the taut skin, removes a tissue-thin layer with a to-and-fro sawing motion. In its new site, this tissue must lie flat but not taut. It is stitched in place and a pressure dressing applied. The dressing is changed every few days and the area kept covered for about three weeks.

Problems arise on occasion when, some

Left A dramatic demonstration of the results of skilled cosmetic surgery. This picture was taken after one half of the woman's face had been given a facelift to remove her wrinkles.

Rex Features

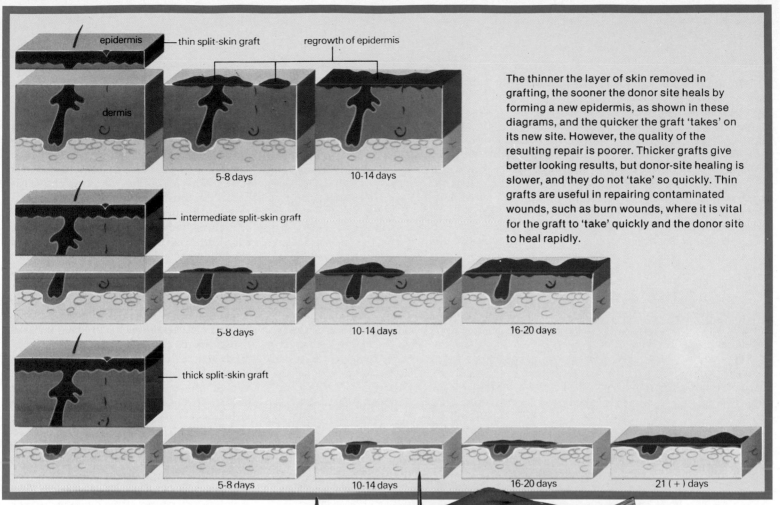

epidermis — thin split-skin graft

regrowth of epidermis

dermis

5-8 days

10-14 days

intermediate split-skin graft

5-8 days

10-14 days

16-20 days

thick split-skin graft

5-8 days

10-14 days

16-20 days

21 (+) days

The thinner the layer of skin removed in grafting, the sooner the donor site heals by forming a new epidermis, as shown in these diagrams, and the quicker the graft 'takes' on its new site. However, the quality of the resulting repair is poorer. Thicker grafts give better looking results, but donor-site healing is slower, and they do not 'take' so quickly. Thin grafts are useful in repairing contaminated wounds, such as burn wounds, where it is vital for the graft to 'take' quickly and the donor site to heal rapidly.

Right This cross-section through the skin shows the two major layers, the epidermis and dermis, and four different thicknesses of skin graft. Split-skin grafts include only a part of the dermis, but a full thickness graft involves removal of the full depth of skin.

time after the operation, the tissue attaching the graft to its new base contracts—to result in a puckered appearance much like the surface of an orange. This risk is greater if the graft has been placed over a deep wound where the lower layers of the skin have not regenerated.

Risk of such contraction is reduced if the surgeon takes, with the outer *epidermis,* more of the skin's inner layer, the *dermis,* crowded with nerves, blood vessels, hair roots and sweat glands. This is, however, the vital layer—its total removal will leave the donor area considerably scarred. When a full thickness graft does 'take' well to its new site results are better than with partial grafts. But the thicker the graft, the more difficult it is to get it to 'take' in the first place. This is because the large numbers of elastic fibres in the thicker grafts may contract, and lead to total rejection of the skin.

Techniques are being tested to improve the chances of success. Experimentally, a

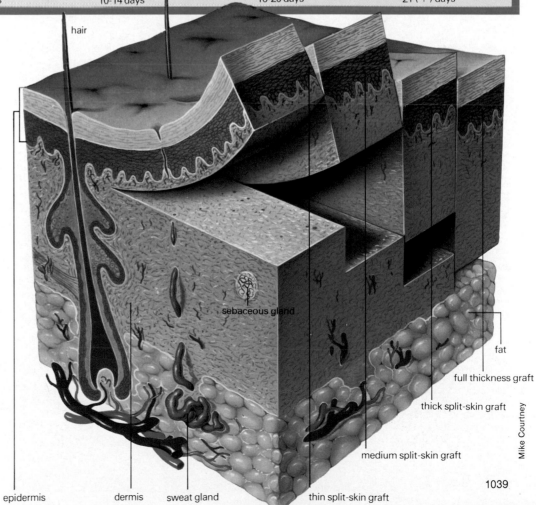

hair

sebaceous gland

fat

full thickness graft

thick split-skin graft

medium split-skin graft

thin split-skin graft

epidermis

dermis

sweat gland

Mike Courtney

natural protein material called collagen has been implanted as a filler between the damaged tissue and a newly grafted epidermis. Collagen gives the skin its normal elastic properties and, if successful, the technique should prevent the orange-skin disfigurement of secondary contraction.

It may soon be possible for collagen extracted from skin removed from one patient during the course of surgery to be used in the treatment of another patient undergoing plastic surgery. The problem is that the collagen has to somehow be chemically altered to make it 'take' in its new site on its new owner.

In other efforts to improve the chances of grafts surviving, doctors at one major plastic surgery centre have been using pieces of the membrane which surrounds an unborn baby in the womb as a layer between the wound and the graft. It is believed that the membrane promotes the growth of blood vessels which supply the site with nutrients and so enhance the chances of graft survival. The

technique has been shown to aid patients with chronic leg ulcers.

Where there is extensive loss of skin, as with large burns, skin grafts can be treated with a machine that cuts the graft to produce a net-like mesh. A meshed graft can prove to be a life-saver in covering a large area of damage with a relatively small piece of skin, and is also useful in areas such as the scrotum where it is difficult to get ordinary grafts to take.

In a successful free skin graft, blood vessels grow and spread between grafted and underlying tissue within about three days of surgery. Next, cells proliferate from the edges and deeper surfaces of the graft itself. After about two weeks, the graft is firmly attached to its new base by connective tissue, and some seven days later a thin layer of fatty deposit forms beneath the graft. First pain and then temperature changes are felt and finally the patient is aware of contact with external objects, though it may be a year before full feeling is achieved.

Free skin grafts can only serve the surgeon's case where the patient has sufficient undamaged skin to supply to the damaged area. Although research continues, no way has been found to transfer skin between patients longer than ten weeks before the donor's tissue is rejected.

Skin-flap grafting

To aid the patient whose wound is too extensive or too deep for a free skin graft, surgeons have developed the technique of *skin-flap grafting*.

In a deeply damaged area, the blood supply to the injured site has been destroyed. If a piece of skin alone was attached to the remaining bone and tissue, it would soon die and fall off, deprived of nutrients. The skin-flap technique grafts skin to a new site, while leaving it attached to its original site by a small amount of connective tissue and blood vessels. The surgeon constructs the flap from the fatty tissues lying immediately below the skin, generally in areas of the abdomen or back.

First, he makes an incision in the skin large enough to draw out the underlying tissue. The lower layers of the skin and underlying fat are then dissected away from the surrounding tissue, taking care to keep the blood supply to the piece of tissue intact. This chunk of tissue is then carefully withdrawn through the incision, then manipulated and stretched so that it can be pulled through 180 degrees to reach the damaged area in, say, the chest or shoulder. Alternatively, if the incision is made lower in the abdomen and the tissue pulled out from underneath the skin covering the stomach, then a 180 degree manipulation will bring the tissue down toward the groin and enable it to be used for repairing injuries to the lower part of the trunk or upper parts of the legs.

The flap of tissue thus formed, still with its attachment internally to the blood supply, is then sutured (stitched) in place in much the same way as if it were a free graft. Once the surgeon is sure that the graft has 'taken' and has put out its own new blood supply to the surrounding tissue, its original internal link can be severed much like an umbilical cord. Any remaining tissue which does not directly contribute to the graft is removed.

A variation on this technique is the formation of a *tube flap*. In this case two parallel longitudinal incisions are made in the skin and the skin between them freed from the underlying tissues so that the length of flesh is still attached to the body at either end.

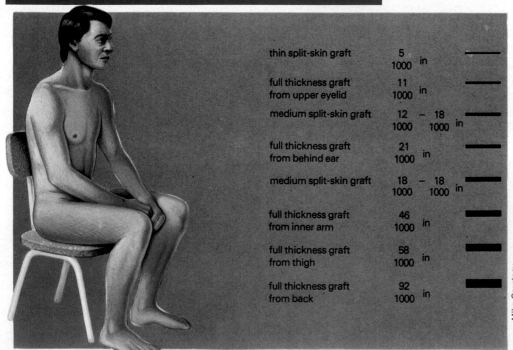

Plastic Surgery Unit, Stoke Mandeville Hospital

Left This special knife, with an adjustable depth guide for the razor-sharp blade, is being used to cut a graft with a to-and-fro sawing action.
Below Common donor sites for free skin grafts, showing their relative thickness.

thin split-skin graft	$\frac{5}{1000}$ in	
full thickness graft from upper eyelid	$\frac{11}{1000}$ in	
medium split-skin graft	$\frac{12}{1000} - \frac{18}{1000}$ in	
full thickness graft from behind ear	$\frac{21}{1000}$ in	
medium split-skin graft	$\frac{18}{1000} - \frac{18}{1000}$ in	
full thickness graft from inner arm	$\frac{46}{1000}$ in	
full thickness graft from thigh	$\frac{58}{1000}$ in	
full thickness graft from back	$\frac{92}{1000}$ in	

Mike Courtney

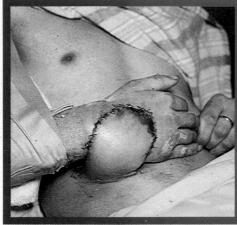

Skin flaps are deeper than grafts and are at first still joined to their donor site *(right)*. This flap from the abdomen has been manipulated and stitched to cover a wound on the patient's hand. When the flap has 'taken', its original connection to the donor site can be severed. The tube flap *(far right)* is attached to the donor area at both ends, either of which can then be moved to a damaged site, often via an intermediate site. Skin grafts can be meshed to form a living net *(below right)* that can be expanded to cover a large wound. It also allows fluid to drain naturally from the wound.

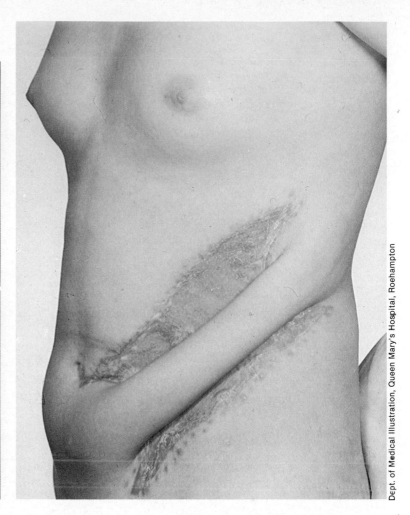

Dept. of Medical Illustration, Queen Mary's Hospital, Roehampton

This is then rolled around and stitched into a tube while the underlying gash which is left is stitched up.

In a later operation, one end of the tube is cut away from its base and rotated upwards towards its new site. It is then sewn on to the damaged tissue. Once it has taken root, the other end of the tube is also transferred to its new site.

The tube can then be opened up and laid flat on its new home.

Alternatively, where a piece of tissue is being transferred from one end of the body to another—from the abdomen to the face for example—a half-way stage is included. One end of the flap is temporarily attached to the wrist and then the other end moved up into place on the patient's face. Once the latter has 'taken', the end temporarily fixed to the wrist is moved up to the face as well, and the tube is opened out and stitched in place as before.

There are major advantages to this technique, although it may seem highly traumatic and restricting to the patient. There is a decreased risk of infection because the tube is closed off. Finally, there is no raw surface on the outside, so superficial scarring is kept to a minimum degree.

Breast reconstruction

Flap techniques are especially useful in repairing facial damage and whenever it is advantageous to have a fatty pad beneath the skin. They are increasingly being used to aid cancer victims who have had an operation to remove one or both breasts (mastectomy). Some plastic surgeons, however, still believe that breast reconstruction, which involves pulling a flap of tissue from the abdomen, is dangerous in post-mastectomy patients because of the risk of transferring the cancer to the new tissues. However, it has been established that about four patients out of every 10 suffer severe depression for a year or more following mastectomy even if a prosthetic breast (artificial breast) is provided, so reconstructive surgery is becoming more and more popular. In addition, there is no clear-cut evidence that such surgery does spread the cancer; indeed, with some patients it may even raise their chances of a healthy recovery, either because of the psychological effect or for some other reason.

Once the flap has been constructed in the area where the breast has been removed, a thin-walled silicone bag containing silicone gel (which is of similar density to normal breast tissue) is inserted into the flap to create an artificial breast. Skin grafts are then used to cover the new breast and, in some cases, the original nipple can be inserted.

Artificial implants can also be used in women requesting breast enlargement operations—now one of the commonest types of cosmetic surgery requested.

Other commonly requested types of breast surgery include breast reduction and breast lifts. In the latter case, a section of the breast is removed and the skin reformed around the remainder, leaving the nipple higher up. There are always obvious scars although these in time may fade and sensitivity of the nipple is lost in fifty per cent of patients. In addition, the refashioned breast will begin to droop again in time.

Micro-surgery

One exciting recent development in reconstructive surgery has been in the field of micro-surgery. Using low-powered microscopes, teams of surgeons are able to rejoin minute blood vessels and nerves (using needles and nylon thread almost invisible to the naked eye), so that amputated fingers, or even whole hands or limbs, can be reconnected in a single operation.

Other expanding new areas include the

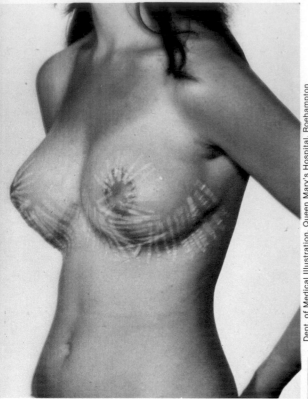

increasing use of flaps involving the transfer of the full thickness of skin together with the underlying muscle. Unlike the flaps described above, these are free flaps, cut away completely from the donor site and moved to the recipient site, where they are re-connected by means of micro-surgery. They provide an excellent method of moving a large bulk of tissue in cases of serious injury, but they are not really suitable for plastic surgery of the face, as the results would be generally unacceptable cosmetically.

Cosmetic surgery

There are now few areas of the body which cannot be changed in some way by plastic surgeons. But much more depends on the expectations of the patient before the surgeon will attempt a face lift, nose remodelling or fat reduction requested. All too often, the scarring which results is grossly in excess of what the patient had expected.

But facelifts, however, can be remarkably successful. Although they involve extensive dissection of the face, scarring is generally limited to above the hairline. A full facelift lasts about ten years.

Similar techniques are used for reducing wrinkles ('crows feet') around the eyes, though an acid peel may also be needed to get rid of all of them. An acid solution is painted into the face and literally burns away, in a controlled manner, the top layer of the skin. The complete peeling and healing process takes about a week, though the skin may be red and unsightly for some time afterwards.

Buttock reduction and stomach flattening can also be achieved quite easily by removal of superfluous fat, though with extensive scarring. In some cases mishaps have occurred. An American woman was awarded heavy damages in 1979 because her surgeon displaced her navel during a stomach-flattening operation.

Plastic surgery has made enormous progress in recent decades, and in the area of reconstruction a great deal can be done to make life bearable for those whose appearance is deformed by injury, disease or one of the many hereditary defects. Cosmetic surgery, although it obviously cannot give an assurance of instant beauty or eternal youth, can transform a person's appearance and relieve much unhappiness caused by self consciousness.

Dept. of Medical Illustration, Queen Mary's Hospital, Roehampton

Breast reduction involves careful preliminary planning *(right)*. The lines indicate where the surgeon will make his incisions, while the semicircular marks show the new sites of the nipples. The appearance of the breast is dramatically and permanently improved by this operation *(above)*.

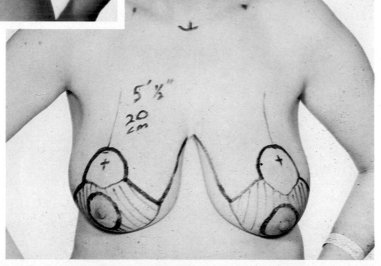

Remodelling of this unattractively shaped nose *(below left)* involves fracturing the nasal bones, then removing excess tissue and repositioning the bones. Next the nose must be immobilized by packing the nostrils and splinting it externally *(below centre)*. The packs and splints are removed after several days; later, when the bruising has subsided, the patient can admire her new nose *(below right)*.

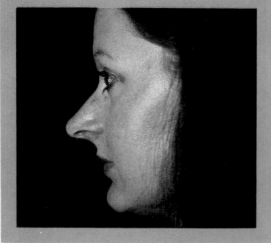
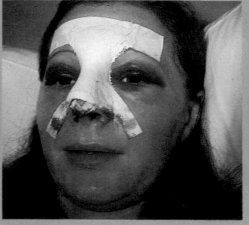
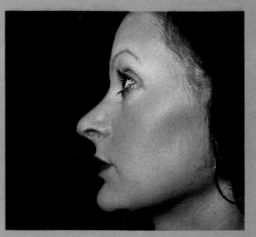

Spectrum Colour Library

Life before the womb

In 1978, after many years of research, the world's first 'test-tube' baby was born, to a woman who had been infertile for many years. The technique of removing an egg cell from the ovary, fertilizing it outside the body, and then implanting it in the womb sounds straightforward, but many obstacles had to be overcome before success was finally achieved.

A common cause of infertility is blockage of a woman's fallopian tubes, the two ducts that normally provide a passage between the ovaries and the uterus along which an egg can travel. The 'test-tube' baby technique is, quite simply, a means of getting round this blockage.

The launching of a potential human life at the microscopic stage takes place every 28 days in a woman of childbearing age. It begins when the pituitary, a gland at the base of the brain with an important influence on growth and body functions, sends a surge of hormone into the bloodstream. This hormone prompts the rapid ripening and release of one or more of the eggs contained in the woman's ovaries, her two 'sacks' of egg cells which are on each side of the uterus. The ripe egg travels from one or other ovary down the fallopian tube connecting it with the uterus, which is where, if fertilized, it will grow into a baby.

In the fallopian tubes

The journey from ovary to uterus takes about three and a half days, and much of this journey is spent passing through the 10–13 cm (4–5 in) long fallopian tube, the inside of which is only as wide as a piece of thread. The tube is lined with *cilia,* tiny hairs that waft the egg very slowly along its course, ensuring that it travels neither too fast nor too slowly. This is important, because it is while the egg is still in the tube, generally about half-way down, that fertilization has to take place.

Within an hour of fertilization occurring, the nuclei of the two cells, the egg and the sperm, fuse together. Another hour later the egg divides into two cells. Then, repeatedly

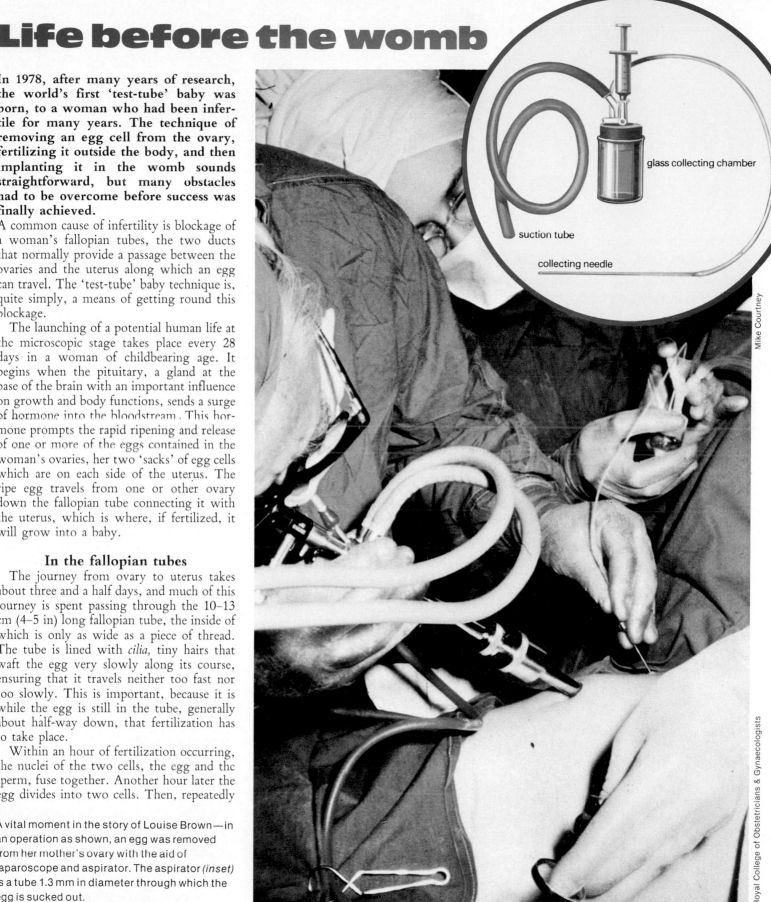

A vital moment in the story of Louise Brown—in an operation as shown, an egg was removed from her mother's ovary with the aid of laparoscope and aspirator. The aspirator *(inset)* is a tube 1.3 mm in diameter through which the egg is sucked out.

glass collecting chamber

suction tube

collecting needle

Mike Courtney

Royal College of Obstetricians & Gynaecologists

1043

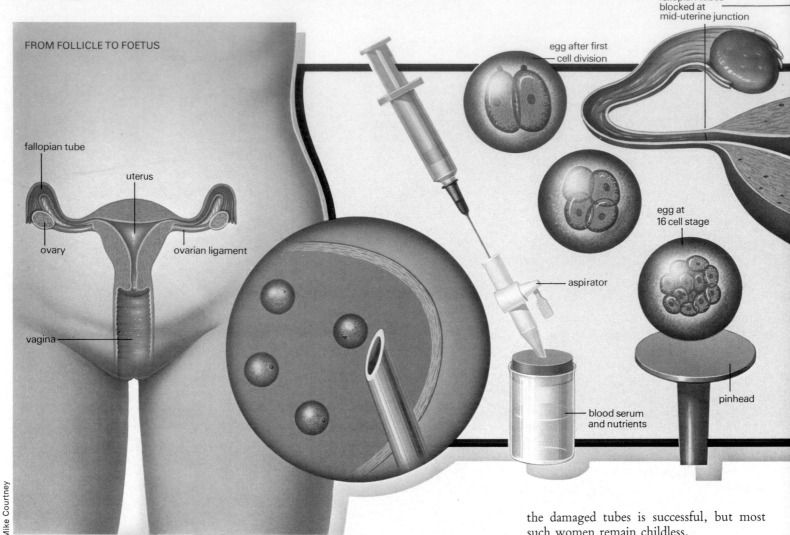

FROM FOLLICLE TO FOETUS

fallopian tubes
blocked at
mid-uterine junction

egg after first
cell division

fallopian tube

uterus

ovary

ovarian ligament

vagina

aspirator

egg at
16 cell stage

blood serum
and nutrients

pinhead

Mike Courtney

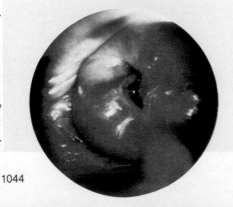

Royal College of Obstetricians & Gynaecologists

Left Views through a laparoscope. *Top* Women who have had operations like this—the removal of fallopian tubes—will gain hope from the Steptoe technique. *Below* The blurred tube in the foreground is the cannula about to enter the cervix and implant the fertilized egg in the uterus of the mother-to-be.

doubling the number of its cells, the egg enters the womb and implants itself in the lining, where if all goes well it will stay for the rest of the nine months of pregnancy.

As the fallopian tubes play such a vital part in this process, the slightest damage to them can break down the whole reproductive mechanism, by making it impossible for egg and sperm to meet. Roughly one in ten couples in the western world is unable to bear children, and in about one third of these the cause is blocked or damaged fallopian tubes. The damage is usually caused by infection or occasionally by abdominal surgery. Sometimes an operation to try to unblock

the damaged tubes is successful, but most such women remain childless.

In the early 1960s an idea developed in the mind of Patrick Steptoe, a gynaecologist and obstetrician at a hospital in Oldham, Britain, who had seen many such cases among the patients attending his clinics. It occurred to him that the blockage might be bypassed in a fairly straightforward way if four steps could be achieved. These were, first, to remove an egg from the woman surgically and, second, to fertilize it in the laboratory with sperm from her partner. (This is not done in a test tube at all, in fact, but in a shallow vessel called a *petri dish. In vitro* fertilization is the correct technical term, but 'test-tube' fertilization is the term that has stuck with the research since its earliest days.) The third step was to keep the egg alive in the laboratory for a few days until the beginnings of an embryo were achieved, and the final step was to implant this embryo directly in the woman's uterus, to grow there as in any other pregnancy. In essence, this is the 'test-tube' baby technique as it was ultimately developed after years of research.

Patrick Steptoe already had at his disposal an instrument for examining the ovaries and

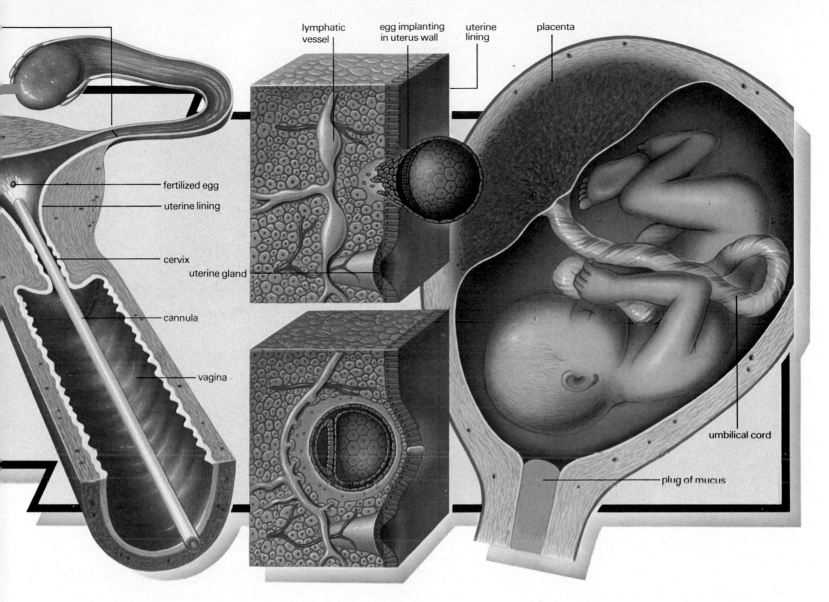

lymphatic vessel · egg implanting in uterus wall · uterine lining · placenta

fertilized egg
uterine lining
cervix
uterine gland
cannula
vagina
umbilical cord
plug of mucus

fallopian tubes, called a *laparoscope*. This is a kind of endoscope which can be inserted through a small cut in the abdomen, allowing the surgeon to look inside the abdominal cavity. He had pioneered the use of this instrument to find out why some women were infertile. It was clear to him that it could also be used, together with a needle-like suction rod, to remove eggs from an ovary of a woman with blocked fallopian tubes.

In the mid-1960s Steptoe found in Dr Robert Edwards, a Cambridge physiologist who had spent years researching the problems of laboratory fertilization, the very person he needed to help him take his idea further. Dr Edwards had learned that to repeat in the laboratory the apparently straightforward way fertilization takes place naturally required very much more than mixing egg and sperm together.

For fertilization to occur, the egg—about the same size as a full stop on this page—has to be at exactly the right stage of maturity

when it meets the sperm. The timing of its removal from the ovary therefore has to be perfect. The maturing process continues during the journey down the fallopian tube, where special secretions are present. Dr Edwards discovered how to reproduce these conditions in the laboratory. He created a mixture of nutrient chemicals, kept at body temperature, that allowed the ripening process to continue.

His own work took a big step forward after he met Patrick Steptoe who, unlike other gynaecologists he had approached, was willing to provide him with a steady supply of eggs. At first the eggs were donated by patients undergoing operations at Oldham General Hospital for the removal of their ovaries. Steptoe would travel to Cambridge with the eggs inside a rabbit's uterus, which kept them at the right temperature until they reached the laboratory bench.

By 1966, Edwards knew how to recognize under the microscope the exact moment

The miracle of the test tube baby. Any woman hoping to conceive by the Steptoe technique is monitored round the clock to catch the hormone surge, which is the best time to take the egg from the ovary. To remove the egg, the surgeon makes an incision in the abdominal wall to insert a laparoscope and see the ovary. Through another incision he inserts the hollow needle of an aspirator to gently suck out the minute egg. The egg is then transferred to a dish of nutrients kept at body temperature and the father's sperm is introduced. Soon one of the sperm penetrates the egg. After about 8 hours an egg chromosome and the sperm fuse into a single cell that divides and sub-divides over the next 2–3 days until there are 8 or 16 cells within the egg. At this point, the fertilized egg is implanted in the womb very simply via the cervix through a narrow polythene tube or cannula. Pregnancy can then develop normally as the egg becomes attached to the thickened uterus lining, works its way in and grows into a healthy foetus.

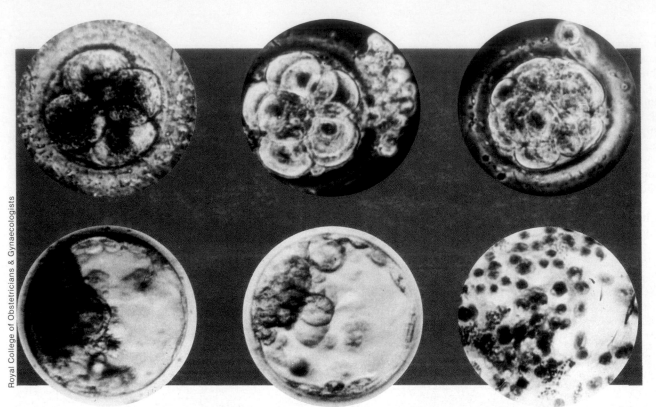

Royal College of Obstetricians & Gynaecologists

The developing embryo. After the egg has been fertilized, egg and sperm nuclei unite to form a single cell. This splits first into two, then into four cells *(top left)*. By the time the egg had reached the 8 cell stage *(top centre)*, it was ready for replacement in Lesley Brown's womb, although Grace Montgomery did not receive the egg until the 16 cell stage *(top right)*.
Bottom left and centre Dark areas are embryonic discs that will form the foetus.
Bottom right The egg seen from outside.

when the egg is ripe and capable of being fertilized: it changes its appearance slightly when it is ready to fuse with the male cell. However, he was still working on a similar problem with the spermatozoa, 20,000 times smaller than the eggs. They too undergo an obscure change, known as *capacitation*, during their journey up the oviduct. Unless 'primed' in this way they will not fuse with the egg.

The first real breakthrough came in February 1969: Steptoe, Edwards and another Cambridge scientist announced in the scientific journal *Nature* that for the first time they had achieved the fertilization of several human eggs outside the body. At a press conference the following year they announced that, given the facilities, the funds and the patients, it should be possible to produce a baby 'within one year'.

By January 1971 they had grown fertilized eggs to the point where they would normally have reached the uterus, but a major problem, that of persuading the uterus to accept the potential embryo, remained. This was to be their main preoccupation in the years that followed, and is still a difficulty in the process. There is no way of grafting a fertilized egg into the uterus surgically; it simply has to be placed there, usually when it has reached the eight-cell stage. In a simple ten-minute operation that does not require an anaesthetic, the egg and culture fluid are expelled from a cannula (tube) into the uterus.

Hormone interference

Between 1973 and 1977, a total of 77 women were subjected to the full treatment, but only three of them became pregnant. One of these pregnancies was *ectopic*, that is, the egg continued to grow in the oviduct instead of implanting itself in the uterine lining. This is very dangerous for the mother, so it had to be ended. The other two ended in very early spontaneous abortions.

Eventually Edwards discovered the reason for the failure. In all these patients, hormones had been administered to ensure rapid ripening of more than one egg —'superovulation'—to make the job of recovery and fertilization easier. Biochemical studies showed that these hormones were not producing normal ripening of the eggs.

In addition, superovulation was interfering with the action of another pituitary hormone which stimulates the lining of the uterus into forming a 'bed' ready to receive an egg if it should become fertilized. Superovulation was delaying the preparation of the womb, making it impossible for implantation to occur.

In 1977 it was decided to abandon that method, and instead to work out ways of recovering an egg at just the right moment in the natural cycle. Timing is obviously crucial, and the techniques developed demand that a team of theatre nurses be on constant standby, while hormone levels in urine and blood samples are monitored round

the clock to watch for the hormonal surge that shows that the egg is ready.

When the surge occurs, sperm from the male partner is collected and prepared, and 18 to 22 hours after the onset of the surge an egg is collected from the woman by laparoscopy. Egg and sperm are then mixed in a dish containing a culture medium, and kept at body temperature. When fertilization takes place, the egg starts to divide and grow, and after two or three days is placed in the uterus. Further hormone monitoring is maintained to find out whether pregnancy has been established.

These stages were the ones followed in a new experimental programme that was set up, for which a further 77 patients were accepted. This time Steptoe and Edwards were to be successful. Louise Brown was born at 11.47 pm on 25 July 1978. The first human being to be conceived outside the womb, she was perfectly normal and healthy. Another healthy baby, Alistair Montgomery, was born from this group of women the following January, demonstrating that Louise Brown's birth was not an unrepeatable achievement. There were two other pregnancies in the group, a total of four out of 32 patients from whom an egg had been removed, fertilized and placed in the uterus. One of the pregnancies ended in a miscarriage at 20 weeks, and the other ended at 11 weeks, because of a genetic defect in the embryonic child.

Even so, there is no definite evidence to suggest that babies conceived in this way carry a higher risk of abnormalities than normal births. *Amniocentesis,* drawing off a sample of the amniotic fluid surrounding the foetus in order to check it for abnormalities, and other techniques such as ultrasound are used to make sure that the unborn baby is developing normally.

With the precaution of removing or blocking off entirely the damaged fallopian tubes to ensure that a woman does not develop an ectopic pregnancy, the technique seems to be very safe for the mother.

The chance of a child

Steptoe and Edwards now feel that, with further refinements of their technique—for example, they noticed that all four of the patients who became pregnant received their embryo at night, a factor requiring investigation— they will be able to offer infertile women a reasonable chance of success. They plan to open a clinic for this purpose, and two London hospitals have also announced 'test-tube' baby programmes. There have already been two more births using the same techniques, in India and Australia.

In vitro fertilization could also be used to help couples with other problems than blocked fallopian tubes. As relatively few sperm are needed to fertilize an egg in a laboratory dish, a man with a low sperm count who would be otherwise infertile could father a child in this way.

There is also cause to hope that, if the procedure can become routine, it could be used to help avoid the birth of some children with inherited disorders. The embryos of couples at risk would be examined in the laboratory, and the abnormal ones discarded.

The technique could be further extended to help women who cannot conceive because of damage to their ovaries, preventing the release of eggs. They could instead receive an egg from a donor, to be fertilized by their partner. The woman would still carry the pregnancy and give birth.

One refinement of the technique that could lead to a significant improvement in the success rate, and would certainly make it easier for doctors to try repeatedly to make a woman pregnant if first attempts fail, would be to store fertilized embryos in deep-freeze before transfer to the uterus up to a year later. Superovulation could once again be used, to produce several ripe eggs at once which might all be fertilized in the laboratory. None would be returned to the mother immediately, so that the effects of the hormone used to induce superovulation could have time to wear off. Then, at appropriate times in successive months, the fertilized embryos could be brought out of storage, slowly thawed, and implanted singly.

So far no human patient who has had a previously stored embryo implanted in the womb has become pregnant, but the technique has been used on animals, with some success, even after the egg has been in store for up to five years. Therefore there is no reason to doubt that it may one day be possible with human beings too, though many will debate the ethics of the technique.

Before the *in vitro* fertilization method leaves the experimental stage and is put into widespread use, the doctors and scientists involved still have to improve their understanding of why so many attempts fail, and adjust their methods to ensure greater success in future.

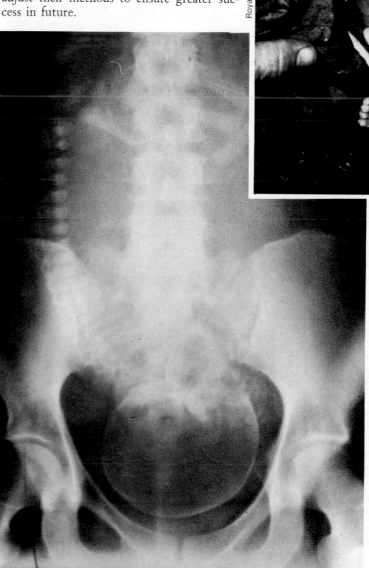

Royal College of Obstetricians & Gynaecologists

Royal College of Obstetricians & Gynaecologists

Left An X-ray of Lesley Brown during her 36th week of pregnancy. Although the baby, with its skull visible bottom centre, looks fairly large, it had not at this stage grown as much as expected and doctors had a few anxious moments. However, a last minute spurt allayed fears. This was just one of the many checks made during the course of the pregnancy. *Above* Louise Brown, just 30 seconds old, emerges into the world as the first ever 'test tube' baby—a great day for childless women everywhere.

Rewriting the genetic code

Since his imagination created such fabulous hybrids as the Sphinx, which had the head of a woman and the winged body of a lion, man has dreamed of shaping new forms of life. Now, using the techniques of genetic engineering, he is able to change the fundamental properties of living organisms. It may one day be possible to create many new organisms with properties that will benefit mankind: for example, bacteria that could clear up oil spills by consuming the oil, and self-fertilizing crops to help solve the world food problem.

The nuts and bolts of the conventional engineer are easy to understand, but what of the raw materials of his genetic counterpart? The basic units the genetic engineer has to work with are cells, the microscopic building blocks of which all living things are made. In the cells are housed thread-like structures —the *chromosomes*. Situated in a row along the length of each chromosome are the *genes*, whose combined action defines the characteristics of their owner—from the humble bacterium to the human being— right down to the finest detail.

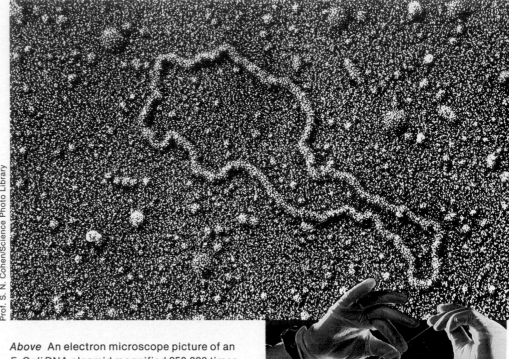

Prof. S. N. Cohen/Science Photo Library

Above An electron microscope picture of an *E. Coli* DNA plasmid magnified 250,000 times.

Ralph Crane/Life © Time, Inc./Colorific!

DNA

At an even more minute level, a single gene is constructed of the chemical *deoxyribonucleic acid* (DNA), shaped like a twisted ladder, the so called 'double helix'. Each ladder 'upright' consists of a chain of alternating sugar and phosphate units—the sugar is called *deoxyribose*. Attached to each sugar unit in the chain is a biochemical compound called a *base* which may be one of only four possibilities: *adenine, thymine, guanine* or *cytosine*. The bases stick out sideways from the sugar-phosphate chains and they link together between the chains to form the rungs of the double helix ladder. Adenine links only with thymine, and guanine only with cytosine, so each rung consists of two linked bases, either adenine-thymine or guanine-cytosine.

The sequence of bases along the sugar-phosphate chains is called the *genetic code,* and it tells all the other chemicals in the cell what to do, rather as a message in Morse code can pass information to a human (dots and dashes being equivalent to the four DNA bases). A gene is simply a sequence of bases along a DNA molecule that carries the instructions for a particular characteristic such as blue eyes or brown hair.

When the gene goes to work in the cell, the two 'uprights' of the DNA ladder break apart and the exposed bases 'read off' in groups of three—triplets—to form a carrier substance known as *messenger ribonucleic acid* (mRNA). This is also in the form of a double helix and is chemically very similar to DNA but it contains the base *uracil* in place of thymine.

The vital function of mRNA is to organize the manufacture of proteins, biochemical molecules vital to every bodily action, by directing the stringing together of smaller units called *amino acids*.

Virus diseases

Many of the most serious diseases afflicting man are caused by viruses. These are the simplest living things known, consisting of little more than pieces of DNA or RNA which invade the living cells of a plant or an animal and force the host cells to produce more viruses within a very short time, harming or even killing the host in the process. The common cold, rabies, polio and yellow fever are all virus diseases.

When a living cell is invaded by a virus such as that of the common cold, it reacts by making a protective protein called *interferon.* Made, like other proteins, through the combined action of DNA and mRNA, interferon is the subject of enormous interest, for limited clinical trials suggest that it may be able to cure not just the common cold but also some forms of cancer (there is evidence that viruses may cause some human cancers). So in 1980 it was with great excitement that the medical world greeted the news that genetic engineers had managed to create a source of interferon, albeit a limited one. The interferon story is a tale that deserves telling in detail, for it exemplifies the complex yet elegant processes of genetic engineering.

The 'hero' is a single-celled bacterium called *Escherichia coli (E. coli* for short) which inhabits the human intestine in large numbers, normally innocuously, and which has solved many a mystery for geneticists. The genetic

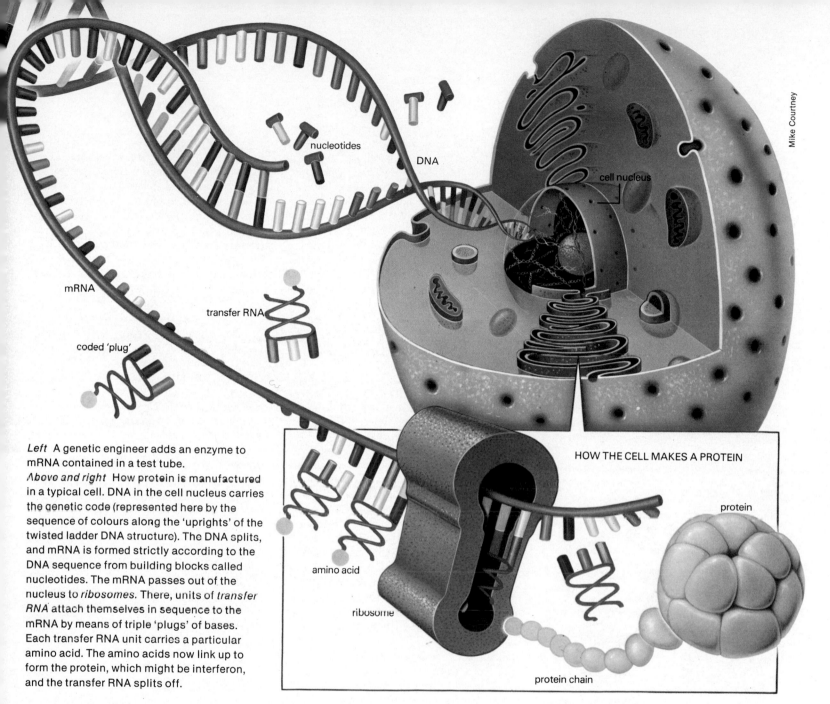

nucleotides

DNA

cell nucleus

mRNA

transfer RNA

coded 'plug'

HOW THE CELL MAKES A PROTEIN

protein

amino acid

ribosome

protein chain

Left A genetic engineer adds an enzyme to mRNA contained in a test tube.
Above and right How protein is manufactured in a typical cell. DNA in the cell nucleus carries the genetic code (represented here by the sequence of colours along the 'uprights' of the twisted ladder DNA structure). The DNA splits, and mRNA is formed strictly according to the DNA sequence from building blocks called nucleotides. The mRNA passes out of the nucleus to *ribosomes*. There, units of *transfer RNA* attach themselves in sequence to the mRNA by means of triple 'plugs' of bases. Each transfer RNA unit carries a particular amino acid. The amino acids now link up to form the protein, which might be interferon, and the transfer RNA splits off.

material of *E. coli* consists of two parts, a single large circle of double-helix DNA plus several more circular double-helix DNA units about a tenth the size which are known as *plasmids*. Although only about a hundredth of a millimetre in diameter, the plasmids can confer characteristics on the cells they inhabit through the action of their genes.

When the cell of *E. coli* reproduces, it does so simply by duplicating all its DNA and splitting into two, so the plasmid DNA is carried on from generation to generation. The aim of the interferon programme was to transfer the human gene carrying the instructions for interferon manufacture into one of the plasmids, and to let the bacterium multiply, thus creating a new source of in-

terferon protein.

The first step in the process was to make cultures of *E. coli* and examine them for suitable plasmids. The chosen *E. coli* was then kept alive in a solution of sugars, mineral salts and vitamins at a steady temperature of 37°C (98.6°F). Step two was to get the plasmids out of the bacteria. This was done by spinning the bacterial solution in a centrifuge (a machine that spins liquids round in a tube very fast to separate constituents of different densities) and then exposing the resulting deposit to chemicals that eat their way into the bacterial cell walls and burst them open. Now the mixture was treated with a very cold chemical solution and the whole lot centrifuged again so that only minute white threads of DNA were left

in the bottom of the centrifuge tube. The plasmid DNA could then be separated from the rest simply on the basis of its density: it was placed on a very thin slice of gelatin, stained and subjected to an electric current, a technique called *electrophoresis*. As a result, the low-density plasmid DNA separated out to end up at the top of the gelatin strip.

Now that the plasmid DNA was ready for treatment, the biochemists had to obtain the human interferon-manufacturing gene. This was far more difficult than the isolation of the plasmid DNA. Nevertheless, using many sophisticated biochemical techniques, the genetic engineers were able to isolate the interferon gene from the white blood cells of humans, and to analyze its coding sequence. There were, however, severe drawbacks to

Mike Courtney

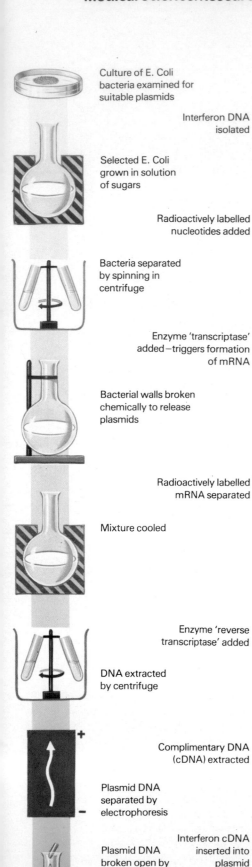

Culture of E. Coli bacteria examined for suitable plasmids

Interferon DNA isolated

Selected E. Coli grown in solution of sugars

Radioactively labelled nucleotides added

Bacteria separated by spinning in centrifuge

Enzyme 'transcriptase' added – triggers formation of mRNA

Bacterial walls broken chemically to release plasmids

Radioactively labelled mRNA separated

Mixture cooled

Enzyme 'reverse transcriptase' added

DNA extracted by centrifuge

Complimentary DNA (cDNA) extracted

Plasmid DNA separated by electrophoresis

Plasmid DNA broken open by enzyme

Interferon cDNA inserted into plasmid

New plasmids infect E. Coli culture

Infected E. Coli culture grows

mRNA extracted

mRNA injected into frog's eggs – human interferon produced by eggs

this method, for the gene contains not only the code for stringing the bits of interferon together and controlling the way the gene works, but also regions of 'nonsense'. When the mRNA is made in the cell by reading off the codes from DNA, all the 'sense', including the general instructions, are translated, but the 'nonsense' is somehow ignored, and this proved a stumbling block to getting the whole human gene for interferon to work in bacteria.

Rather than using the whole gene as their starting point, the genetic engineers worked backwards. Instead of extracting the whole interferon gene they started with mRNA, the intermediary between DNA and interferon. All the interferon so far made in bacteria has been produced in one of two experimental ways.

In the first method, the interferon DNA was isolated and put in a test tube with a 'sludge' of *nucleotides,* the building blocks of DNA and RNA molecules, which were radioactively labelled so that they could be easily traced. An enzyme (a biological catalyst or trigger) called *transcriptase* was then added to the mixture; this triggered the DNA into making mRNA. Because it was radioactively labelled, the mRNA could be separated from the DNA. The mRNA was then added to another mixture of nucleotides and a different enzyme called *reverse transcriptase* added. This enzyme acts by making the mRNA produce a copy of itself in DNA form, but, most importantly, this so-called *complementary DNA, cDNA,* does not contain the 'nonsense' of the original gene.

A schematic representation of how human interferon is prepared by the techniques of genetic engineering. The two starting points *(top left)* are firstly a culture of *E. Coli* bacteria and secondly interferon mRNA. Set out as a series of steps, the process looks fairly straightforward, but in fact the teams of scientists involved took many years to perfect the techniques. The final extraction of human interferon from frogs' eggs for laboratory tests *(right)* was a considerable triumph.

In the second method, scientists started from the interferon protein itself. They took a gamble on the exact sequence of amino acids in the interferon protein, worked out the DNA code, and actually made several pieces of the gene in the laboratory, one of which, they hoped, would be the one they wanted. They then put the gene pieces into a preparation of mRNA, and from this obtained the complete mRNA message for interferon. As in the other method, it was then a comparatively simple step, again with the help of reverse transcriptase, to make a complete double-stranded cDNA for interferon.

Biochemical 'scissors'

Once the cDNA for the interferon gene was prepared, the next step was to break open the bacterium's plasmid DNA in order to insert the new instructions. This was done with biochemical 'scissors', enzymes called *restriction endonucleases* or *ligases* that occur naturally in the cells of bacteria and viruses and have the ability to slice up DNA into little bits. These restriction enzymes are extremely specific. They always make their snips in the same part of the DNA and the 'frayed' ends produced as a result of their actions are 'sticky'. Because the restriction enzymes *are* so specific, any DNA, whatever its source, will join up with any other that has been cut with the same enzyme.

Having been severed by the same enzyme, the circular plasmid DNA from the *E. coli* developed breaks at specific points, as did the human cDNA. The two were finally mixed together, the temperature lowered to help

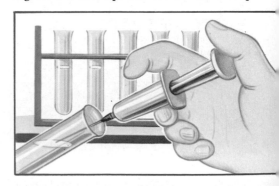

joining of the 'sticky' ends and the DNA from the two sources combined. With luck, the interferon-producing DNA would now have become part of the *E. coli* plasmids, but how did the biochemists test this out? First they had to make the plasmids infect a new selection of *E. coli* bacteria. These new *E. coli* cells hungrily grabbed the plasmids, encouraged to do so by a high calcium, low temperature environment. Again, with good fortune and a considerable degree of trial and error, the living culture of *E. coli* would contain plasmids capable of making human interferon.

To test the manipulated *E. coli,* the scientists extracted the mRNA made by its new genes and injected it into frog's eggs. When the right RNA was present, the eggs quickly started making interferon.

Even such a success story has its failings. So far, the interferon output has been very low. Also, the interferon itself does not match its human counterpart exactly, because it lacks certain sugar molecules that are attached to human interferon and are thought to play an important part in the way it works.

'Recombinant DNA technology' is the phrase used to describe the making of interferon by genetic engineering, but this is by no means the first human protein to have been inserted into bacteria. Insulin and human growth hormone are among a growing list of substances whose genes have been transferred into plasmids and cloned in culture—with enormous implications for the future of medicine. For that future, recombi-

nant DNA technology is at a kind of crossroads. The destination scientists are aiming for is the replacement of defective genes with their normal counterparts, but there are several possible routes open.

One route is to find bacteria that naturally carry genes that are normal equivalents of the deficient ones, or to synthesize normal genes in the laboratory, and to use these in genetic engineering processes. Another is to use normal human genes and to attempt to transfer them from one person to another.

On route one, some progress has been made, in particular with the hereditary disease *galactosemia.* Babies born with this disease cannot use the sugar galactose which is found in milk and other dairy foods. Unless they completely avoid foods containing galactose they soon start to become mentally retarded.

Using a virus

The ever-obliging *E. coli,* however, possesses normal genes for galactose utilization. What the genetic engineers did was to incorporate the galactose gene from *E. coli* into a virus that normally infects bacteria, then used the virus to infect abnormal cells taken from human skin. Incredibly, the technique was successful, and the skin cell culture began to make its own enzymes for breaking down galactose. As far as making genes is concerned, this has been done, but there is not yet any way to get them into human chromosomes.

On route two, progress is more conjectural than concrete. It has been applied to the disease diabetes. In diabetics, the gene gover-

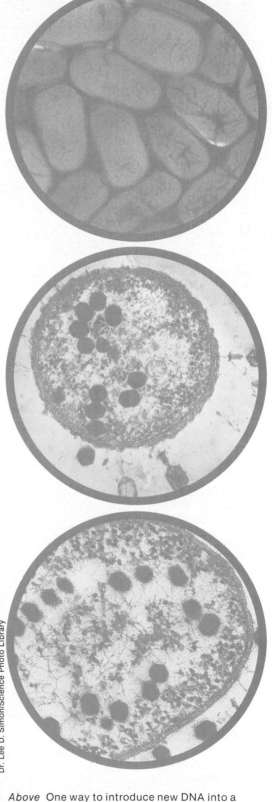

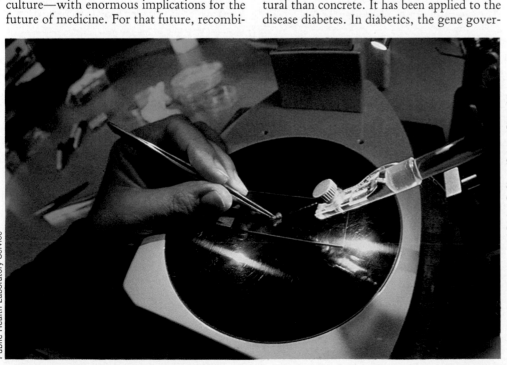

Above One way to introduce new DNA into a cell is to use a modified virus. A group of *E. Coli* cells *(top),* for instance, may be attacked by a virus called a *phage.* The phages invade the *E. Coli* cell *(centre),* and the cell begins to follow the coded instructions of the phage DNA (bottom) to produce more phages. *Left* Injecting mRNA into a frog's egg.

Public Health Laboratory Service

Dr. Lee D. Simon/Science Photo Library

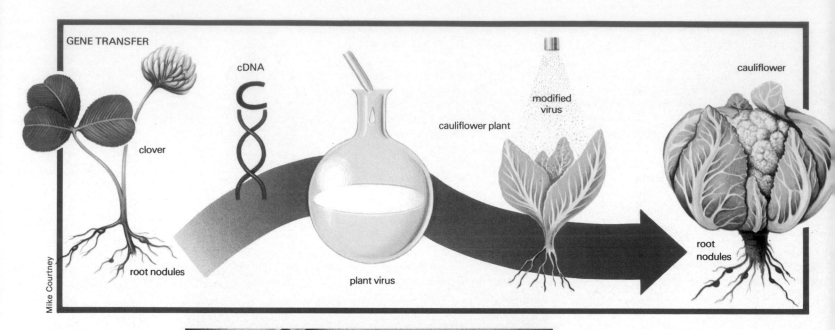

GENE TRANSFER

cDNA

clover

root nodules

plant virus

cauliflower plant

modified virus

cauliflower

root nodules

Mike Courtney

Clover is one of the few plants that can extract (fix) nitrogen from the air. The process is carried on in nodules on the plant's roots *(right)*. It might be possible to prepare cDNA corresponding to the nitrogen fixing gene, insert it into a virus and infect another plant, such as a cauliflower, with the virus so enabling that plant to fix nitrogen for itself.

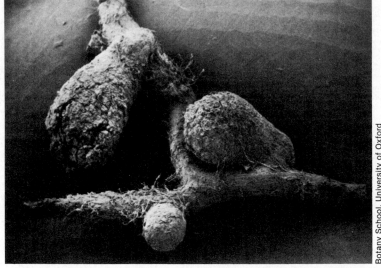

Botany School, University of Oxford

ning the production of the vital hormone *insulin* is defective. Theoretically, instead of a daily injection of insulin, a diabetic could be injected with a virus carrying the normal gene which would then do the work within the body. This is a problematical area indeed. Success would depend not only on getting the virus to infect enough body cells to be effective, but to prevent it from having a malignant effect and to stop it from being destroyed by the body's highly efficient natural defences. The danger is that such attempts to repair defects might have catastrophic rather than curative results.

Although they have received the most urgent attention, medical problems are not the only ones to be attacked by the genetic engineer. As far as plants are concerned, genetic engineering goes back to the Stone Age, for it was then that man first started to

select and breed crop plants with identifiable and desirable traits. Plants are, in many ways, much more malleable material for genetic engineering than animals, because their genes and chromosomes act in a much less rigid way. Unlike animals, plants can, for example, be easily persuaded to regenerate themselves from just one single cell.

In the field of genetic engineering, scientists are working on plants in various ways. One approach is to mix two cells from completely different plants—say wheat and soya bean—and from the resulting cell grow a completely new plant that might revolutionize the world food problem. So far, however, this particular combination has reached no further than a poor-definition 'blob' of cells.

Because they contain between 50 and 80

per cent protein, dried bacteria are widely used as an animal feedstock. One of the bacteria used for this feedstock is called *Methylophilus methyltrophus*. Scientists working on ways to make energy savings in feedstock production found that one of the biochemical steps by which the bacterium makes its proteins is really rather ineffecient. So, using the techniques of recombinant DNA technology, they first found a suitable plasmid, broke its DNA open and inserted a new, more efficient gene into its circle of DNA. The plasmid was then transferred into the *M. methyltrophus* bacteria. The resulting yield of protein was significantly increased.

Safety

Since the prospect of genetic engineering turned, in the 1970s, from a dream to reality, it has raised vital questions of both safety and morality. The alarm was first raised in 1973 when it was feared that strains of *E. coli* that had been altered by genetic engineers could have disastrous effects, perhaps creating havoc by proving resistant to antibiotics. Equally alarming, a manipulated virus might turn out to have tumour producing properties.

Through innumerable working parties and committee on both sides of the Atlantic, the argument has raged about whether genetic engineers should be allowed to 'play God'. The protagonists on one side insist that society will be best served by scientists truly committed to their abilities to sound an 'early warning'; those on the other demand that every citizen should have the right to influence the progress of a science with such momentous potential.

Contraception: the hormone balance

The huge population 'explosion' of the late twentieth century results largely from the fact that modern medical care has cut the death rate dramatically in the developing countries. As a result, many scientists have explored ways to cut birth rates—nowhere more controversially than in the development of the oral contraceptive pill.

Since the 1960s, the pill has proved to be the most popular and effective method of reversible contraception, offering almost 100-percent protection. Nevertheless, as with many other revolutions, the contraceptive 'revolution' heralded by the appearance of the pill disappointed those whose expectations of its benefits were unrealistically high.

Today, doctors generally recognize that for some women at least, the pill does have undesirable side effects. Many doctors believe that there can never be an 'ideal' contraceptive suitable for all women at all stages in their lives. Instead, a whole spectrum of methods is needed for women living under different circumstances in different countries and having differing health problems, requirements and preferences. Each method has its own advantages and disadvantages, and a balance must be struck between effectiveness and risk. This ratio of benefit to risk is at the heart of the problem of developing new contraceptives.

Nearly all the 60 million women taking oral contraceptives in the early 1980s are given the type of pill known as the *combined pill*. It contains two synthetic female sex hormones (chemical messengers)—oestrogen and progestagen. Oestrogen is also produced naturally by the ovaries, as is progesterone, the naturally occurring form of progestagen. Both act on the pituitary gland to produce a surge of gonadotrophic hormones known as follicle-stimulating hormone (FSH) and luteinizing hormone (LH). These trigger off ovulation—the production of an ovum or egg. When the pill is taken for 21 days of the menstrual cycle, the amounts of oestrogen and progestagen are maintained at a constant high level. As a result, the gonadotrophic hormone surge does not occur. Thus there is nothing to trigger off ovulation. With no ovum, there can be no pregnancy.

Top Crystals of female sex hormone make a colourful pattern under the microscope. Discovery of their action paved the way for the development of contraceptive pills.

The oestrogen and progestagen also have important direct effects on the mucus in the cervix and on the endometrium (lining of the womb). The cervical mucus becomes scanty but dense, with its cells tightly packed together, providing a physical barrier against the entry of sperm into the womb. The endometrium becomes thin and unfavourable to the implantation of an ovum. Both these mechanisms make pregnancy unlikely even if ovulation did occur.

During the one week out of every four when the woman stops taking the pill, she will experience *withdrawal bleeding,* which is just like a period, but is caused by stopping taking the pill rather than by the natural menstrual cycle.

The first pills to be marketed during the 1960s contained far higher doses of the hormones than was necessary to prevent conception. Today's pills contain much smaller amounts of hormones in order to minimize the side effects which they produce.

These side effects can be divided into two categories: the first is that of the so-called 'minor' effects, and includes weight gain, nausea, changes in libido (sexual desire), headaches and inter-menstrual (between periods) bleeding. These can usually be avoided by changing the pill for one with a different formulation.

The second group of more serious or potentially serious effects includes an increased risk of various circulatory disorders —blood clots, strokes, high blood-pressure, and heart attacks, increased risk of gall-bladder disease (thought to double after four to five years of pill-taking), a decrease in glucose tolerance which places diabetic women at risk, depression, and a possible increase in birth defects if oral contraceptives continue to be taken during early pregnancy.

The 'pill' is probably the most extensively tested drug in the world. Large-scale studies in the late 1970, investigating the connection between oral contraceptives and mortality, indicated that the risk of dying from circulatory disorders was five times more likely in women taking the pill compared with those not taking it. However, a detailed recent re-evaluation of the data, both for the USA and 21 other countries, published in 1979 by the World Health Organisation (WHO), failed to confirm these results, suggesting that this particular risk may be considerably smaller than had been originally assumed. Nevertheless, for some groups of women the dangers are much greater than for others; these are women over 35, women who are overweight, and women over 30 years of age who smoke.

Although these risks do exist, not all the side effects of the pill are harmful. Some of the potential benefits that taking the pill can bring include the suppression of premenstrual tension and menstrual pain, the inhibition of benign (non-cancerous) breast lumps, a reduction in the chance of developing ovarian cysts and a reduced incidence of rheumatoid arthritis.

And the risks from taking the pill are, on balance, fewer than those from the preg-

Contraceptive pills are manufactured under carefully monitored conditions. The journey from laboratory to consumer involves rigorous quality control *(bottom left)* as well as highly sophisticated modern methods of packing *(bottom right)* in which scrupulous attention to hygiene is a prime consideration.
Below The end product on the conveyor belt.

Left A stage in the manufacture of long-term birth-control devices in the USA.
Centre In Thailand, as in other Third World lands, the 'pill' is sold in general stores.

the body, acting on a suspicion that this could be responsible for some of the side effects. They have been studying women taking a range of different pills, containing the same dose of oestrogen with varying amounts of progestagen. The results of one such study, reported in 1980, confirmed suggestions that some circulatory diseases, particularly strokes, may be related to the amount of progestagen in the body.

Variable pills

Such research has led to the marketing of very low-dosage *variable pills*. These are of two sorts: *triphasic* and *biphasic*. With the triphasic pill, the ratio of progestagen to oestrogen is altered three times during the 21-day cycle of pill taking. The developers of the pill think that this reflects the natural hormonal fluctuations more closely than the conventional pill, which supplies a constant dosage of oestrogen and progestagen in each pill for the entire 21 days. The triphasic pill gives a lower total dosage of progestagen per month than was previously possible, while still giving less than 0.05 mg oestrogen. Triphasic pills were first marketed in West Germany in 1979 and in Britain in 1980. Biphasic pills work in a similar way but give only two different ratios of the hormones.

An alternative approach is that of the *tricycle pill routine*. This method eliminates the need for a monthly withdrawal bleed. A research programme led by Dr Nancy Loudon in Edinburgh showed that by taking combined oral contraceptives continuously for 84 days, followed by 6 pill-free days, one group of women experienced only four withdrawal bleeds.

Progestagen-only pills, sometimes called minipills, are taken on a continuous daily basis, even during menstruation. Containing no oestrogen, they prevent conception by altering the cervical mucus and the lining of the womb, but rarely prevent ovulation. For this reason, they are not as effective at preventing conception as the combined pills, and may also cause irregular periods.

Nevertheless, they are particularly useful for older women at risk from the combined pill, for breast-feeding women (they do not reduce the amount of breast milk as do combined pills) and for women who cannot tolerate oestrogens for various reasons.

Birth control by progestagen is not

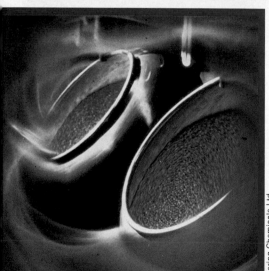

Above Contraceptive pills being coated with dye and sugar. The dye colour identifies the individual pill, and the hard sugar coating stops it breaking up and improves its taste.

nancies and births which they prevent.

Most of the harmful side effects of the pill seem to be caused by its oestrogen content, so the amount of oestrogen has been progressively reduced since the trials with the first pill, Enavid, in 1956. In 1969 the UK Committee on the Safety of Medicines recommended that the amount of oestrogen in combined oral contraceptives should not exceed 0.05 mg. None of the pills in use today contains more than 0.05 mg oestrogen.

Unfortunately, reducing the oestrogen content of the pill may lead to problems in the large number of women who are prescribed other drugs as well. Some antibiotics, tranquillizers, barbiturates and anti-epilepsy drugs are now known to interfere with the pill. Taken with such drugs, the low-dosage pills may give rise to intermenstrual bleeding or even unwanted pregnancies.

Reducing the oestrogen content of the pill preoccupied researchers up to the mid-1970s. Recently, however, researchers have investigated the action of the progestagen on

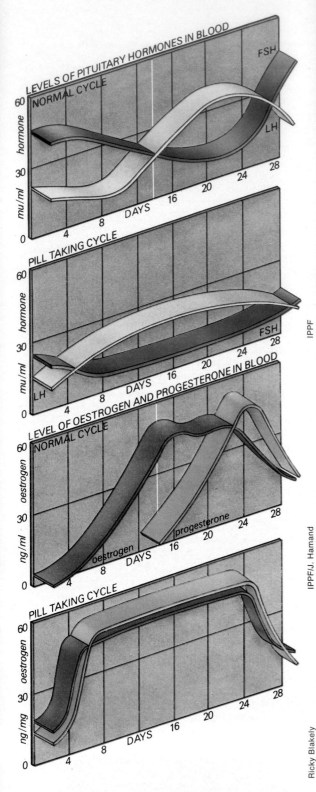

Above During the normal cycle, variations in oestrogen and progestagen levels act on the pituitary gland to alter levels of the gonadotrophic hormones LH and FSH, and these trigger ovulation (white line). The pill keeps oestrogen and progestagen at high levels. This acts on the pituitary gland to level out LH and FSH, preventing ovulation.

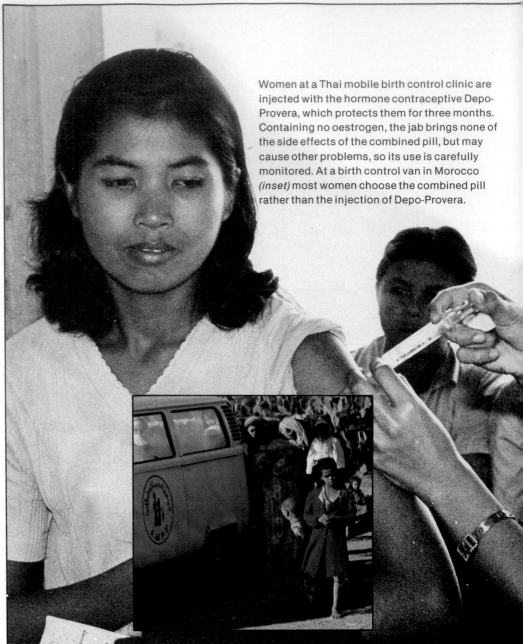

Women at a Thai mobile birth control clinic are injected with the hormone contraceptive Depo-Provera, which protects them for three months. Containing no oestrogen, the jab brings none of the side effects of the combined pill, but may cause other problems, so its use is carefully monitored. At a birth control van in Morocco *(inset)* most women choose the combined pill rather than the injection of Depo-Provera.

IPPF

IPPF/J. Hamand

Ricky Blakely

restricted to pill-taking; injectable progestagens have been used for over 15 years by about 10 million women. Today, about $1\frac{1}{4}$ million women use injectable contraceptives, but a good deal of controversy surrounds their use. The most widely used injectable is that containing depot medroxy-progesterone acetate (DMPA), commonly known by its trade name Depo-Provera. It is currently approved for use in more than 80 countries, including West Germany, France, Belgium, Denmark, Holland, Switzerland and New Zealand. It is at present banned in the USA.

In the UK, Depo-Provera is approved for short-term use for two groups of women only: those recently vaccinated against German measles and those whose partners have just had a vasectomy and are waiting confirmation of the end of sperm production.

One of the worries concerning Depo-Provera has been that it may cause cancer, but the results of the many studies of its possible effects have not definitely established any such connection. There are unpleasant side effects that can be attributed to Depo-Provera, however; these are weight gain and disturbances in the menstrual cycle. And since each injection gives contraception for up to 6 months, there is the problem of temporary infertility for women who want to conceive. On the other hand, Depo-Provera does not carry the possible risk of blood-clotting effects associated with the combined pill. Also, unlike some combined pills, it does not reduce the supply of breast milk.

New contraceptive devices currently being tested include silastic capsules, implanted either subdermally (under the skin) or in be-

| 1 | 2 | 3 | 4 | 5 | 6 | 7 | 8 | 9 | 10 | 11 | 12 | 13 | 14 | 15 | 16 | 17 | 18 | 19 | 20 | 21 | 22 | 23 | 24 | 25 | 26 | 27 | 28 |

28 DAY MENSTRUAL CYCLE
menstruation

days 1-13 egg develops in a follicle in ovary

day 14 follicle bursts at ovulation releasing egg from ovary

days 15-28 egg travels along fallopian tube to womb

| 1 | 2 | 3 | 4 | 5 | 6 | 7 | 8 | 9 | 10 | 11 | 12 | 13 | 14 | 15 | 16 | 17 | 18 | 19 | 20 | 21 | 22 | 23 | 24 | 25 | 26 | 27 | 28 |

ORAL CONTRACEPTIVES
menstruation

21-day course beginning on fifth day of cycle

pre-menstrual phase: no pills

① ② ③ ④ ⑤ ⑥ ⑦ ⑧ ⑨ ⑩ ⑪ ⑫ ⑬ ⑭ ⑮ ⑯ ⑰ ⑱ ⑲ ⑳ ㉑

Ricky Blakely

Left Uppermost chart shows normal cycle. Second chart depicts sequence of events for a woman taking the pill. Ovulation is prevented, but a form of menstruation still occurs.

tween muscle layers. These release progestagens into the body over a period of between one and three years. The long protection period, the ease of removal should a woman wish to become pregnant, and potentially fewer side effects are the expected advantages of this approach.

A synthetic microcapsule, less than 0.2 mm ($\frac{1}{125}$ in.) in size, containing hormone contraceptives, can be injected into the bloodstream to give a programmed release of the drugs for about six months before being completely absorbed by the body.

Silicone vaginal rings, which have been under investigation since the late 1960s, offer yet another approach. At present they can give contraceptive protection for one to three weeks only, but if this method proves safe and effective, the period could be extended.

Much research is being devoted to testing intra-uterine and intra-cervical devices for storing and releasing hormones. The former makes the environment of the womb hostile to implantation of an ovum, while the intra-cervical device produces a local contraceptive action in the cervix and uterus.

'Morning after' contraception

All the methods of hormonal fertility control discussed so far are preventative, in the sense that they work only if taken before intercourse rather than afterwards. However, since the hormones also affect the subsequent development of a fertilized egg, post-coital methods are also being investigated. Relatively high doses of oestrogens will inhibit the implantation of an ovum in the womb. This approach has been tested by giving women tablets containing the oestrogens within 24–36 hours of unprotected intercourse, and continuing the dose for a further five days. This 'morning-after' contraception is unlikely to be adopted as a routine birth-control method, however, because of the risks from high doses of oestrogen and the short period of protection it offers.

An alternative that seemed more promising was to give low doses of progestagen after intercourse, but tests have since proved disappointing. Although few side effects

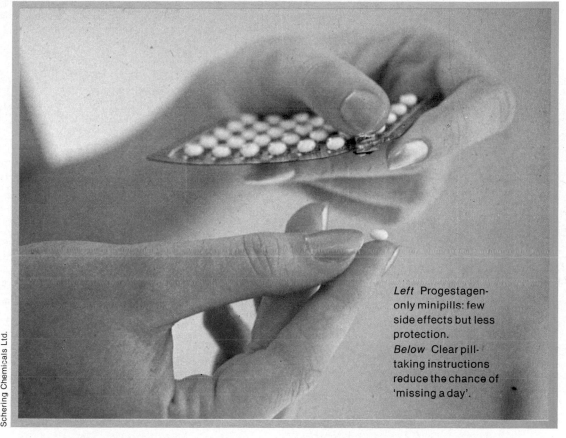

Schering Chemicals Ltd.

Left Progestagen-only minipills: few side effects but less protection.
Below Clear pill-taking instructions reduce the chance of 'missing a day'.

were apparent, the tests showed that if the tablet was taken more than three hours after intercourse, the pregnancy rate actually rose.

The World Health Organization is investigating once-a-month tablets containing low doses of hormone stored in body fat and released slowly to bring on a period whether or not fertilization has occurred. Already marketed in some countries, these work in a similar way to the morning-after methods.

Although much research has already been done in testing various chemicals with a view to suppressing sperm production, a satisfactory male pill is a long way off. The process leading to sperm formation is a continuous and complex one, taking more than two months, and it is still not fully understood. Sperm suppression would require much higher amounts of hormones as well, leading to much greater risks.

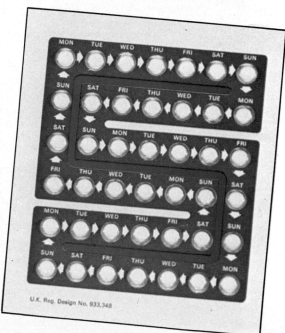

© Nuffield Dept. of Obstetrics & Gynaecology

U.K. Reg. Design No. 933,348

Nuclear medicine

Nuclear medicine has turned radioactive materials—normally considered highly dangerous—to the service of health. Developed largely over the last 25 years, nuclear medicine is a marriage between the technical skills of pharmacist, biochemist, physicist and the medical expertise of the doctor.
Gamma radiation (short-wavelength X-rays) can pass through several centimetres of human tissue. Thus, a trace of gamma rays from a source taken into the body will 'shine' through the body, and the escaping radiation can be detected by a camera sensitive to gamma radiation outside the body.

The gamma camera is used to 'look into' the body and map out the sites where a gamma-emitting *radionuclide* (radioactive material) has been deposited. This simple mapping procedure is the keystone to diagnosis by nuclear medicine. Its major advantage is that most of the gamma radiation escapes from the body. The radiation dose to the patient is often much lower than from a conventional chest X-ray.

There is a major difference between nuclear medicine and other methods of making images of the body's interior (such as conventional X-rays, computerized scans, or the use of high-frequency sound). Whereas these other imaging methods produce a picture of the anatomy of an organ, nuclear medicine finds out how the organ functions.

For example, X-rays of a patient's left and right kidneys might both appear normal even though one is, in fact, performing normally whereas the other is grossly abnormal. This additional physiological, functional information can, however, be obtained by a nuclear medicine imaging technique. It uses radioactively labelled chemicals that are tailor-made for the specific test. The radiopharmaceutical's specific chemical characteristics determine the way it moves through and is taken up by the body, whereas its radionuclide label enables it to be seen inside the body by a suitable camera.

For example, to test whether a kidney is functioning, a pharmaceutical which will be filtered out of the blood is used, coupled to a radionuclide whose radiation is readily detectable. If, and only if, the kidney is working will it filter out the chemical which will then show up as a hot spot on a radionuclide scan. In the same way, since the repair of bone involves the uptake of phosphate, a phosphate-labelled pharmaceutical can be used to show up the site of new bone formation or repair.

In a static scan, an image of the radiopharmaceutical distribution is obtained at a particular point in time. A dynamic study, on the other hand, involves the collection of a time sequence of consecutive images (usually beginning at the time of administration). The images can be viewed one after the other, rather like a strip of movie film. The shape of the image will indicate where the radiopharmaceutical is situated, and the brightness of the image indicates the rate at which the radiopharmaceutical is being assimilated by the body.

The most common static scan—and indeed the most common investigation carried out in a nuclear medicine department—is the bone scan. In a normal scan, the manufacture

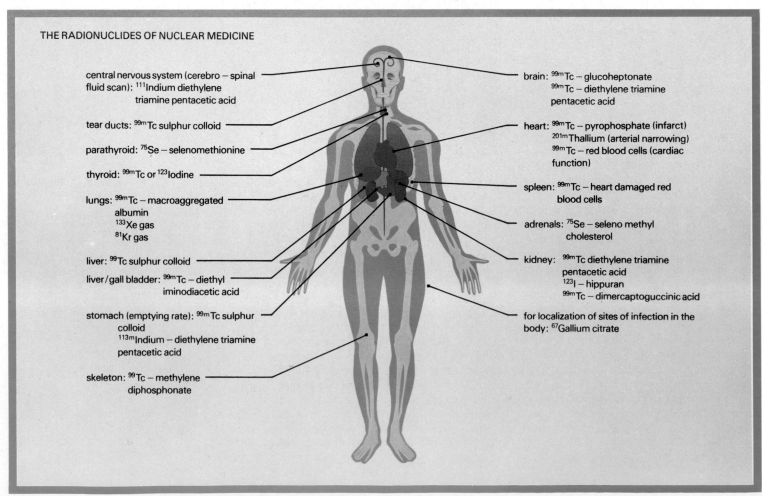

THE RADIONUCLIDES OF NUCLEAR MEDICINE

central nervous system (cerebro – spinal fluid scan): ^{111}Indium diethylene triamine pentacetic acid

tear ducts: ^{99m}Tc sulphur colloid

parathyroid: ^{75}Se – selenomethionine

thyroid: ^{99m}Tc or ^{123}Iodine

lungs: ^{99m}Tc – macroaggregated albumin
^{133}Xe gas
^{81}Kr gas

liver: ^{99}Tc sulphur colloid

liver/gall bladder: ^{99m}Tc – diethyl iminodiacetic acid

stomach (emptying rate): ^{99m}Tc sulphur colloid
^{113m}Indium – diethylene triamine pentacetic acid

skeleton: ^{99}Tc – methylene diphosphonate

brain: ^{99m}Tc – glucoheptonate
^{99m}Tc – diethylene triamine pentacetic acid

heart: ^{99m}Tc – pyrophosphate (infarct)
^{201m}Thallium (arterial narrowing)
^{99m}Tc – red blood cells (cardiac function)

spleen: ^{99m}Tc – heart damaged red blood cells

adrenals: ^{75}Se – seleno methyl cholesterol

kidney: ^{99m}Tc diethylene triamine pentacetic acid
^{123}I – hippuran
^{99m}Tc – dimercaptoguccinic acid

for localization of sites of infection in the body: ^{67}Gallium citrate

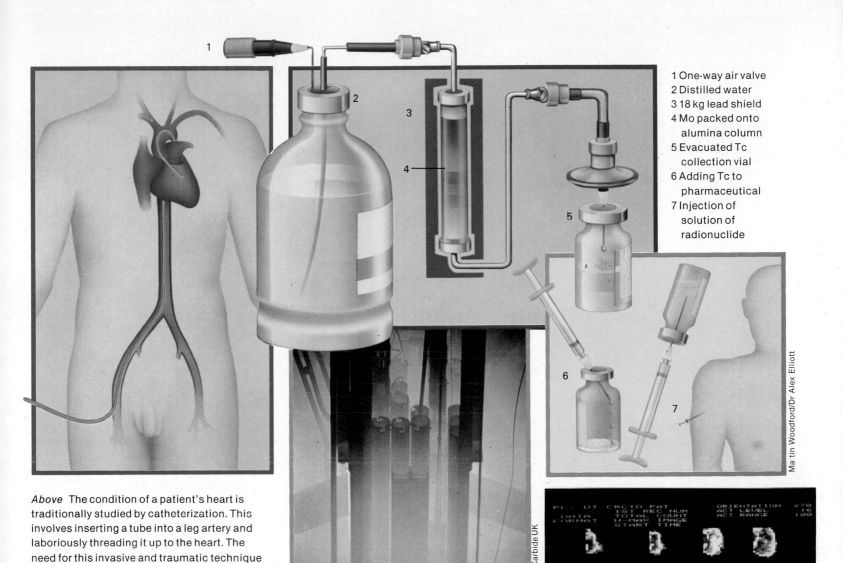

Martin Woodford/Dr Alex Elliott

Union Carbide UK

Martin Woodford/Dr Alex Elliott

1 One-way air valve
2 Distilled water
3 18 kg lead shield
4 Mo packed onto alumina column
5 Evacuated Tc collection vial
6 Adding Tc to pharmaceutical
7 Injection of solution of radionuclide

Above The condition of a patient's heart is traditionally studied by catheterization. This involves inserting a tube into a leg artery and laboriously threading it up to the heart. The need for this invasive and traumatic technique has been greatly reduced by the recent advances in the field of nuclear medicine.

Dr Alex Elliott

of bone can be seen to be occurring throughout the skeleton, and particularly in the hips, spine and shoulders. But the bone scan can show up a particular problem such as an ill-fitting artificial hip. The artificial hip causes increased wear on the surrounding bone, and leads to a local increase in the rate of bone remodelling. The increase in remodelling is revealed as a tell-tale 'hot spot' on a nuclear scan since it leads to an increase in the uptake of radiopharmaceutical.

Lung test

Another example of a static scan involves injecting very small spheres of human albumin labelled with technetium. The size of the spheres is chosen so that they will be carried in the bloodstream as far as the capillaries (the smallest-diameter blood vessels) in the lung, where absorption of oxygen takes place. The spheres are, however, too large to pass through the capillaries and lodge there in the lung.

In a normal image, the uniform deposition of radioactive spheres indicates that the blood supply to the lungs is intact. If, however, there is no radioactivity present in, say, the mid-zone of the right lung, the test indicates that there is no blood supply to this region.

Such a clearly defined defect results from a pulmonary embolism (in which a large blood clot blocks one of the arteries supplying blood to the lung). If the patient is treated with anti-coagulants, the clot may be broken up and the blood supply restored. Thus the lung scan can be used, first to help make the diagnosis, and then to monitor treatment.

A dynamic study is typified by the *renogram*—the investigation of kidney function. The kidney is a collection of about a million tiny filter units, called *nephrons*. These remove certain waste products and foreign materials from the bloodstream, concentrate them, and pass them to the bladder in urine. In order to test the filtering function, a chemical, such as hippuric acid labell-

Top A radionuclide being prepared by drawing distilled water over a column packed with radioactive molybdenum, which is obtained from the core of an atomic reactor (*above centre*). Technetium, a radioactive decay product of molybdenum, is washed out and collected in a bottle. The technetium is then drawn into a syringe and injected into a bottle containing the pharmaceutical to be labelled. The labelled chemical is then administered to the patient by injection—a simple, safe non-invasive technique.

Above Sixteen successive nuclear medicine images that show how the heart is functioning.

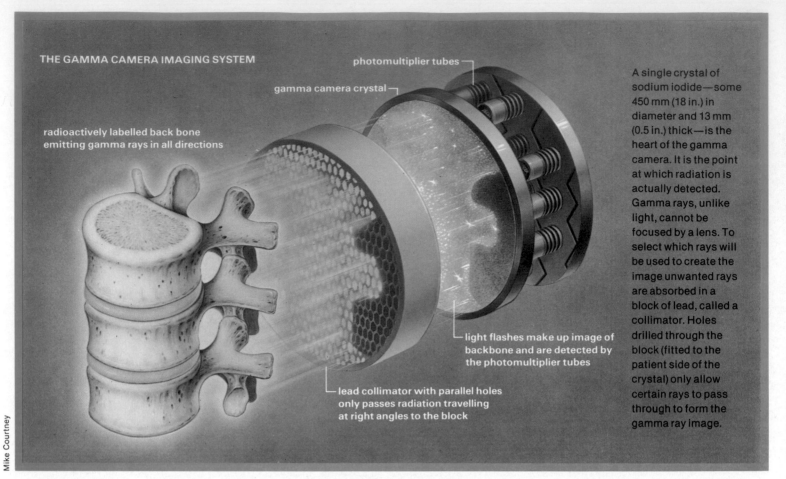

Mike Courtney

International General Electric Company of New York Limited

THE GAMMA CAMERA IMAGING SYSTEM

photomultiplier tubes

gamma camera crystal

radioactively labelled back bone emitting gamma rays in all directions

A single crystal of sodium iodide—some 450 mm (18 in.) in diameter and 13 mm (0.5 in.) thick—is the heart of the gamma camera. It is the point at which radiation is actually detected. Gamma rays, unlike light, cannot be focused by a lens. To select which rays will be used to create the image unwanted rays are absorbed in a block of lead, called a collimator. Holes drilled through the block (fitted to the patient side of the crystal) only allow certain rays to pass through to form the gamma ray image.

light flashes make up image of backbone and are detected by the photomultiplier tubes

lead collimator with parallel holes only passes radiation travelling at right angles to the block

ed with radioactive iodine, is injected into the blood stream.

A series of images is collected at 20-second intervals starting at the time of injection, and the gamma camera is connected to a computer which stores the images. For each image in the series, the computer calculates the amount of radioactivity present within the kidney and displays the result in the form of curves known as activity-time curves.

The curve rises initially, corresponding to the extraction of the radiopharmaceutical from the blood, before output into the urine takes place. Then the curve reaches its peak (as the rate of extraction is balanced by the rate of outflow into urine). Finally, the curve falls (corresponding to a phase in which the radiopharmaceutical is excreted from the kidney). Any variation from this pattern—such as when the kidney is not filtering material from the blood, or when the filtering function is intact but the kidney not excreting the material into the urine—produces its own distinctive curve.

The blood flow to the brain can be investigated by studying the time of appearance, and the amount of radioactivity present in various regions following the cerebral injection of radio-labelled red blood cells. A small sample (2-3 ml) of the patient's blood is withdrawn and put into a centrifuge

to separate out the red cells. These are labelled with technetium and re-injected into the patient. The gamma camera, again linked to a computer, is positioned to look down vertically on the top of the patient's skull. From this angle, a series of images is obtained at 0.3 second intervals, and activity-time curves are generated. After some mathematical processing, it is possible to calculate the rise of blood flow through the brain's arteries.

Blood flow and senility

The total blood flow can be used as a measure of the severity of senility *(senile dementia)*, and variations in the pattern of flow can show up narrowing or blockage of an individual artery, or its branches. Another method of presenting the data is to make a plot of the arrival time of the radioactivity at each point in the brain. A late appearance time is indicative of a narrowed or blocked artery. The rest can be used to assess the success of an arterial operation to circumvent a cerebral blockage—comparing results before and after surgery.

The fastest-growing area in nuclear medicine is the radionuclide investigation of heart function. Nuclear cardiology offers the cardiologist a much less disturbing method of obtaining data than the standard technique, which is called *contrast angiography*. This

traditional method involved inserting a hollow tube (catheter) into an artery at the thigh and passing it along until it reached the chambers of the heart. A radio-opaque dye is then injected into the heart via the catheter and a series of X-ray images obtained. This technique of angiography involves a small risk of serious complications, requires hospitalization, and is traumatic.

Radionuclide tests cannot always supply all the data that can be obtained by conventional angiography, but they do still have a wide range of applications. The most widely used radionuclide tests give information on heart chamber wall motion. They are particularly useful for indicating damage to the heart muscle following a heart attack (myocardial infarction), poor blood supply to the heart muscle (perfusion defects) and abnormalities of blood flow between the chambers of the heart (cardiac shunts).

Nuclear cardiology is mainly concerned with the health and functioning of the left ventricle and the arteries which supply it, since it is this chamber of the heart that is most prone to damage and disease. A commonly used index for heart function is the 'left ventricular ejection fraction'. It is a measure of the heart's range of ability to fill and empty—the greater the range of movement the better the state of health of the

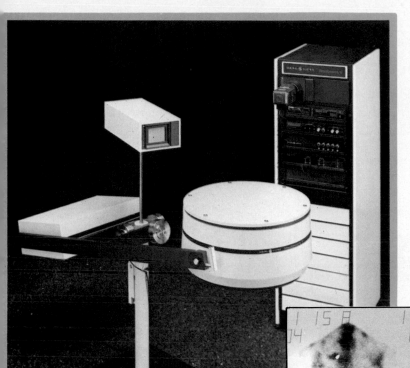

Left MaxiCamera 11 utilizes a computer for consistent and sharp imaging. Inside the camera the gamma ray image is converted into an electrical signal by a bank of 37 photomultipliers.
Below left Gimbal suspension allows precise manual setup for any orientation.
Below Nuclear image—dark areas, sites of radionuclide uptake, indicate that there is a cancer.

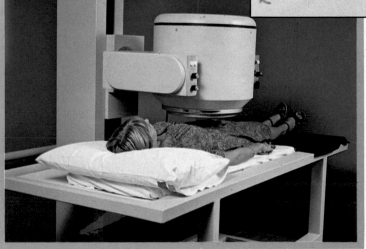

heart muscle. Another aspect of healthy heart function is that the wall motion should be smooth and uniform, and all of the wall should move in a co-ordinated fashion.

A nuclear angiogram can be performed by passing a pulse of labelled material through the heart (the 'first-pass' technique). With this method, a gamma camera is set up to record 25-100 images per second of the passage of the radionuclide in the heart.

The gamma camera is positioned so that it records activity in both chambers of the heart. From the gamma camera image it is then possible to calculate the time taken for the blood to travel from one side of the heart to the other. It is also possible to feed into a computer information from successive beats

Above The nuclear scan is a sensitive indicator of cancer spreading to bone. At an early stage, cancer shows up as a 'hot spot', so the condition can be detected quicker than with a conventional X-ray of the skeleton and treatment can be started at once.
Left Gamma camera for whole body scan.

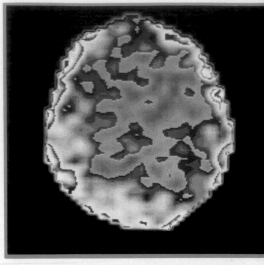

Far right The state of health of the brain is monitored by the inhalation of radioactively labelled xenon gas. Detectors at the side of the head monitor the rate of clearance of the gas—thus measuring the level of blood flow in the brain. Results are displayed on a video monitor *(right)*. Brain scans can also be obtained by injecting radioactively labelled red blood cells *(centre left* and *right)*. A portable gamma camera *(below right)* can be used to pick up signals emanating from a radioactive tracer in the patient's head. The technique is used to detect signs of a possible brain tumour at an early stage.

John Watney

Dr Alex Elliott

of the heart. The computer can then be instructed to combine the images to produce an accurate representative cardiac cycle.

The representative cycle images can be displayed in the form of a moving pattern on a television screen to help doctors analyze heart-wall motion. The pattern of movement can be used to identify a patient with an aneurysm—a segment of heart muscle which has died and hardened following a heart attack. The dead fibrous segment is weak and may bulge out into the ventricular wall of the heart, which will then move out of sequence with the rest of the ventricle.

The ability of nuclear medical techniques to distinguish between the various heart conditions is a major contribution to diagnosis. It removes the need for the traumatic, and potentially dangerous, traditional process of contrast angiography.

Choice of treatment

If the radionuclide procedure reveals that a patient's condition is normal then no further heart investigation is required, whereas patients in whom a diseased heart muscle has been identified can be treated with drugs. However, patients with aneurysms may require surgery. They will go forward to traditional contrast angiography, since the nuclear angiogram does not provide sufficient anatomical detail for surgery.

Another method of performing a nuclear angiogram is to introduce a radioactive label into the body so that it becomes uniformly distributed throughout the bloodstream (the 'equilibrium-gated' method). Again, the gamma camera information is fed into a computer which then builds up a typical, representative cycle.

It is also possible to conduct nuclear imaging tests after the patient has exercised on a bicycle ergometer or treadmill so that the patient's heart rate is raised to a predetermined level. The purpose is to unmask abnor-

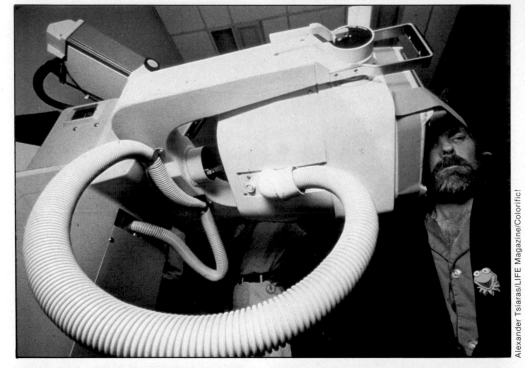

Alexander Tsiaras/LIFE Magazine/Colorific!

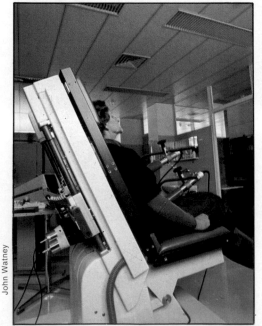

John Watney

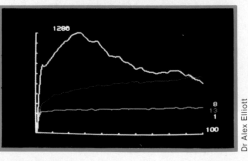

Dr Alex Elliott

Left Kidney function analyzer. There are two detectors at the back of the seat, one for each kidney. The detectors at the front measure the radioactivity in the aorta—which gives a measure of the background radiation to be deducted from the results.
Above Activity time curves indicate the rate of uptake elimination of radionuclide—a measure of the health of the kidney.

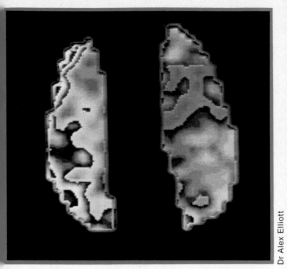

malities of wall motion that are not present at rest. For example, the procedure might be used in the case of a patient suffering from poor blood supply from a coronary artery due to narrowing or obstruction.

At rest, the quantity of blood delivered to the heart muscle is capable of supplying sufficient oxygen for the muscle to function. But whilst exercising, the oxygen demand cannot be satisfied and the portion of muscle supplied by the artery may cease to contract.

If the ability of the left ventricle to pump blood is found to fall instead of rise when the patient exercises, then undergoing coronary artery bypass surgery may relieve the problem. Nuclear angiography is extremely valuable since it may be unwise to use traditional angiography in this type of case—where heart function is impaired and the patient is already in a delicate condition.

Nuclear angiography is also used to assess the heart's response to various classes of drug, a process for which the traumatic and disturbing technique of contrast angiography would not be ethical or acceptable.

Magic bullets

Imaging of patients suspected of a heart condition that has led to scarring of the heart muscle (acute myocardial infarction) is employed in addition to traditional techniques (such as ECG and blood enzyme measurements). Nuclear imaging is particularly useful in those instances where traditional methods give unclear or misleading results.

Nuclear medicine also uses beta radiation, a short-range killer of cells. It comprises a stream of electrons, which are rapidly absorbed in tissue (having a maximum range of a few millimetres). Because the electrons give up their energy over such a short distance, they deliver a high-radiation dose that kills the tissues through which they pass.

But this cell-killing ability can also be ex-ploited for safe and non-traumatic therapy. If the radionuclide is aimed so that its lethal effects are restricted to the site of a tumour, the beta radiation will destroy the cancer cells without damaging the normal healthy surrounding cells. For example, if radioactive iodine is administered, the body will tend to concentrate it in the thyroid gland just as it concentrates normal iodine. Thus—like a 'magic bullet'—the radionuclide finds its way to this small target where it kills cancer cells. In this way, radioactive iodine can be used for the treatment of cancer of the thyroid gland, which is located in the throat.

Nuclear medicine procedures are not traumatic for the patient, most requiring only a small injection into an arm vein. Virtually all tests can be performed without having to hospitalize the patients, and the procedures have no side-effects. Nuclear medicine offers a safe and convenient method of obtaining reliable information about the function, as well as the structure, of a wide range of body organs.

In quest of interferon

Research into a mysterious substance called interferon is one of the most active lines of medical investigation in the 1980s. Interferon, the generic name for a group of proteins involved in the body's defence against disease, is released naturally by cells in response to attack by viruses. The proteins seem to prevent further infection—probably acting similarly to antibiotics. But it is the suggestion that interferon may also be valuable in the treatment of cancer that has caused the excitement.

The interferon story began in 1957, when Alick Isaacs and Jean Lindemann, working at the National Institute for Medical Research in England, isolated a substance which seemed to prevent virus infections. They had artificially infected cells in the membrane of a chicken's egg with a certain type of virus, and were trying to discover why it was impossible to infect the same cells with other kinds of viruses introduced afterwards.

Origin of the name

They found that an extract taken from the infected cells and added to a separate group of completely healthy cells caused the second group to become resistant to a wide range of virus infections. It was soon shown that the active ingredient in the extract was a protein, and because it seemed to interfere with a virus's ability to infect cells, Isaacs called the protein *interferon*.

The discovery of a substance that was potentially a cure for all virus-caused diseases was greeted with excitement, and many laboratories began experimenting with interferon. However, it soon became clear that there were obstacles to its immediate use.

For a start, scientists found that the precise chemical structure of the substance varies critically from species to species. As a result, interferon from animals simply would not work in humans. In addition, a way of obtaining human interferon to treat patients was needed. This was a daunting task, partly because interferon is present only in extremely small concentrations, and partly because it is far from easy to purify.

When these problems became generally known, enthusiasm for interferon research evaporated as quickly as it had appeared. The story would probably have ended there if it

Bottom left At Imperial College, London, this flask contains *Escherichia coli* bacteria genetically engineered to produce human interferon. Flask culture serves as inoculum for a small-scale 50 litre fermenter *(below)*.

Niall McInerney

had not been for a number of advances in cell biology, biochemistry and genetics which came together in the late 1960s. The result was that it became possible (though still difficult) to produce pure interferon from cells of a number of species, including Man, on a scale which made further research practical. In addition, techniques became available for discovering how interferon manages to protect cells against viruses.

Viruses work by invading a cell, taking over genetic control of its biochemical systems, and using those systems to produce identical copies of themselves. The normal functioning of the infected cell is so disturbed that it rapidly disintegrates, releasing all the newly manufactured viruses which can infect other cells. However, one of the last actions of the dying cell is to release a minute quantity of interferon into the surrounding fluid and bloodstream. Molecules of interferon then attach themselves to special receptor sites on the surfaces of healthy cells. Acting very like a hormone, interferon causes the production of enzymes which exert the anti viral effect within healthy cells.

Fighting virus invasion

The most recent research indicates that interferon protects healthy cells not by destroying viruses or by preventing them from entering, but by blocking the mechanism whereby a virus is able to take over control. Specifically, it seems that the enzymes which are released when interferon binds to the surface receptors of healthy cells act to destroy the messenger-RNA that the viruses first produce when they infect a cell.

From a clinical point of view, it is significant that once the body has been 'alerted' by interferon, its cells become resistant to a wide variety of viruses—not just to the particular infection which sparked off the reaction. The earliest samples of human interferon studied confirmed that the substance could indeed help control viral infections.

A single nasal spray containing interferon gave protection for up to two weeks against the common flu virus, and doctors at Moorfields Eye Hospital showed that it could protect victims of shingles against blindness.

Hoping to demonstrate the same effect against a cancer-causing virus in mice, American scientist Ion Gressor tried injecting the mice with mouse interferon. He expected the interferon to prevent the further spread of the tumours, but to his astonishment, the tumours actually shrank—and in some cases disappeared altogether.

Again, funds became rapidly available,

enabling research teams to follow up this promising lead. Soon it was shown that in animals, and in large doses, interferon appears to have a two-fold effect against cancer. In the first place it activates the scavenging white cells in the blood stream which normally destroy foreign material in the body. Secondly, the surface coating of tumour cells which have been exposed to interferon undergo a slight change. The effect of the change is that cancer cells become more recognizable to the body's natural immune system as foreign material, and this encourages their destruction.

Before interferon could be tried on human cancer victims, it was necessary to establish a steady supply of the substance. Dr Kari Cantell pioneered a technique for obtaining limited quantities of interferon from human blood. His Finnish laboratory was to be the world's major source of human interferon throughout the 1970s. Cantell's method was to separate off all the white cells from blood donated to the Helsinki Red Cross. By keep-

Below The Imperial College 3,000 litre production fermenter, into which the interferon-producing bacterial culture is transferred from the 50 litre fermenter. High yields of interferon are expected. Inset shows instruments for monitoring and controlling growth of the bacteria during production.

Niall McInerney

INTERFERON IN ACTION

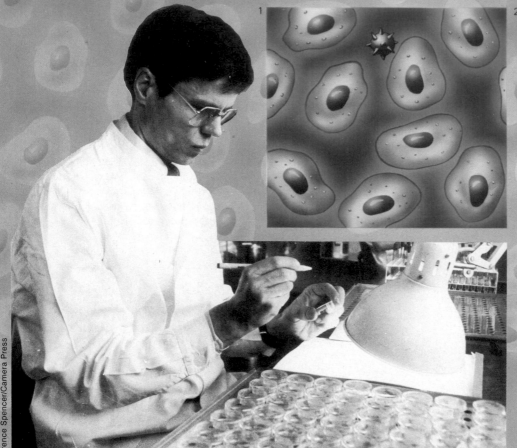

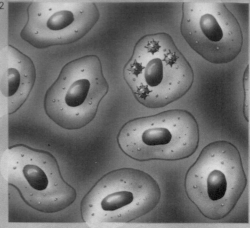

Above Sequence of diagrams shows the natural process by which interferon protects cells against attack by viruses.
1 A virus prepares to enter a cell.
2 The virus enters the cell and divides.
3 The cell disintegrates, releasing the new virus particles, which infect other cells.
4 Infected cells respond by manufacturing anti-vital protein to attack the viruses.
5 Infected cells also produce interferon.
6 The interferon is released to surround other, unprotected cells.
7 These then react to the interferon by producing anti-viral protein, and thus become resistant to attack by viruses.

Terence Spencer/Camera Press

Above At the Central Public Health Laboratory in Helsinki, Finnish virologist Dr Kari Cantell cultures selected virus strains that help to produce interferon. The viruses are used to stimulate production of interferon by white blood cells, which must first be separated from the liquid blood *(right)*.

ing the white cells alive and treating them with a specially selected virus which stimulated maximum production of interferon, it became possible to extract sufficient quantities for use in hospital trials.

The first systematic experiments took place at the Karolinska Hospital in Sweden during the mid-1970s. There, 35 patients with a rare form of bone cancer were treated three times a week for 18 months.

The results of the interferon therapy were encouraging, and while the size of the sample was too small to draw general conclusions, more than twice the expected number of patients showed long-term recovery. Similar small-scale treatment programmes followed in a number of hospitals worldwide, again with apparently promising results for future treatment.

Many doctors were sceptical. Identical

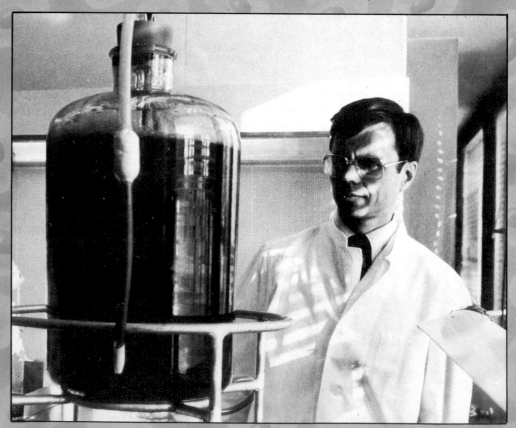

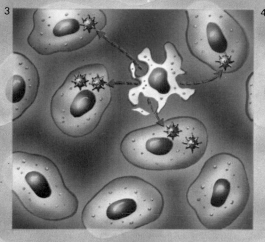

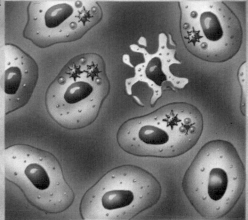

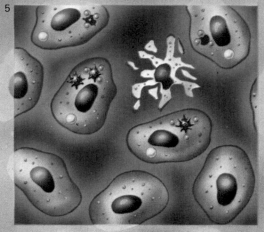

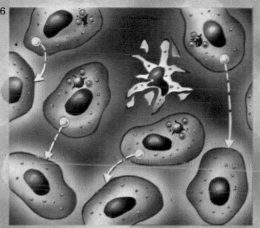

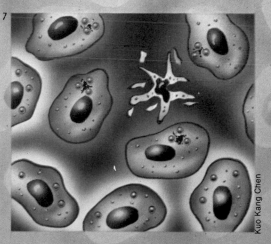

claims have been made for a number of 'cures' now regarded as without any real effect. In small, uncontrolled studies, the sample of patients receiving treatment may not represent all sufferers of the disease.

Inconclusive trials

Nevertheless, the American Cancer Society (ACS) felt justified in funding a multi-million dollar series of clinical trials. This two-year study, the largest to date, involved 150 patients with four different kinds of cancer. The interferon for their treatment, which lasted up to three months, cost $6,000,000 and used up almost the entire world supply.

The ACS trials, despite their scale and cost, have proved inconclusive. Undoubtedly, some cancers responded better to the therapy than others. Breast cancer, for example, seemed to show signs of receding, while multiple myeloma (a disease of the bone marrow) showed little evidence of improvement. There were other useful observations. For example, all patients received fairly large amounts of interferon, and this produced 'flu-like side effects.

Since it is a natural body substance, interferon seems to have a less disruptive effect on non-diseased body systems than conventional chemotherapy. It seems, too, that patients can tolerate the treatment for long periods without loss of effectiveness. On the other hand, where improvement was noted, the disease often returned when the interferon injections ceased.

Above all, the ACS clinical trials were still too small and inadequately controlled to rule out the possibility that something other than the interferon was involved.

Other studies take place quietly and sometimes in conditions of considerable secrecy. Results from two separate trials in major London hospitals are just beginning to

be known. In one, a group of twelve cancer patients were treated over a year. Alarmingly, three of those receiving the drug died suddenly and inexplicably. It is not yet known whether these unfortunate deaths were related to the treatment and, if so, whether they were due to impurities in the interferon or caused by the drug itself.

It is clear that the main restraint on interferon research is the lack of an adequate method of production. Current supplies suffice for only a fraction of the reputable research projects seeking support. In 1978, Kari Cantell's laboratory required 50,000 litres of human blood to produce a total of one-tenth of a gram of pure interferon. This would have been enough to treat about 10,000 people with minor viral infections, but only about 200 people with a chronic viral disease. Alternatively, a mere 80 cancer victims could have been put on a year-long course—each at a cost of about $25,000.

As a result, hope for the future lies in finding a large-scale, cheaper source of the drug. The aim is to produce interferon artificially in quantities which are not restricted by the need for human raw materials. So far three different approaches to the problem look promising.

The method which is furthest advanced, and which has already supplied some human interferon for use in the American Cancer Society trials, is being developed by the Burroughs-Wellcome group in the UK. Their technique is to culture human white-blood cells that have lost their normal growth control and multiply readily. The Burroughs-Wellcome group has a 1,000-litre fermentation facility, and is already producing human leukocyte interferon from a line of such transformed white cells.

Serious difficulties remain, however. For example, the best way of making the blood cells multiply is to expose them to a special

virus. This virus has been implicated in certain types of human cancer, so it is vital that the interferon produced is uncontaminated with an agent that will cause the very disease that the drug is designed to cure. There will also be difficulties in scaling-up production to large-scale commercial cultures.

A variation of the Burroughs-Wellcome approach is being pursued by several commercial laboratories in the USA. They have avoided the contamination problem by dispensing with white cells and using fibroblast cells instead. These cells, which can easily be obtained from human foreskins removed during circumcision, will multiply in culture without being treated with dangerous viruses.

While the interferon produced by this method is slightly different from white-blood cell interferon, it seems to work just as well. Indeed, some researchers claim that a mixture of types is most effective.

Genetic engineering

A quite-different artificial method of producing interferon in large quantities has become possible as a result of recent advances in genetic engineering. A section of DNA, representing the gene coding for human interferon, has been successfully inserted into the DNA of a bacterium, *Escherichia coli*, which can readily be grown in submerged culture. As the bacterium grows and divides, it produces interferon as an intracellular protein.

Potentially large amounts of interferon could be produced by this method, provided that a suitable means can be devised for extracting and purifying the interferon protein. Many teams are active in this field, including commercial firms. One European company, Biogen, has obtained *E. coli* strains producing human leucocyte interferon, and is collaborating with scientists at Imperial College, London, to produce interferon in 3,000 litre fermenters in the College's fermentation pilot plant.

The Genetic Manipulation Advisory Group, a body formed in the UK to control safety precautions employed in genetic engineering work, has stipulated that the bacterium must be grown under containment conditions. It has, therefore, been necessary to modify the fermenters to provide for containment and to devise techniques for killing the bacterial cells at the end of the fermentation without destroying the intracellular interferon protein.

Although interferon yields were initially low, progress has been rapid. Biogen claim that yields are now comparable, if not superior, to the best yields that can be obtained by the growth of human cell lines. The major problems concern the extraction and purification of the interferon so produced, but these problems seem likely to be overcome.

Biogen hope to produce sufficient pure interferon by this technique for clinical trials in mid-1981. Of all the methods for producing interferon in quantity, this is potentially the most exciting. Further considerable increases in yield are theoretically possible by the application of genetic engineering, mutation and strain selection techniques.

Whilst it may be possible to produce adequate interferon for research purposes and clinical trials by other methods, such as large-scale tissue culture, if the hopes for interferon are eventually realized and production on a commercial scale is required, the use of genetically engineered bacterial strains appears to offer the best chance of success.

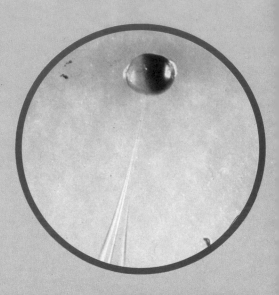

Below In experiments to insert the human gene for interferon into bacteria, a Biogen researcher operates a tiny syringe under a microscope. In close-up *(right)*, the needle penetrates a frog's egg to inject RNA in a test to see if the genetic engineering worked.

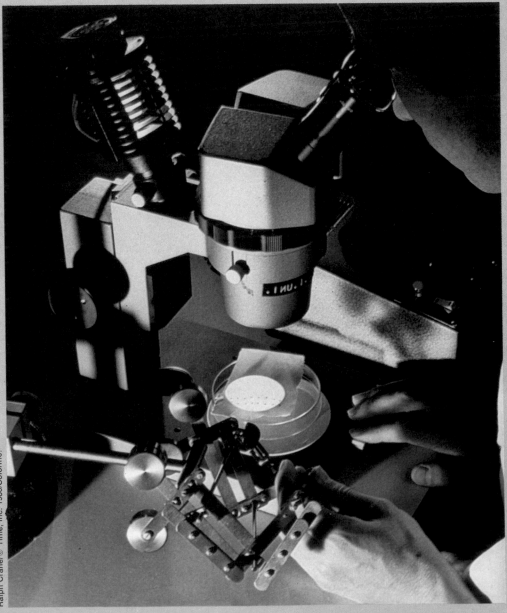

Ralph Crane/© Time, Inc. 1980/Colorific!

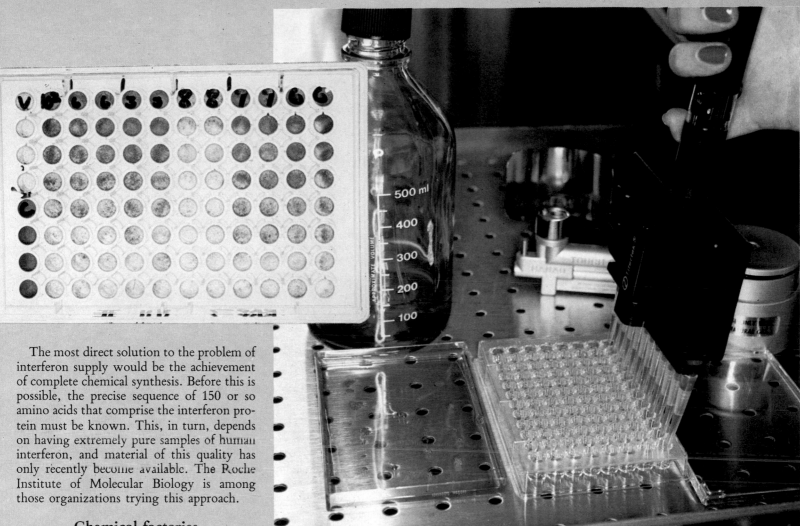

The most direct solution to the problem of interferon supply would be the achievement of complete chemical synthesis. Before this is possible, the precise sequence of 150 or so amino acids that comprise the interferon protein must be known. This, in turn, depends on having extremely pure samples of human interferon, and material of this quality has only recently become available. The Roche Institute of Molecular Biology is among those organizations trying this approach.

Chemical factories

Both the direct synthetic production of interferon and the use of bacterial 'chemical factories' to achieve the same result may well come up against a serious obstacle. Interferon is not simply a protein—it also contains carbohydrate components known as sugar residues. The exact nature of these residues is not determined by the gene for interferon. Instead, the chemical constitution of the interferon molecule is partly affected by the type of cell that makes it.

It is possible that interferon from bacteria which have undergone genetic engineering and interferon from straightforward chemical synthesis will both be critically different from natural human interferon. Whether this difference significantly reduces the power of artificial interferon to help destroy cancer is unknown. It may be responsible for the inability of interferon from one species to act properly in the body of another.

Other ways of using the interferon system to treat human illness are being developed and investigated. Of these, one of the most promising is artificial stimulation of the body's own cells to produce their own interferon, rapidly and in large amounts. This

Above A bioassay of interferon in progress at Imperial College. Wells in tray *(above left)* are filled with eleven samples containing interferon. The further the blue dye extends down each column the greater the amount of interferon in the sample. Column on extreme left is for virus and human cell controls.

can be achieved by infecting the body with a carefully chosen virus. But in practice this may be difficult because a viral infection which strongly stimulates the interferon system may also kill a patient weakened from the primary disease. So far, no way of separating the two effects has been found.

Although the use of interferon to treat cancer has drawn attention away from its original anti-viral effect, it is still valuable in its old capacity. In particular, interferon is used to treat virus-caused diseases which do not respond to conventional treatment. A worker at the Centre for Applied Microbiological Research, UK, almost certainly owes his life to interferon treatment after accidental exposure to the deadly Green Monkey virus.

Less exotic examples of dangerous diseases which may be cured by interferon include hepatitis B and acute chicken-pox in children. The drug may also play a useful role in the aftermath of organ transplants, when administered in dosages which boost the body's defences against infection without stimulating the immune mechanisms that lead to tissue rejection.

Future hopes

For many thousands of cancer victims and their families, the false hope that interferon might rescue them has caused enormous extra suffering. Television and the press have been as much to blame as some of the research teams who, excited by initial results and keen that funding should continue, have been over-optimistic.

It is easy to cite a few cases of dramatic improvement, but much more difficult to explain why such results mean very little. In the light of this unfortunate history, it would be particularly sad if the scepticism that has now arisen should prevent continued research.

Nature's own painkillers

For over two thousand years opium, and the compounds such as morphine that can be made from it—known collectively as opiates—have been one of the most powerful drugs in the fight against pain and suffering. Yet because they are addictive, the opiates themselves have also been the cause of untold misery. We are just beginning to discover how these ancient drugs work in the body to produce their effects. The story that is emerging suggests that the brain contains its own natural opiate system. If we learn how that system works then it should become possible to develop drugs which have all the advantages of morphine without the problems of its addictiveness. Understanding this complex system will also give us an insight into the way that our experiences of pain, pleasure and other emotions are controlled in the brain.

Brain cells communicate with one another by the use of chemical messengers called *neurotransmitters*. These substances are released at places called *synapses* where two nerve cells almost connect one another. The molecules of transmitter have to cross a very narrow gap until they reach another nerve cell. Here they combine with receptor molecules located in the membrane of the cell. When they do so, the second cell will respond. The majority of drugs that have an effect on the mind act specifically at the synapse, either encouraging or suppressing the ability of the neurotransmitter to effect the nerve cell on the receiving side.

Mirror images

There are several pieces of evidence to suggest that the opiate drugs also work in this manner. The main clue is that all the various substances which can produce opiate *analgesia* (analgesia is relief from pain) have a smilar molecular structure. A transmitter substance fits its receptor molecule like two pieces of a jigsaw puzzle. Provided a critical section of the molecule is the right shape, the rest can vary considerably. But there are other clues which suggest that the opiates work directly on receptors in the brain. It is possible to produce *isomers* of the opiate drugs—isomers are chemical compounds whose constituent elements are present in equal parts but arranged differently, giving them different properties. The isomers of the opiate drugs have molecular structures which are exact mirror images of the normal arrangement (just as your right hand is a mirror image of your left). These wrong isomers have no ability to remove pain and this, too, suggests that opiates work by having the

Bottom left A section through the base of the brain shows the pituitary gland (pale object) in which natural painkillers have been found.
Right The high concentration of opiate receptor sites and of the natural painkiller enkephalin in the limbic system of the mid-brain suggests that this is a key site involved in natural pain control.
Below Nerve pathways carrying information from the body's pain receptors to the brain have been found in the spinal cord.

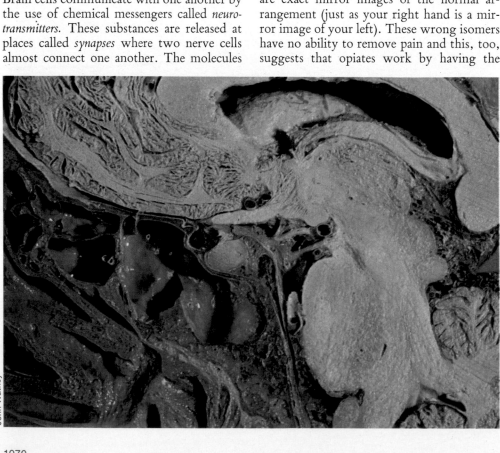

John Watney

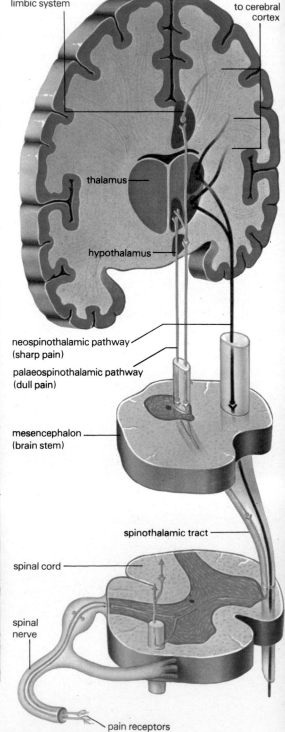

limbic system

to cerebral cortex

thalamus

hypothalamus

neospinothalamic pathway (sharp pain)

palaeospinothalamic pathway (dull pain)

mesencephalon (brain stem)

spinothalamic tract

spinal cord

spinal nerve

pain receptors

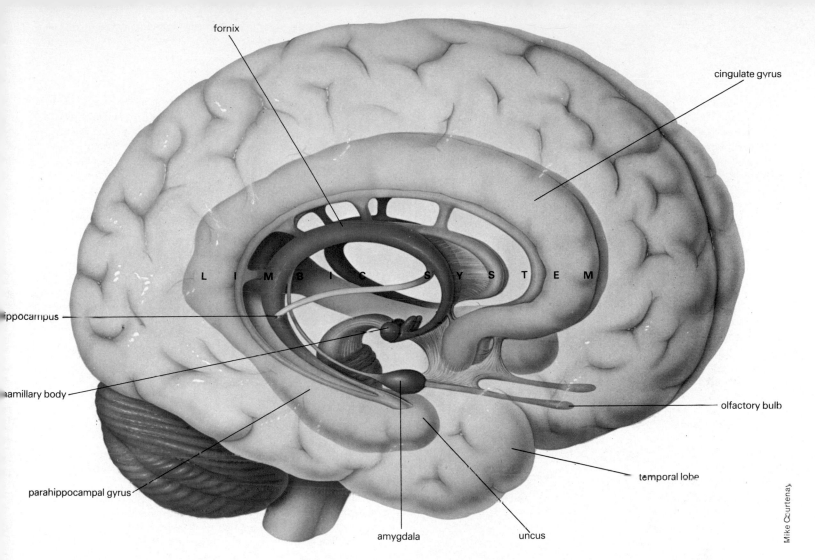

fornix

cingulate gyrus

LIMBIC SYSTEM

hippocampus

mamillary body

olfactory bulb

parahippocampal gyrus

temporal lobe

amygdala

uncus

Mike Courtenay

right shape to fit into receptors already in the brain.

During the last decade, techniques have been developed which demonstrate the existence of these receptors. Workers at the Johns Hopkins University School of Medicine in the United States used a drug called *naloxone* which, despite having a molecular structure closely resembling morphine, has the power to prevent opiates from working. Someone on the point of death from morphine poisoning can be revived within seconds by an injection of just a little naloxone. It seems that naloxone will 'bind' with the opiate receptors without causing an analgesic response. In other words, it will block the receptors, and because it binds to them more strongly than morphine, it will tend to prevent any morphine that is also present in the brain from working.

The American workers used very small amounts of radioactively 'labelled' naloxone which they added to fragments of cell membranes from rat brains. They found that even after washing, some radioactivity remained on the membranes. To prove that this was due to the naloxone binding with specific opiate receptors rather than simply sticking

to the membrane in general, they carried out a further test. This involved adding various opiates, such as heroin and morphine, and measuring the extent to which the radioactive naloxone was displaced. Conclusive proof that they had indeed located the receptors came from the fact that the extent to which the added opiate removed the naloxone matched the power of that opiate to relieve pain in human patients.

A natural opiate

By giving radioactive opiates to Man and to animals it is possible to show that the receptors are concentrated in the synapses of nerve fibres and brain regions thought to be associated with the perception of pain. But a question remains: what are they doing there? Presumably there is a natural brain substance, a neurotransmitter, which is released at the synapses sensitive to the opiates. The task of finding this morphine-like substance was extremely daunting, since it is present in almost inconceivably minute quantities. Furthermore, unless the researcher knows exactly what to look for, any extract will be contaminated a thousandfold by unwanted impurities. Nevertheless, in an experiment

conducted in 1974, John Hughes and Hans Kosterlitz, of the University of Aberdeen in Scotland, succeeded in producing an extract of pig brains that exhibited morphine-like properties.

The starting material was several tonnes of pigs' brain from slaughterhouse animals. Only a small part of the brain was used, and thousands of pigs were needed to produce the final product. Hughes and Kosterlitz used a variety of sophisticated purification techniques, such as gel filtration and high pressure liquid chromatography, to produce a large assortment of extracts—any of which might have contained the substance they were searching for. A key breakthrough was their development of a sensitive and specific test for the presence of a natural opiate. They used a strip of living muscle taken from mouse vas deferens (the contracting tube that connects the testes to the penis). For some unknown reason, this unlikely structure contains sensitive opiate receptors. Hughes and Kosterlitz found that electrically induced contractions of the vas deferens were weakened by the presence of opiates, even in very low concentrations. They were able to check that the effect was definitely due to an

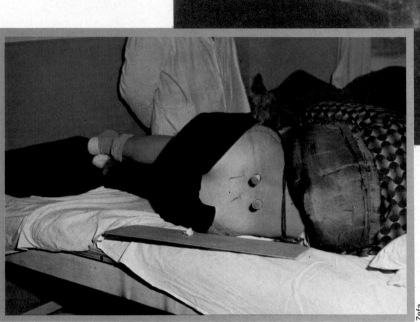

Right Electrical stimulation of the mid-brain seems to make the body produce its own painkillers. Here electrodes have been implanted in the patient's brain, activated by a portable transmitter. Electrical stimulation of a particular part of the spinal cord also provides speedy natural pain relief.

Left It is likely that the anaesthetic effect of acupuncture depends on the body's natural painkilling system, though major operations performed in China without anaesthetic turned out to be propaganda.

opiate by adding a little naloxone, which blocked the opiate receptors and restored the contractions.

The two researchers tested their extracts using this technique and showed without doubt that one of them contained a natural opiate, present in the brain. They called their substance *enkephalin* from the Greek for 'in the head' and they found that it was a type of compound called a peptide, composed of a sequence of amino acid units. At first they had difficulty in determining the exact sequence of amino acids. But then came the discovery that they had isolated not one but two, similar, peptides, each made up of five amino acids. One of them, methionine-enkephalin, has the sequence tyrosine-glycine-glycine-phenylalanine-methionine. The other peptide, leucine-enkephalin, is the same, except that the amino acid leucine replaces methionine at the end of the chain.

The balance of the two types of enkephalin seems to vary not only between different species of animal, but also between different parts of the brain. Whether the two enkephalins have different functions has yet to be discovered.

Once the structure of the natural opiate was known, biochemists were able to synthesize it in a laboratory. The first sample took two weeks to produce, but as soon as it was ready it was tested, using Hughes and Kosterlitz's mouse vas deferens technique. Fortunately, it showed exactly the same properties as the natural substance that had been extracted with such effort from pigs' brains. Within weeks of the results being made known, enkephalin research teams were being formed all around the world.

However, the final step in the demonstration that there is a natural opiate system in the brain was yet to come. It was still

necessary to show that the distribution of the receptors matched the distribution of the enkephalins. The researchers found the location of the receptors by using a technique known as *autoradiography*. They gave a test animal a low dose of radioactively labelled opiate which 'bound' selectively at the receptors. Then they pressed a flat section of the brain against a photographic plate. After some time, they developed the plate, when the location of the opiate (and hence the receptors) was revealed in the form of dark areas where radioactivity had affected the silver in the emulsion.

Chemical messengers

The enkephalins are not the only opiate-like natural substances in the brain. At the invitation of a fellow worker in the field of biochemistry who suspected a possible link, Dr H. R. Morris, of London University, who had just established the amino acid sequence of the enkephalins, was present at a lecture on an obscure *hormone* (chemical messenger) from the pituitary gland. To his great delight, he noted that five amino acids of this hormone, which was a much longer peptide called beta-lipotropin, had exactly the same sequence as one of the enkephalins. This new discovery soon led to an investigation of the opiate-like properties of the pituitary hormones. As a result several potent morphine-like peptides have been

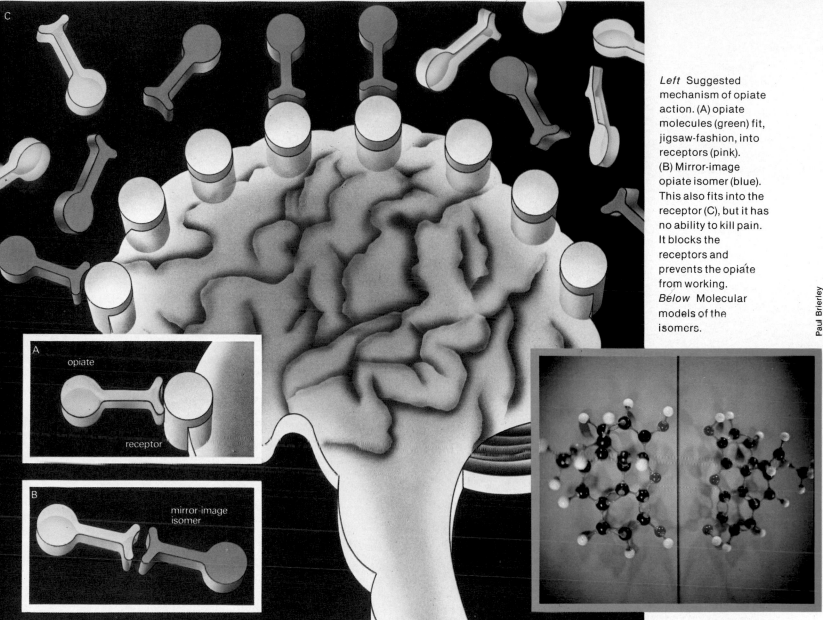

Paul Brierley

Left Suggested mechanism of opiate action. (A) opiate molecules (green) fit, jigsaw-fashion, into receptors (pink). (B) Mirror-image opiate isomer (blue). This also fits into the receptor (C), but it has no ability to kill pain. It blocks the receptors and prevents the opiate from working.
Below Molecular models of the isomers.

isolated from the pituitary. They have been given the general name *endorphin* from a combination of 'endogenous' and 'morphine'. The term endorphin has come to include all opiate substances found naturally in the body.

The pituitary gland is a pea-like structure which lies just underneath the brain about 5 cm (2 in.) behind the bridge of the nose. The pituitary is the conductor of the hormonal 'orchestra' and it communicates with the brain and the rest of the body by releasing substances into the bloodstream. Because there is no nervous connection between the pituitary and the brain, the analgesic effects of morphine cannot be explained by its effect on the pituitary receptors. This fact, together with the finding that the endorphins are only present in the brain itself in extremely low concentrations, has led some researchers to speculate that the analgesic

effect of the endorphins is entirely fortuitous and has nothing to do with their normal function. Instead, they suggest that the endorphins may be involved in controlling internal pituitary changes relating to hormone levels in the blood.

However, other scientists believe that the endorphins have a much more significant role directly related to pain. The lifetime of enkephalin once it has been released in the brain is only a few seconds. This is because it is rapidly destroyed by enzymes in the blood. Endorphins, however, can remain active for several hours. It is therefore possible that while enkephalin is a fast-acting neurotransmitter, endorphin is a *neuromodulator*, having a longer-lasting effect on the enkephalin system. Support for this idea comes from the discovery that endorphin is an extraordinarily potent substance: thirty times as much morphine is required to pro-

duce equivalent effects. So the low concentration of endorphin in the brain is understandable.

There may be another reason why endorphin levels seem to be so low. If a healthy human volunteer is given a large dose of naloxone, the drug that blocks the opiate receptors and prevents morphine from working, then nothing very much happens when he or she is injected with morphine. Naloxone will also prevent any natural opiates from working. So this implies that enkephalin and endorphin are normally held in reserve. However, some dramatic new experiments have demonstrated the body's natural painkillers in action.

There are certain occasions when people seem able to function normally in the face of injuries and other afflictions which one would expect to be unbearably painful. Soldiers with appalling wounds sometimes

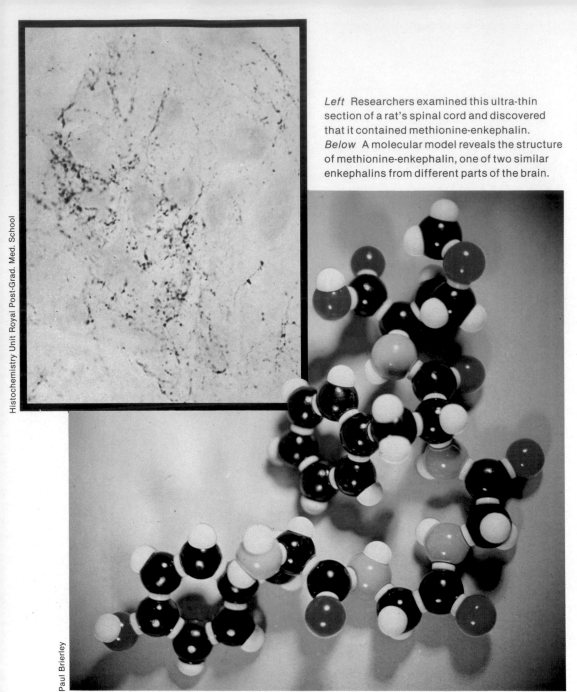

Histochemistry Unit Royal Post-Grad. Med. School

Left Researchers examined this ultra-thin section of a rat's spinal cord and discovered that it contained methionine-enkephalin. *Below* A molecular model reveals the structure of methionine-enkephalin, one of two similar enkephalins from different parts of the brain.

Paul Brierley

side-effects of the opiates is to depress the breathing centre in the brain, so it is possible that endorphin from the placenta may help to stop the foetus breathing before birth.)

A recent American study suggests that endorphins may also be responsible for the so-called *placebo effect:* pills given to a patient to reduce pain will often work even if they contain no drug at all. In a careful experiment in which even the doctors who administered the drug did not know whether they were giving morphine, saline solution or naloxone, it was found that only naloxone produced no easing of volunteers' pain after their wisdom teeth were removed. Presumably, because the volunteers all expected to receive painkillers, endorphin was released in their brains. This accounted for·a lessening of their pain even when all they actually received was saline solution. But those who received naloxone did not feel any improvement because it blocked the action of any endorphin that was released.

The discovery that endorphins can be produced as a result of an expectation (as with the volunteers above who expected to receive a pain-killing drug) suggests that the enkephalin system may be involved in many aspects of our emotional life. Some scientists have even speculated that our experience of pleasure may be related to the release of enkephalins or endorphins. Almost certainly, some mental illnesses are due to some kind of disturbance in the natural opiate system.

Non-addictive opiates?

The drug companies are interested in the enkephalins and the endorphins because they promise to lead to the development of a powerful non-addictive analgesic. Unfortunately, however, none of the substances so far discovered hold the key to this breakthrough. It seems that the natural opiates are just as addictive as morphine, heroin and the rest! If enkephalin is injected directly into the brains of rats, it produces the expected analgesic effects, but the rats also develop all the symptoms of tolerance and physical dependence that occur in opiate withdrawal.

The key question is whether the addictive nature of the opiates is an inevitable consequence of the way in which they produce analgesia. The answer will come from knowledge of how enkephalin works, at the level of the nerve cell, to reduce the experience of pain. This is one of the fastest growing areas of pharmacological research. In years to come we can expect greatly improved control of pain and a much better understanding of the brain.

carry on fighting for long periods before realizing how badly they are hurt. Women in labour do not as a rule require an anaesthetic. The ancient Oriental technique of acupuncture has long been recognized as a method of pain control. When the endorphins and enkephalins were discovered, it seemed possible that they might be involved in these circumstances. Evidence soon began to mount up showing that they were.

Doctors treating chronic pain caused by damage to the spine had known for a long time that electrical stimulation to a particular region of the spinal cord could provide rapid relief. There was argument over how this worked, until researchers analysed the fluid surrounding the spinal cord. They found that a few minutes after the electrical

stimulation, the spinal fluid contained significant levels of an opiate-like substance—endorphin. This suggested the possibility that a similar mechanism might be at work in acupuncture, which can also involve the use of electric needles. Strong evidence that this theory was correct came from the finding that the anaesthetic effects of acupuncture could be reversed within seconds by giving a small dose of the opiate antidote naloxone.

The endorphins and the enkephalins have now been found to be involved in many other situations where the body must control pain. It seems that the placenta produces endorphin as labour nears, thus not only reducing pain for the mother, but also possibly keeping the baby quiet! (One of the

Probing the inner man

One of the most remarkable tools at the disposal of the modern doctor is an optical instrument which allows him to look right inside the patient's body. He can see into the intestine, stomach, lungs, bladder, womb or even the knee joint. The precision instrument which makes this possible is called an endoscope.

Examining the vocal chords using a small mirror held at the back of the patient's mouth is a simple form of endoscopy (the word comes from the Greek and means 'to look inside'). But this technique is obviously very limited — it takes more than a mirror to look into someone's stomach — and for over 100 years scientists have been developing special instruments for looking at the interior of living organs.

The first endoscopes were open, rigid tubes which could be illuminated by candles. Although a few rigid metal endoscopes are still used today, the majority of modern instruments are flexible 'fibre optic' types. The key component of a fibre optic instrument is a bundle composed of thousands of long, thin glass fibres. Each filament is as thin and flexible as a human hair, and light entering one end is reflected repeatedly along the length of the fibre to emerge at the other end. The fibres, 9 to 12 microns in diameter (a micron is a millionth of a metre), are covered with a reflective coating so that however much they are bent or curved, very little light is lost as it travels from one end to the other.

Fibre optic bundles

A modern endoscope has at least two fibre optic bundles. The first one transmits light from the operator's end of the instrument to the end which is inserted into the patient, the *distal* end. This lights up the area to be inspected. A second fibre optic bundle then transmits an image of the illuminated area back up the endoscope shaft to an eyepiece.

The arrangement of fibres in the illuminated fibre bundle is not critical — each fibre simply transmits a small amount of light from the lamp to the organ being examined. But the organization of fibres in the image bundle is crucial. Because each fibre is transmitting the light responsible for a small bit of the image, rather like a dot on a printed picture, it is very important that the arrangement of fibres is the same at each end of the bundle. A fibre which starts, say, at the top left of one end of the fibre bundle must end at the top left of the other end, otherwise the image seen in the eyepiece will be hopelessly jumbled up. A fibre bundle whose filaments are identically arranged at each end is called a *coherent* bundle.

In most fibre optic endoscopes light is transmitted through a randomly arranged (non-coherent) bundle from a powerful lamp, usually a quartz-halogen or xenon arc lamp, to the operator's end of the instrument. From there a second non-coherent bundle transmits it down the flexible shaft to the distal end. A coherent bundle carries the image back up the shaft to the eyepiece. Lenses at each end of the

Olympus KeyMed

Dr R. Zeegen

Above A modern endoscope is used to locate and remove a polyp from the patient's colon. This is a common disorder which can cause bleeding from the bowel—the doctor's view of the polyp is seen in the lower picture. The doctor manipulates and removes the polyp by means of tiny forceps and a snare at the tip of the instrument which he operates using the white handles seen in the picture. The patient remains conscious throughout the operation which may take no more than a few minutes.

coherent fibre bundle bring the image to a focus in the eyepiece.

As well as the glass fibre bundles, several channels pass through the shaft of the endoscope. One of these allows air to be passed down the shaft. If the organ being examined is a hollow one, such as the stomach, the air channel allows it to be inflated slightly to make thorough examination possible. And as the organ often contains liquid, a suction channel is provided so that it can be drained through the endoscope. Another duct carries fluid for washing mucus or small particles from the viewing lens at the tip.

Simple operations

Modern endoscopes are much more than just viewing instruments. Tiny forceps can be passed through yet another channel in the instrument's shaft to take specimens of tissue for analysis. And simple operations, such as the removal of a polyp from the colon, can be performed.

The tip of a modern endoscope is remarkably flexible and can typically be moved through 100° to the left or right,

180° upwards and 90° downwards. All the controls for these various manoeuvres are conveniently positioned near the eyepiece within easy reach of the operator's fingers. All the channels are compactly contained within the shaft, and in spite of the complexity, the shaft diameter may be as small as 8.8 mm (0.35 in.), little more than the width of a pencil.

Images seen down the endoscope can be photographed with conventional, Polaroid or movie cameras. And more recently a television system has been perfected, so the possibilities for display are virtually limitless.

The detailed design of the endoscope depends on its intended use. For stomach examination, for instance, the shaft must be long and flexible so that it can easily pass down the patient's throat into the stomach, whereas for knee joint examination the shaft is much shorter, and usually rigid. Endoscopes designed for specialized tasks are named accordingly — instruments for looking into the fluid filled spaces of the brain are called *ventriculoscopes,* for examining the lungs *bronchoscopes,* for

Below An X-ray picture shows the tortuous path followed by an endoscope shaft during a full examination of the colon. The shaft seen here is about 80 cm (31 in) long.

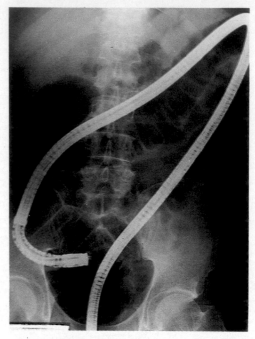

Dr R. Zeegen

HOW FIBRE OPTICS WORK

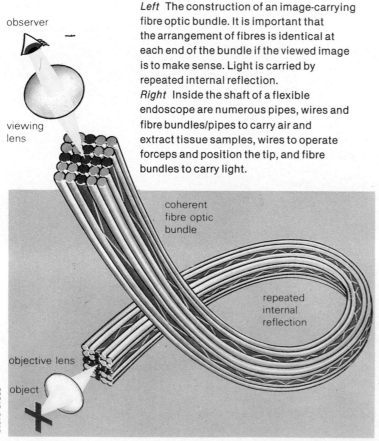

observer

viewing lens

Left The construction of an image-carrying fibre optic bundle. It is important that the arrangement of fibres is identical at each end of the bundle if the viewed image is to make sense. Light is carried by repeated internal reflection.
Right Inside the shaft of a flexible endoscope are numerous pipes, wires and fibre bundles/pipes to carry air and extract tissue samples, wires to operate forceps and position the tip, and fibre bundles to carry light.

coherent fibre optic bundle

repeated internal reflection

objective lens

object

Steve Cross

INTERNAL CONSTRUCTION OF ENDOSCOPE TUBE

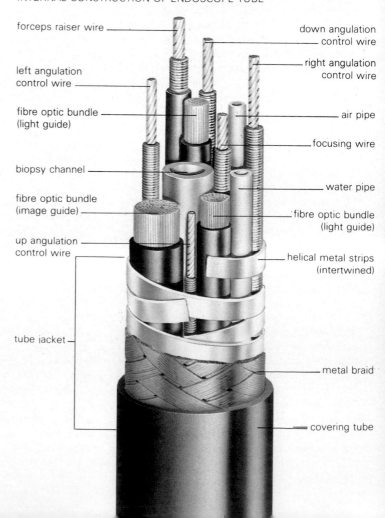

forceps raiser wire

left angulation control wire

fibre optic bundle (light guide)

biopsy channel

fibre optic bundle (image guide)

up angulation control wire

tube jacket

down angulation control wire

right angulation control wire

air pipe

focusing wire

water pipe

fibre optic bundle (light guide)

helical metal strips (intertwined)

metal braid

covering tube

Kuo Kang Chen

CONTROLS END

1 combined air and lens-washing control
2 suction button
3 small-channel non-return valve
4 large-channel non-return valve
5 viewing lens
6 eyepiece
7 focusing ring
8 brake for up/down control
9 up/down angulation control for distal end
10 left/right angulation control for distal end
11 brake for left/right control
12 forceps raiser control

DISTAL END

13 forceps raiser for small channel
14 objective lens
15 air and water outlet
16 fibre optic viewing lights
17 forceps raiser for large channel

ENDOSCOPE ACCESSORIES

surgical scissors

rat tooth forceps

alligator jaws forceps

biopsy forceps

looking into the stomach, *gastroscopes,* for looking into joints *arthroscopes* and so on.

By far the commonest subject for endoscopic examination is the gastro-intestinal tract, the tube which runs from stomach to anus and through which our food passes as it is being digested and absorbed. To view the gullet or stomach, the endoscope is inserted through the mouth, and the patient must fast from the night before the examination. If the large bowel, the *colon,* is to be examined, the instrument (a *colonoscope*) is inserted via the rectum, excreta being removed beforehand by means of enemas and washouts. The doctor inserts the endoscope under local anaesthetic with minimal discomfort to the patient, and after a full examination, which is complete in a few minutes, he has an excellent picture of just what is going on in the suspect organ.

Diagnostic tool

The endoscope is used most often as a diagnostic tool. For instance, if X-ray pictures are difficult to interpret or show nothing unusual, the doctor may turn to the endoscope for a closer look. And if a tumour is revealed by an X-ray picture, the endoscope may be used to take a sample of tissue for analysis before the doctor decides on any treatment. Now that very slender

Above An endoscope for looking at the stomach. The various controls are for focusing, moving the tip, introducing air or water, sucking out fluids and so on.
Right All kinds of accessories are available for modern endoscopes. This picture shows a selection for use with the instrument shown above. The rat tooth forceps are for grasping while the lower two are for taking tissue samples.

endoscopes are available, some clinics use them as the main diagnostic tool for patients whose symptoms suggest disease of the upper gastro-intestinal tract. And where patients have had an operation on the stomach or colon and their symptoms recur, or where the operation has been for cancer and a regular watch is essential, endoscopy is normally the method chosen.

Endoscopy is sometimes used to examine apparently healthy patients. In Japan, for instance, the incidence of stomach cancer is much higher than in other countries, and regular endoscopic examination helps to detect the disease in its early stages and so makes a cure more likely.

Jaundice is an unpleasant ailment where the skin and the white of the eye become

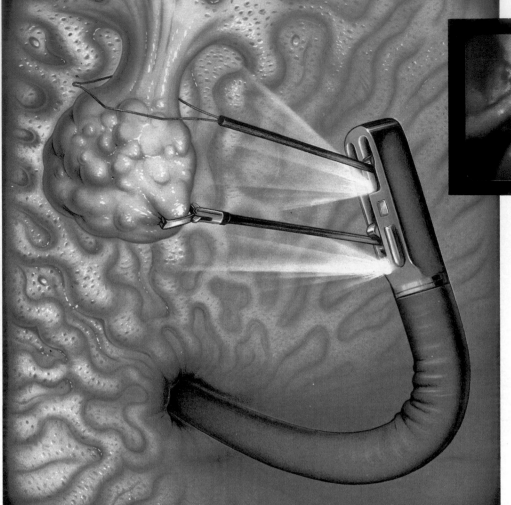

Left A polyp is removed by means of a side viewing endoscope. The polyp is snared and a high frequency current is passed to remove it. It is extracted by the forceps.
Above left A doctor's eye view as the root of a polyp is snared.
Above right The tip of an endoscope. As well as forceps and snare, two light channel lenses can be seen (one channel for light from the lamp and one to carry the image to the eyepiece).
Below left Looking at the interior of a patient's stomach using an endoscope. A teaching attachment enables a second doctor to view the image simultaneously.

discoloured with a yellowish tinge. It has many causes and the doctor may find it hard to pin down the right one. Endoscopy can help him diagnose the underlying problem. A tiny tube is passed through the endoscope and into the outlet of the patient's biliary and pancreatic systems and a dye is injected. When an X-ray is taken, the bile ducts and the pancreatic ducts show up clearly. (In fact this is the only satisfactory way of examining the pancreatic duct system, short of an operation). The X-ray picture may show a cancer of the pancreas or a gall stone, either of which can give rise to jaundice.

The endoscope is a very useful tool in the diagnosis of patients suffering from severe bleeding in the upper or lower gastro-intestinal tract. In such cases endoscopy of the gullet, stomach and duodenum will often be carried out as an emergency procedure and it can tell the doctor not only exactly where the site of the bleeding is, but also what is causing it. This information is vital because treatment depends very much on the cause. For instance, an operation may be needed for a duodenal ulcer, but inflammation and bleeding caused by aspirin will only very rarely require such drastic action. The cause of chronic blood loss from the large bowel may escape detection by all

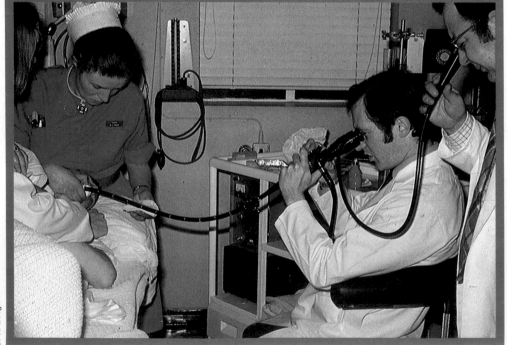

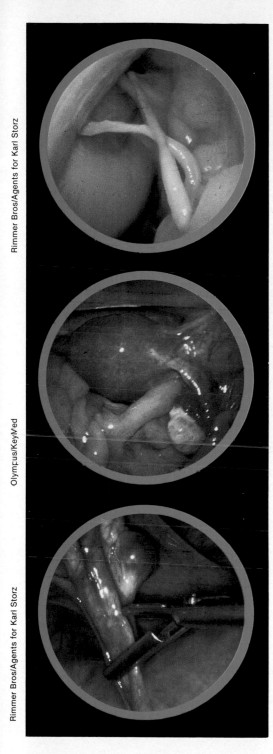

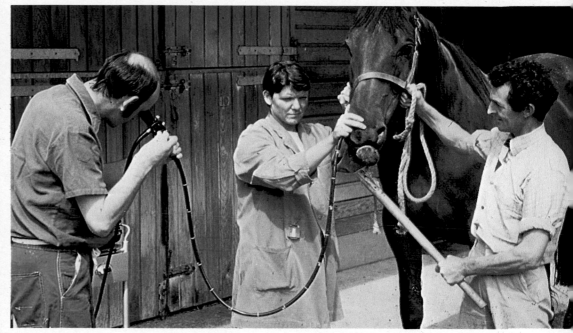

Above A thin, flexible endoscope is used to locate a nasal blockage in a racehorse. A rigid endoscope would once have been used, but the newer instrument allows the vet to keep his distance.

Top An inside view of a knee joint. Operations on the knee can now be done using an endoscope, and a stay in hospital is usually unnecessary. The patient's knee ligaments, which are crossed, are clearly visible in this picture.

Above centre Seen through a laparoscope—a type of endoscope—are the womb, a fallopian tube and ovary (lower right).

Above bottom Forceps extending from the tip of a operating laparoscope grasp a fallopian tube prior to a sterilization operation. Once again endoscopic methods make the procedure much simpler.

other methods except endoscopy — a small bleeding polyp may look like a particle of faeces on an X-ray picture, and abnormal, chronically bleeding blood vessels may not show up at all on an X-ray picture taken after a barium enema.

But the endoscope is not just an instrument of diagnosis — it can be used to treat patients as well. For example, the traditional method of treating a patient found to have a polyp in the colon involved an operation and a 10 to 14 day stay in hospital. Nowadays the polyp can be removed by means of an endoscope. With the instrument suitably insulated and the patient earthed, a wire is passed down the shaft. A high frequency current is then applied to get rid of the offending polyp. The patient can leave hospital the next day and the procedure is both cheap and safe. A similar wire may be used to enlarge the opening of the bile duct and so allow gall stones contained in it to be removed with a small basket or probe passed down the endoscope. Alternatively, the gall stones may now pass of their own accord into the bowel; either way an operation is avoided. The technique is especially useful for elderly patients unfit for conventional surgery.

All kinds of ingenious modifications of the endoscope and tools for use with it are available so that swallowed foreign bodies may be removed from the gullet and stomach. In the past, if the doctor decided that a particular object would not pass naturally through the gut, he would have no choice but to proceed with an operation.

Constrictions of the gullet can be widened by using an endoscope, and it is possible to pass a tube through an inoperable cancer of the gullet to allow the patient to swallow normally despite the disease.

New treatment

One of the latest developments in endoscopy is a new treatment for patients suffering from a major haemorrhage in the upper gastro-intestinal tract — all too often a cause of death among the elderly. The technique uses a special endoscope fitted with a quartz fibre extending down one of the channels in the shaft. At the operator's end of the instrument the fibre is coupled to a laser light source. First of all the doctor pinpoints the site of the bleeding using the endoscope in the normal way. Then he positions the tip of the quartz fibre over it and triggers the laser for a carefully determined time. The laser light, directed precisely on to the point of the bleeding by the quartz fibre, coagulates the blood and the bleeding stops.

Developments like this will ensure that the endoscope remains a powerful weapon in the doctor's armoury. And in the future it can be expected that cheap, fast endoscopic treatment will increasingly replace conventional surgery.

Probing the living brain

Lifted from the skull, the human brain sags like a waterlogged, grey, three pound blancmange. But within this soft lump of tissue, in life, some 14 thousand million cells function as the body's most delicate and complex organ. Surgery of the brain is, technically, the most demanding and difficult of medical activities.

Though prehistory is strewn with skulls that bear evidence of efforts to probe the secrets within, surgeons shunned the brain until the 1900s. Its functions were a mystery. Interference with it meant death by infection or bleeding. Then, in 1905, an American surgeon, Harvey Cushing, removed a brain tumour and began a new era. Today, brain surgery is still, on occasion, hazardous, sometimes controversial and in general a last resort. But three in four operations are successful.

Harvey Cushing's success was due largely to his 'bloodless' technique. He ensured that blood vessels, disturbed by the operation, did not bleed into the brain and put fatal pressure on it. He maintained the brain's normal blood supply—without which, starved of oxygen, brain cells die within minutes.

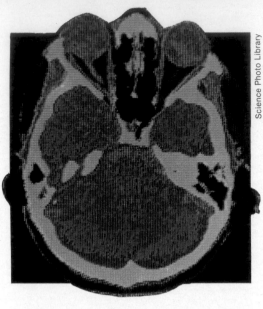

Above This tomography 'scan', showing a cross-section through the lower part of the brain, is made up of several X-ray images linked together using a computer. It is a safe and accurate method of diagnosing brain damage.

Haemorrhages and blood clots remain major hazards in brain surgery: in other respects today's surgeons have advantages denied to the first pioneers. With antibiotics they can keep infection at bay and cure brain abscesses previously untreatable. Microscopes aid work on tiny blood vessels. X-ray equipment pinpoints targets deep in the brain. Tomography, or 'scanning', has improved diagnosis and maps the precise size and site of lesions.

But, in opening the human skull to venture a scalpel on the tissue within, modern surgeons still face a task demanding a concert pianist's sureness of touch—and one wrong note may cost a life, or destroy a vital function.

Removing tumours

Cushing's target tumour was a *meningioma,* a growth of tissue known as 'benign' since its cells do not break away to invade other parts of the body—but still dangerous as its growth pressures the brain and impairs it. First stage in a meningioma removal involves the surgeon in skills recognizable to any carpenter.

He begins by turning down a flap of the scalp to bare the part of the skull above the tumour. With a brace and bit he drills partly through the skull, then continues with a burr (a conical drill). Its shape prevents the rest of the drill from slipping straight through the membranes around the brain, once the tip has cut through. The surgeon makes five burr holes, in a pentagon pattern. Next, four sides of the pentagon are cut out by slipping a thin, flexible saw (much like a cheesecutter's wire with teeth) in through one hole and out through the next. The saw blade is held diagonally, so that the blade cuts upwards and outwards. Thus the piece of skull which has been cut out can be safely replaced without any risk of it slipping through the hole. The fifth side of the pentagon is not sawn, but cracked. It stays attached to the living muscle beneath the scalp,

In this computer tomography scan, a type of tumour known as a glioma is clearly visible (white area) in one of the brain's cavities.

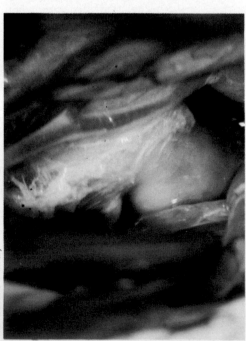

Here an artery in the brain has bulged out to form an aneurysm (pink egg-like object, *centre right*) which may burst and damage the brain.

This scan shows a blood tumour (haematoma) between the brain and the skull of a patient who has suffered a head injury.

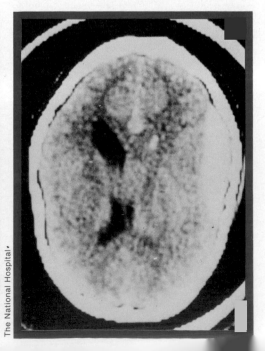

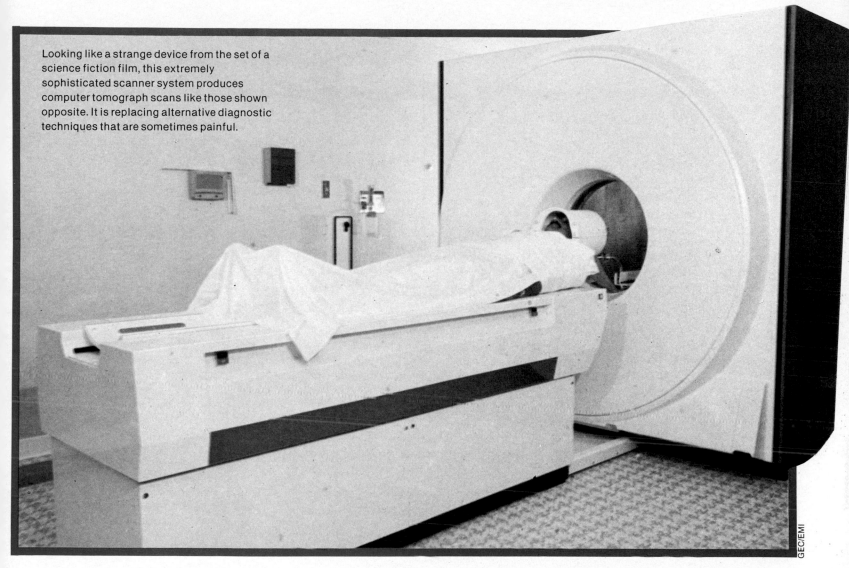

Looking like a strange device from the set of a science fiction film, this extremely sophisticated scanner system produces computer tomograph scans like those shown opposite. It is replacing alternative diagnostic techniques that are sometimes painful.

which, acting as a hinge, allows the surgeon to pull back the section of skull like a door. This door is called an 'osteoplastic flap'. The bone remains alive because it is still getting its blood supply through the attached muscle.

To get at the tumour the surgeon cuts through the brain's outer membranes to expose the cortex of the brain—a layer of tissue fissured like a walnut. The tumour, usually quite firm, can be separated from the cortex by inserting little *pledgets* of cotton wool around it. Once it has been packed round with cotton wool it can be safely winkled out. As soon as the tumour is gone, the brain, which has been compressed, will expand to fill the cavity left by the tumour.

Other types of tumour are less easy to deal with surgically and most surgeons regard them as inoperable and better treated with drugs or radiation. One common malignant tumour is the *glioma,* a tumour of the connective (glial) tissue between nerves. Though

the neurones, or nerve cells, themselves are not involved in the tumour they may suffer when the connective tissue tumour around them begins to press. Some surgeons have tried to cut out gliomas, but this can only be done by excising all the neurones in that area—which may leave the patient speechless, paralyzed or lacking other important functions. The decision to operate must weigh the damage caused by surgery against the symptoms of the cancer.

Occasionally a brain tumour will start to decay from the middle and form a cyst. Such tumours can be destroyed by draining the fluid from the cyst and then introducing a fragment of radioactive material into the cavity. Having found the tumour with the help of X-rays, the surgeon opens the skull, makes a small slit in the cortex and uses an instrument to splay the fibres of the cortex so that he can see—with the aid of a small lamp—right down into the brain. The fluid is drained mechanically from the cyst and

then the radioactive material is inserted on the end of a needle. The radioactivity dies away after a few days, but, if the operation has been judged correctly, it is sufficient to destroy the tumour from the inside without affecting the surrounding brain tissue.

Brain abscesses

Abscesses in the brain, once untreatable and often fatal, can now be treated in a similar way. An abscess is a pool of pus resulting from an inflammatory reaction to an infection. The body's natural healing mechanisms try to isolate the pus by forming a fibrous wall of glial tissue around it, but they may not be strong enough to mop up the bacteria in the pus itself. The abscess is drained by inserting a hollow needle through a burr hole. Antibiotics can then be introduced into the fibrous capsule to kill remaining bacteria.

This direct attack is more effective than antibiotics given by mouth or by injection

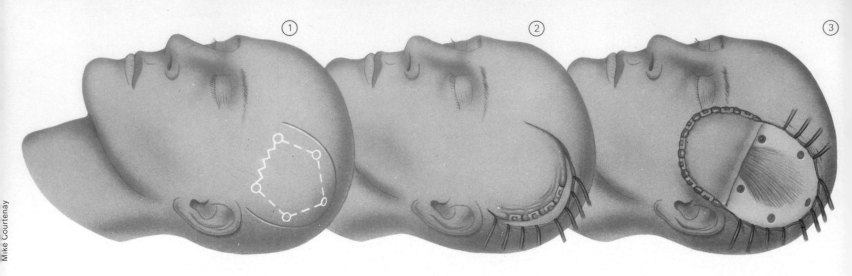

In an operation to remove a tumour, the brain surgeon first determines the pattern of the 'door' *(above left)* he will cut into the skull to allow him entry into the brain beneath. Next *(above centre)* he cuts a flap of skin and pulls it away from the scalp. Clips along the side of the flap and along the cut edges of the scalp prevent any bleeding. The flap is then turned right back *(above right)* so that the surgeon can drill the five burr holes that will enable him to lift up a flap of bone. A stalk of muscle is left uncut so that blood can still reach the bone and keep it alive during operative procedures.

because the fibrous wall of the abscess stops the antibiotics from entering the capsule. An antibiotic put straight into the abscess will not escape, and will confine its activity to the abscess, so it can be given without risk to the surrounding brain tissue.

Such techniques have neutralized the killing power of *meningitis,* an inflammatory infection of the meninges (membranes of the brain) which until 30 years ago took many lives. Now, very severe cases can be treated by injecting antibiotics directly into the deeply located ventricles of the brain which

contain the cerebrospinal fluid. As this fluid bathes the membranes all round the brain, the injected antibiotic will very rapidly reach its target.

Damage caused by blows to the head, by accident or violence, occupies most of the brain surgeon's time. Fortunately, nine in ten people who are brought to hospital with head injuries suffer bruising only and recover without surgical intervention. Of the remainder there will be some who have suffered such severe damage to the parts of the brain which control their vital functions that nothing can be done for them surgically. They remain severely disabled or die, sometimes after a long period of coma. Patients the surgeon can help are those with a small amount of superficial brain damage and at risk from bleeding in the brain. A knock on the head can damage a blood vessel slightly so that the effects of the bleeding only begin to show up after several hours or even days after the incident.

Diagnosis

An experienced surgeon can very often diagnose internal bleeding without any sophisticated equipment but simply by testing reflexes, coordination and other nervous system functions. Before scanning was invented, there were a variety of X-ray techniques—still used in hospitals which do not have scanning equipment—which could identify bleeding in cases which were difficult to diagnose by observation alone. *Angiography,* for instance, is a technique that involves injecting an opaque liquid into the carotid artery which feeds the brain. This liquid's passage will show up under a series of

Left Surgery has exposed this large tumour in the right temporal lobe, at the front of the patient's brain. Without treatment, such patients may survive for only a few months after diagnosis, but drug or radiation therapy is often preferable to surgery.

CUTTING THE BURR HOLES

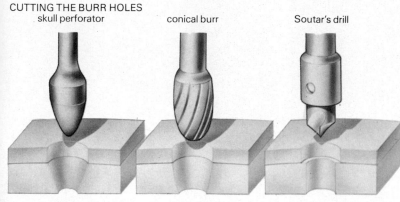

skull perforator   conical burr   Soutar's drill

Left These drill bits are used by the surgeon to make the burr holes in the skull which enable him to cut out the bone flap. The conical burr or Soutar's drill is used to enlarge the initial hole made by the skull perforator. This ensures that fragments will not slip through.

TREPHINING

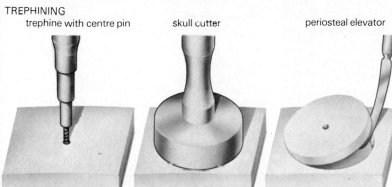

trephine with centre pin   skull cutter   periosteal elevator

Trephining is used to cut out a small hole in the skull for localized brain operations. First a small hole is drilled in the bone *(far left)*, then a trephine is used to cut out a circular groove *(centre left)*. Finally the bone disc is lifted out *(left)*. The surgeon can now operate.

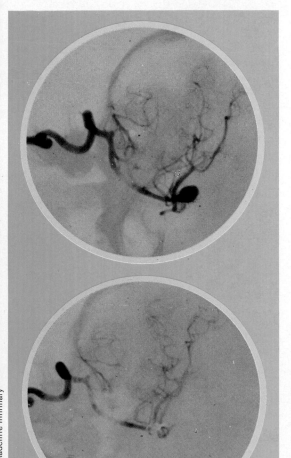

Above left This angiogram of blood vessels in the brain, made by injecting them with opaque dye and then taking X-ray pictures, reveals an aneurysm as a black blob.
Left This angiogram was taken after repair of the aneurysm with a clip (white object).

X-ray pictures, enabling the surgeon to see whether the blood flow is blocked or flowing where it should not. But angiography is not without risk of damage to the blood vessels by filling them with a foreign fluid.

Scanning is much safer. A scan is basically a series of X-ray images which are linked together with the help of a computer to give a three-dimensional view. Whereas an ordinary X-ray film will only reveal relatively dense tissue such as bone, or an introduced opaque material, a scan can reveal a broad spectrum of tissues, including blood clots and tumours. Once identified, a clot can be sucked out through a burr hole, or, if it is large and solid, via an osteoplastic flap.

Patching up the skull

Damage to the skull itself is less serious than the bleeding which may occur underneath. Like any bone the skull can repair itself and will regain its former shape if the fracture is pinned into position. If it is so badly smashed that the fragments of bone are beyond self repair or missing, the hole can be covered by a shaped metal or acrylic plate. Silver was the first metal to be used for replacing lost pieces of skull, but it tended to oxidize and become pitted. Tantalum, a metal both strong and malleable, replaced silver, and has been replaced in turn by acrylic, which is cheaper and can be moulded very quickly.

Acrylic is also used by brain surgeons for another purpose: to prevent *aneurysms* from bursting. Aneurysms are bulges in arteries, caused by a weakness in the lining of a blood vessel, and tend to occur at points where an artery bends or branches. The pressure of the blood blows the weak lining out into the shape of a balloon or berry. Aneurysms can exist for years without provoking any symptoms, unless they become so large that they press on a nerve. But as time passes there is an increasing likelihood that the balloon will burst. If it does burst, the blood will spill into the space between the brain and the meninges and sometimes into the brain itself.

Aneurysm repair

Controversy still surrounds surgical attempts to repair burst aneurysms. Many physicians believe that a burst aneurysm is best left alone and claim that surgical intervention is only likely to make matters worse. On the other hand there are brain surgeons who have developed great skill at repairing burst arteries. They point out that even if the wound manages to heal, that part of the artery will remain weak and liable to burst again if it is not strengthened. Tiny holes in slippery artery walls are difficult to see unaided, let alone manipulate and repair. Modern success rates are very largely due to the microscope.

Having cut an osteoplastic flap to reveal the damaged artery, the surgeon can either wrap the aneurysm or clip it. Wrapping is rather like repairing a puncture in a bicycle tyre; the idea is to block the hole and strengthen the artery wall. Aneurysms can be wrapped with acrylic, with a 'muscle stamp' or with a rubber-like material. Acrylic, when heated, becomes soft and flexible, but sets hard when it cools. A muscle stamp is a small piece of facial muscle which is beaten with a little surgical hammer. Beating makes the tissue produce sticky substances called *thromboplastins* which help the stamp adhere to the artery. Both these materials have disadvantages: acrylic, unlike the artery wall, is rigid and cannot pulsate in the same way as the artery does when blood is being pumped through it. And muscle stamps tend to form scar tissue which can

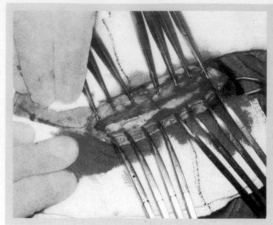

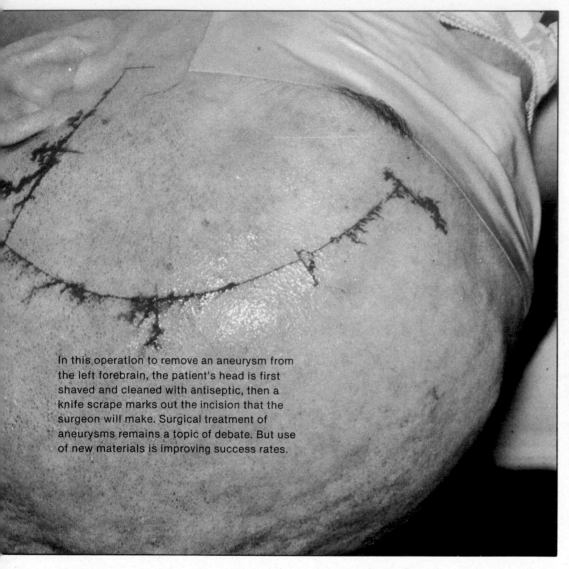

In this operation to remove an aneurysm from the left forebrain, the patient's head is first shaved and cleaned with antiseptic, then a knife scrape marks out the incision that the surgeon will make. Surgical treatment of aneurysms remains a topic of debate. But use of new materials is improving success rates.

Above The surgeon has started to make the first cut into the scalp. By pressing down on the cut and inserting clips along each side of the incision, he can control the bleeding.

Above Now the surgeon has turned back the skin flap to reveal the bony skull beneath. The stalk of temporalis muscle, via which the skull can still receive blood, is visible.

eventually stretch and loosen. New rubber-like materials which expand and contract are now being tried with apparent success.

Another aneurysm repair technique involves placing a little clip around the neck of the aneurysm balloon. This can be done as a preventive measure before the aneurysm has burst or afterwards to prevent further bursts. The surgeon must avoid setting the clip too close to the artery because he could easily pinch the vessel and obstruct the blood flow. The clip is therefore placed about a millimetre up the neck of the aneurysm.

Psychosurgery

If brain surgery for aneurysms is debatable, nothing has equalled the controversy which surrounds psychosurgery. In 1935 a Portuguese neurosurgeon, Egas Moniz, found that patients with severe mental disturbances, notably schizophrenia, could be relieved of their misery by cutting

the fibres leading to the prefrontal lobes of the brain at the forward end of the cortex. Monitz won a Nobel prize, and his many patients who had been in mental institutions for years were free to take up new jobs. But *leucotomy* is an irreversible operation—once cut, the fibres never regenerate—and it causes personality changes. Many patients became not only carefree but irresponsible. Medical views of schizophrenia have changed since the 1930s. A popular view today is that schizophrenia (which is a rather vague term for a variety of mental disturbances) is a biochemical disorder, possibly caused by an inborn error of metabolism, and better treated by drugs than the knife.

Although many neurosurgeons and a large body of public opinion reject operations which can permanently change a patient's personality, some psychosurgery undeniably relieved a great deal of suffering. The first leucotomies were crude, massive operations,

Above Burr holes are now being drilled into the skull to enable the surgeon to cut out the bone flap (osteoplastic flap). He does this by sawing the bone between each hole, except for the two on the side forming the hinge; here the bone is broken, not sawn.

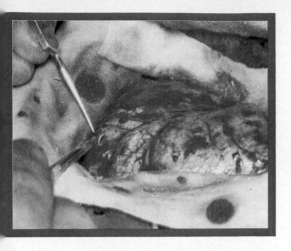

Above With the bone flap lifted up like a trap door, the surgeon is now able to cut into the dura mater, the tough outer membrane which envelops the brain.

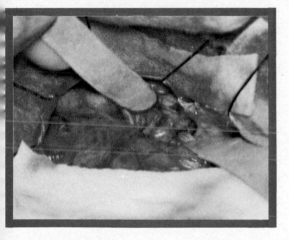

Above The brain is now exposed, and by moving the temporal lobe to the left and the frontal lobe to the right, the surgeon can at last see the aneurysm.

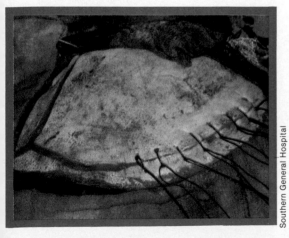

Above After repairing the aneurysm by wrapping it with acrylic, specially treated muscle or a new rubber-like material, or by clipping it, the surgeon has closed the cut in the dura mater, replaced the bone flap, and is now stitching the bone flap back into place.

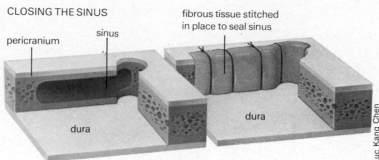

CLOSING THE SINUS

pericranium sinus

fibrous tissue stitched in place to seal sinus

dura dura

Kuc Kang Chen

When the bone flap is cut out, the exposed sinus *(far left,* dark area) must be closed with a piece of fibrous muscle lining. This is held in place *(left)* by sutures through the dura and the fibrous layer around the bone of the skull.

Southern General Hospital

but in the 1950s and 1960s British surgeons found that they could relieve chronic pain, obsessional states and agitated depression by destroying a very much smaller area of the frontal lobes. This was the so-called 'Area 13', which measures about one cubic centimetre and lies in the orbital cortex, about 4.5 cm (1.8 in) back from the very front of the brain.

It can be destroyed in a variety of ways, some of them most ingenious. It can be cut with a blade, coagulated by heat or high frequency electrical stimulation, or frozen with a pressurized gas such as nitrous oxide. Sometimes a small metal ball was inserted into Area 13 (under X-ray control it could be placed to within two millimetres of the centre of the target), then an induction coil would be wrapped round the patient's head. A small current would be passed through the coil and the electromagnetic field it created would heat the metal ball so that it gradually burnt out the brain fibres around it. Alternatively a small 'seed' of radioactive material such as yttrium could be introduced into Area 13 to destroy the target fibres. Destruction of Area 13 is claimed to have less effect on personality and intellect than full leucotomy, though it was only recommended when all other forms of treatment had failed.

Similar restraints are applied to the treatment of epilepsy by neurosurgery. Epilepsy may often be a congenital problem, but it can also result from brain injuries, infections or brain tumours. It is not certain whether all epilepsy is caused by a lesion or scar in the brain, but neurosurgeons did find lesions in the temporal lobes (the part of the brain nearest the ears) of a large number of their patients.

During the 1950s and 1960s the surgeon Murray Falconer treated epilepsy by disconnecting parts of the temporal lobe. The operation was usually successful but tended to bring about personality changes. Patients who had previously been aggressive would become docile. Some years earlier, the Canadian surgeon Wilder Penfield, whose pa-

tients were almost all accident victims, looked for distinct lesions which he could remove. His work also confirmed the idea that the temporal lobe plays an important part in memory. Using only local anaesthetic he would stimulate different parts of the temporal lobe and this would often arouse distant recollections in his patients' minds. One woman 'heard' Beethoven's Fifth Symphony on the operating table. But despite these intriguing findings, surgical treatment for epilepsy is now relatively uncommon and only used when drug treatment has failed.

Implanted electrodes

Under active development, however, is an innovative technique for pain control: the implanted electrode. Gold or platinum electrodes are set into the grey matter of the midbrain. The emerging wires are fixed under the scalp and lead to a receiver device in the chest which has an aerial protruding through the skin. The electrodes are activated by a portable transmitter which the patient can keep in his pocket. It seems that electrical stimulation of the midbrain makes the brain produce its own painkilling chemicals, called *endorphins* and *enkephalins,* which have a similar effect to morphine. It is an expensive operation and not fully proven—but appears to relieve severe pains which cannot be controlled by drugs.

Techniques in brain surgery regarded as valuable breakthroughs in one decade may be discarded a decade later. Some of the controversy is due to the traditional rivalry between surgeons, physicians and psychiatrists, who mutually mistrust other specialists 'trespassing' on their field. Sometimes techniques are abandoned because experience shows that they do harm as well as good. And the success of a technique depends as much on the individual surgeon's skill as on the method itself. Nevertheless, in the years since Harvey Cushing first removed a tumour, brain surgery has developed to a highly refined technique which, when other means fail, can save lives and relieve pain.

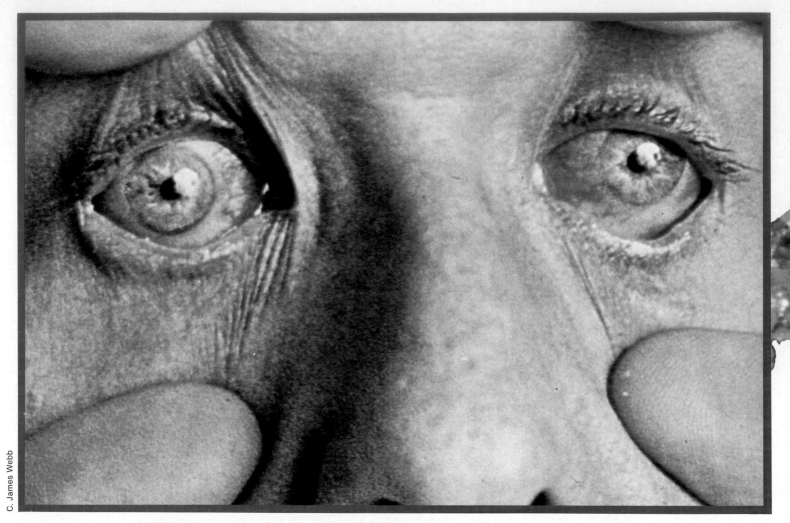

C. James Webb

Why antibiotics work

Of all the advances in medical science achieved over the past hundred years, perhaps none has had more impact on Western society than the development of antibiotics. Indispensable weapons in the war against killer diseases, they also have an important role in the treatment of minor infections. Few people in the developed world now go through life without undergoing a course of antibiotics at some time. Admittedly, the emergence of increasing numbers of resistant strains of disease-causing organisms has recently put the widespread use of the drugs under scrutiny. But even so, the discovery of antibiotics remains one of the great miracles of modern medicine.

Since Alexander Fleming discovered penicillin almost by accident in 1929, over a thousand different antibiotics have been found. They are classed as a sub-group of the range of drugs known as 'chemotherapeutic agents' (substances that kill or inhibit the growth of micro-organisms responsible for a disease). Antibiotics are antimicrobial substances extracted from living organisms. The term was invented by the German scientist Paul Ehrlich as long ago as 1906, following the realization that every human disease might be cured by the use of a 'magic bullet'—a chemical that would seek out and kill the microbe leaving the host unharmed.

Penicillin, from the fungus *Penicillium notatum,* is by far the best known and still one of the most commonly used antibiotics. But the list of new entrants to the field is growing all the time. The majority of these natural antimicrobial agents are produced by a form of bacteria-like fungi called *Actinomycetales.* Almost 60 per cent of the antibiotics now in use come from this little group of organisms.

Nobody knows quite why these micro-organisms produce antibiotic substances, though various suggestions have been put forward. The substances are referred to as 'secondary metabolites' because they are produced not in the first—growth—stage of the organism's life cycle but in the second—resting—stage. It has therefore been suggested that they help to conserve energy during the resting stage.

Synthesis of these substances for medicinal purposes has always relied on the culture of fungi and bacteria—only a few of the simpler varieties, such as chloramphenicol, can be synthesized artificially. Nevertheless, a number of chemical changes can be made to the natural product to get closer to the required substance. In commercial production

Infectious bacteria can produce toxins hundreds of thousands of times more poisonous than arsenic or strychnine.
Above Eye infection caused by staphylococci.

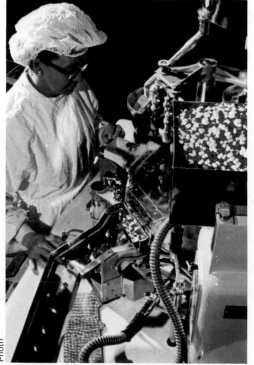

Above left Pseudomonas microbes identified by an oerie blueish glow under ultra-violet light. Pseudomonas causes urinary tract infection. The complaint is treatable with tetracycline (crystals shown above and left).
Below left Under sterile conditions drug-company production lines produce thousands of tons of antibiotics each year.

of penicillin, for example, phenylacetic acid is generally added to the culture medium in order to stimulate the development of side chains on the basic molecules to give penicillin G. New varieties of penicillin that are effective when penicillin G fails are being manufactured semi-synthetically. This involves the removal of the side chain and the substitution of others.

One of the strange things about antibiotics is that the producing fungus or bacterium clearly does not suffer from the toxic effects of the substance, while very closely related disease organisms do. Recent research is beginning to show why this may be. Experiments conducted on the antibiotic Thiostrepton have shown that while other streptomyces are fully susceptible, Thiostrepton is not toxic to its producer organism *Streptomyces aureus*. This is because the organism has a particular substance that guards it from attack.

The key to antibiotic action is that the drug is only toxic to certain organisms while leaving others unaffected—this is why anti-

biotics are so valuable to medicine. If they simply killed off every organism without discrimination, they would be little better than antiseptics and could not be used as drugs. But because of their selective toxicity they can be safely administered to patients throughout the body, whilst the antibiotics only exert their toxic effect on the pathogenic organism.

Drugs such as penicillin and cephalosporin, for instance, work by inhibiting the synthesis of a substance called mucopeptide—an important component of the bacteria cell walls. Mucopeptide is only found in bacteria, so only bacterial cells are subject to the toxic effects of penicillin. With so few side effects, penicillin is still a very popular drug. The adverse and occasionally fatal reactions that can occur with penicillin are due not to the toxicity of the drug but to a hypersensitive response in the patient. Only certain people have this adverse response and naturally these people will not be prescribed penicillin.

The mechanism that makes an antibiotic selectively toxic may seem relatively delicate, but even so the difference between two organisms must be fairly clear if the antibiotic is to have any chance of success. This is why antibiotics have never been a great deal of use against cancer. Nor are they effective against viral infections. For when a virus takes over the host cell, there is little to point out the difference between the infectious

1087

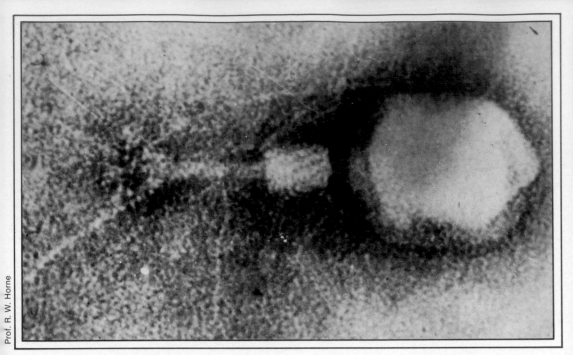

Prof. R. W. Horne

agent and the host. Susuma Hotta and his co-workers in Japan have extracted substances from the bacteria *Streptococcus faecalis* that inhibited growth of human pathogenic viruses. But this success against a virus, and a few earlier examples, are isolated and probably not significant.

It may be simpler to improve targeting of the drug rather than alter its selective toxicity. This would involve adding substances to the drug that help it to 'home in' on specific infected cells. It has been observed, for example, that cells invaded by a virus have slightly more permeable membranes than normal cells. Thus antibiotics could, perhaps, be administered in a form which permeates infected cells but cannot enter normal cells. Such targeting techniques will probably become increasingly important.

Different antibiotics work against different organisms. If a patient is given a particular drug but fails to respond after a full course, it will not be because the drug has not 'worked' but because the organism causing the illness was not susceptible. A second course of a different antibiotic may well succeed. This is why it is so important to establish which organism is to blame.

Unless the doctor can identify the organism accurately, he can never be sure which drug he should use. Unfortunately, there is rarely the time or the money available for establishing the pathogenic organism in every single infection—a doctor cannot wait for a detailed laboratory report before starting to treat the patient. He must therefore make an informed guess as to the identity of the pathogen and prescribe on that basis.

However, the availability of so-called 'broad spectrum' antibiotics, effective

Above and right Bacteriophages—viruses that attack bacteria—are shaped like tiny tadpoles a 5,000th of a millimetre long.
Below The phage acts as a self-propelled hypodermic syringe. After attaching to the bacterium, the spring-like sheath contracts, driving its tip into the cell wall. The polyhedral head then releases its contents—viral DNA which induces the bacteria to produce new phages in due course.

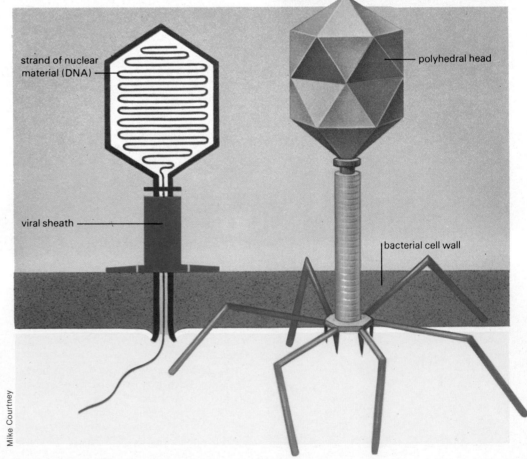

strand of nuclear material (DNA)

viral sheath

polyhedral head

bacterial cell wall

Mike Courtney

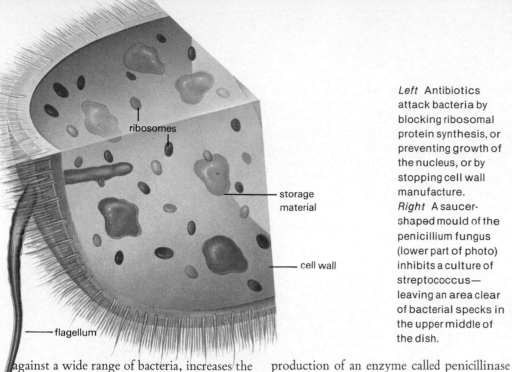

ribosomes

storage
material

cell wall

flagellum

Left Antibiotics
attack bacteria by
blocking ribosomal
protein synthesis, or
preventing growth of
the nucleus, or by
stopping cell wall
manufacture.
Right A saucer-
shaped mould of the
penicillium fungus
(lower part of photo)
inhibits a culture of
streptococcus—
leaving an area clear
of bacterial specks in
the upper middle of
the dish.

Paul Brierley

against a wide range of bacteria, increases the chances of success in the treatment of an unidentified organism. Such drugs are widely used, particularly in the Third World where the facilities for accurate diagnosis are simply not available.

Even if the pathogen can be recognized, choosing the appropriate antibiotic is rarely simple. One problem is that for any one organism there are often a considerable number of apparently appropriate drugs. Also, minute differences in an apparently susceptible organism may make it completely resistant to a particular antibiotic.

One of the major problems with the use of antibiotics has been the emergence of resistant strains of bacteria. A patient suffering from infection by an initially susceptible species may fail to respond to treatment because the bacteria have changed very slightly to become resistant. Very soon after the introduction of penicillin, for instance, a resistant variety of staphylococci emerged.

Resistance in this case was not due to the bacteria becoming invulnerable to attack by penicillin. What happened was that varieties of the bacteria resistant to penicillin were able to survive and reproduce in great number thereby defeating the drug. The resistant varieties owed their success to the production of an enzyme called penicillinase which inactivates penicillin.

Increasing resistance

Since then, many other instances of resistance in a wide variety of species of bacteria have been found. For virtually every bacteria treated by antibiotics, strains have emerged that are resistant to the same drugs to which they were formerly susceptible.

However, the emergence of resistant strains does not mean that an antibiotic will altogether fail to work against a particular infection. It simply means that in some cases the infection will not respond to treatment by a particular antibiotic—but it may, however, respond to others. In fact, if an infection appears resistant to one form of antibiotic, it will generally respond to another. Tuberculosis, for instance, has traditionally been treated by streptomycin. But as strains resistant to streptomycin have appeared, doctors have been turning increasingly to the antibiotic rifampicin for both initial and maintenance therapy. In very few cases is there no suitable alternative—even though the alternatives may be expensive.

Unfortunately, common use of any antibiotic increases the chances of a resistant bacterial strain emerging. Because rifampicin is now used more frequently, it is inevitable that strains of tuberculosis resistant to rifampicin will become increasingly common. Rifampicin is usually used in combination with another drug to reduce the speed at which these resistant strains emerge. But eventually a replacement will have to be found for rifampicin, just as rifampicin is now replacing streptomycin. Antibiotic research is a constant battle to keep ahead of bacterial resistance—finding new substances to combat micro-organisms as they develop resistance to existing antibiotics.

New antibiotics have generally been found in soil by elaborate screening techniques, but this source is rapidly being exhausted. Soil screening is yielding new substances at an ever diminishing rate. Now scientists are beginning to look at alternative sources of natural antimicrobial agents. One of the most important may well be the sea.

The idea of using a marine source of antibiotics is now a new one. Dried powders and syrups extracted from the Red Sea alga *Digenia simplex* have been used for centuries in Asia and the Mediterranean as antihelminthics—drugs active against worms. But with the need for new sources of antibiotics the possibilities are being re-examined.

One drug company, for instance, has set

viral DNA

virus particle

new viral DNA

new virus
particles

cell
ruptures

Left Viral attack:
1 virus attaches to the cell membrane.
2 injected viral DNA reproduces itself.
3 new viruses are assembled in the host.
4 the cell ruptures, spreading infection to other host cells.

1   2   3   4

Mike Courtney

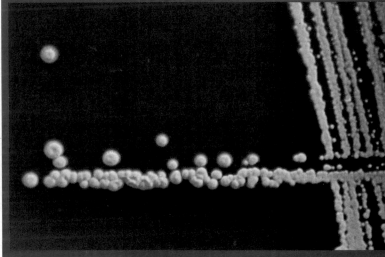

up an Institute of Marine Pharmacology in Australia, simply to research this field. That there are suitable substances has already been demonstrated. Marine bacteria, fungi, phytoplankton, and the higher algae, all contain substances with antibacterial, antifungal and antiviral activities.

In a study in 1962, Kraslinkova found that 124 out of 326 cultures collected from marine organisms produced substances active against *Staphylococcus, Escherichia coli, Myobacteria luteum* and *Saccharomyces cervisiae.*

penicillinase inactivator. Such an inactivator might well be the naturally occurring enzyme clavulanic acid. Recent experiments suggest that clavulanate inhibits the penicillinase from *Staphylococcus aureus.*

Undoubtedly, adjusting existing antibiotics is the simplest approach to finding new drugs to combat resistance and fight susceptible bacteria more effectively. It is cheaper and it reduces the need for extensive long-term trials before the drug is brought into general use.

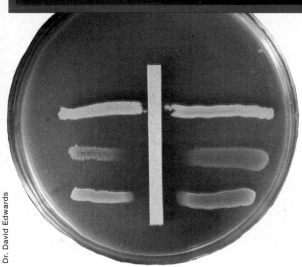

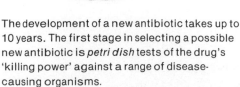

The development of a new antibiotic takes up to 10 years. The first stage in selecting a possible new antibiotic is *petri dish* tests of the drug's 'killing power' against a range of disease-causing organisms.
Top A coded range of antibiotics on test against klebsiella aerogenes (G-ve) microbe.
Centre left Penicillium chrysogenum—the mould producing the penicillin antibiotic.
Centre right Mould producing streptomyces.
Above Strip test: streptomycin *vs* 3 microbes.

He also found an actinomycete with all the characteristics of a broad spectrum antibiotic. Cephalosporin came from the sea too. (It was found in a sewage outfall off Sardinia!)

Alternatively, new antimicrobial drugs may be made synthetically. Synthetic drugs have always been used against viral and fungal infections since antibiotics are generally ineffective against these, but they may be used in bacterial infections in the future.

A third approach, and one that has already brought many rewards, is to synthetically adjust natural antibiotics to overcome the bacteria's resistance mechanism. A penicillinase-resistant form of penicillin, called cloxacillin, was made years ago by substituting the benzyl side chain of penicillin G with a chain that protects the basic penicillin molecule from the effects of penicillinase. Unfortunately, without the benzyl side chain, the penicillin is much less active. And although cloxacillin works against penicillin-resistant strains (that is, strains that produce penicillinase), it is less effective than penicillin G against susceptible strains.

However, another possibility is to use penicillin G against penicillinase-producing organisms by combining it with a synthetic

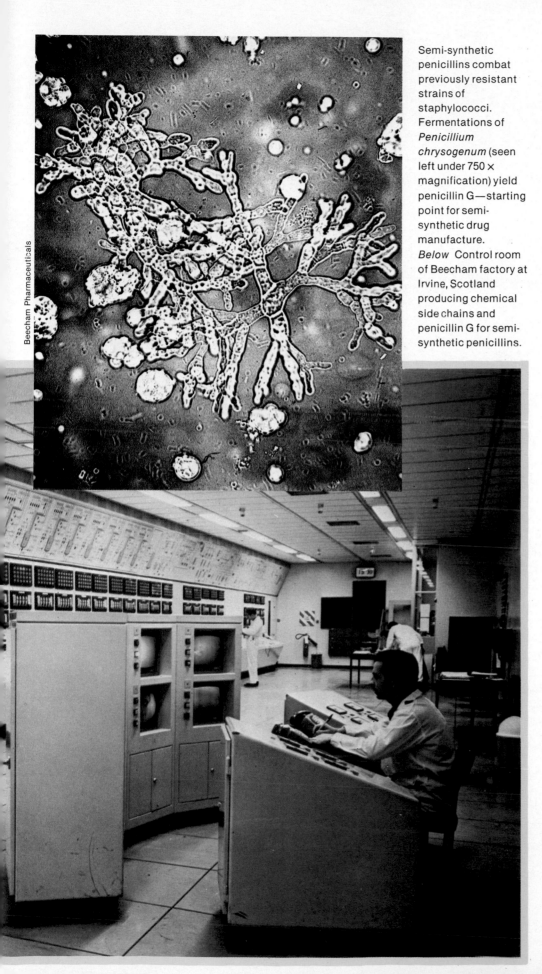

Semi-synthetic penicillins combat previously resistant strains of staphylococci. Fermentations of *Penicillium chrysogenum* (seen left under 750 × magnification) yield penicillin G—starting point for semi-synthetic drug manufacture.
Below Control room of Beecham factory at Irvine, Scotland producing chemical side chains and penicillin G for semi-synthetic penicillins.

However, it has been suggested that the race to find new antibiotics may ultimately be unrewarding because eventually resistance to every new antibiotic will emerge. Therefore, it is argued that such drugs should be used sparingly to slow the development of resistant strains to a minimum. Although scientists do not suggest that antibiotics should be withheld from anyone seriously ill, doctors may be asked to think twice before prescribing them for minor infections when there is little risk of serious complications. In particular, it is suggested that prophylactic (preventative) use of antibiotics should be avoided whenever possible.

One of the big controversies of recent years has been the large-scale use, in many countries, of antibiotics as a growth additive in animal fodder. This antibiotic additive has now been banned in Britain for the following reason: in 1958, two Japanese scientists, Akiba and Ochiae, discovered that resistance can be transferred between two micro-organisms by a process known as conjugation. Resistance to many drugs could be carried from cell to cell by a 'bridge' synthesised by RTF (resistance transfer factor). It was shown that this transfer of resistance is due to the existence of a special genetic element called the R-factor (resistance-factor).

Resistance is infectious

R-factors mean that multiple drug resistance could be transferred by conjugation to any other bacteria in the vicinity. Resistance is therefore infectious.

All this means that the use of antibiotics against intestinal infections could lead to the natural selection of resistant strains. And once this has occurred, the resistance could be transferred by R-factors to many other organisms. The level of R-factor resistance in some farm animals fed on antibiotic-supplemented diets was found to be very high. In one sample of 1,000 animals, only 2.4 per cent were susceptible to drug treatment of Salmonella. The disturbing feature is that such resistance can apparently be very easily transferred to humans. There are many routes of contact—meat, dairy products, etc—and R-factors have been found on the skins of sausages. If this is indeed so, the feeding of antibiotics to animals in their normal diet would appear to involve a number of dangers.

One day a substance that prevents bacteria conjugating (and so stops the spread of R-factor) may be found. But until that day, the search for new and more effective antibiotics must go on.

Cancer, kill or cure

Most of us fear cancer—with good reason, since it is the second of the two great killer diseases of the modern Western world (the first being heart disease). Nevertheless, recent advances in cancer research mean that more than a third of people who develop cancer can now be cured. Mostly, these are people whose cancer has not spread, and can be eradicated by surgery or radiotherapy applied to a small area. A major challenge to cancer research is to find treatments which will help those patients whose cancer had already spread beyond the reach of surgery or radiotherapy when first diagnosed.

To appreciate the difficulties in treating cancer, it is essential to know a little of how the disease arises. The ten million million or so cells of which our bodies are composed are organized to form various tissues. Different tissues—such as those of the eye, liver and skin of the hand—have their own distinctive properties and organization. The cells within them 'know' where they belong and how they should behave to make each specific tissue; and they maintain this pattern throughout life.

Skin tissue, for example, is constantly being renewed by the shedding of dead cells from the surface and their replacement by new ones beneath. The production of new cells is precisely geared to loss, and the new cells never stray beyond their appointed territory, so the form of the skin is maintained.

In cancer, this organization breaks down. More and more cells are made, and they accumulate in the tissue, distorting it and forming a lump, or *tumour*. More important, the cells seem no longer to recognize their surroundings. They lose their sense of territory and begin to stray, first within the tissue and then beyond it to distant parts of the body.

This description of how a cancer starts has two important consequences for attempts at treatment. The first arises because the description is so incomplete. Biochemists can describe in general outline the processes which maintain healthy tissues, but virtually nothing is known of the mechanisms which regulate these processes. So when the system breaks down, as it seems to do in a cancer, doctors do not know how to restore the body's own controls. They are left with the alternative of simply destroying the cancerous cells.

If the cancer is localized in a non-essential part of the body, such as a leg or a breast, it can be cut out. Sometimes, however, an operation is technically too difficult. Then, especially if the cancer is sensitive to X-rays, *radiotherapy* may be an alternative. This kills all the cancer cells (and many of the normal cells) in a particular area. But if the cancer has spread further afield, the patient needs a treatment which will reach all parts of the body. In most cases, this involves administering an anti-cancer drug which will be carried in the blood to the cancer cells, wherever they may be. This approach is called *chemotherapy*.

Cancer cells are not totally new cells. They remain very similar to normal cells, but out of their proper context. In most cases, their chemical machinery, which is what most anti-cancer drugs attack, is not altered very much.

So it seems that the drugs that kill the cancer are quite likely to kill the normal tissue too. (This is in sharp contrast to the situation with bacterial infections and antibiotics. A streptococcal sore throat can be cured with penicillin without harming the patient at all. The penicillin attacks the chemical machinery of the bacteria by preventing the assembly of their protective cell walls, and they die. The cells of the patient's throat on the other hand, do not have this type of cell wall, and are unharmed.)

The second problem with cancer treatment, then, is one of selectivity. Researchers

Far left Cocooned against infection, a leukaemia patient waits in a sterile room.
Below left Graph shows typical events during chemotherapy. As well as killing cancer cells, drugs damage healthy tissue, including the blood-making bone marrow. Blood count falls as chemotherapy continues, until it reaches a level at which infection can kill.
Below One of the effects of some anti-cancer drugs is the temporary loss of hair.

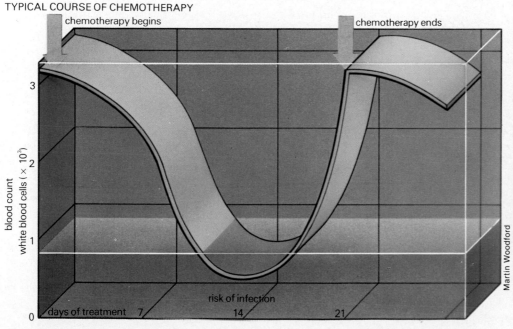

TYPICAL COURSE OF CHEMOTHERAPY

chemotherapy begins — chemotherapy ends

blood count
white blood cells (× 10³)

3

2

1

risk of infection

0   days of treatment   7   14   21

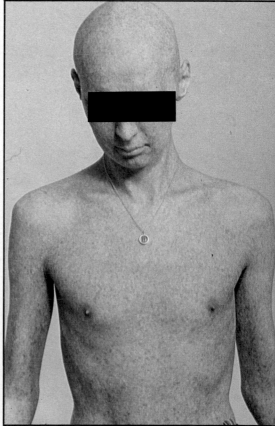

Institute of Cancer Research

Martin Woodford

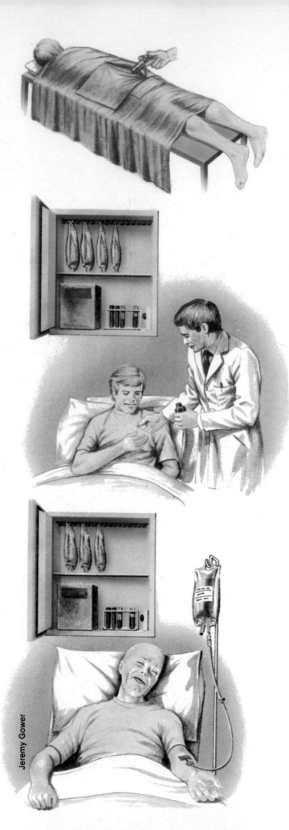

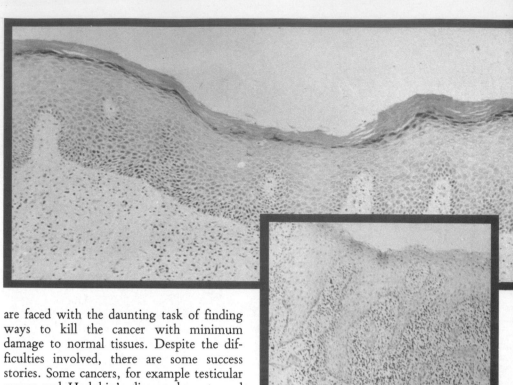

Cancer transforms the organized pattern of healthy skin (microscopic section, *top*) into an amorphous area *(above),* where the cells have lost their sense of territory.
Below Detail of marrow transplant technique. Marrow cells are removed from the richly supplied marrow cavity of the pelvic bones via a hollow needle. Nearly a pint of marrow may be taken in a half-hour session *(right).*

are faced with the daunting task of finding ways to kill the cancer with minimum damage to normal tissues. Despite the difficulties involved, there are some success stories. Some cancers, for example testicular cancer and Hodgkin's disease, have turned out to be very sensitive to particular combinations of anti-cancer drugs.

Normal tissues vary in their sensitivity to the killing effects of anti-cancer drugs. Different drugs have their main effect on different tissues. The most familiar is the temporary loss of the patient's hair caused by certain drugs which damage healthy hair follicles. Overall, however, the tissue which is at greatest risk of serious effects from most anti-cancer drugs is the bone marrow.

The bone marrow contains cells which make blood. Damage to these cells eventually results in anaemia because of a shortage of red cells in the blood, extreme susceptibility to infection because of shortage of white blood cells, and a tendency to bruising and bleeding because of a shortage of blood platelets—which form an important part of the clotting mechanism.

A risk of infection

Generally, anti-cancer drugs are given in short courses, causing some damage to the bone marrow, followed by periods of two to six weeks with no chemotherapy to allow the bone marrow to recover. At the time when the blood count is lowest, the patient is at greatest risk, especially from infection.

As the treatment continues, more and more of the reserves of the normal bone marrow are used up repairing the damage caused by each successive course of chemotherapy, and recovery becomes slower and less complete. The bone marrow thus sets a limit to the amount of chemotherapy possible.

What is particularly frustrating to cancer researchers, but at the same time a spur to

Above In the simplest type of transplant, some of the patient's own marrow cells are removed (1), then stored in a deep-freeze in liquid nitrogen (2). Meanwhile, the patient begins a course of chemotherapy (3). Thus, the removed marrow escapes the harmful side-effects of the drugs. When the drugs cease to act, the marrow cells are returned via a drip (4). They travel in the blood to the bone marrow, where they make new blood cells, speeding recovery.

REMOVING BONE MARROW

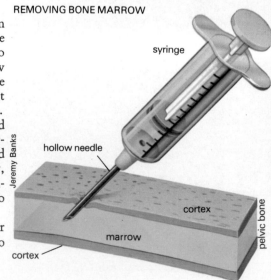

their further efforts, is that this limit seems to be only a little short of the amount of chemotherapy that might produce real success in some of the commoner cancers, if only it could be given safely. For example, in one variety of lung cancer, chemotherapy will regularly cause the disease to shrink until it is too small to be detected in half or more of the patients. But only ten per cent or so are still alive three years later. In the others, the disease will have inexorably grown back. At this point, the body's reserves cannot stand another programme of treatment at the doses needed to control the cancer and the patient dies.

A vital question for cancer researchers to answer is whether the patients whose cancer 'disappears' and then returns a few months later are nearly cured. Would just a little more chemotherapy get rid of the last few thousand cancer cells and prevent the inevitable relapse? There is no theory at present which can provide the answer to this question. It seems that if higher doses of chemotherapy are to work, they should probably be given in only a small number of intensive courses.

Meanwhile, bone marrow transplants, which have been carried out for a number of years in the treatment of a variety of non-cancerous blood diseases, are now being applied to the treatment of cancer. In the simplest case, a portion of the patient's own marrow is simply removed before chemotherapy to escape the effects of the drugs. Replaced as soon as the drugs are no longer active, it speeds up the recovery of the marrow subjected to chemotherapy.

This strategy is adopted in the treatment of cancers, such as lung cancer, which do not primarily involve the bone marrow. In patients with leukaemia, the marrow from someone else must be used.

The technique of marrow transplantation is relatively straightforward. The bone marrow cells are taken from the donor by sucking them from the marrow cavity through a hollow needle. Almost a pint of marrow may be removed in a half-hour session.

Deep-frozen marrow cells

Depending on when they are needed, marrow cells can either be deep frozen and stored for months in liquid nitrogen, or they can be kept cool in an ordinary refrigerator and used the same day. Transplanting cells into a patient is also extremely simple. They are passed into the vein, just like a blood transfusion. The marrow cells circulate in the blood until they pass through the bone marrow. There, by unknown processes, they recognize their 'home' and resume their normal function of making new blood cells.

However, a complication with some cancers which tend to spread to the bone marrow is that the 'normal' transplant might put back cancer cells as well as bone marrow, defeating the whole purpose.

Small-scale trials on volunteer patients with one variety of lung cancer will soon tell researchers whether the extra chemotherapy works. Even if it does, the costs of marrow transplants may prohibit widespread use.

Nonetheless, some encouragement for the use of marrow transplantation in the treatment of cancer has come from the results so far with the other group of patients, those with some types of leukaemia. Technically, the problems here are greater. All of the patient's own marrow must be destroyed by chemotherapy and radiotherapy to eradicate the leukaemia. The grafted marrow has therefore to start from nothing in building up the new marrow, and the period of risk from infection and bleeding while the patient is without adequate blood cells is greater—generally about three weeks.

Another obstacle is that host and donor may not be compatible. This problem is most familiar with kidney transplantation, where it is well known that, despite careful 'matching' (rather like the matching of

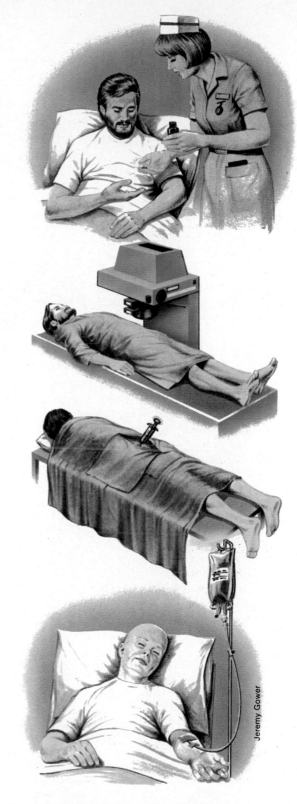

Above The infected marrow of a leukaemia victim destroyed by drugs (1) and radiotherapy (2). Marrow cells taken from donor (3) are transplanted into patient (4). Problems arise when immune cells in graft attack patient, causing 'graft-versus-host disease'. Drugs that suppress immune cells in kidney transplants cannot be used because they slow down the growth of the marrow, but a new drug, Cyclosporin A, is giving good results.

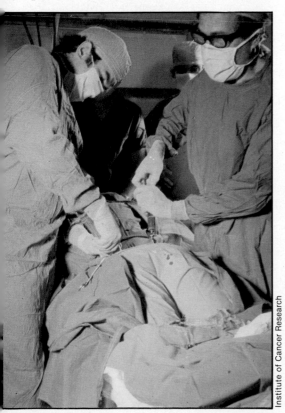

Jeremy Gower

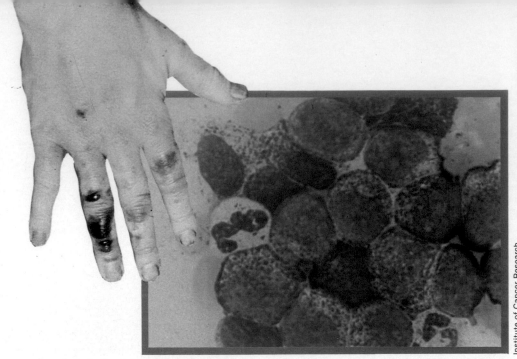

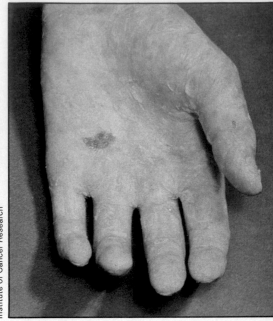

blood groups for transfusion) a proportion of kidneys are 'rejected' by their host, and die. This rejection is part of the natural defence of the body against foreign substances, carried out by the cells of the immune system. Unless the immune system is suppressed by special drugs, the kidney will be destroyed.

With marrow transplants, the immunity problem has a new twist. Many of the immune cells are derived from the bone marrow. The patient's own bone marrow and immune system have been destroyed by the treatment for the leukaemia, so the patient does not react to the graft. The marrow graft, however, contains active immune cells from another person, and these attack the patient because they 'see' him or her as foreign. The drugs which are used to suppress these cells in kidney patients are a handicap here because they also slow down the growth of the new bone marrow. *'Graft-versus-host disease'*, which causes skin rashes, severe diarrhoea, liver failure and lung damage, has been a major obstacle to success with these transplants. Until recently, it led to the death, a few weeks after the transplant, of more than one-third of the patients.

Encouragement has come from two directions. First, a new drug, Cyclosporin A, seems able to suppress the immune system and prevent 'graft-versus-host disease' without damaging the new bone marrow. There are still problems and side-effects from this drug, but with its use the number of deaths is dropping dramatically. This in turn has allowed the impact of the extra treatment for the leukaemia to become clear.

Whereas with the best 'standard' chemotherapy only about 15 per cent of adults with the commonest kind of acute leukaemia can expect to survive five years,

the expected results with marrow transplantation are nearly 50 per cent. Some patients do die of leukaemia which has not been eradicated despite the increased level of drug treatment. It is clear, though, that the higher dosages made possible by marrow transplantation have killed the leukaemia cells in many patients where 'standard' doses would have failed. Finding the best way to exploit this lead and extend these first results obtained with acute leukaemia will prove an exciting task.

An alternative approach is to try to design an anti-cancer drug that really is specific for cancer cells. Such a drug is an attractive prospect because it would not damage normal tissues, and so should require none of the elaborate support of marrow transplantation, and involve little risk or discomfort.

Using antibodies

How is such a drug to be designed, if it is true that most cancer cells are still closely related to their normal counterparts? A clue lies in the very feature that sets cancer cells apart—their seeming failure to communicate effectively with their normal neighbours.

It is reasonable to suppose that cells recognize their neighbours where their surfaces come into contact, and there is a good deal of evidence that the surfaces of cancer cells are different from those of their normal counterparts. These surface changes are probably not completely specific to the cancer cells—they may, in particular, be shared by some of the early cells from which the adult tissue is derived—but even so, attempts have been made to make use of them, both in diagnosis and in treatment.

Although some cancers are probably recognized as foreign by the body's immune

Above far left Chemotherapy and radiotherapy leave a leukaemia patient with no marrow cells. One of the effects is that the blood does not clot properly and the patient bleeds readily. Infection is also a serious risk.
Above left White blood cells affected by leukaemia do not develop properly.
Above One of the distressing symptoms of 'graft-versus-host disease'—skin rashes appear on a patient's hands.
Right In this computer-drawn scan of a rat's heart (H) and right lung (L), radioactively labelled antibodies reveal a tumour (T) in the lung. Diagnostic techniques such as this are a great aid in the fight against cancer.
Below Left-hand X-ray shows tumour (white bulge) in a patient's lung. X-ray on right shows lung after tumour has been removed.

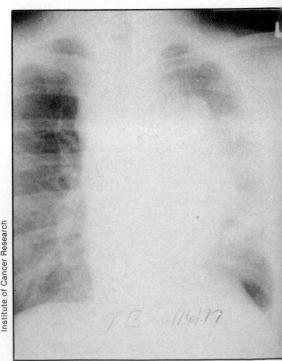

system, efforts to boost the immune system, to help it destroy the cancer, have been almost entirely unsuccessful. Recent interest has focused instead on one component of the immune system, the *antibodies.*

Antibodies are proteins made by cells of the immune system in response to foreign substances. Each antibody has the ability to combine specifically with the substance that provoked its formation. This specificity is an important feature of the immune mechanism. Researchers have injected cancer cells into an animal of another species (for example, a mouse or rabbit), analyzed the antibodies produced in the animal's blood, and purified those which react only with cancer cells and not with the normal cells as well.

Rather than work out for themselves the

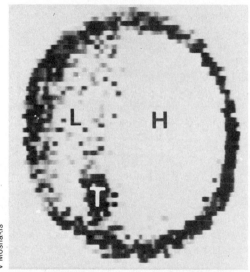

V Moshakis

differences that distinguish cancer cells from normal cells, the cancer researchers have let the animal's immune system do it for them. The antibody that is provided may not be uniquely specific for cancer cells, as was pointed out above, but it will at least distinguish cancer cells of a particular type from their normal fellows.

The first step in treatment with antibodies was to see if injection of the antibodies alone would kill the cancer cells. This met with little success, probably because the antibodies were not of a type that can kill cells as a simple consequence of combining with them.

Search for a 'magic bullet'

The next step was to link antibodies with anti-cancer drugs, to see if 'homing' the drug in this way would allow a smaller *total* dose of drug to be delivered more effectively to the cancer cells. These experiments continue, but the results so far are not dramatic.

This and other similar tests suggest that the anti-cancer drug reaches the target area of the cancer, but fails to work properly once it is there. There is some evidence, for example, that the drug may not enter the cancer cells effectively.

Accordingly, the cancer specialists are developing a refinement of the antibody-drug technique which may solve this problem. Certain natural poisons, such as ricin (implicated in the 'umbrella murder' of Bulgarian broadcaster Georgi Markov in London, 1978), kill cells very efficiently by punching a hole in the cell surface as soon as they make contact. Antibodies joined to the

active part of ricin would, in theory at least, make very effective 'bullets' for killing specific cells. Much work remains to be done, but this system has proved highly effective in the very simple test of killing specific cells in laboratory cultures.

The limited success of the antibody-drug approach has encouraged cancer doctors in their search for a 'magic bullet' which will kill cancer cells, but the realization of such a breakthrough is still a long way off. In the short term, the use of techniques such as marrow transplantation, which will extend the scope of existing anti-cancer treatments, is likely to be more successful. However, doctors must bear in mind that cancer is a common disease, and predominantly a disease of the elderly. High-technology treatment such as bone marrow transplantation is costly in resources and also makes heavy demands on the patient's strength. So its use is likely to be confined to a select minority.

An altogether different approach also involves the use of antibodies. Production of large quantities of highly specific, highly purified antibodies has been revolutionized in the past five years. Cancer researchers are beginning to use these antibodies as specific probes to 'dissect out' the sequence of events that occurs when the organization of a normal tissue breaks down and a cancer appears. It may take the researchers five years or, more probably, fifty; but if they can understand these early stages well enough to predict precisely their course in people found to be at risk of cancer, they may be able to prevent the cancer from appearing.

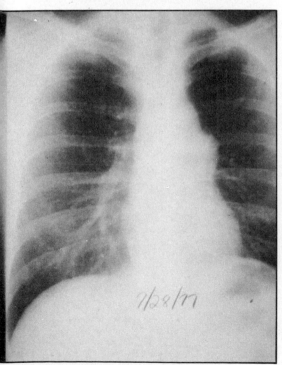

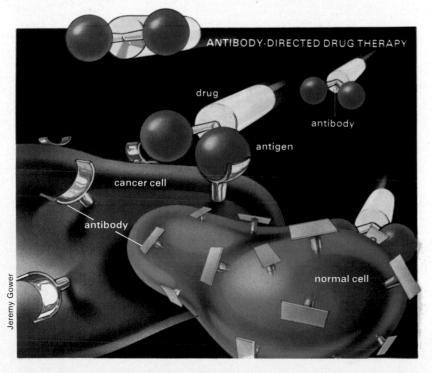

Right The technique of linking antibodies to anti-cancer drugs has so far met with only limited success, but tests still continue. The antibodies carry the drug to the target area of the cancer, where they 'home in' to bind with antigens on the surface of the cancer cells. Normal cells have a specific type of surface to which the antibody does not bind. Hopefully, 'homing' anti-cancer drugs in this way may enable smaller drug doses to be effective.

Jeremy Gower

ANTIBODY-DIRECTED DRUG THERAPY

drug

antibody

antigen

cancer cell

antibody

normal cell

Drugs to change the mind

The pressures and pace of daily life in modern industrialized society have been held responsible for the vast increase in prescriptions for tension-relieving medication during the 1970s. In 1980 there was no let-up in the apparent need worldwide for 'psychotropic' or mood affecting drugs, and in the UK alone it is estimated that eight per cent of the adult population take some form of psychotropic medication each day.

The costs involved, as well as the obvious clinical necessity, therefore make it essential that these drugs are only used when they have proved their effectiveness in treating a particular condition. Casual prescribing, although a tempting way out to the hard-pressed practitioner, is expensive and counter-therapeutic.

Used correctly, however, psychotropic drugs have proven value. As a group, the chemical compounds involved are highly complex, and their actual method of inter-action with the body's chemistry has only lately been defined in detail.

'A drug', according to a report of the World Health Organization, is 'any sub-stance which, when taken into the living organism, may modify one of its functions'. Most psychological distress is of a fairly minor degree, manifested in the form of anx-iety. The search to find drugs to relieve this unpleasant state has produced a wide choice including *benzodiazepines*—Valium, Librium and the like—and barbiturates. There are currently available 28 phenothiazine derivatives (the major tranquillizers), nine antidepressants, and more than 25 non-bar-biturate sedatives. These three categories can be examined side by side with the *psychodysleptics*, so-called pleasure-attaining drugs, which are of more recent origin.

The anxiolytics—or 'minor tranquillizers' —are the mood drugs most likely to be prescribed to an anxious patient. They are in-creasingly used as the means of evading the real problems of stress, disappointment, un-satisfactory home life, and general unhap-piness which so many people encounter. Studies of populations consulting their doc-tors have shown a high percentage seeking consultation for symptoms which reflect underlying anxiety. A Scandinavian study recently showed that one third of the adult population had overt symptoms of anxiety, whilst five per cent had symptoms severe enough to diagnose as an 'anxiety state'.

The psychophysiology of anxiety (the relating of physical events to psychological phenomena) is the key to understanding its treatment. When we experience awareness of anxiety, there is a general 'arousal' of the central nervous system (the CNS) and

Mood-changing drugs include depressants and anti-depressants. *Chlorpromazine*, a depressant, calms the agitated schizophrenic's hallucinations *(below)*. Drugs shown are tricyclic anti-depressants.

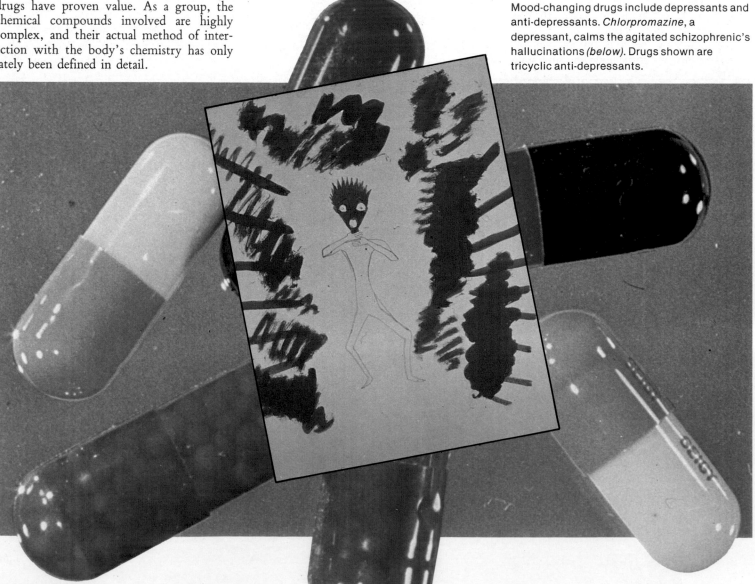

Adamson Collection

C.James Webb

peripheral nervous discharge (hands shaking and dry mouth, for instance). Objects or situations which are seen as threatening produce activation of the *reticular activating* system which is in the lower end of the brain—the 'brainstem'. Messages have passed here via the cerebral cortex, which 'decides' whether or not the stimuli are threatening. The reticular activating system also acts to inhibit overactivity from the cerebral cortex, which might lead to incapacitating anxiety. And between these two areas—the cortex and the reticular activating system—balance of arousal is maintained so that the individual is sufficiently stimulated to respond—for instance, to defend himself appropriately if attacked.

Brain messages

There are other specific areas in the brain which are involved in anxiety production, notably the *amygdala* and the *hippocampus* which seem to control a further structure —the *hypothalamus*. The hypothalamus in turn relays the brain messages to the autonomic nervous system.

The autonomic nervous system controls the basic functions of the body and, in a state

of anxiety, the autonomic changes mainly affect the circulatory system and the sweat glands. A quickening heart-beat has long been associated with increasing anxiety. Measurements of peripheral blood flow (usually taken in the forearm) show that it too is markedly increased. These 'peripheral' responses reinforce the central awareness of anxiety and compound the severity of the symptoms. The palmar sweat glands are also activated by the autonomic nervous system. The level of sweating, as measured by electrical resistance of the palmar skin, produces the so-called 'psychogalvanic response' and this is a useful way of measuring actual levels of physiological anxiety.

The autonomic manifestations of anxiety —the pounding heart and sweating hands —can so affect the individual that they, rather the central feeling of anxiety, cause him to seek help. Drugs used in the treatment of anxiety can be divided broadly into two groups: those which act primarily on the central nervous system, and those which block peripheral nervous activity.

Centrally acting drugs are known as the anxiolytic *sedatives* (since there is no pure compound they all have some sedative pro

perties). Barbiturates, although most commonly used to induce sleep are effective anxiolytics in small doses, especially phenobarbitone and amylo-barbitone. However, it is difficult to obtain a satisfactory anxiolytic effect without some reduction of alertness—or even drowsiness—and as the dose increases, the respiratory centre becomes depressed. The dangers involved in overdosing are therefore great and these, together with the tendency to produce dependency and the harmful effect on the liver, led to the search for alternative anxiety-reducing medication.

Easing the symptoms

The *benzodiazepines* were manufactured in response to the need for an anxiolytic drug which would not produce sedation. There are several available for the treatment of anxiety—chlordiazepoxide (Librium), diazepam (Valium), medazepam (Nobrium), oxazepa (Serenid), lorazepam (Ativan) and clorazepate (Tranxene). They all have basically the same structure, but, despite initial claims by drug companies, it is now clear that they too cause physical and psychological dependency, and that withdrawal after long-term use can lead to convulsions.

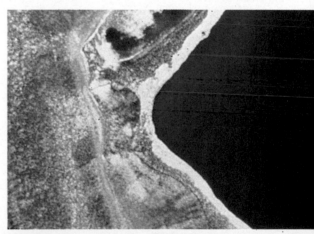

Excessive anxiety arouses the central nervous system. The group of drugs known as *benzodiazepines* act by blocking the nerve endings which anxiety-arousing events have stimulated. Two widely prescribed members of the group are *chlordiazepoxide* (Librium) and *diazepam* (Valium). Crystals of both under polarized light are shown *(right)*. Valium *(top)* is more active than Librium *(below)* in causing muscular relaxation and sleep. But, structurally, all members of this group are closely related. They act within an hour, as a rule. A special advantage is that overdoses are rarely lethal.

C.James Webb

Overdosage, however, is not usually fatal and in this respect benzodiazepines are superior to the barbiturates. Although relatively safe when taken alone they do increase the effects of alcohol and barbiturates, and any combination of a benzodiazepine with these can be dangerous. Essentially, the benzodiazepines do not solve the problem of anxiety, only ease some of its symptoms.

The peripherally acting drugs work by blocking the adrenergic nerve endings. These are the ones which receive stimuli when anxiety is aroused, via the release of adrenalin. Propanalol (Inderal) is effective in reducing anxiety symptoms such as tremor and palpitations and, when used in combination with a small dose of a benzodiazepine, seems to be an effective, rational treatment.

Neuroleptic drugs are those which have therapeutic effects on the psychoses (of which *schizophrenia* is a main example) and they are thus also known as 'anti-psychotics' or 'major tranquillizers'. Their advantage over earlier drugs is that they induce sedation without causing sleep.

The most common perceptual disorders are auditory hallucinations (hearing voices) when the patient usually hears involved conversations between two or more voices, frequently making disparaging references to himself. Disturbance of thought processes

Dr.David Jacobewitz/Science Photo Library

Above Treated with formaldehyde, a tangle of nerve endings glows green and reveals the presence of the body's chemical messenger, *noradrenaline. Below* A typical course of treatment by tricyclic drugs shows that dosage may require variation to keep the blood level of the drug within the optimum range that relieves anxiety in an individual.

means that patients are unable to put their thoughts together in a logical way. The characteristic emotional disturbance in schizophrenia is of 'flattening', when the patient appears lacking in emotional responsiveness. His behaviour is incongruous and he may respond inappropriately by laughing in the wrong places, or giggling to himself. Lastly, physical activity in schizophrenia can become less and less, lapsing eventually into an apparently stuporous state (catatonia). This may be interrupted by an outburst of violent activity (catatonic excitement) which ends as suddenly as it begins.

Treating schizophrenia

The neuroleptics in current use—drugs which aim to treat schizophrenia—are mainly the *Phenothiazine* group and the *Butyrophenone* group. Chlorpromazine is the most widely used of the phenothiazines and was the prototype of this drug series. It was initially used as a booster for anaesthetics, but was soon found to calm disturbed psychotic patients.

One of the problems of administering these powerful drugs is their side-effects. The main form of these side-effects are 'extrapyramidal' symptoms which mimic those of Parkinson's disease, causing stiffness, tremor, inability to initiate movements, and excess salivation. Other side-effects are abnormal pigmentation (when exposed areas of skin acquire a grey-blue colour) and jaundice, which occasionally appears after a few weeks of treatment.

Although the butyrophenones are struc-

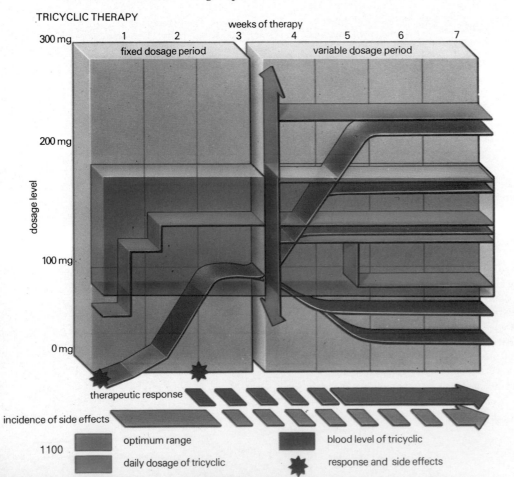

TRICYCLIC THERAPY

weeks of therapy

1 2 3 4 5 6 7

300 mg

fixed dosage period   variable dosage period

dosage level

200 mg

100 mg

0 mg

therapeutic response

incidence of side effects

Ian Stephens

1100

optimum range

daily dosage of tricyclic

blood level of tricyclic

response and side effects

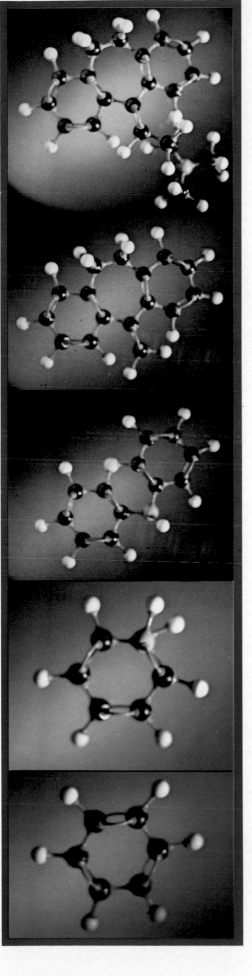

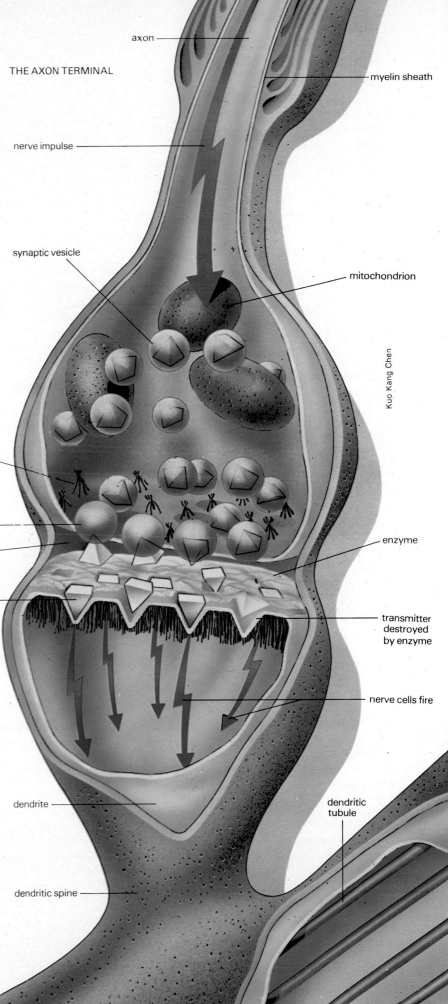

THE AXON TERMINAL

axon

myelin sheath

nerve impulse

The tricyciic anti-depressants are so named because their chemical structure involves three joined rings of atoms with an attached tail or side-chain. 1 *Amitriptyline* is among the most widely used. It is a derivative of 2 *Dibenzocycloheptene*. 3 The *phenothiazine* group of drugs also has a characteristic chemical structure. 4 *Aniline*, a coal-tar derivative, is an intermediate stage in phenothiazine production. 5 The *benzene* ring is the starting point in synthesis.

synaptic vesicle

mitochondrion

protein fibres

vesicles discharge transmitters through membrane

enzyme

synaptic cleft

transmitters impinge on dendritic spine

transmitter destroyed by enzyme

nerve cells fire

Right *Noradrenaline* transmits an impulse from the axon of one nerve cell to the dendritic spines or cell body of another, through a synapse between two cells. Enzymes degrade the noradrenaline, clearing the way for the next impulse. Certain drugs, notably the tricyclic group, inhibit the uptake of noradrenaline from nerve cells.

dendrite

dendritic tubule

dendritic spine

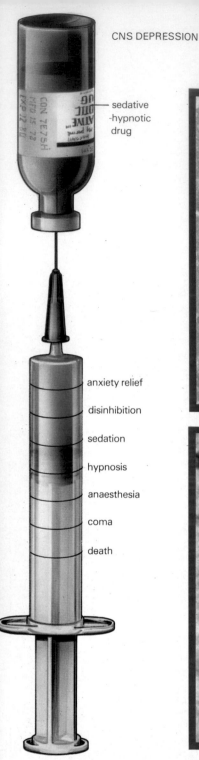

sedative
-hypnotic
drug

anxiety relief

disinhibition

sedation

hypnosis

anaesthesia

coma

death

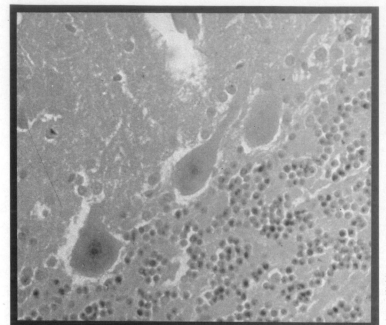

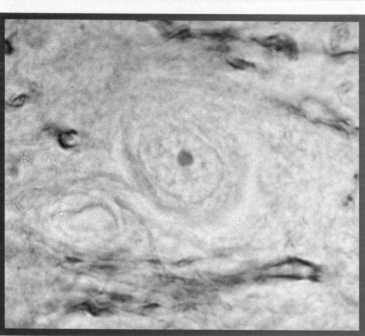

Barbiturates, effective relievers of anxiety when taken in small doses, are sedatives that act on the central nervous system, and risks from over doses are high. *Above* Progressive stages of increasing dosage. Barbiturates have a depressant action on arousal mechanisms in the brain stem reticular formation. Cell structures here *(top)* and in the cortex *(above)* regulate the degree of arousal a living creature must maintain in order to relate to his surroundings and protect himself. Hyperactivity at such brain centres overstimulates the body system and triggers exaggerated response to disturbing events.

turally different from the phenothiazines, they share many of their chemical properties. They are potent anti-emetic drugs and also produce a loss of control, similar to the symptoms of Parkinson's disease. In addition, they increase the sedative effects of alcohol and other anaesthetic drugs—which means they should be used with caution.

It is clear that these drugs can only be used when treating severe mental illness. Their powerful effects would be devastating to the normal individual. How, then, can they be usefully prescribed? There is evidence that continued medication with phenothiazines

reduces the incidence of relapse in schizophrenia. To ensure that schizophrenic patients really do receive their medication, many centres now use injectable long-acting phenothiazines and there are reports of a significant reduction in hospital readmission rates and in the duration of each illness relapse. However, there are also reports of depression developing in some long-term patients, and greater hope may lie in less stressful life situations for this type of patient. Studies, particularly that of Birley and Brown in 1970, have shown that severe breakdowns are precipitated by traumatic life-events, and the use of anti-psychotic drugs can only be successful if such stresses are reduced.

Elation and depression

The anti-depressants are a group of drugs which are encountered more often in everyday life than neuroleptics, but unfortunately, this means that they are sometimes abused, either by the prescriber or the patient. Depression can be a symptom of a reaction to adverse circumstances. Everyone gets depressed from time to time when things go wrong and this is a perfectly natural reaction. A few people, however, develop a much more serious condition—a true 'depressive illness' which usually comes on completely unexpectedly and for no apparent reason. When states of depression and elation ('mania') alternate, it is termed 'manic-depressive psychosis'.

This depression can be described as persistent alteration of mood and is usually accom-

C.James Webb

Daily Telegraph Colour Library

A—Z Collection

panied by one or more of these symptoms —self-deprecation and a morbid sense of guilt; sleep disturbance (typically early morning wakening); retardation of thought or action; agitated behaviour; suicidal ideas or attempts, inability to concentrate, or lack of interest in surroundings; severe anorexia (reluctance to take nourishment) with consequent weight loss. The symptoms are not altered by any change in circumstances and usually require treatment with either drugs or, less often, electro-convulsive therapy.

Mania occurs less frequently than depression and is characterized by over-activity during day and night, loss of social inhibitions and lack of judgement. This imbalance often leads to self-assertiveness, over-generosity and recklessness. The manic patient also has a sense of well-being and talks non-stop in a continuous stream of jokes, puns and personal remarks, making him entertaining, if exhausting to the listener.

Brain transmitters

Depression is due to an absolute or relative decrease in monoamines (brain 'transmitters') at receptor sites in the brain, while mania is due to an excess. Successful chemical treatment of these conditions, therefore, depends on restoring normal monoamine activity at the central receptor sites in the brain. Two main types of anti-depressant are in common use. They are the *Monoamine Oxidaze Inhibitors* (MAOI) which work by blocking the breakdown of brain amines, and the *tricyclic antidepressants* which achieve the same effect—increasing the total amount

Drugs which produce hallucinations include the hemp plant *Cannabis sativa (left)* and the alkaloid, mescaline, derived from a Mexican cactus *(right). Above* Phobias have been cured by exposing patients to the phobic object whilst relaxed by drugs.

of brain monoamines—by blocking the actual uptake of monoamines into the axon terminal—the area where brain cells interconnect.

Lithium carbonate is increasingly used as a preventive drug. It was introduced as a treatment for mania, but chemical trials in the late 1970s have shown that it is useful in manic-depressive illness (the 'up and down' illness) as well. Two important trials in the 1970s proved that prophylactic lithium given to patients known to have suffered previous episodes of mania or depression actually prevented them—in a significant number of cases—from having a relapse.

It is becoming increasingly clear that drugs will not remedy the problems with which modern society confronts the individual. Just as the final and complete 'cure' for tuberculosis lay in the eradication of poverty and squalor as much as in the use of anti-tuberculosis drugs, so the key to schizophrenia must be found in appropriate living conditions for the individual afflicted. Flattening him with a shot of largactil is no lasting cure. Treatment involving behavioural regimes as well as drugs, and docters using psychotherapeutic skills rather than their old prescribing patterns—these are the areas of true innovation for the 1980s.

John Watney

The schizophrenic may hear voices, see visions and reflect the disorder of his inner world in agitated, chaotic paintings. Neuroleptic drugs such as *chlorpromazine* have revolutionized the therapy of the disease by stabilizing the sufferer's anxieties and tensions.

The poison epidemic

Every year in modern industrialized societies, more and more people are poisoned. Although criminal poisoning is out of fashion and accounts for very few cases today, the death toll from accidental and suicidal poisoning is still unacceptably high. The statistics make grim reading: over 5,000 people die from poisoning each year in the USA, and five million cases of poisoning are reported there annually. In the UK, over 3,000 people die from poisoning each year.

The incidence of acute poisoning, mainly self-administered, has increased so steadily over the past 30 years that it now takes on epidemic proportions. For instance, the number of patients admitted to hospital in England and Wales with acute poisoning was 33,600 in 1962, whereas by 1972 this figure had trebled, and continues to increase annually. On average, poisoning accounts for 15 per cent of all admissions to casualty departments of UK hospitals.

The young at risk

In the UK, the age group most at risk from acute poisoning are the 16-25 year-olds, and it is three times as common in girls as in boys. For example, in Edinburgh, Scotland, one in every 1,000 adults poisons themselves annually, but the figure for teenage girls is one in every 600.

Not only the incidence of poisoning but also the choice of poison appears age related. In the 16-20 age group, analgesic (pain killing) drugs (including aspirin) account for most cases; between 20 and 40, antidepressants and tranquillizers are more often taken; while in older age groups, barbiturates and hypnotic drugs are the commonly available ones.

Complicating this picture is the fact that in about 40 per cent of poisonings mixtures of drugs are involved, and this is becoming increasingly common. This obviously makes the task of the doctors who must diagnose and treat poisonings much more difficult.

In the USA, acute poisoning is now the fourth most common cause of accidental death, topped only by automobile accidents, drownings and burns. About 90 per cent of

reported cases of accidental poisoning involve children—mainly in the home.

Similar statistics have been obtained in most other Western societies, though the death rate from poisoning varies widely from one country to another, reflecting different standards of diagnosis and treatment. Also there is a wide difference between countries as to how poisonings are recorded.

Against these considerable odds, medical experts on poisons—*toxicologists*—strive to improve our knowledge of toxic substances and how to deal with their effects, with the encouraging result that although more people are poisoned each year, a higher proportion survives.

Toxicologists have classified acute poisoning into four groups: suicidal, accidental, homicidal and 'parasuicidal'. This last category is relatively new, and reflects the large proportion of patients who have taken

poison as a conscious, often impulsive, attention-seeking act with little or no emphasis on thoughts of self-destruction.

'Parasuicide' accounts for much of the growing increase in hospital admissions for poisoning. For example, in the UK, as many as 80 per cent of poisoned patients referred to a hospital may be parasuicides.

It is likely that the single most important cause of the sharp increase in parasuicide is the ease with which people can obtain drugs from state medical services, and the widespread acceptance of the idea that a course of pills can cure all manner of physical and psychological troubles. Few households today are without a potentially lethal supply of drugs.

In the UK, specific information on poisons and treatment of poisoned patients has been available since 1961, when an advisory telephone service was established.

Then, in 1963, four government-sponsored information centres were set up in London, Cardiff, Edinburgh and Belfast, in addition to the unit at Leeds and another in Manchester. Eire has a centre in Dublin.

Co-ordinated from the National Poisons Information Service at Guy's Hospital in London, the UK government centres provide a round-the-clock telephone information service, which answers over 30,000 queries annually on over 10,000 toxic or potentially toxic substances used in medicine, agriculture, industry or for domestic purposes. It can draw on an impressive array of data on the toxicity, constituents, clinical symptoms of poisoning and suitable treatment for each substance. These data are constantly being up-dated as experience increases and new approaches to treatment become available.

Unlike the situation in most other countries, the information provided by the centres is available to the medical profession only, since it is felt that manufacturers of drugs and other potentially hazardous substances are likely to act more responsibly in divulg

Left About 90 per cent of reported accidental poisonings involve children. Modern 'child-proof' containers make access more difficult, but too many people hoard drugs, leave them within easy reach, or discard them in waste bins rather than by flushing away.
Below Potentially lethal pills, coloured for identification, compare easily with a well-known brand of sweets.
Right Graph shows the changing pattern of poisoning over the years.

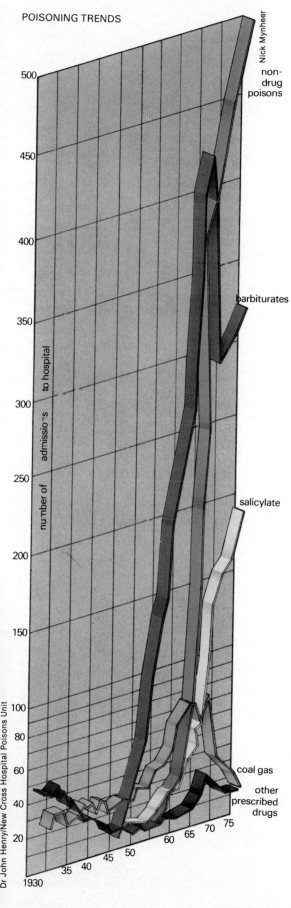

POISONING TRENDS

ing information if they do so solely to doctors and toxicologists for the specific purpose of patient care.

In the USA, the first Poisons Information Centre was set up in Chicago in 1953, and was rapidly followed by others; there are now 660 such centres, though they differ widely and there is considerable duplication of resources: in the State of Illinois alone, for example, there are over 100 centres, and several may exist in a single city.

This has led to a situation where some centres are without adequate facilities or staff, and the information they hold may even be out of date. Hopefully, this is being remedied by the establishment of fewer but better-equipped and more expertly staffed Regional Centres. Early results appear encouraging.

Toxicological tapes

In fact, as far as information gathering and dissemination go, the US system is potentially the best in the world, making extensive use of computer and microfilm technology. One source of information is the Clinical Toxicology of Commercial Products (CTCP), stored on computer tapes. This provides information on over 25,000 household products.

The National Clearinghouse of Poison Control Centres holds index-card data on more than 16,000 consumer products, linked to a computer system which can relay information almost instantly to a number of regional centres. The National Library of Medicine supplies quarterly abstracts of papers relating to poisons from 2,500 journals, as well as a data service of computer-stored extracts from reports by federal government and industry. This data service,

called Toxline, has also been stored in a British Computer since 1976 and is used by UK doctors.

Some US Poison Control Centres answer queries from the medical profession only, as with their UK counterparts, but most also provide detailed data and advice to members of the public. This approach is justified on the grounds that many accidental poisonings, especially in children, are of a minor nature, and can be dealt with in the home. Also, more seriously poisoned patients may be less at risk if they receive prompt attention before being admitted to hospital. The centres also serve an educational function in warning parents of possibly hazardous substances so that they can prevent a tragedy from occurring.

Despite the evidence put forward by proponents of this approach that the dramatic reduction in the death rate from poisoning in US children in recent years is due to the ready availability of data, many toxicologists feel that information should not be provided so freely, because of the danger of encouraging drug abuse and parasuicide.

The first task in dealing with a poisoned patient is that of trying to establish a correct

diagnosis. A doctor can use various clues to help determine what substance or substances are involved in a poisoning. For example, the combination of loss of consciousness with absence of bowel sounds, irregular heartbeat and widely dilated pupils, together with changes in nervous system reflexes, suggests an overdose of tricyclic antidepressants.

Unfortunately, few poisons produce specific diagnostic features, and the doctor must also rely on circumstantial evidence. Size, shape, colour and marking of tablets are important, but it is not safe to assume that an empty bottle contained what was written on its label. There have been cases where patients were found to have taken an overdose of different pills from the ones clutched in their hands as they lay unconscious.

A recent invention by a Welsh doctor should help the ambulance and medical staff to identify unknown tablets. The pill is dropped into a slot and a machine measures its dimensions, shape and colour accurately. All these details are transferred into digital form and relayed to a data bank at the hospital, where a computer print-out gives the name of the drug within seconds. For instance, a pill with diameter 33, thickness 1,

Right A kit enabling quick and simple screening of patients for paracetamol poisoning. Yellow colour indicates paracetamol. Checking against a card *(below)* allows estimation of its concentration to withing 50 ug per ml.

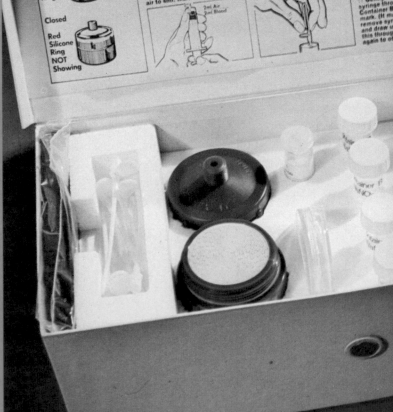

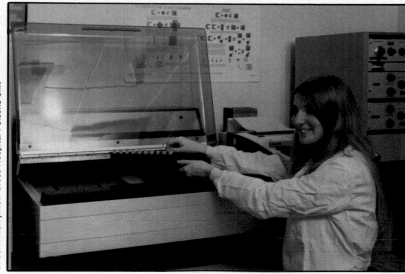

Dr John Henry/New Cross Hospital Poisons Unit

Above Examining a liquid chromatogram.
Above right Radio-immuno assay. Samples of test solution are incubated with a series of radioactively labelled antibodies. If a drug is present, it will combine with its specific antibody; amount is estimated by measuring radioactivity with a scintillation counter.
Below right Thin-layer chromatogram. Large, reddish spot at top of third line is Prothiaden; faint blue spot below, nicotine from smoking.

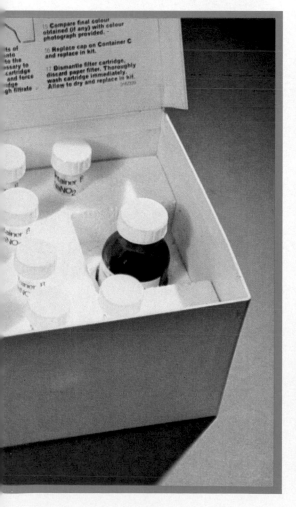

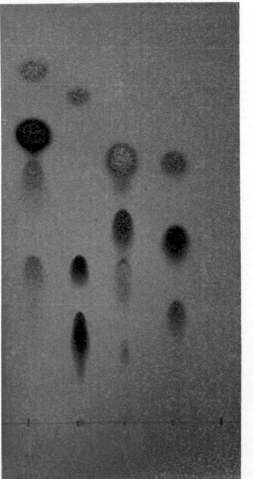

shape 1 (circular) and colour 1 (white) turns out to be aspirin, while one that looks identical is found to have a diameter of 33, shape 1, colour 1, but a thickness of 4 rather than 1, and this proves to be paracetamol.

Such knowledge is vital. An aspirin overdose can usually be successfully treated by pumping the stomach (and maybe giving diuresis), but this is not enough with paracetamol—an antidote must be given promptly to prevent irreversible and possibly fatal damage to the liver.

The most reliable method of obtaining a precise diagnosis, however, is by laboratory analysis of samples of blood, urine or stomach contents. Simple and rapid screening methods have been developed for the analysis of about 90 per cent of common poisons, using the technique of *thin-layer chromatography*. This depends on the distance travelled by the drug or drugs in the sample when spots of the latter are placed on a glass plate covered with an absorbent material and immersed in some kind of solvent.

The plate is then sprayed with various reagents that produce distinctive colour changes with particular poisons, creating a 'picture' known as a *chromatogram*. For example, in screening urine samples for dangerous drugs, the chromatogram may be sprayed with ninhydrin, which produces a red or pink colour if amphetamines are present.

EMIT

There have been important developments recently in the technique of *immunoassay*, which relies on the fact that antibodies will attach themselves firmly to molecules of a particular drug, but only indifferently or not at all to those of other substances in the sample. *Enzyme multiplied immunoassay technique* (EMIT), for example, is a relatively simple procedure which permits rapid screening of specific poisons and also indicates how much poison is present. Results are obtainable within a few minutes. EMIT is used to detect many drugs, including opiates, cocaine, barbiturates and amphetamines.

Complex screening techniques, using a combination of thin-layer and gas-liquid chromatography with ultra-violet spectro-

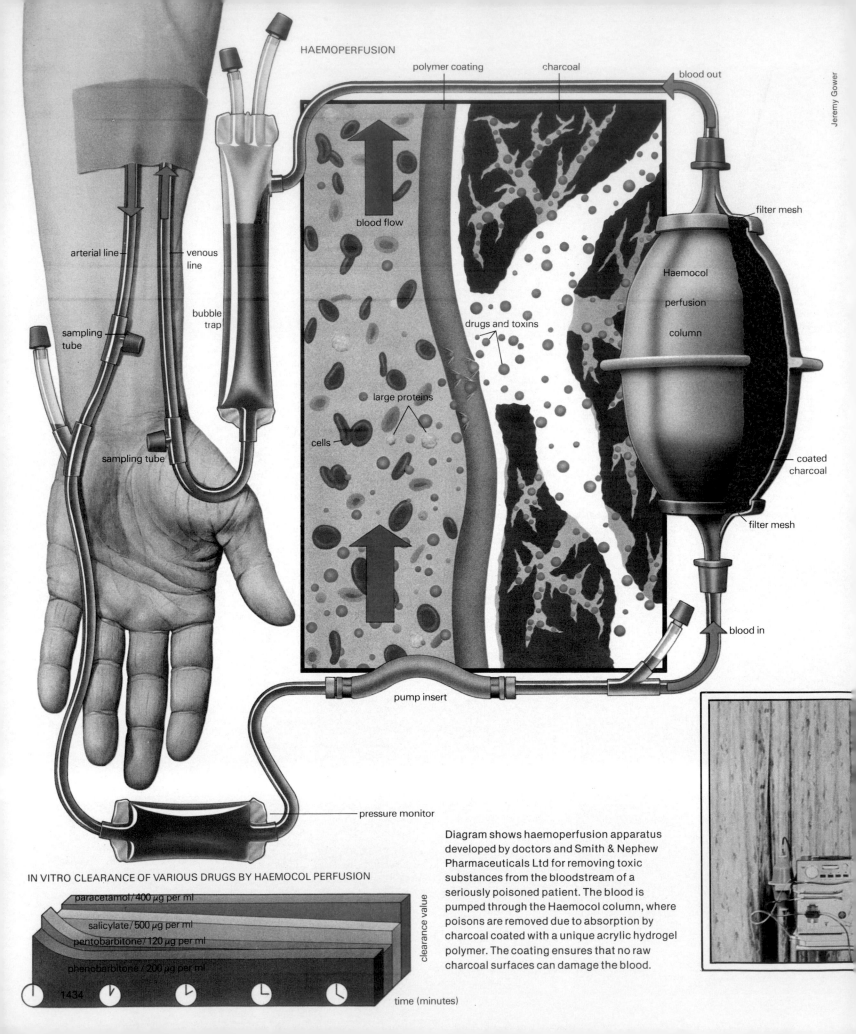

HAEMOPERFUSION

polymer coating   charcoal   blood out

blood flow

arterial line   venous line

sampling tube

bubble trap

drugs and toxins

sampling tube

large proteins

cells

Haemocol perfusion column

filter mesh

coated charcoal

filter mesh

blood in

pump insert

pressure monitor

IN VITRO CLEARANCE OF VARIOUS DRUGS BY HAEMOCOL PERFUSION

paracetamol / 400 µg per ml

salicylate / 500 µg per ml

pentobarbitone / 120 µg per ml

phenobarbitone / 200 µg per ml

1434

clearance value

time (minutes)

Diagram shows haemoperfusion apparatus developed by doctors and Smith & Nephew Pharmaceuticals Ltd for removing toxic substances from the bloodstream of a seriously poisoned patient. The blood is pumped through the Haemocol column, where poisons are removed due to absorption by charcoal coated with a unique acrylic hydrogel polymer. The coating ensures that no raw charcoal surfaces can damage the blood.

Jeremy Gower

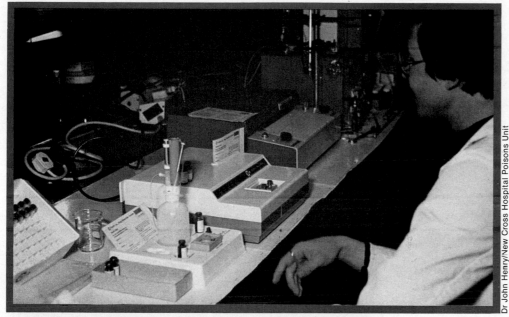

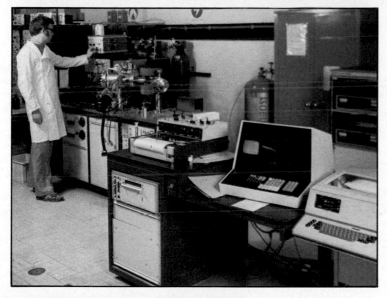

Above Rapid screening tests for drugs can be made on this EMIT (enzyme multiplicd immuno-assay) device. *Left* Mass spectrometry is a technique for determining the molecular structure of drugs and it is especially useful for identifying the breakdown products of drugs in both blood and urine samples. *Below* Haemoperfusion in progress.

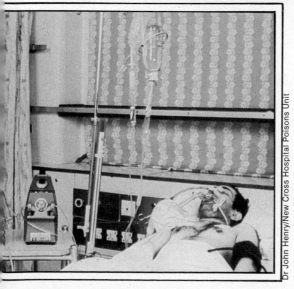

towards reversing any respiratory failure, dealing with problems of blood circulation, and preventing further absorption of the poison as quickly as possible.

Gastric lavage (washing out of the stomach cavity) is often carried out. It is vital that the patient should be in the semi-prone position with the head lower than the feet and turned to one side. Warm water is usually all that is necessary for washing out the stomach, though normal saline solution is used for babies, and for some poisons other fluids are more effective (for example, sodium thiosulphate for cyanide poisoning).

Gastric lavage is of little use, however, for most poisons unless the patient is given the treatment within four hours of having ingested the poison. Important exceptions to this are salicylate poisoning, when all patients should have the stomach washed, and poisoning with tricyclic antidepressant drugs, where a delay of 12 hours between ingestion and treatment is possible.

In addition to this emergency treatment, the hospital staff must ensure that the patient does not suffer from hypothermia, water imbalance or infection.

Diuresis and haemodialysis

Special methods for encouraging drug elimination include *forced diuresis*—stimulating the excretion of urine with an intravenous infusion of sodium bicarbonate, dextrose and other substances. This is used in cases such as acute salicylate poisoning. *Haemodialysis,* using a kidney machine, is another possible treatment: this is the purification of the blood by passing it across a semipermeable membrane. This requires specialized techniques and expert supervision. A simpler method is that of *peritoneal dialysis,* by which the poison can be removed from the abdominal, or peritoneal, cavity, in which the major body organs lie. A recent advance, for removing toxic substances from the body after absorption has occurred, has proved to be by the technique of *haemoperfusion,* which involves passing the patient's blood through a column containing charcoal coated with acrylic hydrogel (a colloid containing water that solidifies in a gelatinous form). Charcoal forms chemical bonds with many drugs and poisons.

Toxicologists continue to make great advances in both their knowledge and treatment of poisoning. Proof of their success is indicated by the fact that in 1945 no fewer than 25 per cent of patients hospitalized with acute barbiturate poisoning died; whereas today the death rate is less than 1 per cent.

photometry, are used when a specialized laboratory manned by a large team is available—at considerable expense.

Even more specific is the combination of gas chromatography and mass spectrometry. Its precision makes it particularly useful to forensic toxicologists. Computers are used to compare results against known standards, increasing the speed and accuracy of this method of analysis.

Treatment of poisoned patients in hospital has changed in important ways over the last 25 years. Contrary to popular belief, there are remarkably few specific antidotes against poisons. Even if there is an antidote, it may be unwise to administer it unless the poisoning is severe, because the antidote itself may have toxic effects.

Instead, treatment is generally directed

INDEX